Fire and Ecosystems

PHYSIOLOGICAL ECOLOGY

A Series of Monographs, Texts, and Treatises

EDITED BY

T. T. KOZLOWSKI

University of Wisconsin
Madison, Wisconsin

T. T. KOZLOWSKI. Growth and Development of Trees, Volumes I and II – 1971

DANIEL HILLEL. Soil and Water: Physical Principles and Processes, 1971

J. LEVITT. Responses of Plants to Environmental Stresses, 1972

V. B. YOUNGNER AND C. M. McKELL (Eds.). The Biology and Utilization of Grasses, 1972

T. T. KOZLOWSKI (Ed.). Seed Biology, Volumes I, II, and III – 1972

YOAV WAISEL. Biology of Halophytes, 1972

G. C. MARKS AND T. T. KOZLOWSKI (Eds.). Ectomycorrhizae: Their Ecology and Physiology, 1973

T. T. KOZLOWSKI (Ed.). Shedding of Plant Parts, 1973

ELROY L. RICE. Allelopathy, 1974

T. T. KOZLOWSKI AND C. E. AHLGREN (Eds.). Fire and Ecosystems, 1974

J. BRIAN MUDD AND T. T. KOZLOWSKI (Eds.). Responses of Plants to Air Pollution, 1975

Fire and Ecosystems

EDITED BY

T. T. Kozlowski

Department of Forestry
University of Wisconsin
Madison, Wisconsin

C. E. Ahlgren

Quetico-Superior Wilderness Research Center
Ely, Minnesota

ACADEMIC PRESS   New York San Francisco London   1974

A Subsidiary of Harcourt Brace Jovanovich, Publishers

ACADEMIC PRESS, INC.
111 Fifth Avenue, New York, New York 10003

United Kingdom Edition published by
ACADEMIC PRESS, INC. (LONDON) LTD.
24/28 Oval Road, London NW1

Library of Congress Cataloging in Publication Data

Kozlowski, Theodore Thomas, Date
    Fire and ecosystems.

    (Physiological ecology)
    Includes bibliographies.
    1.   Fire ecology.   2.   Fire ecology–United States.
I.   Ahlgren, Clifford Elmer, Date   joint author.
II.   Title.
QH545.F5K68   574.5'222   74-5695
ISBN 0–12–424255–3

Contents

4. Effects of Fire on Birds and Mammals
J. F. Bendell

5. Effects of Fire on Grasslands
Richard J. Vogl

6. Effects of Fires on Temperate Forests: North Central United States
C. E. Ahlgren

7. Effects of Fire on Temperate Forests: Northeastern United States
Silas Little

12. Effects of Fire in the Mediterranean Region
Z. Naveh

13. Effects of Fire in Forest and Savanna Ecosystems of Sub-Saharan Africa
John Phillips

14. Use of Fire in Land Management
A. J. Kayll

List of Contributors

Numbers in parentheses indicate the pages on which the authors' contributions begin.

C. E. Ahlgren (*1, 195*), Quetico-Superior Wilderness Research Center, Ely, Minnesota and Wilderness Research Foundation, Chicago, Illinois

Isabel F. Ahlgren (*47*), Quetico-Superior Wilderness Research Center, Ely, Minnesota

J. F. Bendell (*73*), Faculty of Forestry, University of Toronto, Toronto, Ontario, Canada

Harold H. Biswell (*321*), School of Forestry and Conservation, University of California, Berkeley, California

Robert R. Humphrey (*365*), Department of Biological Sciences, The University of Arizona, Tucson, Arizona

A. J. Kayll (*483*), Fire Science Centre, University of New Brunswick, Fredericton, N. B., Canada

E. V. Komarek (*251*), Tall Timbers Research Station, Tallahassee, Florida

Silas Little (*225*), Northeastern Forest Experiment Station, Forest Service, United States Department of Agriculture, Pennington, New Jersey

Z. Naveh (*401*), Faculty of Agricultural Engineering, Technion-Israel Institute of Technology, Haifa, Israel

John Phillips (*435*), University of Natal, Pietermaritzburg, South Africa

P. J. Viro (7), The Finnish Forest Research Institute, Helsinki, Finland

Richard J. Vogl* (139), Department of Biological Sciences, The University of Arizona, Tucson, Arizona

Harold Weaver (279), 236 S.E. 155 Place, Portland, Oregon

* Present address: Department of Biology, California State University, Los Angeles, California.

Preface

Over the years the importance of fire to mankind has been a subject of turbulent debate. Early man appreciated fire and used it in many ways for his well-being—for hunting, grazing for domestic stock, clearing of forests for agriculture, producing ash to fertilize fields, favoring certain plants over others, assisting in harvesting of crops, and eliminating undesirable plant materials. Then suddenly, in the 1930's, the attitude toward fire changed drastically. Within a few years fire was widely considered an insidious enemy rather than a useful friend. This change of attitude was catalyzed by the reckless burning by early settlers and by extensive publicity which dramatized only the harmful effects of fire. The new philosophy of almost complete condemnation of all fire reflected rather wholesale rejection of the historically demonstrated beneficial effects of fire. However, in recent years the pendulum has begun to swing again, and it has become fashionable to reexamine the beneficial effects of fire. Research data are appearing which show that some plants and animals depend on fire for good health. Many land managers are now willing to acknowledge that complete exclusion of fire in many areas has caused dangerous fuel accumulations which may be expected to result in catastrophic fires, disease and insect problems, deterioration of range, reduced wildlife-carrying capacity, and decreased watershed yield. Superimposed on the changing prevailing views of the majority at various times have been lack of objectivity and overstatement in favor of or against fire in natural ecosystems. It is against such a historical background that this book examines in depth the influence of fire on ecosystems.

We recognize that fire affects all the interdependent components of the ecosystem. However, the strong interest of investigators on various parts of the ecosystem makes it feasible to discuss separately a number of aspects of fire ecology. Therefore, while the book is comprehensive in scope, its chapters deal separately with both harmful and beneficial effects of fire on soils, soil organisms, birds and mammals, and plants. One chapter treats the effects of fire on grasslands, and four chapters consider the role of fire in temperate forests and related ecosystems. The latter four chapters are handled on a regional basis to highlight variations in responses, especially plant succession, to fire. Chapters are included on effects of fire on chaparral and on North American deserts. Two chapters discuss the influence of fire on Mediterranean and African ecosystems. The final chapter is devoted to the use of fire in land management. Although the book does not overlook the important role of fire in tropical ecosystems, it nevertheless concentrates on temperate zone ecosystems, particularly in the United States. Such coverage was planned because of the extensive body of literature on fire ecology in the United States and because the editors were most familiar with the work of qualified contributors in the United States.

The book was planned to be interdisciplinary and to be of major interest to researchers, teachers, and land managers. It should be useful to agronomists, bacteriologists, biochemists, botanists, climatologists, ecologists, foresters, geneticists, horticulturists, meteorologists, microbiologists, mycologists, ornithologists, mammalogists, plant physiologists, plant pathologists, range managers, soil scientists, wildlife managers, and zoologists.

In planning this volume, invitations to prepare chapters were extended to investigators of demonstrated competence in the United States and abroad. We express sincere thanks to each contributor for sharing his knowledge and experience and for his patience during the production phases. We deeply appreciate the help and cooperation of the Wilderness Research Foundation. Dr. Isabel F. Ahlgren helped with editing and manuscript revision, and Mr. P. E. Marshall assisted in the Subject Index preparation.

T. T. Kozlowski
C. E. Ahlgren

Fire and Ecosystems

. 1 .

Introduction

C. E. Ahlgren

In recent years our environment, which once was taken for granted, has become a subject of great concern to society. Various natural forces have shaped the biotic community over time and will continue to do so. Fire has been one of the most dramatic of these natural forces. On a global basis, fire and man—separately and together—have had a tremendous impact in shaping or altering world vegetation. Fire, therefore, has become a subject of great interest, not only among biologists and foresters, but also among conservationists.

Despite this recent upsurge of interest, fire probably has been an important ecological factor for as long as flammable vegetation has existed on earth, and man has long taken note of this. The Bible records evidence of man's consciousness of the effect of fire on vegetation [Joel 1:19; Isa. 9:18, 6:13; Job 1:16 (Naveh, 1973)]. Throughout this book, attention is called to the prewhite settlement use of fire by American Indians in different parts of North America, as evidence of man's inherent knowledge of the potential of fire.

In modern times our recognition of the role of fire in the ecosystem has come from studies of past vegetation, identification of soil charcoal layers, fire scars on trees, even-aged character of some forests, records of explorers, etc. These also indicate that fire has been a factor in shaping forests and grasslands for centuries (Soper, 1919; Maissurow, 1935). Muller (1929) stated that in southeastern Europe:

> Virgin forests require fires for fullest fruition just as the Phoenix arose only from fire. There is, in short, no virgin forest that has not resulted from fires; for fires are as "natural" as the forest itself.

1

While the history of fire in the various ecosystems of the world is indeed fascinating, it does not reveal the biological principles involved. Because of man's intervention and the ways in which he has altered world vegetation, the lessons learned from past fire history cannot be used alone in predicting the future of any individual modern ecosystem. Agriculture, disease, harvesting of timber, use of herbicides and pesticides, settlement, man-caused changes in animal populations, etc., as well as man's use and suppression of fire, have all set a different stage on which fire may play a somewhat varying role in modern ecosystems. Our consideration of fire, therefore, cannot be based solely on retrospection. Rather, it must include careful evaluation of present conditions and influences.

As recognition of the importance of fire grew, several popular ideas became accepted by laymen, conservationists, and biologists. These included attempts at exclusion of fire. Suppression of uncontrolled fire in the ecosystem has been necessary in the face of increased population, land use, demands for forest products, and the damaging effects of fires caused by man's carelessness in some places. Public awareness of the need for prevention of such man-caused holocausts was a necessity. Such efforts, which included extensive publicity for fire control in the United States and elsewhere, were aimed at prevention of damaging, dangerous fires. The originators of such campaigns did not intend the public to infer that fire had no place in nature nor that every fire was a "bad" fire. Public enthusiasm for such campaigns led to overapplication, a fear of any fire in the ecosystem, and a hesitancy to recognize the dependence of some ecosystems on fire for renewal.

At the same time a growing recognition of the role fire has played in perpetuating certain desirable natural ecosystems has led some ecologists and conservationists to believe that fire could be used as a cure-all for most or all problems of forest and grassland regeneration. Proponents of this school feel that fire should be allowed to play its role in the natural environment with a minimum of restriction. This school of thought ignores the fact that not all ecological niches within an environment are fire-adapted.

In some cases this school developed from a broad application of the historical approach to the future. This neglects the fact that fire today is acting on a different stage than it did in past times. We must consider use of fire not only in terms of what it has done in the past but also in terms of what it will do in modern, man-altered ecosystems. Even the most remote forests are influenced by man's manipulation. Introduction of diseases and insects, destruction of large supplies of seed of certain plants, especially of some tree species, and utilization of increased acreages in agriculture and urbanization, etc., have all changed the environment surrounding and influencing such areas. In many cases, the past cannot be repeated naturally; in others it can. However, each situation, ecological niche, and species complex must be considered separately and evaluated carefully.

Another dangerous approach to fire involves generalization from one geographical area to another. The beneficial effect of fire in pine forests of the southern United States, for example, does not necessarily mean that fire can be used in the same way with beneficial effects in the pine forests of the north central United States or in Scandinavian countries. Similarly, the response of grass or shrubs to fire varies with the ecosystem and species involved.

Because of growing interest, pressures exerted by public and conservation groups, and conflicting schools of thought among biologists regarding the role of fire, the editors decided that the time was appropriate to bring together in one volume a summary of available information on the role of fire in the ecosystems of the modern world. It was felt that the most realistic approach would be to obtain detailed summaries of the effect of fire on various aspects of the environment: soils, soil organisms, birds and mammals, and vegetation. However, special emphasis has been placed on vegetation, since it provides the fuel for fire, determines conditions of fire hazard, is most immediately and directly affected by fire, and varies greatly from region to region. Plants of grassland, temperate forests, tropics, chaparral, and desert are considered separately. Further division of the temperate forest was necessary because of the variable role fire plays in different regions. For example, response to fire by pines of the western United States is very different from that of pines in the north central region. In various areas, the ground vegetation, especially shrubs, may play completely different roles in postfire succession. Therefore, temperate forests are treated regionally by biologists who are thoroughly familiar with postfire response in each area. Many of the examples of vegetation types are taken from the United States, because of the current emphasis on fire ecology in that country and the editors' familiarity with qualified contributors.

In subsequent chapters similarities in postfire responses of seed-reproducing and vegetatively reproducing herbs, shrubs, and trees in different ecosystems will be noted. These undoubtedly are all related to certain ecological factors in common among these areas. While there are these common factors in fire-oriented ecosystems, their effect on the total vegetation and other parts of the environment is tempered by their interaction and the potential of the particular species, land form, soil, etc., involved. The reader's attention is frequently called to responses of native vegetation to fire-caused changes in seedbed, soil moisture and nutrients, climate, temperature and light conditions, plant competition, and animal food and habitat. Only after thorough study of the variations in response by region and species involved can there be complete understanding of the effect of fire on the ecosystem in general.

Adaptations of species to fire vary greatly, and consequently the kind of fire environment in which these species thrive must be considered. Some tree species with fire-resistant bark and other characteristics that make them able to survive ground fires are particularly adapted to the frequent, periodic burning characteristic of certain ecosystems. This is true of pines of the southeastern United States. In this area, growth of pine as well as of associated forbs, shrubs, and grasses, is profuse and rapid, and fire plays an important role in speeding up the recycling process for litter and ground vegetation.

Other species are best adapted to fire periodicity of only once per tree species rotation. In these cases the mature trees are destroyed by fire, and conditions are established in which a new generation of seedlings will result. This is true of the *Pinus banksiana–Picea mariana* ecosystems of the north central United States. Frequent periodic burning would destroy the young forests of this ecosystem. Litter accumulation is slow and its frequent, rapid destruction would be disadvantageous. With these species further adaptation to the once-per-rotation fire sequence can be seen in the serotinous cones which are opened by the heat of fire. Mutch (1970) presented evidence in support of the hypothesis that flammability of litter and needles of fire-adapted vegetation is greater than that of nonfire-adapted vegetation. This would set the stage for future fires necessary for perpetuation of such fire ecosystems.

In the absence of fire, however, nonfire-adapted species will invade an area, usually compete successfully, and eventually replace the fire-adapted species.

It becomes clear, then, that the factors involved in fire ecology must be identified and carefully considered for each region and species involved. Only after such consideration can these factors be fully understood and utilized in the use or suppression of fires. In this book it is our aim to bring together such regional and specific consideration in the hope that it will stimulate more careful evaluation of the role of fire in the future. In the words of Holloway (1954), fire ecology must be . . .

> founded in a deep appreciation of the forest, of its origins and structure, and of the complex inter-relationships of all its component parts. And, since the forest is an ever-changing, living community, understanding also demands unrelenting study of all trends in forest evolution, man-made

or natural. For a single forest, the complex actions and interactions of all plants, soil, animals, microorganisms, and man together with all factors of climate, topography, lithology, and history must be studied together and in synthesis.

Further, caution in the approach to fire was expressed by Haig (1938) and is relevant today:

> In considering the whole question of fire use and its effects, it might be well to . . . remind ourselves of the natural prejudice introduced by the rightful recognition of the need for a more extensive and more effective protection from indiscriminate and destructive fires, and approach the entire question in the scientific spirit so aptly described by Francis Bacon: 'A mind eager in search, patient of doubt, fond of meditation, slow to assert, ready to reconsider, not carried away either by love of novelty or admiration of antiquity, and hating every kind of imposture, a mind, therefore, especially framed for the study and pursuit of truth.'

References

Haig, I. T. (1938). Fire in modern forest management. *J. Forest.* 36, 1045–1051.

Holloway, J. T. (1954). Forests and climate in the South Islands of New Zealand. *Trans. Roy. Soc. N. Z.* 82, 329–410.

Maissurow, D. K. (1935). Fire as a necessary factor in the perpetuation of white pine. *J. Forest.* 33, 373–378.

Muller, K. M. (1929). "Aufbau, Wuchs, und Verjüngung der Südosteuropäischen Urwälder." Schaper, Hannover.

Mutch, R. W. (1970). Wildland fires and ecosystems—an hypothesis. *Ecology* 51, 1046–1051.

Naveh, Z. (1973). Fire ecology in Israel. *Proc. 13th Annu. Tall Timbers Fire Ecol. Conf.* (in press).

Soper, E. K. (1919). The peat deposits of Minnesota. *Minn., Geol. Surv., Bull.* 16.

. 2 .

Effects of Forest Fire on Soil

P. J. Viro

Fire is a good servant but a poor master.
Finnish proverb

I. Introduction

In discussing the effects of fire in forestry, a distinction must be made between wildfires, swaling, and prescribed burning. Wildfires, which occur during the driest parts of summer, are not caused deliberately and usually only devastate. In the past, wildfires were caused mainly by lightning. Swaling is an old form of agriculture. In addition to burning the area, it included some primitive tilling of the soil. Swaling is planned and prepared in advance. Prescribed burning is a means of preparing and improving a forest site for a new generation of trees. This chapter emphasizes prescribed burning.

Swaling was formerly practiced on a very large scale in agriculture in the forested parts of the world. In Germany and Austria, for instance, it was still used to some extent at the end of the last century (Heikinheimo, 1915). In the Finnish hinterland it was the chief form of cultivation as recently as the latter half of the nineteenth century, and even at the beginning of the present century it was quite common in remote provinces. Usually one to five crops—and sometimes more than ten—were taken from a swaled area. The area was then left for natural reforestation. On the most fertile sites swaling was sometimes repeated as frequently as every eighth year; on the most barren sites it could be repeated only once in 30 years or more. It follows that this form of agriculture required vast areas of land. In some regions more than four-fifths of the productive upland might have been recently swaled.

As a rule the swaled site was naturally taken over by pine, birch, and alder; the more frequently it was swaled, the more hardwood-dominated the forest became. If a site had been swaled only once, the result was often a very beautiful stand of pines. During the latter half of the nineteenth century, these fine stands led foresters to regard fire as a means of converting stands, having a stunted growth of spruce or worthless species of deciduous trees, into more valuable pine stands. Such burning planned in advance for purposes of forestry is called "prescribed burning." The effect of wild forest fires on the soil is much more severe than that of prescribed burning because wildfires usually occur during the driest parts of the summer.

In the latter half of the nineteenth century, the State Forest Service prohibited unrestricted swaling on state-owned lands, but forest sites could still be rented for swaling provided the cropper had the burned site sown with pine seed. In this way, a large number of low-value spruce and alder thickets were converted into stands of valuable species. The first mention of sowing pine seed on swaled land probably dates back to 1864.

The purpose of this chapter is primarily to describe the effects of prescribed burning on nutrient status of the soil and to consider whether burning causes losses of nutrients. The advantages and disadvantages of burning will also be summarized.

The discussion on the effects of burning is based mainly on material and experience from Finland, but the conclusions are probably valid for almost the whole of Sweden and for Fenno-Karelia up to the Lake Ladoga–Lake Onega–White Sea area. This entire region will be termed "Fenno-Scandia." The climate, bedrock, and tree species of these areas west and east of Finland are very similar to those in Finland. The author's own material was collected in Finland between the 61° and 63° parallels North (Viro, 1969), and the Finnish and Swedish studies quoted extend to the northern timber line (67°–69° N).

To study the duration of the effects of burning, the plots were divided into groups according to the length of time elapsed since burning. The groups covered short periods of time in sites burned recently and longer periods in older burned sites. Every burned plot had a control plot in the nearby stand, and the average of all the unburned sites was used as a control. The material comprised 92 pairs of sample plots.

Fenno-Scandia has a very limited number of tree species. The most important are the conifers, pine (*Pinus silvestris*) and spruce (*Picea abies*). Of the deciduous species, birches (*Betula verrucosa* and *B. pubescens*) are the most important, although aspen (*Populus tremula*) and alders (*Alnus incana* and *A. glutinosa*) also appear to some extent. In the most southerly parts there are a few continental European species, but only in minute proportions.

Of the main Finnish tree species, pine needs high light intensity for good growth but is modest in its demands for nutrients. Spruce is exactly the reverse: it requires abundant nutrients but only a moderately high light intensity. Birch requires both a high light intensity and abundant nutrients for good growth, but it can also be grown on quite barren sites.

Forest site classification in Finland is based on ground vegetation, according to Cajander (1949). For example, OMT (*Oxalis–Myrtillus* type) denotes a very fertile site for pine and a good site for spruce, MT (*Myrtillus* type) site is good for pine and fair for spruce, and VT (*Vaccinium* type) is a rather barren site, fair for pine but not fit for spruce.

The earliest prescribed burning in Finland was performed in the last decades of the nineteenth century. However, as a regular procedure in forestry it only got under way in the 1920's, and expanded on a larger scale in the late 1940's. It almost ceased in the late 1960's, the main reasons being the fire hazard and the dependence of burning on the weather. Various methods of mechanical soil treatment have been introduced to compensate for prescribed burning, but none of them has given as good results in reforestation.

II. Reasons for Burning

A. Growth Factors

Growth depends on both external and internal factors, but only the important external factors—water, temperature, and nutrients—will be discussed here. The climate of Fenno-Scandia is temperate. The rainfall is mostly high enough to ensure good growth in normal years, but, owing to their poor water-holding capacity, esker gravel and coarse sand soils (about 10% of the soils) are sometimes too dry during spells of drought. The temperature is reasonably high in the southern part of the area, but the northern part of the area is tundra. The average annual temperature in the Finnish part of Fenno-Scandia varies from 4.8° to —2.0°C, and in the warmest month from 17.8° to 11.8°C. Annual rainfall varies from 680 to 310 mm. The higher figures refer to the south, the lower ones to the north of the country.

The bedrock of Fenno-Scandia consists of crystalline rocks, mainly Precambrian granites and gneisses. Their plagioclase is rich in albite and the proportion of dark minerals is low; the mica is mainly biotite, and these granites always contain small amounts of apatite (0.1%). Owing to this mineral composition, the soils are always acid. The degree of acidity depends largely on the percentage of basic minerals—the amphiboles and pyroxenes—and on the calcium content of the plagioclase. The two first-mentioned minerals and the micas are also important sources of magnesium. Potassium is derived almost totally from biotite, and phosphorus entirely from apatite. The bedrock of Fenno-Scandia is very old, dating from the Precambrian era, but, owing to the glacial periods, the soils in this area are very young, only 9000–12,000 years old. The prevailing soil class accounts for a total of 73%, comprising a variety of moraines, the bulk of them fine-sand moraines. Of the water sediments, the most abundant are sand soils, 13%. Fine sand and clay soils account for only 9% (Aaltonen, 1941).

Forest sites on mineral soils have a humus layer that is composed of remnants of dead vegetation—in coniferous forests mainly of mosses. The humus layer in Finnish forests is usually not very thick: in pine stands it is 2–3 cm, in spruce stands 3–4 cm thick. The thickness of the humus layer has been kept within these reasonable dimensions by the earlier very frequent wildfires and the mixture of deciduous trees that usually are the first species to rise after fire. However, in places where spruce has been the only tree species for centuries on rather barren sites, the humus layer may well be 10 cm thick and even thicker around the base of the trees. Spruce has a long living crown that prevents solar radiation from penetrating into spruce stands, and aeration is poor. Consequently, these stands have a cool and moist microclimate, known as a "basement climate." The formation of a thick humus layer is favored by a cool climate, acid soil, and coniferous species of trees, especially spruce. All these factors slow down decomposition of humus, and the humus layer grows continuously thicker if the natural course of development continues undisturbed.

Wet humus needs considerable heat to warm up, and dry humus is an effective heat insulator. For these reasons soils with a thick humus layer remain cool for a long time in the spring and warm in the autumn; of these conditions, the coolness in spring has a greater effect on growth of trees. The shallow root systems of dwarf shrubs and spruce keep the humus layer porous, further promoting its temperature-insulating capacity. The drier, thicker, and more porous the layer is, the more effective is the temperature insulation.

The main purpose of prescribed burning in forestry is to improve the factors that regulate the growth of trees—water, temperature, and nutrients—while the immediate aim is to assure reforestation, often to change species of trees, and to facilitate the work of reforestation. One of the most serious obstacles to natural regeneration is a thick layer of raw humus. As a consequence of poor forestry, dry or fairly dry sites (*Vaccinium* type), which are most suitable for pine, are often occupied by spruce, completely preventing regeneration of pine. Burning has been found to be the most effective means of establishing productive stands under these circumstances.

B. Development of Virgin Forests

All the virgin forests remaining in Fenno-Scandia grew after early wild forest fires. The general trend in virgin forests is as follows: on an open site pine and deciduous trees, usually birch, are pioneer species. The shade-tolerant spruce rises underneath, and gradually the regeneration of birch and then pine is impeded. The birches die and rot, and the final result is a spruce-dominated stand of stunted growth with a few very old pines. Spruce litter decomposes more slowly than that of pine and slower still than that of birch. As the stand deteriorates the humus layer becomes thicker, more raw, and more acid. Simultaneously, the nutritional status of the stand worsens continuously because the bulk of the available nutrients is bound by the humus layer in a nonavailable form. As time goes on it becomes more and more difficult to regenerate spruce, and the spruce stand itself begins to degenerate.

Fig. 1. Typical final stage of degradation of a stand: 120- to 160-year-old pure spruce stand on a moist *Vaccinium*-type site. Volume of the stand 125 m³/ha, annual growth 2.1 m³/ha. Humus plus moss layer 30–50 cm thick. No signs of ancient forest fires nor of cuttings. Burning object of first order.

Owing to the very slow development of spruce, the result is an unproductive spruce stand of trees of all ages. On poor sites this process from burning to final degeneration of the spruce stand may take up to 200 years. Good sites are fairly resistant to this process so it takes a longer time. After a new fire the process starts again from the beginning. Figure 1 shows the stage of degeneration on a *Vaccinium*-type site.

The process described above occurs especially on sites that are unquestionably suitable for pine—on *Vaccinium*-type and the more barren *Myrtillus*-type sites. The basic reason for the succession is that spruce as a shade-tolerant tree can regenerate under a full canopy, but, as it needs 2 to 3 times as much nutrients as pine does, it reaches dimensions of commercial timber very slowly on soils low in available nutrients.

Every now and then, wildfires interrupt the degenerative process in virgin forests. They always occur during the driest periods of summer, and their effect on stand development is often drastic. They kill all trees and other vegetation and usually burn the bulk of the humus layer over an extensive area. The area is seeded from the surrounding stands, and the cycle is renewed. Pine and birch are again the pioneer species because regeneration of spruce is very slow on burned sites (Hesselman, 1934). The good growth of the new generation of pine and birch stands has given foresters the idea of prescribed burning.

III. Physical Effects of Burning

A. SOIL TEMPERATURE

Temperature is an important growth factor in the Fenno-Scandian climate, and its effect is greater in the north. Soil temperature depends primarily on air temperature, but it can also be affected artificially. In Fenno-Scandia, soil temperature is generally low or moderate, only in exceptional circumstances excessively high, and an improvement in the thermal conditions of the soil usually benefits fertility of the site. For instance, raising the temperature from 6° to 12.5°C doubled the CO_2 evolution of forest humus in our incubation experiments in the laboratory, and raising it to 20°C quadrupled it.

Most of the matter that is consumed during prescribed burning is live moss or other surface vegetation and slash, and their amounts vary widely from site to site according to the site type and tree stand. In our material, the humus layer became slightly thinner as a direct consequence of burning, decreasing from 5.3 to 4.1 cm. However, it became distinctly thinner during the following years, too. The thinning after burning averaged 1.6 cm but ranged up to 20 cm at the base of trees and was due mainly to compression: When fire kills dwarf shrubs, especially heather, their decaying roots can no longer keep the humus porous. The thinning of the humus layer after burning is well demonstrated in Fig. 2. The

Fig. 2. A stump on *Vaccinium*-type site 16 years after burning. The bleached part illustrates the compression of the humus layer: Right after burning it was still covered by humus. Stump diameter 26 cm.

humus layer was thinnest between the eighth and twentieth years after burning, subsequently becoming thicker again. To benefit the soil temperature, it is extremely important to ensure that at least all the living moss is destroyed by burning. Optimum burning is that which consumes about half of the humus layer, in addition to killing the living vegetation, and the minimum requirement is that the moss layer will be totally destroyed.

Thinning the humus layer improves the thermal conditions of the site, which is one of the main purposes of prescribed burning. Our unpublished studies shed light on the effect of burning on soil temperature. Table I shows the average temperatures, reported at 1400 hrs, during the period of observation (June 17–August 23, 1959) in the year of burning. The thickness of the humus layer was 3 cm at the burned sites, that of the humus and moss layers 9 cm in the unburned clearings and under tree canopies. It can be seen that prescribed burning and clear cutting had a great effect on soil temperature, and the effect was clearly seen down to a depth of 20 cm in the mineral soil. The dark surface of a burned site effectively absorbs solar radiation; consequently, the surface layers of the burned sites were warmer during the growth period than those of unburned sites, especially those under the canopy. The highest temperature measured at the surface of burned humus was 52.8°C, that at the surface of the humus in the clearing was 36.0°C, while the simultaneous air temperature was 29.9°C. It was also found that the humus layer was a good insulator: The temperature under

TABLE I
Effect of Burning and Shading on Soil Temperature
between June 17 and August 23, 1959[a]

	Burned area (°C)	Unburned area	
		Clearing (°C)	Under canopy (°C)
Air, 5 cm above ground	23.6	24.8	20.6
Humus surface	31.3	24.4	18.0
Mineral soil surface	16.0	12.8	9.1
Mineral soil			
Depth 10 cm	12.5	10.6	8.7
Depth 20 cm	12.0	10.2	8.3

[a] Spruce stand, 100 years old, South Finland.

the humus layer was quite low, especially under the canopy. The vital processes depend greatly on soil temperature, particularly its influence on microbial activity and nutrient uptake. Even the thin humus layer of the burned area proved to be a good insulator: The highest temperature measured at the interface of the mineral soil and humus layer of the burned site was 18.3°C.

The thickness of the humus layer has a great effect on soil temperature in midsummer. The average daily temperature of June to August at the surface of the humus was 15.3°C in the summer of 1956 and 20.6°C in the summer of 1959. A humus layer 10 cm thick lowered the temperature by 4.4°C in 1956 and by as much as 8.9°C in the warmer summer of 1959. Especially on very hot days in the latter summer, differences of as much as 12°C were recorded between the air and the mineral soil surface. The humus layer also decreased the temperature gradient in the mineral soil, and in every case a dry humus layer was a more efficient heat insulator than a moist one.

B. Soil Moisture

The importance of the moisture factor is not as conspicuous in Fenno-Scandia as in the more southern regions. Even in Fenno-Scandia, however, water may be the minimum growth factor in exceptionally dry and warm summers and on soils with a poor water-holding capacity. The most usual soil class in Fenno-Scandia is morainic soil; only in exceptional cases is water scarce in these soils. The lack of water is much more evident on esker gravel and coarse sand soils.

In forest soils with a deep groundwater table, the moisture conditions depend largely on the humus layer. The effect of the humus layer on soil moisture is twofold: When it is dry, the living moss and humus layers sometimes imbibe large amounts of water, so that only heavy rains can reach the tree roots; during dry periods, the humus layer effectively hinders evaporation from the soil. Both water-holding capacity and the ability to hinder evaporation are much smaller for charred humus than unburned humus. In addition, the higher temperature of a burned site tends to increase evaporation. On the other hand, the humus layer of a burned site dries quickly, and evaporation is diminished as the capillaries are broken.

The hindrance of water penetration and the higher evaporation are detrimental to growth on coarse soils with poor water-holding capacity and in dry summers, but these influences are mainly favorable on common morainic soils of Fenno-Scandia and in rainy summers. Experience from the field has shown that young stands on burned sites do not usually suffer from a lack of water if the humus layer has not been totally burned.

IV. Organic Matter

A. Ground Vegetation

The ground vegetation of an old spruce stand is very monotonous if the vegetation of an average or poor forest site has developed undisturbed for centuries. The most common plants are mosses, followed by dwarf shrubs. Grasses and herbs are scarce, and lichens also appear in minute amounts on sites where prescribed burning is recommended. It should be borne in mind, however, that under natural conditions no species of plants disappear from the site as a consequence of competition from climax species. They remain in dwarf size, and, when the circumstances become more favorable, they react rapidly. In Fenno-Scandian countries it is very important to consider the development of ground vegetation because evaluation of site fertility is based principally on ground vegetation.

Above-ground vegetation is completely destroyed by prescribed burning, but species with subterranean regenerative organs may survive. Examples are certain herbs, most grasses, dwarf shrubs, and deciduous trees. The more thoroughly the humus layer is burned, the more completely the subterranean organs of regeneration are destroyed. In our studies the development of vegetation on burned sites revealed largely the same trends as those on the wild-burn sites in North Finland described by Kujala (1926) and Sarvas (1937), and the spread of the various species of plants can generally be attributed to the same causes.

The spreading capacity of seed, however, seems to have a greater effect on development of vegetation than was claimed by the above authors; the first species found in our burned areas were dandelion (*Taraxacum officinale* Web.), hawksweed (*Hieracium* spp.), and fireweed [*Chamaenerion angustifolium* (L.) Scop.]. Every seed kernel of these species has a pappus that enables it to be transported for long distances. The seeds of certain other species of plants are spread by birds; for instance, those of raspberry (*Rubus idaeus* L.) and of several grasses. At least of equal importance for development of vegetation in a burned area are the subterranean regeneration organs of several species. These may be largely destroyed by burning, but they are hardly ever killed off entirely.

In the year of burning only a few scattered plant specimens could be found at the site, but the vegetation increased rapidly in subsequent

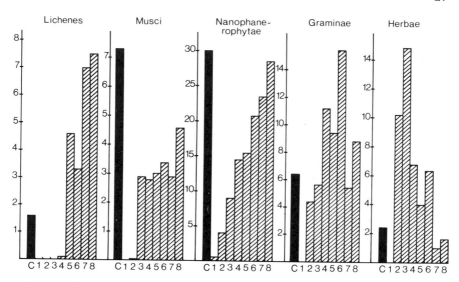

Fig. 3. The percentage coverage of different groups of ground vegetation. On x axis time after burning: 1 < 1, 2 = 2–3, 3 = 4–6, 4 = 7–9, 5 = 10–12, 6 = 13–19, 7 = 20–32, 8 = 33–50 years; C = controls.

years (Fig. 3). Grasses and herbs were the first to appear on the site; owing to lack of competition, they spread rapidly in a few years. The maximum abundance of grasses was 2.5 times and that of herbs 5 times as great as on the control sites. Consequent to shading by the growing tree stand, the quantity of herbs began to fall off after 6 years and that of grasses 20 years after burning. For a short time fireweed was the dominant herb species, but 20 years later only a few specimens could be found. Most abundant among the grasses were species of *Calamagrostis;* their coverage was double that on the control sites between the seventh and twentieth years after burning.

The first dwarf shrub to appear was raspberry; its coverage was 0.5% in the year of burning and 1.4% 5 years later. The total coverage of dwarf shrubs reached the normal level 50 years after burning. On the control sites lingonberry (*Vaccinium vitis-idaea* L.), blueberry (*Vaccinium myrtillus* L.), and heather [*Calluna vulgaris* (L.) Hull] were the most common dwarf shrubs. Of these, heather increased its coverage fastest, followed by lingonberry.

Mosses first appeared in substantial amounts on the burned sites after 3 years. Between 4 and 6 years after burning their coverage rapidly increased to 30%, but it increased above this level only after 30 years. In 50 years the coverage of mosses reached 60% of that on control sites.

Initially the most common moss was fire moss (*Ceratodon purpuraeus*), but its coverage was never extensive. *Polytrichum* spp., *Pleurozium schreberi*, and *Dicranum* spp. were the first mosses to appear in substantial amounts. The most common species on the old burned sites were the latter two; the coverage of *Pleurozium* alone was 20%.

Lichens were slower to appear on the burned sites than any other species. They became clearly apparent in the seventh year after burning, but then they increased rapidly; on the 50-year-old burned site they were 5 times as abundant as on control sites. Perhaps the culmination point had just been reached.

It was earlier believed that fireweed and raspberry appeared abundantly on recently burned sites because they needed nitrogen in nitrate form (Hesselman, 1917). However, Tamm (1956) has shown that fireweed is able to utilize ammonia nitrogen too. According to Uggla (1957), raspberry is so abundant on burned sites because its seeds are carried everywhere by birds, and the heat of burning promotes their germination.

The histograms in Fig. 3 make it clear that vegetation takes a long time to stabilize after burning. This is a crucial fact to be borne in mind when site fertility is classified according to ground vegetation. The slow stabilization of lichens and mosses is of special significance. Sarvas' (1937) studies indicate that stabilization is much slower in North than in South Finland.

The development of herbs, grasses, and dwarf shrubs after burning is reasonable: Herbs and grasses have effective means of spreading their seed, they are fast-growing, and they grow best on fertile sites. Dwarf shrubs are slower to develop, but being evergreen they gradually displace the herbs and grasses, especially when the young trees begin to shade the ground at the same time.

The moss cover was slow to increase. The most likely reason is that the microclimate on the control sites in the old spruce stands was different from that in the young pine stands of burned sites. In the former, the dense spruce canopy prevented both aeration and light penetration into the interior of the stand, thus making the climate under it moist and cool. Judging by their abundant growth, mosses evidently favor this kind of microclimate. The rapid development of lichens after burning is largely due to the fact that they need high light intensity.

Once again it must be emphasized that grasses and herbs, which denote high fertility in conventional forest site classification, had a denser cover for as much as 20 years after burning. Of the indicators of a poor site, only lichens were found in greater abundance on the burned sites than on the controls, and even here the excess was only 7%. Dissimilarities in the development of different species of plants must be borne in mind in classifying sites on the basis of ground vegetation. When determined on this basis, the site quality of young stands on burned sites is often overestimated on moraine soils and underestimated on coarse water sediments.

B. Quantity of Humus in Soil

The prevalent soil type in Fenno-Scandian forests is podsol; brown soils are very rare exceptions. The humus layer is usually raw mor; on fertile sites moder humus is prevalent. Mull humus is very rare, appearing only on the most fertile sites. These are mainly on calcareous soils and occasionally under deciduous trees. Burning only benefits forestry on mor–humus sites.

Podsol soils have a humus layer of varying thickness over the mineral soil and a living moss layer over the humus layer. The thickness of these layers depends on several factors, especially the rate of humus decomposition. Because the moss is a more effective heat insulator as it becomes thicker, the decomposition of organic matter in the humus layer below it slows down at the same rate as the increase in thickness of the moss layer. For this reason the thermal conditions continuously develop in an unfavorable direction, leading to continuously retarded decomposition, and thus to an accumulation of humus. One unfavorable factor leads to another, and if natural progression is allowed to continue undisturbed, site fertility deteriorates slowly but surely. This course of development is most typical of rather poor sites under spruce.

Spruce favors fertile soils; fertile soils are moist, and moist soils are cool. However, the humus on fertile sites decomposes rapidly, and on the most fertile sites thick layers of raw humus do not accumulate, not even under spruce. Both spruce and pine grow on average sites, but barren sites are good for pine only. As a result of poor forestry, large areas of mediocre and barren sites have turned into almost pure spruce stands, with a large part of the nutrients locked in a nearly unavailable form into the thick humus layer. These stands grow slowly; the lower the site fertility, the faster they deteriorate.

The average weight of organic matter in the humus layer before and after burning is shown in Fig. 4. On the average, the control sites contained 15 metric tons of ash-free organic matter per hectare in the living ground vegetation and 33 tons in the humus layer. The mineral soils contained a much greater amount of humus than the humus layer proper. The percentage of humus in the soil decreased rapidly with increased depth, and, down to a depth of 60 cm, the control mineral soils contained

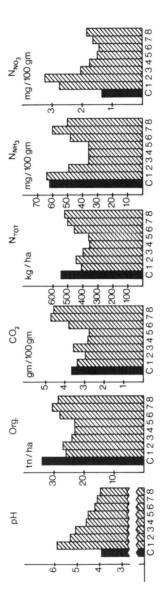

Fig. 4. Effect of burning on certain properties of the humus layer. Acidity was measured in water, organic matter as loss on ignition, CO_2 evolution gravimetrically after 3 weeks' incubation (Viro, 1955), total N by the Kjeldahl method, mineral nitrogen was extracted with KCl and measured by Nessler's reagent. See Fig. 3 for key to x axis.

an average of 78 tons of humus per hectare. The bulk of this humus was almost undecomposable.

C. Changes in Humus Quantity

Decomposition of humus is one of the most important phenomena of nature. Upon decomposition, nutrients that have been bound by organisms again become available. In virgin forests the decomposition of old humus and formation of new organic matter are in equilibrium. Under unfavorable conditions, however, decomposition of organic matter may become slower than formation of new material. This leads to an accumulation of organic matter, which simultaneously causes deterioration in the thermal conditions in the soil. Such a development is typical of spruce stands on poor soils and is more frequent toward the north. On the other hand, the litter of deciduous trees decomposes rapidly, and no accumulation of humus occurs in such stands, as the litter of these species—like grasses and herbs—promotes humus decomposition.

The weight of the humus layer proper (F + H layers) diminished during our burnings by about one-fourth, from 33 to 25 tons per hectare. The difference between burned and control sites did not subsequently decrease very much. In the first 12 years it fell by about 12%. After that, formation of new humus exceeded decomposition of old humus, but it took 50 years for the sites to return to the same state as before burning.

In mineral soil, the burned sites contained an average of 10% less humus than the control sites. The difference was much greater in the first 10 cm layer (17%) than in the deeper layers (10–7%). Burning cannot have a direct consuming effect on the amount of humus in mineral soil. The diminution of the humus in mineral soil was probably due mainly to the increase in decomposition caused by decrease of acidity. On the other hand, alkali and alkali earth metals were present in the ash as oxides or carbonates. As their solution is basic, they may have dissolved some of the humus while seeping through the soil.

V. Burning and Nutrients

A. Mineral Nutrients

1. Total in Humus Layer

Most of the material destroyed by prescribed burning is logging waste and ground vegetation, whose mineral substances penetrate the soil with rainwater. Our samples from the humus layer on control sites did not include living vegetation or logging waste. For this reason the total amount of nutrients measured on the burned sites was usually larger than that measured on control sites because humus samples from the former also included the ash of all the burned matter. The rate at which the quantities of nutrients returned to their normal level differed widely from one nutrient to another. Figure 5 depicts the total amounts of nutrients in the humus layer in the year of burning and for 50 years after burning. Histograms also show the amounts of nutrients in the humus layer of the control sites as an average of all the controls.

Unfortunately, the nutrients in the living vegetation on control sites were not analyzed, so we have to use approximations. According to Mork (1946), Aaltonen (1950), Tamm (1953, 1964), and Viro (1955), the quantities of calcium and magnesium in slash and ground vegetation can be estimated as being similar to the total quantities in the humus layer. Potassium and phosphorus are probably more abundant in the humus layer. This means that the total amounts of these nutrients in the living and dead organic layer were 620 kg CaO, 110 kg MgO, 85 kg K_2O, and 120 kg P_2O_5 per hectare.

Calcium and magnesium in ash are mainly in oxide and carbonate forms, with small amounts of phosphates. The oxides are water soluble but are rapidly converted into carbonates that are soluble under acidic conditions only. Calcium and magnesium were not leached from the charred humus layer at all in the year of burning, and only very slowly in the first years after burning. The control level of the humus layer was reached in about 20 years. There were no increases in the total amounts of calcium and magnesium as there was for potassium in the oldest burned sites. The reason is that new vegetation takes up relatively more potassium than calcium and magnesium, and the liberation of potassium from decomposing new vegetation is faster than that of the latter two nutrients.

Potassium in fresh ash is also in either oxide or carbonate form, and the oxide rapidly changes to carbonate. All the potassium compounds formed by burning are water soluble; this can be clearly seen in Fig. 5. Even in the year of burning, the amount of potassium in the charred humus layer fell below that of the humus layer of the controls, notwithstanding the nutrients from the living vegetation, and it continued to decrease in the following 5 years. After that, the amount of potassium decreased very little during the next 6 years. From the twelfth year on the total amount of potassium in the humus layer increased, probably due to formation of rapidly decomposing new organic matter, but not even at the end of the 50 years had it attained the level of the controls.

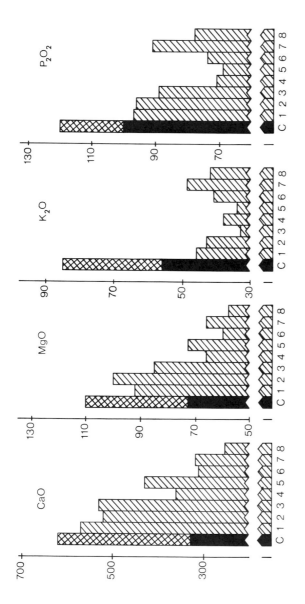

Fig. 5. Total amounts of nutrients in the humus layer, kg/ha (cross-hatching represents living vegetation). Ignited at 550°C. See Fig. 3 for key to x axis.

After burning, phosphorus also was found to a large extent as water-soluble alkali phosphates, and it followed roughly the same course in leaching as potassium. But when the amount of potassium fell by about 0.8 kg equivalent per hectare in the year of burning, the decrease in the amount of phosphorus was only 0.6 kg equivalent. As the ash also contained some sodium, a considerable part of the alkalis were evidently leached in the form of carbonates.

According to these figures, considerable quantities of nutrients migrated from the organic layer to the mineral soil as a consequence of burning. The leaching of cationic nutrients corresponded to their order of retention in organic matter. Net losses of the most soluble nutrients, phosphorus and potassium, from the humus layer were most evident during the first 6 years after burning. Equilibrium prevailed between the seventh and twelfth years: The amounts leached away or taken up by vegetation were about as large as those liberated from the decomposing new vegetation. From the twelfth year on, liberation from new organic matter exceeded the uptake and leaching. In the case of calcium, uptake and leaching exceeded the liberation throughout the period of observation. Magnesium reacted in almost the same way as calcium.

2. *Exchangeable Nutrients*

Exchangeable nutrients are usually considered to be available to vegetation. Under normal conditions there is a balance in the amount of exchangeable nutrients. They are released continuously by decomposition of organic matter and weathering of minerals, and they are taken up by micro- and macrovegetation. In addition, a small quantity of the liberated nutrients is lost by leaching. Burning drastically upsets this balance for a long period. Some of the mineral nutrients released by burning are dissolved and rapidly leached out of humus, and some are bound in the soil in an exchangeable or nonexchangeable form. After burning, there are large amounts of available nutrients in the surface soil, and losses by leaching are then possible, owing to absence of vegetation. It takes a long time for the balance between liberation and uptake of nutrients to be restored, with length of time varying from one nutrient to another.

The solution used for extraction of the exchangeable nutrients and easily soluble phosphorus consisted of 1 N acetic acid that was adjusted to pH 4.65 with ammonia. Organic matter was determined as loss on ignition at 550°C and total nitrogen by the Kjeldahl method. For an overall picture, the quantities of exchangeable nutrients of the control soils per unit of area are shown by fertility classes in Table II. The quantities of nutrients and organic matter were generally largest in the

TABLE II

EXCHANGEABLE MINERAL NUTRIENTS AND TOTAL ORGANIC MATTER AND
NITROGEN IN CONTROL SOILS[a]

Site type	Soil layer (cm)	kg per hectare				tons per hectare	
		CaO	MgO	K$_2$O	P$_2$O$_5$	Organic	N
Vaccinium	Humus	123.0	22.8	34.6	20.3	28.4	0.434
	0–10	95.5	25.9	47.4	18.8	25.9	0.793
	10–20	57.7	17.8	36.2	12.9	18.9	0.654
	20–30	54.8	17.1	32.8	8.8	12.2	0.460
	30–60	129.9	48.6	88.2	27.3	13.8	0.822
Myrtillus	Humus	164.6	38.6	45.9	28.0	35.4	0.623
	0–10	111.6	30.3	52.2	17.0	29.0	0.893
	10–20	76.9	18.3	38.8	10.3	20.9	0.759
	20–30	62.4	21.3	34.0	7.4	14.1	0.561
	30–60	132.0	53.7	94.5	22.8	17.1	0.972
Oxalis-	Humus	202.6	43.4	44.9	22.9	29.3	0.565
Myrtillus	0–10	372.5	57.8	61.8	12.8	38.8	1.265
	10–20	150.4	29.5	47.9	8.5	23.7	0.907
	20–30	98.2	25.7	44.3	8.0	19.4	0.755
	30–60	285.6	87.6	115.3	17.1	29.0	1.257

[a] Eighty- to 120-year-old conifer stands of different tree species composition. Extracting solution: 1 N acetic acid adjusted to pH 4.65 with ammonia. The coarse soil ($\phi >$ 2 mm) is also taken into account when calculating the amounts per area. Humus layer does not include the living moss or other vegetation.

most fertile soils; the only exception was phosphorus, which in each mineral soil layer was most abundant on the most barren sites and least abundant on the most fertile sites. Evidently, the balance of phosphorus in Fenno-Scandian mineral soils is fairly good in general because only in exceptional cases is there an increase of growth with phosphorus fertilization (Viro, 1967).

Exchangeable nutrients after burning are presented in Fig. 6. In the autumn of the year of burning, the amount of exchangeable nutrients in the humus layer was greater on burned than on control sites; only the amount of potassium was equal in both groups. The amounts began to decrease because of leaching, and the speed of leaching varied considerably for the different nutrients. The poor fixation of potassium by the humus layer was the most conspicuous feature; although burning added about 50 kg per hectare, the amount of exchangeable potassium fell to 50% below the level of the controls in 9 years. Contrary to all other mineral nutrients, no increase of potassium in the humus layer could be detected even in the autumn of the year of burning. The

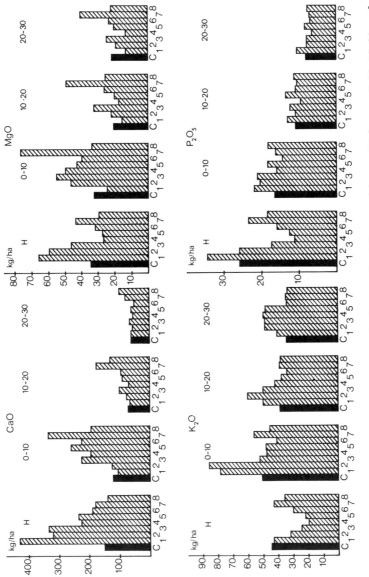

Fig. 6. Amounts of exchangeable nutrients (kg/ha) in the different soil layers: H, humus; 0–10, 10–20, and 20–30 in centimeters. Nutrients were extracted with 1 N ammonium acetate (pH 4.65). See Fig. 3 for key to x axis.

potassium that was leached from the humus layer was partly bound by the mineral soil, but there too its adsorption was rather weak. After burning, the surface layer of the mineral soil contained increased amounts of exchangeable potassium, and even in the year of burning a distinct increase of potassium was noted as deep as 30 cm in the mineral soil. A further increase of potassium, probably deriving from the new vegetation, was found in the humus layer 10 years after burning.

After burning, the humus layer contained large quantities of exchangeable calcium—3 times the amount on control sites. Calcium was leached from the humus layer very slowly, only reverting to the control level after 50 years. In the year of burning, there was no increase of calcium in the mineral soil. Most of the leached calcium was adsorbed by the surface layer of the mineral soil, and only after 20 years was a small increase detected in the subsoil. The effect of burning on the amount of exchangeable calcium lasted even longer in the surface layer of the mineral soil than it did in the humus layer. Perhaps the increase of calcium had just culminated 50 years after burning.

After burning, the humus layer contained twice as much magnesium as the controls, but 6 years later, the amounts were equal. As in the case of calcium, most of the magnesium lost from the humus layer was adsorbed by the surface of the mineral soil through ion exchange.

Leaching of magnesium from the humus layer was faster than that of calcium. There were also small decreases in the amounts of both calcium and magnesium in the surface layer of mineral soil during the year of burning, and in the case of magnesium a decrease was also found deeper in the soil. This was most likely due to potassium which partly displaced calcium and magnesium on the surface of soil particles. It can be seen that the leaching of all the cationic nutrients followed the general laws of chemistry: Bivalent elements displaced monovalent elements on the surface of the soil particles. Among the bivalent elements, calcium displaced magnesium according to the law of mass action. Monovalent potassium displaced bivalent elements to some extent, owing to its abundance after burning.

The amount of easily soluble phosphorus in the humus layer was practically doubled by burning. However, part of this phosphorus was so easily soluble that even in the year of burning noticeable amounts were found at a depth of 30 cm in the mineral soil. During this year the humus layer had lost half of the additional phosphorus (about 25 kg P_2O_5 per hectare), probably mainly as potassium phosphate. In the following 6 years, the amount of phosphorus in the humus layer rapidly fell to 11 kg. Nevertheless, no increase of easily soluble phosphorus was detected in the mineral soil. Although it would seem as if 24 kg phosphorus per hectare were leached beyond the reach of vegetation, this cannot have been the case. Phosphates are so sparingly soluble in soils that contain great amounts of calcium and magnesium that the leaching of phosphate is very slight.

The rapid decrease of phosphorus in the year of burning was probably due to an abundance of potassium in the ash. The easily soluble phosphates were leached out before equilibrium between the different electrolytes was restored in the soil. After that, the mineral phosphorus in the soil was mostly contained in sparingly soluble magnesium and calcium phosphates.

There was a highly significant single correlation between exchangeable magnesium and easily soluble phosphorus in the humus layer of the burned sites. The early decrease and later increase of magnesium and phosphorus after burning occurred simultaneously (see Fig. 6); evidently they were signs of the same occurrences: the rapid formation of sparingly soluble magnesium phosphate in an almost neutral soil (pH 6) and much slower dissolving of phosphate in a fairly acid soil (pH 4).

Since the sparingly soluble phosphates formed by calcium resemble those formed by magnesium, the ratio of calcium in the retention of phosphorus was also studied by variance analysis. Although the solubility of calcium phosphates is about equal to that of magnesium phosphates and although the concentration of the exchangeable calcium was 4 times that of magnesium, the single correlation between calcium and phosphorus ($r = 0.583$) was much lower than that between magnesium and phosphorus.

The single correlation between magnesium and phosphorus proved to be almost as high as the multiple correlation between magnesium and calcium and phosphorus ($R = 0.857$). The partial correlation between calcium and phosphorus was very low (0.02), indicating that phosphorus was bound in the humus layer of the burned sites almost entirely by magnesium (partial correlation coefficient 0.765).

At first, the quantities of several nutrients in the humus layer decreased after burning but began to increase again about 10 years later. This was clearest in the cases of potassium and phosphorus, and it was similar to the increase of ammonia (Fig. 4). The available amounts of potassium and phosphorus had just about reached normal levels in all soil layers 50 years later, but the amount of calcium still exceeded it and probably that of magnesium also, although not so clearly.

Stabilization of the nutrient status was a result of increase in amounts of new vegetation, especially profuse growth of grasses and herbs in the early years and later the vigorous growth of the young pine–birch stands. These effectively took up dissolved nutrients from the mineral soil and thus decreased leaching due to burning. The litter from these plants decomposes rapidly, and the nutrients immediately reenter the organic cycle.

B. TOTAL NITROGEN

Forest soils contain large amounts of nitrogen. The nitrogen content of the organic matter in the humus layer of control sites varied with fertility of the site: It was higher on fertile than on barren sites, and the same held good for the surface layers of mineral soil. The nitrogen content of organic matter as calculated from the ash-free humus increased on moving downward from about 1% in the living moss and 1.6% in the humus layer to 5.7% in the 30–60 cm layer of mineral soil. This means that the rapidly decomposing nitrogen compounds disappeared quickly; afterward the nitrogen compounds decomposed more slowly than those not containing nitrogen.

In the humus layer the control sites contained 540 kg nitrogen and in the lower vegetation and slash 180 kg nitrogen per hectare. The amount of nitrogen was much greater in mineral soil than in the humus layer, and it decreased in the 10 cm soil layers on moving downward. Down to a depth of 30 cm the mineral soils contained 2350 kg nitrogen per hectare, and the total amount on the sites down to the same depth was 3070 kg per hectare.

Burning caused a clear decrease in the carbon–nitrogen ratio of the portion of the humus layer that remained unburned. The C/N ratio in Fenno-Scandia is usually quite high, especially in the humus layer (see Table III). It is evident that C/N ratios were somewhat lower on burned than on control sites. They were very high in the humus layer but much lower on the surface of the mineral soil, and they decreased rectilinearly toward deeper layers. Burning itself hardly produced any effect on C/N ratios of different mineral soil layers. This ratio is generally assumed to indicate a tendency of nitrogen to be mobi-

TABLE III
AVERAGE C/N RATIOS IN TOTAL MATERIAL

Site	Humus	0–10 cm	10–20 cm	20–30 cm	30–60 cm
Control	38.9	18.2	16.6	14.6	10.6
Burned	35.1	17.5	16.1	14.6	9.8

lized: the lower the ratio, the greater the tendency. However, though C/N ratio of the control mineral soil humus decreased from 18.2 in the surface layer to 10.6 in the subsoil, mobilization of nitrogen in the latter was much slower than in the former. The humus of the subsoil had been leached from the humus layer and by then it was bound in inert complexes that were almost undecomposable.

VI. Biological Effects of Burning

A. Acidity

Podsol soils are acid by nature and tend to become more so under coniferous tree species, especially under spruce. According to the literature, the degree of acidity has a marked effect on site fertility because biological activities, such as nitrogen mobilization and decomposition of soil humus, largely depend upon it (Waksman, 1936). After burning, the herbaceous vegetation and deciduous trees that comprise the pioneer species on a burned area help to keep the acidity down and promote biological activity of the site.

When organic matter is burned, the mineral substances in it are released in the form of oxides or carbonates that usually have an alkaline reaction. In our experiments these compounds reduced acidity of the remaining humus layer by 2–3 pH units. The decrease in soil acidity did not last long: Because of leaching away of alkalis and earth alkalis, ion exchange, and the formation of new organic matter, the acidity reverted to its original level 50 years after burning. In mineral soil, changes in acidity were small owing to its great mass as compared to the humus layer. During the first 20 years after burning, however, acidity of the surface of mineral soil was on the average 0.4 pH unit lower on burned sites than on unburned sites, and a difference of 0.2 unit persisted for at least 50 years.

It should be noted that in our material the acidity of the humus layer, as measured from a mixed sample of the plot, never became alkaline; the highest pH measured in the humus layer of the burned site was 6.8. However, the reaction of the soil surface, measured in the field immediately after burning, was sometimes clearly on the alkaline side, the highest reading being pH 8.0. The acidity of control sites varied from pH 4 to 5.5; the humus layer was always the most acid and the subsoil the least.

The most active part of mineral soil is the clay fraction in which ion exchange takes place. In the virgin forest, the amounts of the various ions in the clay fraction are in equilibrium, but after burning large quantities of easily soluble electrolytes enter the soil and the balance is disturbed.

In our material the correlation coefficient between the active and exchange acidity (c_H measured in water and in potassium chloride, respectively) was highly significant. The coefficient was 0.81 for control sites and 0.95 for burned sites. These coefficients indicate that, in the control soils especially, equilibrium prevailed in the content of the various ions in the humus layer. The unusually high correlation in the humus layer of the burned sites was due to large amounts of electrolytes in the soil surface after burning and to their slow but steady replacement by hydrogen ions. In the mineral soil there was no correlation between different hydrogen ion concentrations on the burned sites, but a highly significant one ($r = 0.49$) in the surface layer of control sites.

Acidity of control sites was rather independent of the amount of exchangeable cations ($r = 0.37$). On burned sites the correlations were clearer, but even there the coefficients were low ($r = 0.50$ for exchange acidity, 0.56 for actual acidity).

B. Humus Decomposition

The tendency of the humus layer in podsol soils is to become thicker. It was seen earlier that the humus layer may contain large amounts of nutrients, but these are primarily in an unavailable form. Under unfavorable conditions this poor nutrient status may continue for centuries or even deteriorate. The principal cause of this trend is the high degree of acidity of the soil, mainly owing to the kind of bedrock. Other major reasons are cover by coniferous species of trees, especially spruce, and low soil temperature, which in turn is due to the thickness of the humus layer. The decomposition of old humus is very slow. Owing to the loss of the most rapidly decomposing surface parts by burning, the remaining humus consists largely of lignins. Such lignins are mainly decomposed by certain actinomycetes, and in the absence of oxygen their decomposition is slow.

Such a development toward a site with a thick humus layer and an almost unproductive tree stand does not occur on calcareous soils or under deciduous trees. To a considerable degree, prescribed burning gives us the advantages of these two factors, at least for one generation of trees.

The temperature of the flames is very high in prescribed burning, but a high temperature does not occur below the burning humus. In a soil with a 3 cm layer of humus, the temperature at the surface of the mineral soil did not exceed 100°C in Uggla's (1957) experiments. In Ahlgren's experiments (1970), in which the thickness of the humus was 2–3 inches initially and 1 inch after burning, the temperature at the surface of the mineral soil rose above 300°C for only 14 min during a burning that lasted 5 hr. The temperature of the soil during burning is important chiefly in relation to survival of microbial and mycorrhizal populations. In the light of the above figures it is obviously rare for all the soil microorganisms to be killed at burning as these can also be found quite deep in the soil (Mikola et al., 1964). In our experiments, too, a lively CO_2 evolution was noted in all the humus samples from the burned sites, which indicates good survival of microbes.

It was pointed out earlier that burning releases large amounts of nutrients, primarily from logging waste, living vegetation, and the surface of the humus layer. Without fire this release through natural decomposition would take a very long time, and perhaps the bulk of the matter would never decompose. For if there are no forest fires the thickness of the humus layer of coniferous stands tends to increase continuously under Fenno-Scandian conditions.

The speed of decomposition of organic matter depends mainly on its origin, the temperature of the soil during the growing season, and soil reaction. It is impossible to construct an accurate scale for the rate of decomposition of organic matter of different origins, but, in general, herbs are the quickest to decompose, followed in order by grasses, leaves of deciduous trees and blueberry, then lingonberry, heather, and conifer needles. The slowest to decompose are mosses and lichens (Mikola, 1954; Viro, 1955).

Among the leaves of trees, alder and birch leaves decompose the most quickly. Decomposition of pine needles is much slower and that of spruce needles is slowest. Roughly speaking, herbs and alder leaves, blueberry and birch leaves, grasses and pine needles, and mosses and spruce needles decompose at about the same speed. The speed also varies widely for the same kind of litter, depending on its nutrient content, which in turn depends on site fertility.

It is a well-known fact in chemistry that the speed of a reaction increases two- to threefold when the temperature rises by 10°C, and biological activity probably increases even more. From the point of view of site fertility, any increase in soil temperature is almost entirely favorable under Fenno-Scandian conditions. Only on a very coarse soil, after the entire humus layer has been burned from a site, may the soil temperature be too high on a warm summer day. Under such circumstances, Vaartaja (1950) has measured temperatures of over 50°C in the surface soil.

C. Nitrogen Mobilization

1. Biological Aspects

The humus layer of Fenno-Scandian forest sites is usually acid. Particularly as a consequence of several successive generations of spruce it has a tendency in virgin forests to become continuously more and more acid. Microbial activity depends largely on acidity and temperature of the soil, and burning has a favorable effect on both of these. The earlier fear that burning was unfavorable for soil microbes seems to be exaggerated. Only the microbes in a thin surface layer are destroyed; those lying somewhat deeper survive and soon repopulate the surface soil. As a result of burning, the number of microbes in the surface layers of the soil increases multifold (Ahlgren and Ahlgren, 1965).

The humus layer contains large amounts of nitrogen but only a minute part of it is in a form that the vegetation can use. The nitrogen must first be mineralized e.g., it must be converted into ammonia or nitrate nitrogen. From the point of view of forest site fertility, nitrogen mobilization is one of the most important phenomena in the soil. The hypothesis that burning has a favorable effect on nitrification was put forward by pioneer researchers such as Heikinheimo (1915), Hesselman (1917), and Eneroth (1928).

Figure 4 shows the effect of burning on the amount of ammonia and nitrate nitrogen in the humus layer (leached with 1 N KCl after 6 weeks' incubation at 20°C). The histogram shows no increase in the amount of ammonia nitrogen as a consequence of burning. On the contrary, it indicates that the amount of ammonia dropped sharply during the first 3 years after burning. An increase in the amount of ammonia was noted later but only after 12 years, and the normal level was not reached until after 50 years. On the other hand, there was already a clear increase in the amount of nitrate nitrogen during the year of burning, and increased amounts of nitrate were found for at least 6 years after burning. The greatest amounts were 3 times those of the controls.

Figure 4 also shows that the ammonia nitrogen content of the humus layer of the controls was about 20 times the nitrate content. The decrease in the amount of ammonia in the early years after burning was probably due partly to the simultaneous rise in nitrate, caused by the increased nitrification of ammonia owing to the drop in acidity. However, when the amount of ammonia fell by 15 mg per 100 gm humus, the simultaneous increase of nitrate was only 2 mg. Owing to the abundance of oxides and carbonates of alkalis and earth alkalis in the humus layer on a recently burned site, the reaction at the surface of the freshly burned humus layer may be neutral or even alkaline. In such a case, part of the ammonia nitrogen may have been lost through evaporation. However, this loss of ammonia is of no consequence, because the alkalinity probably disappeared after the first rains.

In the field the ammonia was partly leached from the humus layer into the mineral soil, as can be seen from Table IV. The principal reason for the apparent decrease of ammonia nitrogen in the incubation experiments, however, is that only the net changes of ammonia and nitrate nitrogen could be measured. Simultaneous phenomena that could not be measured included ammonification, nitrification, uptake of ammonia and nitrate by the microbes and vegetation, and in the control samples perhaps some denitrification also. Overrein's findings (1967) bear out our results. Biological nitrogen fixation by microbes and vegetation was accelerated by the reduced acidity, and it probably consumed a large part of the mineralized nitrogen. Nitrification largely depends on the reaction on the humus, and, as shown by Fig. 4, the nitrate content decreased as acidity increased.

Incubation experiments in the laboratory have shown that lowering acidity with calcium carbonate greatly decreases ammonification and

TABLE IV

MINERAL NITROGEN IN SOIL AFTER BURNING[a]

Soil layer (cm)	kg N per hectare			
	NH_4-N		NO_3-N	
	1st year	2nd year	1st year	2nd year
Burned plots				
Humus	7.8	4.7	0.13	0.38
0–10	11.0	17.2	0.63	1.22
10–20	6.3	7.6	0.68	1.24
20–30	5.3	6.9	0.68	1.20
Total	30.4	36.4	2.12	4.04
Unburned plots				
Humus	2.6	2.6	0.18	0.32
0–10	4.0	5.4	0.50	0.93
10–20	4.5	6.3	0.46	1.22
20–30	5.1	5.1	0.47	1.15
Total	16.2	19.4	1.62	3.62

[a] *Myrtillus* type site. Fine sand moraine soil. South Finland. Samples were taken twice a month from May 15 to September 15 and analyzed fresh. Sampling was begun the day after burning.

increases nitrification. The changes are much greater in fertile than in barren soil. On the most fertile sites, the increase of nitrate is greater than decrease of ammonia, whereas on poor sites, decrease of ammonia can be 6 times as great as increase of nitrate (Kaila *et al.*, 1953; Viro, 1963).

Figure 4 shows the mineral nitrogen content of the humus layer sampled in late summer. Nitrogen mobilization was also followed during the whole growing season. Table IV gives the average amounts of mineral nitrogen in a burned and an unburned site during the first two growing seasons after burning. Samples were taken every two weeks and analyzed while still fresh. The effect of burning on mineral nitrogen shown by the table is similar to that in Fig. 4, and the leaching of ammonia from the humus layer and fixation in the mineral soil is obvious. It can also be seen that very little ammonia had penetrated through the first 10 cm mineral soil layer. On the other hand, nitrate was poorly fixed in both the humus and mineral soil.

Taking the above into account, Fig. 4 evidently does not give a correct picture of changes in the amount of mineralized nitrogen caused by burning. When samples were taken continuously, it was noticed that the ammonia content of the humus layer rose fivefold at the time of burning—from 10 to 50 mg per 100 gm humus. The high ammonia content did not last long. It fell by half during the year of burning and almost reached the preburning level during the third summer.

2. Nonbiological Aspects

The change of organic nitrogen into ammonia by burning has long been a familiar feature in Finnish agriculture. Before the adoption of commercial fertilizers, burning cultivation was quite common on peat land fields. Burning was repeated every third or fourth year, and each time a layer of humus about 5 cm thick was consumed. Svinhufvud (1929) found that when peat was partly burned some of its nitrogen was converted into ammonia, and this became bound to the peat that remained unburned. He concluded that this accounted for good crops on burned peat lands. In our experiments, too, the amount of ammonia increased during burning (see Table IV). In the experiments of Kaila *et al.* (1953), heating up to 65°C was enough to cause a multiple increase in the amount of ammonia nitrogen, whereas it did not affect the amount of nitrate nitrogen. Before invention of techniques for industrial synthesis of ammonia, ammonium sulfate was a valuable by-product of the fuel peat industry (Gissing, 1909). However, the ammonia yield in these processes was quite low: at most 10% of the total nitrogen.

To make sure that formation of ammonia at burning is due mainly to heating, humus samples were slowly heated in a quartz tube. The gases generated were conducted first through a tube filled with damp humus and then through a sulfuric acid solution. The tests showed that 53–74% of the nitrogen in humus was converted into ammonia, and almost all this ammonia became bound to the damp humus. The more fertile the site, the greater was the percentage of mineralization. Only very little nitrate formed.

The nitrogen compounds that are converted by heat into ammonia at prescribed burning are probably the most easily decomposed compounds. Their natural decomposition finally produces the same result without fire. In laboratory experiments perhaps some of the more slowly decomposing compounds were also converted.

D. Measurement of Biological Activity

Carbon dioxide evolution is generally regarded as an index of microbial activity and speed of decomposition (Waksman, 1936). However, CO_2 evolution (measured after 4 weeks of incubation) does not give a correct picture of the speed of decomposition. For instance, it failed to do so in incubation experiments with humus samples fertilized in various ways (Viro, 1963). Similarly, in the present material the rate of CO_2 evolution was smaller in burned than in unburned humus for several years after burning (Fig. 4). Subsequently it increased, and 20 years after burning it was greater in the burned site. The reason for retarded CO_2 evolution after burning was the disappearance of the most easily decomposing surface humus; the later acceleration was mainly due to formation of easily decomposing humus from herbs and grasses.

Biological activity in soil usually increases with decreasing acidity. However, here the acidity increased continuously after burning, but so too did CO_2 evolution. In the incubation experiments mentioned earlier (Viro, 1963), liming the humus decreased CO_2 evolution, but nevertheless the amount of organic matter decreased in the limed plots at the same time. So all our evidence indicates that CO_2 evolution is not a good universal indicator of biological activity in burned samples. On the other hand, there was a highly significant positive correlation between ammonia nitrogen content and CO_2 evolution, and this was higher on burned sites than on control sites. The correlation between nitrate nitrogen content and CO_2 evolution was significant only on control sites. The carbon–nitrogen ratio explained nitrogen mobilization much better than did CO_2 evolution. Both in the control and burned humus greater ammonia and nitrate nitrogen contents coincided with lower C/N ratios.

The effect of burning on biological activity of the soil in our material was smaller than had been anticipated. In particular, a greater increase of nitrogen mobilization owing to decreased acidity had been expected. Fertilization experiments have reliably demonstrated that fertility of poor and average mineral soils is most closely correlated with available nitrogen, and nitrogen mobilization is commonly supposed to be promoted by low acidity. It is true that there was a highly significant correlation between the amount of nitrate nitrogen and the degree of acidity (c_H), but the highest correlation coefficient was only 0.45. Correlations between c_H and ammonia were even lower. One explanation for the low correlation is that only net mineralization could be measured and microbes used up the mobilized nitrogen immediately.

VII. Loss of Nutrients due to Burning

A. CATIONIC NUTRIENTS

Before studies were made on the effects of prescribed burning, the loss of nutrients due to burning was a subject of contention among foresters. The most common fear, apart from the fire hazard, was that burning reduces soil fertility by causing additional leaching. Figures 5 and 6 enable us to estimate the balance between the various nutrients in the soil.

Average upland soils contain rather small amounts of nutrients in an available form (Viro, 1951). On the other hand, these sites have large amounts of nutrients bound to organic matter, e.g., ground vegetation, slash, and humus. By prescribed burning these nutrients are partly freed in an available form, chiefly as carbonates and oxides.

Of the cationic nutrients, calcium losses caused by burning were the smallest. Leaching from the humus layer was very slow, and the amounts leached were mainly fixed by ion exchange in the top layer of the mineral soil so that any loss of calcium due to burning was probably negligible. Magnesium, another bivalent element, was also rather well retained by the humus layer and mineral soil. Nonetheless, much of the increased amount of magnesium found in the humus layer during the first years after burning was quickly leached into the mineral soil. Owing to its precipitation with phosphorus, the chain of events was not as clear as that for calcium. In any case, appreciable amounts of leached magnesium were found as deep as 30 cm in the mineral soil, and small amounts had probably been leached even deeper. Thus, small losses of magnesium are likely. Potassium is a monovalent element, and the histograms of Fig. 6 show that its fixation, especially by the humus layer, was very poor. At least during the first 12 years after burning, potassium was leached deeper than 30 cm, and all the potassium in the burned matter, about 60 kg per hectare, were leached out of the reach of tree roots.

Nutrients in the surface layers of soil, where most roots are found, are most important for tree growth. In mineral soil down to a depth of 30 cm, the burned sites held exchangeable calcium and magnesium in quantities larger than or equal to those in the unburned sites for at least 50 years, while the amount of potassium remained greater for only 12 years.

It can be seen from the foregoing that leaching of cations occurred in accordance with the theoretical fixation capacity of the ions, but the initial abundance of potassium in the soil caused some leaching of calcium and magnesium in accordance with the law of mass action.

The total amounts of exchangeable calcium and magnesium in the topsoil layers evidently do not change much during burning, for the amount taken up by the new vegetation is probably about equal to that released from the burned slash and humus. Some of the magnesium, and perhaps some calcium too, was precipitated as insoluble phosphates. Potassium is leached easily from the humus layer but is bound fairly well to mineral soil through ion exchange. However, some potassium is evidently leached beyond reach of the plants, although the amount cannot be very large. Compared with the available potassium resources of Fenno-Scandian soils, established losses are negligible.

B. Phosphorus

The histograms showing changes in the amount of easily soluble phosphorus in the humus layer greatly resemble those for exchangeable magnesium: The amount decreased during the first 6 years by about 75% and began to rise again 12 years after the burning. However, the diminution in the amount of easily soluble phosphorus does not indicate the real loss. As stated earlier, it is likely that phosphorus was mainly precipitated in the surface layers in the form of sparingly soluble magnesium phosphate.

The histograms in Fig. 4 show that the amount of easily soluble phosphorus was greater in the first two 10 cm layers of the mineral soil for several years after burning, but in the 20–30 cm layer only in the year of burning. Presumably, phosphorus was precipitated in the mineral soil mainly by aluminum and iron oxide hydrates. Taking all this into consideration, the losses of phosphorus due to prescribed burning are probably rather small.

C. Nitrogen

Living vegetation, needles, and smaller branches of the slash are totally consumed by burning, as also are parts of the humus layer and of the coarser slash. The mineral nutrients of the vegetation and slash remain at the site in the ash, whereas their nitrogen is totally lost by burning. The bulk of the nitrogen in the burned humus layer is also lost by burning, but an unknown amount of it is changed into ammonia and bound in the unburned humus. The average loss of nitrogen due to burning could be estimated at 320 kg per hectare—about 180 kg from the slash and vegetation and 140 kg from the humus layer. This is 10% of the total amount of nitrogen on the site. There are no clues as to the amount of ammonia nitrogen that escaped with the waste gases, but it was very probably greater than the amount bound on the site. Such nitrogen returns to the soil with rainwater but not usually to the same site.

These figures indicate a substantial loss of nitrogen from the site at burning, but this loss is unimportant because the nitrogen in living vegetation before burning is totally unavailable to other plants. Mineralization of the nitrogen in the logging waste and humus layer is very slow. On the other hand, burning decreases acidity of the humus layer and in this way encourages the mineralization of nitrogen–nitrification immediately and ammonification somewhat later (Fig. 4). So, despite the loss in the total amount of nitrogen, burning greatly increases the mineral nitrogen.

D. Conclusions

From the foregoing we can attempt to summarize the pros and cons of prescribed burning. The organic matter that burns is slash, living moss, and humus layer which, if unburned, have little effect on the fertility of the site because they decompose and release nutrients very slowly. So, from the point of view of plant nutrition they are only important as a nutrient reserve for the distant future. After burning, the nutrient status of the site remains excellent for a long time. Burning unquestionably results in great losses of total nitrogen from the site, but simultaneously it results in an increase of mineralized nitrogen. The former is practically unimportant, the latter of great consequence.

Particular attention must be paid to the long duration of the large amounts of exchangeable calcium in the mineral soil. Most Fenno-Scandian soils are poor in calcium, and a very high positive correlation has been established between the amount of calcium and fertility of the site (Viro, 1951). It is most probable that this correlation is due to the effect of calcium on nitrogen mobilization. Burning may induce some unimportant losses through leaching of magnesium and especially of potassium, but thanks to the richness of Fenno-Scandian soils in micas, these losses do not reduce site fertility very much. The loss of phosphorus may be more serious. But these studies do not indicate any noteworthy loss, and fertilizer experiments have also shown that most Fenno-Scandian mineral soils are quite well supplied with phosphorus (Viro, 1967).

It was shown earlier that the amount of exchangeable nutrients begins to increase again about 12 years after burning. This rise was assumed to be due to decomposition of the lush vegetation, especially herbs and grasses. These take large amounts of nutrients from the soil and so diminish leaching losses. Their litter decomposes rapidly and thus maintains rapid circulation of nutrients. The period of herb dominance after burning is quite short, and herbs do not hamper development of the new stand of trees to any great extent. The abundance of grasses lasts longer and may sometimes be detrimental to trees for a long time.

As a general observation, it was found that the amounts of nutrients leached from the site after burning are rather small. Magnesium, and especially calcium, are well attached to all layers of the soil. Phosphorus and potassium, especially phosphorus, are attached loosely to the charred humus layer, but quite firmly to mineral soil. In other words, prescribed burning does not lead to any serious impoverishment of the soil—at least if it is not repeated more often than absolutely necessary to remedy the consequences of ineffective forestry. Generally speaking, however, prescribed burning should not be necessary in Fenno-Scandia if good forestry is continuously practiced.

VIII. Burning in Forestry Practice

The climax stage of a Fenno-Scandian forest is an old spruce stand of poor growth; this can rightly be called the stage of degradation of the stand. It will not improve naturally; it calls for strong action, and the most reliable method is prescribed burning.

Natural regeneration of a stand is very difficult on sites with a thick humus layer because young seedlings must compete for water and nutrients with the existing vegetation. If the humus layer is very porous,

Fig. 7. A vigorous seedling stand of pine, 6 years after burning and planting. The originally very lush and now recessive growth of fireweed did not hamper regeneration.

the young seedlings may die from lack of water, especially in dry summers. The living moss may be a carpet several centimeters thick—one that offers no available nutrients. Furthermore, the thick organic layer is an effective heat insulator.

The purpose of prescribed burning is to improve all the external growth factors—nutrition, moisture, and thermal conditions—but the immediate objective is to facilitate regeneration of a stand and change the species. The leading foresters in Fenno-Scandia, in research and in the field alike, have accepted prescribed burning as one of the best methods of restoring good growth in a stand that has stagnated under poor management. Figure 7 shows an excellent result of prescribed burning. The effect of fire on soil moisture is ambivalent. Moss and humus layers hold rainwater well and also hinder evaporation. Thus the effect of burning on soil moisture may be favorable (on a paludified site) or unfavorable (on a site having low water-holding capacity), depending on circumstances. The effect of burning on thermal conditions is almost entirely favorable. Only if the humus layer is completely burned may the temperature of dry soils occasionally rise to a lethal level. The effect of burning on availability of nutrients is entirely positive; especially noteworthy is its effect on nitrogen mobilization. The effect of burning on the possible leaching of phosphorus must undeniably be regarded as unfavorable. The quantities lost, however, probably are quite small.

Owing to the high percentage of mica in Fenno-Scandian soils, the loss of potassium is unimportant.

Fire destroys the worthless stunted trees and other vegetation, the slash, and part of the humus layer. Slash may protect the seedlings against the pressure of snow in winter but burning it cuts down labor requirements and facilitates the work of reforestation machines. There can hardly be any harm in burning the moss layer, while the conversion of its nutrients into utilizable form is beneficial to the new stand.

Humus does not accumulate on the most barren sites; it is common on moister sites only—in Fenno-Scandia mainly on morainic soil. There usually is not any risk of the humus layer of these soils being totally burned; the most common fault is that the burning is not thorough enough. Thus, burning can be recommended on all morainic soils, even on the stoniest ones.

Very dense stands of stunted spruce occasionally appear on barren coarse sand or esker gravel soils. As these soils have low water-holding capacity, they dry up rapidly, and so fire usually consumes their thin humus layer completely. Thermal and water conditions after burning are then unfavorable, and after thorough burning the reforestation of these sites may take decades. For these reasons, prescribed burning is not recommended for sites on coarse water-sediment soils. Owing to the scarceness of fine soil particles, their ion-exchange capacity is poor, and therefore the nutrients liberated by burning are largely lost by leaching. If the purpose is to kill the spruces by fire, the burning must be done fast enough for the humus layer proper to be left almost intact; in addition to the spruces, only the living vegetation, dwarf shrubs in particular, should be killed.

Burning is not necessary on the most fertile soils, except for swalelike cultivation. If fertile sites are burned, they must be burned more intensely than normal so as to effectively destroy the root systems of the grasses.

The favorable effect of burning on reforestation is generally acknowledged. There are no comparative studies on the effect of prescribed burning on growth of stands during a whole rotation. However, observations in young stands show indisputably that reforestation is much easier and surer on a burned site, and the development of the young stand is faster than without burning. The good growth of the young pine stands is due to improvement of all the growth factors. It is very important for the fire to kill the vegetation almost totally so as to reduce root competition during the most critical years in the early growth of pine seedlings, and the burned site must be reforested without delay. Once the seedlings have had a good start and reach a height of 1–2 m, the lush growth of herbs, grasses, and deciduous trees may be even beneficial.

When natural reforestation or broadcast sowing is employed after burning, the abundance of easily soluble salts on the soil surface may destroy the seedlings at the cotyledon stage, especially spruce seedlings (Heikinheimo, 1915). This danger is small with planting, because burning affects the nutrient concentration less in mineral soil than in the humus layer.

Prescribed burning also has its disadvantages. If a site with a thin humus layer is burned too intensely, the entire humus layer may be consumed. There are many examples of this as a result of wild forest fires, for these usually occur during the driest periods of the summer. However, such problems are very rare after careful prescribed burning. The most common mistake in practice is burning while the humus layer is still too wet. The fear of the humus layer being burned too thoroughly probably derives from experience of reforestation difficulties after certain wild forest fires.

In some cases the burned site may produce such a profusion of ground vegetation that this interferes with growth of the seedling stock. A lush growth of grasses is a sign of inadequate burning, even though its detrimental effect on reforestation has obviously been exaggerated. In recent years an abundance of an ascomycete, *Rhizina undulata,* has been found on burned sites, and this may have destroyed the young seedling stock almost entirely (Laine, 1968). Observations of this fungus indicate wide variations in its occurrence, and means of combatting it are still unknown. According to some field observations, the occurrence of *Rhizina* is a sign that the burning was not intense enough.

One of the factors restricting the use of prescribed burning is the weather. In a wet summer prescribed burning may be quite impossible, and in a dry summer the fire hazard is excessive. For these reasons the period suitable for prescribed burning may be very short, and an advance planning of forest work that includes burning is uncertain. Furthermore, prescribed burning with the inevitable subsequent supervision may be rather expensive for small areas.

A comparison between the advantages and disadvantages of prescribed burning leads to the conclusion that the advantages mostly outweigh the disadvantages. The most important advantages are improvement in nutrient and thermal conditions of the site, assured regeneration, and the opportunity to select the species of tree that is most suitable for a site. Some of the benefits of prescribed burning have been sought also through fertilization, scarification, and ploughing. However, none of these alone can provide all the advantages of prescribed burning.

Foresters can greatly influence the duration of favorable conditions in the soil after burning. Everything practicable should be done to prevent a return to the unfavorable conditions.

References

Aaltonen, V. T. (1941). Referat: Die finnischen Waldböden nach den Erhebungen der zweiten Reichswaldschätzung. *Commun. Inst. Forest. Fenn.* **29,** No. 5.

Aaltonen, V. T. (1950). Die Blattanalyse als Bonitierungsgrundlage des Waldbodens. *Commun. Inst. Forest. Fenn.* **37,** No. 8.

Ahlgren, C. E. (1970). Some effects of prescribed burning on jack pine reproduction in Northeastern Minnesota. *Minn., Agr. Exp. Sta., Misc. Rep.* **94.**

Ahlgren, I. F., and Ahlgren, C. E. (1965). Effects of prescribed burning on soil microorganisms in a Minnesota jack pine forest. *Ecology* **46,** 304–310.

Cajander, A. K. (1949). Forest types and their significance. *Acta Forest. Fenn.* **56.**

Eneroth, O. (1928). Referat: "Beiträge zur Kenntnis der Einwirkung des Schlagabbrennens auf den Boden." Skogshögskolan, Stockholm.

Gissing, F. T. (1909). "Commercial Peat." Griffin, London.

Heikinheimo, O. (1915). Referat: Der Einfluss der Brandwirtschaft auf die Wälder Finnlands. *Acta Forest. Fenn.* **4,** No. 2.

Hesselman, H. (1917). Summary: On the effect of our regeneration measures on the formation of saltpetre in the ground and its importance in the regeneration of coniferous forests. *Medd. Statens Skogsfoersoeks anst.* **13–14,** 923–1076.

Hesselman, H. (1934). Referat: Einige Beobachtungen über die Beziehung zwischen der Samenproduktion von Fichte und Kiefer und der Besamung der Kahlhiebe. *Medd. Statens Skogsfoersoeks anst.* **27,** 145–182.

Kaila, A., Köylijärvi, J., and Kivinen, E. (1953). Influence of temperature upon the mobilization of nitrogen in peat. *J. Sci. Agr. Soc. Finl.* **25,** 37–46.

Kujala, V. (1926). Untersuchungen über den Einfluss von Waldbränden auf die Waldvegetation in Nordfinnland. *Commun. Inst. Forest. Fenn.* **10,** No. 5.

Laine, L. (1968). Summary: *Rhizina undulata* Fr., a new forest disease in Finland. *Folia Forest.* **44.**

Mikola, P. (1954). Summary: Experiments on the rate of decomposition of forest litter. *Commun. Inst. Forest. Fenn.* **43,** No. 1.

Mikola, P., Laiho, O., Erikäinen, J., and Kuvaja, K. (1964). The effect of slash burning on the commencement of mycorrhizal association. *Acta Forest. Fenn.* **77,** No. 3.

Mork, E. (1946). Summary: On the dwarf shrub vegetation on forest ground. *Medd. Nor. Skogforsoeksv.* **33** (n9), 269–392.

Overrein, L. N. (1967). Isotope studies on the release of mineral nitrogen in forest raw humus. *Medd. Nor. Skogforsoeksv.* **85** (n23), 541–565.

Sarvas, R. (1937). Referat: Beobachtungen über die Entwicklung der Vegetation auf den Waldbrandflächen Nord-Finnlands. *Silva Fenn.* **44.**

Svinhufvud, E. G. (1929). Referat: Untersuchungen über die Einwirkungen des Brennens auf Moorboden. *Suom. Suonviljelysyhdistyksen Julkaisuja* **10.**

Tamm, C. O. (1953). Growth, yield and nutrition in carpets of a forest moss (*Hylocomium splendens*). *Medd. Statens Skogsforskningsinst.* (*Swed.*) **43,** No. 1.

Tamm, C. O. (1956). The response of *Chamaenerion angustifolium* (L.) Scop. to different nitrogen sources in water culture. *Physiol. Plant.* **9**, 331–337.

Tamm, C. O. (1964). Growth of *Hylocomium splendens* in relation to tree canopy. *The Bryologist* **67** (no. 4), 423–426.

Uggla, E. (1957). Summary: Temperature during controlled burning. *Norrlands Skogsvårdsforb. Tidskr.* **4**, 443–500.

Vaartaja, O. (1950). On factors affecting the initial development of pine. *Oikos* **2**, 89–108.

Viro, P. J. (1951). Nutrient status and fertility of forest soil. I. *Commun. Inst. Forest. Fenn.* **39**, No. 4.

Viro, P. J. (1955). Investigations on forest litter. *Commun. Inst. Forest. Fenn.* **45**, No. 6.

Viro, P. J. (1963). Factorial experiments on forest humus decomposition. *Soil Sci.* **95**, 24–30.

Viro, P. J. (1967). Forest manuring on mineral soils. *Medd. Nor. Skogforsoeksv.* **85** (n23), 111–136.

Viro, P. J. (1969). Prescribed burning in forestry. *Commun. Inst. Forest. Fenn.* **67**, No. 7.

Waksman, S. A. (1936). "Humus." Baillière, London.

. 3 .

The Effect of Fire on Soil Organisms

Isabel F. Ahlgren

I. Introduction

Because of their apparent lack of direct economic importance and the involved procedures sometimes required in their study, soil microflora and fauna have not been investigated as extensively in relation to fire as have larger plants and vertebrates. Consequently, postfire ecology of soil organisms is not well understood. Yet these organisms are in

47

direct contact with their soil environment and are sensitive to changes in season, moisture, nutrients, temperature, and biotics. Many are completely dependent on the soil for survival. As indicators of change in soil conditions and as important contributors to soil texture, recycling of nutrients, and nitrogen availability, the soil organisms must not be overlooked in an overall study of postfire changes in the ecosystem.

As the soil-inhabiting species vary greatly in form, growth requirements, and life cycles, generalizations are difficult and may be misleading. Identification of species is often difficult and time-consuming. Most of the past research on these organisms has been quantitative in nature, identifying the organisms only to major groups. Consequently, the discussion in this chapter is limited to the larger groups with only occasional references to specific organisms.

II. Algae

Few references to the occurrence of algae in soil on burned lands are available. Shields and Durrell (1964), in discussing the sequence of algal species in different successional stages of land development, noted that the first organisms to colonize soil on burned English heath were algae, especially simple green algae such as *Cystococcus humicola* or species with slimy coverings, e.g., *Gloeocystis vesiculosa*. Veretennikova (1963) noted unicellular or simple, filamentous algae as pioneers on wet, burned sites in Russia and believed them to aid in improving conditions for growth of higher plants.

Although no other observations of algae in postfire soil are available to the writer, it is reasonable to assume that these organisms are affected by fire and exhibit detectable postfire successional changes. Undoubtedly, additional information on algal succession will be available as methods of cultivating and studying algae become more widely used in ecological work.

III. Bacteria

Under various conditions investigators have found that fire has no effect on the bacterial population, that fire completely destroys it, or that fire results in a striking increase in bacteria. A closer examination of the soil environment in which the work was done reveals possible reasons for the differences. In addition, experimental techniques and materials used vary widely among investigators and could be responsible for some variation in results.

A. Techniques for Study

In most quantitative studies dilution plating on selective media was used. The limitations of the plating technique as an indication of total microbial population have long been recognized (Paarlahti and Hanoija, 1962). Most workers readily acknowledge that such media do not foster growth of all soil microflora species and can be used only as general indicators of gross changes in the population. There are also many differences in soil-sampling techniques, in the time delay between soil sampling and testing, and between burning date and testing date. These and other factors may contribute to the different and often conflicting results reported.

B. Depth and Intensity of Fire

Changes in microbial populations following fire are most evident in the upper soil layers. The depth to which the effect can be detected increases with increased burn intensity. Bacterial populations in the upper $1\frac{1}{2}$ inches of soil on burned land were higher than on unburned land 7 months after fire in the northwestern United States (Wright and Tarrant, 1957). No changes were evident on light burns. In the southern United States, bacteria in the F and H soil layers increased following burning but did not change significantly below the organic zone (Jorgensen and Hodges, 1970, 1971). Neal *et al.* (1965) found no alteration of bacterial populations 2 inches below the surface following burning of Douglas-fir (*Pseudotsuga menziesii*) slash. In prescribed-burned jack pine (*Pinus banksiana*) areas in the north central United States, there was a small but significant increase in bacteria 2–3 inches below the surface the second and third growing seasons after burning. Since the increase was not detected the first year, it is believed to be related to the gradual leaching of ash minerals to lower soil layers (Ahlgren and Ahlgren, 1965).

C. Temperature during Burning

Heat can cause immediate reduction of bacterial populations. Soil temperatures recorded during fire vary greatly, but most are sufficient to kill unicellular organisms. Surface soil temperatures of 1841°, 1452°, 1260°, and 1013°F have been reported (Isaac and Hopkins, 1937; Iwanami, 1969; Stinson and Wright, 1969; Smith and Sparling, 1966). In Minnesota jack pine prescribed burns, temperatures exceeded 1472°F for 1 min, 932°F for 9 min, and 572°F for 17 min. Between humus and mineral soil at a depth of 2 to 3 inches, temperatures reached 572°F

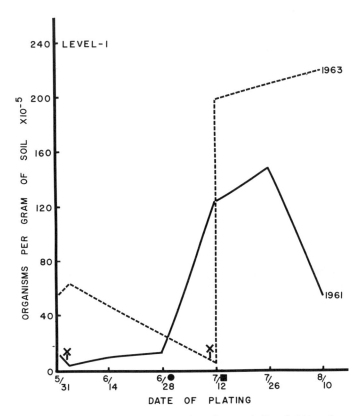

Fig. 1. Number of bacteria per gram of surface soil (level 1) as determined by dilution plate counts on soil extract agar the first growing season after prescribed burning on two jack pine (*Pinus banksiana*) stands in northeastern Minnesota. X, Date of burn; ●, date of first rainfall following 1961 burn; ■, date of first rainfall following 1963 burn.

for 14 min and were above 122°F for 72 min (Ahlgren, 1970). Hall (1921) reported that heating soil for 1 hr at 212°F produced an initial depression in number of bacteria, followed by a sharp increase. Ahlgren and Ahlgren (1965) noted a similar reduction on two burned areas (Fig. 1). The decrease was greater on a more severe burn, where soil temperatures reached around 900°F over much of the area, than it was on a more moderate burn where such temperatures were rare. The population decline was evident only in the upper inch of soil.

In areas where moisture is a limiting factor, these population declines are temporary, lasting only until the first postfire rainfall. Consequently, if this temperature-induced decrease is to be detected, plating must be done before the fire, immediately after fire, and at intervals thereafter for comparison. In those studies in which platings were not made until several days, weeks, or months after fire, such decreases would not be detected.

D. Soil Moisture

The determining effect of moisture may be more widespread, however, than in the cases cited. In moist soil, bacterial populations may respond differently to fire than in drier soil. This response may explain other population differences reported in the literature. For example, reductions in bacterial populations are often recorded in dry, north temperate sites immediately after fire (Fritz, 1930; Kivekäs, 1939; Heikinheimo, 1915; Ahlgren and Ahlgren, 1965). In moist soils of humid, tropical areas of the Malay Peninsula, however, Corbet (1934) reported that microbial populations rose immediately after fire and declined to preburn numbers within 9 months. In contrast, Meiklejohn (1955) found that burning arid bushland in Kenya resulted in long-range reduction of bacteria.

In the Douglas-fir area of the United States, Wright and Bollen (1961) found that microbial populations increased during the rainy period and decreased during drought on both burned and unburned land. In northern Minnesota, after the first rainfall following burning, bacterial populations in the surface inch of soil increased to 3 or 4 times preburn numbers the first growing season after fire (Fig. 2). Such increases above normal levels were not evident at 3 inch depths, on unburned land, and did not occur in subsequent years (Ahlgren and Ahlgren, 1965). The sharp

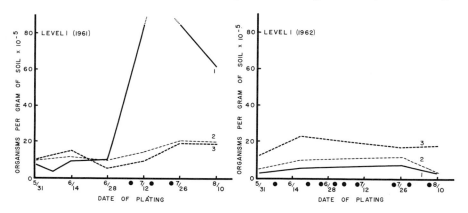

Fig. 2. Number of bacteria per gram of surface soil (level 1) during first (1961) and second (1962) growing seasons following prescribed burning on a northeastern Minnesota jack pine (*Pinus banksiana*) stand. 1, Area cut and burned; 2, area cut, unburned; 3, area uncut and unburned. (Taken from Ahlgren and Ahlgren, 1965.)

first-year increase may be related to availability of ash minerals dissolved into the soil by rain.

E. SOIL CHEMISTRY

The sudden availability of mineral nutrients, along with the other soil chemical changes associated with burning, undoubtedly can have a profound effect on the bacterial population. Early attempts to demonstrate this by adding burned products to the soil were not successful. Tryon (1948) found that adding charcoal to soil altered the moisture-holding capacity but did not affect the microbial population, although Hesselman (1917) had reported that soil on charcoal burning grounds in Sweden was very rich in bacteria. Dügelli (1938) added ash to mineral soil in Swiss spruce forests and detected no change in bacterial numbers.

Soil pH frequently increases after fire because of the addition of ash minerals. There is a close correlation between spring increases in bacterial populations and spring increases in soil pH, temperature, ammonia, and nitrate (Neal *et al.*, 1965; Wright and Tarrant, 1957; Wright and Bollen, 1961). In the Douglas-fir areas, pH is higher after rain on both burned and unburned land. Since bacteria are generally favored on more alkaline soils and the majority of the fungi thrive on more acid soils, pH differences are probably the reason for the many reports that bacteria are favored over fungi on burned lands (Fowells and Stephenson, 1933; Wright and Tarrant, 1957; Fuller *et al.*, 1955; Vandecayve and Baker, 1938).

F. QUALITATIVE DIFFERENCES

Most work on postfire bacterial changes involves only quantitative differences, since species identification in bacteria is time-consuming and sometimes difficult. A postfire succession of microbes would be expected, however, since preliminary studies indicate that soil microflora varies with the ecotype (Bollen and Wright, 1961; Tresner *et al.*, 1954; Vandecayve and Baker, 1938). Wright and Bollen (1961) noted that hard-burned Douglas-fir soil developed complex microflora about 1 year after burning, but that the population became more characteristic of unburned areas about 2 years later. There are few other qualitative studies of bacteria in various ecotypes including burns, and further investigations should be fruitful.

The bacteria involved in the nitrogen cycle are more readily recognized than most others, because of their growth on nitrogen-free media and other very specific media. Burning increases nitrogen-fixing *Azotobacter* and *Clostridium* species in the soil of red pine (*Pinus resinosa*) and white pine (*P. strobus*) areas (Lunt, 1951), in the New Jersey pine barrens (Lutz, 1934), and elsewhere (Sushkina, 1933; Remezov, 1941; Fowells and Stephenson, 1933). Isaac and Hopkins (1937) related similar increases following Douglas-fir slash burning to increases in soil calcium from the ash. Kivekäs (1939) reported a slight increase in nitrification and a decrease in ammonification in burned soil in Finland. Increased soil nitrification may continue for up to 12 years after fire (Hesselman, 1917). Decreases in nitrogen fixation and nitrification were reported after burning only in dry, Kenya bush soils (Meiklejohn, 1955).

G. CARBON DIOXIDE EVOLUTION

Most studies reported above detected bacterial changes by dilution plate techniques or by plating on enrichment media which permitted only growth of organisms capable of utilizing specific compounds. Carbon dioxide evolution has also been used as a quantitative index of microbial activity. It may be a more accurate index of changes, since the activity is measured in the soil and not on media which may be selective and eliminate significant groups of bacteria. The respiratory rate of different species varies, however, so the method can give only broad, comparative information on quantitative population changes. In northern Arizona, Fuller *et al.* (1955) found carbon dioxide evolution lower in burned than in unburned soil, except on experimental slash burns. In quantitative platings of these areas, however, bacterial populations were higher on burned land. Carbon dioxide evolution in soil from burned jack pine lands corresponds closely with quantitative plating results, with a decrease immediately after fire, followed by a marked increase (Ahlgren and Ahlgren, 1965).

IV. Actinomycetes

It is estimated that actinomycetes may constitute up to 30% of the living organisms in the soil. Because their growth requirements and colony appearance on media are similar to those of bacteria, the two groups are often combined in studies of postfire soil microorganisms, or the actinomycetes are overlooked entirely.

Most workers have reported that soil actinomycetes and bacteria behave similarly in response to fire (Bollen and Wright, 1961; Jorgensen and Hodges, 1970, 1971; Lutz and Chandler, 1946). Neal *et al.* (1965) reported that in the Douglas-fir area, actinomycetes make up a relatively constant percent of the total soil microbial population—between 35 and 40%—although they are more stable than bacteria and less sensitive to moisture changes. In northern Minnesota, the author found actinomycetes constituted from 3 to 70% of the total microbial population. The largest percentage occurred in the surface level immediately after fire. The smallest percentage was in the surface level of the severe burn after the first postfire rainfall.

Differences between bacteria and actinomycete postfire behavior have also been noted. Wright and Tarrant (1957) found that the ratio of bacteria to actinomycetes remained the same on unburned and lightly burned land but increased sharply on severely burned land. In northern Minnesota jack pine prescribed burning studies, the author found that the ratio of bacteria to actinomycetes was reduced immediately after fire in the surface soil on both moderate and severe burns, with greater reduction on the severe burn (Table I). The ratio increased strikingly

TABLE I

RATIO OF BACTERIA TO ACTINOMYCETES ON MODERATE AND SEVERE JACK PINE (*Pinus banksiana*) BURNS[a]

	Level 1			Level 2		
	Burn	Cut	Control	Burn	Cut	Control
1961 (moderate burn)						
Preburn	0.74	0.63	0.93	0.66	2.68	0.42
2 Days after fire	0.50	0.74	1.01	1.86	2.91	1.28
9 Days after fire	1.52	1.11	0.91	1.03	4.10	1.67
16 Days after fire	1.20	1.03	0.81	1.49	5.00	1.03
30 Days after fire	12.14	1.50	1.05	1.51	2.30	8.76
1962 (second postfire year)						
June 15	0.77	1.18	2.29	1.48	1.80	4.11
July 15	1.30	1.32	1.88	4.32	2.87	5.50
August 15	0.55	0.33	2.51	2.25	1.73	1.03
1963 (third postfire year)						
June 15	1.83		1.89	2.04		1.59
July 15	9.00		2.20	1.67		4.44
August 15	12.90		7.89	1.32		2.39
1963 (severe burn)						
Preburn	9.16			4.00		
2 Days after fire	2.33			2.91		
9 Days after fire	26.33			2.26		
16 Days after fire	22.80			2.37		

[a] Level 1, top 1½ inches of soil, including burned humus; level 2, 1½ to 3 inches into mineral soil.

after the first postfire rainfall. These results suggest that the actinomycetes are more resistant to heat than bacteria and are less affected by moisture changes.

Two years after burning, the actinomycete numbers rose significantly in the lower, 3-inch soil level on burned land. This increase was much greater the third year after fire and did not occur on cutover or unburned forest. There is a normal, seasonal decline in the proportion of actinomycetes to bacteria from early June through August, in both the surface soil and 3 inches below the surface, regardless of whether the land has been burned, cleared, or is forested. This decline must be taken into consideration in interpreting results.

Thus it would appear that actinomycetes are less affected by heat and drying than are bacteria. In the lower soil levels, actinomycetes may be stimulated to increase in later years, possibly because of the downward leaching of ash minerals.

V. Fungi

Most forest ecologists are familiar with the abundant fruit bodies of certain soil fungi, especially discomycetes, following forest fire. Some of these fungi rarely, if ever, are found on unburned land, and terms implying their association with fire are incorporated in their names—e.g., *Pyronema*, or the large group of ascomycetes to which Seaver (1909) applied the term "pyrophilous discomycetes." As a broad group, soil fungi are usually associated with acid soils, yet it is well known that ash from burning often increases soil pH and makes the soil less favorable to fungal growth. The group "fungi," however, includes a wide variety of species with many highly specialized and diverse physiological needs. Consequently, as with the other two groups of microorganisms previously discussed, there are numerous apparent contradictions in the literature on the behavior of fungi on burned lands. Many of these contradictions can be interpreted in terms of differences in environmental conditions, methods of study, or types of fungi investigated.

A. VEGETATIVE VERSUS REPRODUCTIVE GROWTH

Many soil fungi are molds which complete their life cycle as delicate mycelia or microscopic spore forms. These include the *Penicillii, Aspergilli, Fusaria,* and related genera, and are studied in culture on artificial media. They make up about 45% of the soil fungi (Jorgensen and Hodges,

1971). Other fungi grow vegetatively in the soil and perform most of their metabolic functions as mycelia but are detected primarily when they produce macroscopically visible fruit bodies on the soil surface. These fruiting fungi are the ones usually noted in field studies and include the so-called cup fungi, the discomycetes, and the mushrooms. Of these macroscopically fruiting species found on burned land, 75% are Pezizales (Petersen, 1970). It should be stressed, however, that absence of fruit bodies does not necessarily indicate the absence of these fungi from the soil as mycelia. Other groups of soil-inhabiting fungi, such as chytrids, Mucorales, and semiaquatic phycomycetes, have not been studied in burned soil. Such groups, however, contribute a small fraction to the total microbial soil metabolism and therefore may not be of particular ecological significance.

B. QUANTITATIVE DIFFERENCES

Decreases in fungal populations after fire because of unfavorable conditions were assumed by Fowells and Stephenson (1933). Using culture plating techniques, Wright and Tarrant (1958) found that, in the Douglas-fir area, there were fewer fungi in the upper 1½ inches of soil on burned land than on unburned land 6 months after fire. Only on severely burned soil was this decrease detected at soil depths as low as 3 inches. In the same area, Neal et al. (1965) detected a slight reduction in fungi throughout the first postfire year; shortly afterward, the numbers returned to preburn levels. In the southeastern United States, however, Jorgensen and Hodges (1970) detected no significant difference in number of fungi cultured from either annually or periodically burned lands or unburned land.

In field studies of fungal species detected by presence of fruit bodies, Jalaluddin (1969) found that soil on burned land in England remained free of fungi for some time after fire. Gradual colonization began at the margin of the burned land. Fewer species of fruiting fungi have been noted on pioneer soil, including burned land, than in older, climax forests where the soil is more acid (Tresner et al., 1954; Vandecayve and Baker, 1938).

No postfire decrease in fungi in the upper soil layer was detected in quantitative platings from northern Minnesota jack pine burns (Fig. 3). After the first rain, however, the fungi on the burned areas increased, and the increase continued during the first postfire growing season. In later years, the burned and unburned soil populations were similar (Fig. 3). No significant postfire differences were detected at a 3 inch soil depth (Ahlgren and Ahlgren, 1965).

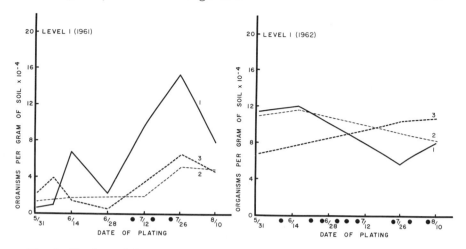

Fig. 3. Number of fungi per gram of surface soil (level 1) as determined by dilution plate counts on Martin's medium, the first (1961) and second (1962) growing seasons following prescribed burning on a northeastern Minnesota jack pine (*Pinus banksiana*) stand. 1, Area cut and burned; 2, area cut unburned; 3, area uncut and unburned. (Taken from Ahlgren and Ahlgren, 1965.)

Cohen (1950) reported fungus spore production to be stimulated by burning in the Transvaal, and Corbet (1934) found that mold populations increased after fire on the Malay Peninsula.

C. Qualitative Differences

As would be expected with such a heterogeneous group of organisms, the fungus species comprising the population on burned land differ from those on unburned land and also vary with the age of the burn. This variation suggests a postfire succession of fungi comparable with the succession of higher plant groups. Since identification of fungi to genus is often less involved than bacterial identification, fungal succession is better known.

Among the early postfire species in many parts of the world, most prominent are the pyrophilous discomycetes. Of these, *Pyronema* sp., *Peziza praetervis*, *P. anthacophila*, *Trichopheia abundans*, and *T. hemisphaeriodes* are frequently mentioned (Sarvas, 1937; Seaver and Clark, 1910; Webster *et al.*, 1964; Moser, 1949; Petersen, 1970). Higher basidiomycetes common on recently burned land include *Boletus* sp., *Trametes* sp., *Pholiota carbonaria*, *Lyophyllum spaerosporum*, *Omphalia maura*, and *Coprinus boudieri* (French and Keirle, 1969; Hintikka, 1960). Jalaluddin (1968) found fruit bodies of *Rhizina undulata*, *Anthacobia*,

Pyronema, and *Peziza* on burned sites the first postfire year. The second year, however, the species present were *Omphalia maura, Pholiota carbonaria, Rhizina*, and *Lyophyllum*.

Wright and Tarrant (1958) found that the occurrence of ectotrophic mycorrhizae, frequently formed by basidiomycetes on the roots of forest trees, decreased on burned land, and that on more severe burns the mycorrhizae were found deeper in the soil. Tree seedlings thrived on the burned land, however, so the decrease in mycorrhizae was apparently compensated for by the increased nutrition of the ash. In Australian pine plantings, fruit bodies of the mycorrhizal *Boletus* occurred within 3 weeks after fire on light burns and 5 weeks after fire on severe burns (French and Keirle, 1969).

Among the molds, Wright and Tarrant (1957) found that species of *Fusarium* were less frequent on lightly burned land than on unburned land. The genera found only on burned land were *Hormodendron, Gliocladium*, and *Cladosporium*. On severely burned sites, no *Aspergillus* was found, *Fusarium* and *Stemphyllium* were rare, and *Botrytis* and *Hormodendron* increased. In the southeastern United States, Jorgensen and Hodges (1970) found that *Gliocladium* and *Mortierella ramannia* increased significantly on land burned annually. The first molds to colonize the center of burned sites in Britain were *Penicillium* and *Trichoderma viride* (Jalaluddin, 1969).

D. Environmental Influences

Among the factors mentioned as related to the occurrence of certain fungi on burned lands are nitrification, sterilization of soil by heat, effect of heat on spore germination, heat formation of nutrient substances, chemical properties of ash, and altered biotic competition. Apparently, no one factor accounts for all cases. Among the higher plants, occurrence of early postfire followers is usually related to seed source, conditions for germination, and lack of competition. The relationships are more complex for the nonphotosynthetic organisms so completely dependent on the soil for all aspects of survival. Investigations of the physiological basis for species selectivity among the fungi on burned land reveal three main factors: temperature, acidity, and nutrient source.

1. Temperature

Although spores of some species of discomycetes are stimulated to germinate by heat, spore viability of the majority, including some pyrophilous species, is reduced by heating above 50° to 60°C (El-Abyad and Webster, 1968; Petersen, 1970; Hintikka, 1960). El-Abyad and Webster (1968) concluded that the overall effect of heat on spore germination was negligible, since the majority of the pyrophilous fungi will germinate on burned land whether the spores were present in the soil before the burn or whether they were introduced by air currents after fire.

Differences in postfire soil temperatures are not determining factors in perpetuation of pyrophilous or nonpyrophilous fungi, since optimum temperatures for their mycelial growth do not differ significantly. The more extreme postfire temperature fluctuations on burned land could, however, alter the sporulation of nonfire adapted species (El-Abyad and Webster, 1968).

Seaver and Clark (1910) found that heating soil, or using an aqueous extract of heated soil in the medium, favored growth of *Pyronema* sp., but that excess heating of soil retarded its growth. Johnson (1919) found that fungi increased in soil heated to 250°C. Tryon (1948), however, reported that addition of charcoal to soil or media did not affect fungal growth.

2. *Acidity*

The occurrence of fruit bodies of pyrophilous fungi is related to soil pH. Many discomycetes and basidiomycetes found on burned land are basophilic. The optimum pH in which fruit bodies of a given species develop is quite specific within the extremes of pH 5.5 to 10.0 (Hintikka, 1960; Petersen, 1970).

Postfire pH varies from an alkaline condition soon after fire to a more acid condition later as the ash minerals are gradually leached out. Consequently, the species of fungi found in the area vary with age of burn. Classifications of fungi based on pH requirements and presence or absence on burns of different ages have been proposed. Moser (1949) classified fungi of burned areas in Austria as either Anthracobionte (burn-limited), Anthracophile (preferring burned land), Anthrakozene (less frequent on burned land), or Anthrakophobe (not on burned land). Similar classifications were proposed by Ebert (1958), Pirk (1950), and Petersen (1970). The similarities between such categories and the ecological categories proposed for behavior of flowering plants on burned land are striking (Vogl, 1964; Ahlgren, 1960).

3. *Nutrient Source*

In Denmark, Petersen (1970) divided fungi with fruit bodies occurring on slash burn sites into four classes. These were based on the occurrence of fruit bodies at different time intervals after fire. Group 1 included those species which appeared first within 7 to 8 weeks after fire; group 2, 10 to 15 weeks; group 3, 20 to 50 weeks; and group 4, 50 or more weeks after fire. These time intervals are associated with a gradually decreasing soil pH. In some cases, optimum pH for fruiting of fungal species found during these time intervals corresponds with actual soil pH found at these times. Other species, however, found within these time intervals, are tolerant of a fairly wide range of basophilic soil reactions. The occurrence of these fungi could also correlate with the availability of various sources of nutrients. Petersen (1970) proposed that the fungi occurring as fruit bodies on burned soil in Denmark may fit into one of three categories on the basis of both time since fire and nutrient source:

1. Species involved in the breakdown of plant roots left after fire. Roots of plants destroyed by fire are available immediately after burning and would be the first nutrient source used.

2. Species involved in utilization of organic matter in the humus. According to Petersen, this partially decomposed material would be utilized by different fungus species after plant roots were broken down by species in category 1.

3. Species which occur only in association with the postfire moss carpet, most of which develops after the first growing season.

It must be emphasized, however, that the data on which these classifications are based include only appearance of fruit bodies and not the occurrence of vegetative mycelium in the soil.

E. PLANT PATHOGENS

Many fungi causing plant disease occur in or on the soil for a portion of their life histories. These species vary in their response to fire as do the nonpathogenic species already discussed.

As early as 1929, Muller (1929) pointed out that fire is of value in purging the forest of disease. A classic example is brown needle spot of longleaf pine (*Pinus palustris*) caused by *Septoria alpicola*. Winter burning eliminates the disease for 1 year, allowing better seedling development. Although the disease recurs 2 and 3 years after fire, damage to seedlings is reduced (Garren, 1943; Siggers, 1934; Chapman, 1932). Leafspot of blueberry (*Septoria* sp.) is also controlled by burning, as is blueberry red leaf (*Exobasidium* sp.) (Markim, 1943). Haig (1938) reported that fire controls nectria cankers.

In other cases, fire seems to increase disease. Fire-damaged trees frequently have scars which provide entry points for fungi, especially heart rot of aspen and pine caused by *Fomes* sp. (Schmitz and Jackson,

1927; Basham, 1957). Fire may stimulate growth of prolific stands of the host plant, thereby multiplying and spreading the pathogen. This is believed to be the case for powdery mildew on blueberry (*Microsphaera alni*) and blueberry rust (*Pucciniastrum myrtelli*) (Demaree and Wilcox, 1947). Similarly, Davis and Klehm (1939) reported that increase of *Ribes* after fire could further stimulate the spread of white pine blister rust (*Cronartium ribicola*).

Perhaps the most serious relationship of plant disease to fire has been the association of *Rhizina undulata* root rot with postfire plantations of Scots pine (*Pinus sylvestris*) and mugho pine (*Pinus mugo*) (Laine, 1968). Viro (1969) reported this disease to be one of the reasons for reduction in the use of prescribed burning in Finland. Germination of *Rhizina* ascospores is enhanced by heat, and mycelial growth is stimulated by the addition of heated extract of pine roots to the medium (Jalaluddin, 1968). The organism grows best, however, on acid soil, as low as pH 3 (Norkrans and Hammarstrom, 1963) and is seldom found in limestone areas (Jalaluddin, 1968) or where burning has been intense (Viro, 1969). Petersen (1970) noted that fruit bodies of this organism occur first at the margins of burned areas, 13 to 95 weeks after fire, when soil pH is 8.0 to 9.5. The difference between optimum pH for mycelium growth and that for fruit body production further points up the inadvisability of using occurrence of fruit bodies as an indication of presence of the fungus in the soil.

VI. Soil Fauna

Information on the effect of fire on soil fauna is limited. Fire has been used since ancient times to control various agricultural insect pests in crops, field borders, and to some extent on range and pasture lands (Komarek, 1970). The few scientific publications dealing with these practices in either forests or grasslands have been primarily of a survey or short-term observational nature. Ecological interpretation from a successional standpoint and the relationship of soil fauna to postfire soil chemistry and other environmental changes is usually lacking. Thus, interpretation of these organisms in terms of the effects of different ecological factors will be fragmentary.

The term "soil fauna" covers a broad area which is not easily delimited. The lesser soil animals, for which at least 15 × magnification is necessary for identification, are usually referred to as microfauna. No literature on effects of burning on these organisms was found. Mesofauna, the intermediate-sized organisms, include mites, collembolans, and other very small arthropods (Metz and Farrier, 1971; Rice, 1932). Some workers include these small arthropods in the microfaunal category. The macrofauna consists of larger insects, snails, earthworms, and spiders. Soil-inhabiting vertebrates are not included in these categories.

It is sometimes difficult to delimit those organisms which can be classified accurately as soil inhabitants. Unlike the microflora which more or less "stays put," the various life stages of a given faunal species may be found in different habitats. Sometimes the habitat itself may move, carrying the fauna with it, as is the case of some stem and twig larvae, beetles, and worms which inhabit the vegetation above the ground at one interval in their lives. As the vegetation dies and begins to break down, it becomes part of the duff, litter, and organic soil and takes the fauna with it. For our purposes, we will include any organisms which normally inhabit either the soil or litter for some portion of their lives and thus could be affected by fire when in the soil.

The macrofauna make up the most diverse group of soil fauna. The mesofauna are the most active in the breakdown of organic tissue into tiny particles prior to bacterial and fungal decomposition (Metz and Farrier, 1971; Bornebusch, 1930). The microfauna probably outnumber the others in total population. All groups of soil fauna play an important role in making the soil more permeable by constantly mixing subsoil with topsoil, incorporating organic matter, and providing media and aeration necessary for bacterial decomposition (Taylor, 1935; Romell, 1935; Heyward and Tissot, 1936).

A. Causes of Population Changes

In areas where fires are moderate, the heat of fire is apparently less important than later environmental changes in reducing insect populations. In Africa, Coults (1945) found that all organisms in the top inch of mineral soil survive fire. Rice (1932) and others have found living organisms under charred boards and debris on the soil surface after fire. Buffington (1967) related the decreases in soil fauna after fire to loss of both incorporated and unincorporated organic matter which reduces food supply for the smaller organisms and in turn for their predators. Xeric conditions which follow fire in the surface soil levels are also responsible for loss of some organisms, as are the greater soil temperature extremes found on burns.

Because of their mobility, the rate at which soil organisms recolonize a burned area must be considered in an interpretation of results in terms of the effect these organisms have in the soil. Metz and Farrier (1971) found mesofaunal recovery to take less than 43 months. In prairie lands, Rice (1932) found that many species declined in number during the prevernal period following burning but increased again during the vernal period when lush recovering vegetation improved habitat.

In most forest areas, the general reduction in soil fauna varies with the area, type of fire, and animal species involved. In loblolly pine (*Pinus taeda*) in South Carolina, Pearse (1943) found soil fauna reduced one-third by fire, with species remaining in about the same proportions as on unburned land except for a greater decrease in earthworms and an increase in ants. In longleaf pine forests, Heyward and Tissot (1936) found 5 times more soil organisms in the A-O layer of unburned soil than in the corresponding layer on burned land. They also found 11 times more organisms in the top 2 inches of unburned mineral soil. These decreases were related to the desiccation caused by xeric postfire soil conditions. The soil was much more permeable with noticeably more tunnels and holes when fire was absent from the land for 10 years. In the New Jersey pine barrens, soil arthropod populations were reduced 50% following fire, with an 80% reduction in taxa (Buffington, 1967). Various species have different life requirements and can be expected to behave differently in response to fire.

B. Annelids and Mollusks

1. Earthworms

In all habitats studied thus far, earthworms are reduced markedly by fire. A 50% reduction resulted from burning off litter in the Duke Forest (Pearse, 1943). In the 0- to 2-inch level of mineral soil in longleaf pine stands, the population was 4 times greater in unburned than in burned soil (Heyward and Tissot, 1936). Although earthworms were not frequent in the A-O layer before fire, some postfire reduction was also noted in this surface layer. In studies of Illinois prairie, Rice (1932) found lower populations of earthworms on burned land. She noted that earthworm populations were similar on both burned and unburned land in early spring when soil moisture content was similar on both. In late April and May, however, as soil moisture increased on the unburned prairie and declined on the burned prairie, the earthworm population in the unburned area rose while that of the burned area dropped markedly. The moisture decline was associated with increased evaporation caused by lack of surface vegetation. Thus, since earthworms are located primarily in the mull and mineral-soil layers, they are affected more by postfire loss of soil moisture than by the actual heat of fire.

2. Snails

In the southern United States, fire reduced the population of snails (Heyward and Tissot, 1936). Snails also disappeared for at least 3 years after burning in jack pine forests of northern Minnesota.

C. INSECTS

1. Grasshoppers and Leafhoppers

Overwintering stages of grasshoppers occur in surface soil or burrows and could be destroyed by spring burning. Burning is often recommended for control of both groups in agricultural practice (Komarek, 1970). Cantlon and Buell (1952) believed that burning could indirectly control certain plant virus diseases by destroying their leafhopper vectors.

In prairies and grasslands, however, grasshoppers and leafhoppers increase after fire (Hurst, 1970; Tester and Marshall, 1961; Rice, 1932). In northern Minnesota jack pine burns, an increase in grasshoppers after fire was also noted. It is possible that the increases were the results of recolonization from adjacent unburned areas or survival in patches of unburned land. Warm daytime temperatures and lush early postfire herbaceous recovery may also contribute to the increase. In African savannah, fire affected various species of grasshoppers differently: Some were almost completely removed by fire; others were relatively unaffected. It is believed that periodic burning contributes to diversity of species within this group (Gillon, 1971).

2. Ants

Ants are less affected by fire than many other groups of insects because of their adaptations to the hot, xeric conditions of early postfire topsoil. Furthermore, their cryptic habits enable them to survive fire below the level of intense heat. Their colonization habits and social organization adapt them to rapid reestablishment on burned land.

In the longleaf pine region, Pearse (1943) reported ant populations reduced one-third by burning. Buffington (1967) also found a reduction in ants on burned soil in the New Jersey pine barrens. The reduction, however, was not as great as for other soil arthropods. He found that two species, Lasius flavus and Solenopsis molesta, were strikingly more numerous on burned land and related their success to a preference for xeric conditions and a dry seed hoarding habit. In Australian Pinus radiata plantations, French and Keirle (1969) found that ants, along with most other insects, were destroyed by fire. Ants were among the first to recolonize, however, and were soon frequent in all burned areas studied.

Reports of increases in ant populations after fire are frequent for other areas and are variously attributed to rapid recolonization and survival of the fire in lower soil layers. In pine forests of the southern United States, Heyward and Tissot (1936) found more ants in the burned 0- to 2-inch mineral-soil layers than in unburned soil. The population in the duff and litter of burned land was quite low. In the grass habitat on burned transmission line right-of-ways in Mississippi, Hurst (1970) found that ants, especially the fire ant, *Solenopsis*, increased following fire. In an Illinois prairie, Rice (1932) found that the ant population was over one-third higher on burned than on unburned soil. She related this to their survival of fire below ground and to the very suitable forage conditions of grassland after fire.

3. Termites

Some termite species can survive fire in lower soil depths. Other species appear to be quite heat resistant. In the New Jersey pine barrens, Buffington (1967) found live *Reticulotermes* under a charred board on severely burned land. Heyward and Tissot (1936), however, reported a reduction in termites in the southern pine (*Pinus palustris*) area after fire.

4. Beetles

In forest areas, most beetle genera are decreased by fire, at least temporarily. The beetle population of grassland and prairies is not affected as much, partly because soil temperatures are lower during grassland fires than during forest fires. In addition, there are many safe harbors in partially burned grass tussocks. A 60% decrease in beetle populations was noted in southern pine areas (Pearse, 1943; Heyward and Tissot, 1936). In the New Jersey pine barrens, 4 times more beetles were found on unburned land than on burned land (Buffington, 1967). The author found fewer beetles on burned than unburned land the first 3 months after prescribed burning in Minnesota jack pine stands (I. F. Ahlgren, unpublished). Tester and Marshal (1961) recorded an increase in beetles in burned Minnesota prairie, and Hurst (1970) noted an increase in both numbers and biomass of beetles on burned transmission line right-of-ways in Mississippi. In an Illinois prairie, following an initial reduction of 15%, beetles recolonized rapidly (Rice, 1932).

In forest areas, burning is sometimes used as a control for bark beetles

(Haig, 1938). The chinch bug has long been controlled in grain- and corn-growing areas by burning (Komarek, 1970). In Belorussia, some species of Elaterids, in their larval stages as wire worms, occur only on burned land; other species thrive on unburned land (Rubzova, 1967). The phytophagus beetles, *Pelycyphorus densicollis* and *Eleodes hispitabris*, were reduced after fire on shrub steppe land in Washington (Rickard, 1970). In Australian *Pinus radiata* plantations, French and Keirle (1969) found that carabid and scarabid beetles were reduced immediately after fire, although they were among the first insects to recolonize burned areas.

D. Arachnids

1. *Spiders*

Spiders, primarily surface dwellers, are drastically reduced by fire in most areas. Reductions of 9 to 31% of the population of unburned land are frequently reported (Buffington, 1967; Rice, 1932; Heyward and Tissot, 1936; French and Keirle, 1969). Only on burned, grassy transmission line right-of-ways in Mississippi was an increase noted after fire, predominantly of ground and wolf spiders (Hurst, 1970).

2. *Mites and Collembolans*

The mesofauna (largely mites and collembolans) are a major part of the total soil fauna and are usually considered collectively. Metz and Farrier (1971) cite a Russian experiment in which naphthalene was placed in oak litter to reduce mesofauna without harming the microflora. After 140 days, the naphthalene-treated litter lost 9% of its weight; the untreated litter lost 55%.

Although their numbers vary with regions, mites can constitute one of the largest groups of animals in the soil. Heyward and Tissot (1936) estimate that in *Pinus palustris* stands they made up between 71 and 93% of the fauna in unburned soil and between 30 and 72% in burned soil, depending on depth. All investigators agree that mite populations are reduced by burning (Rice, 1932; Heyward and Tissot, 1936; Metz and Farrier, 1971; Pearse, 1943). In a detailed study of this group, Metz and Farrier (1971) found a sudden reduction of mites within 24 hr after fire. A much greater reduction occurred in the surface 3 inches of mineral soil than in the forest floor (Fig. 4). The recovery of mites to preburn levels occurred in less than 43 months.

Collembolans, while insects, are usually considered with the mites

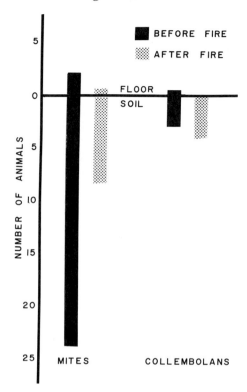

Fig. 4. Average number of mites and collembolans extracted per 20 cm² soil sample, immediately before and 24 hours after an annual summer burn in North Carolina. (Taken from Metz and Farrier, 1971.)

because of size and similarities in habitat. These organisms were also reduced by fire (Heyward and Tissot, 1936; Metz and Farrier, 1971). Metz noted this reduction only on the forest floor, with little effect in the mineral soil (Fig. 4). This group, then, is one of the few in which an immediate postfire reduction has been described, with all indications that it is induced by the heat of fire. Both mites and collembolans recovered to near preburn population levels in 3 to 4 years, as indicated by the high populations in periodic burns (Fig. 5).

E. CENTIPEDES AND MILLIPEDES

Centipedes and millipedes are usually reduced by burning, often by as much as 80% (Rice, 1932; Heyward and Tissot, 1936; Pearse, 1943). The larger numbers of centipedes on unburned land have been related

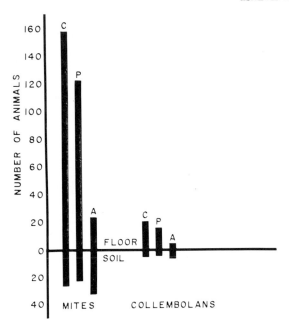

Fig. 5. Average number of mites and collembolans extracted per 20 cm² soil sample on annually burned, periodically burned (38 to 46 months previous to sampling) and unburned areas. C, Control; P, periodic burn; A, annual burn. (Taken from Metz and Farrier, 1971.)

to the greater populations of other insects, since centipedes are predacious in feeding habits (Rice, 1932).

F. OVERVIEW

Studies of soil fauna vary in sampling methods, habitat, time of study, and fire intensity. Despite such diversity, two generalizations are apparent. First, the effect of fire is greater in the forest than in grassland. There are several possible reasons, none of which has been fully investigated. The species in grassland may be more adapted initially to xeric conditions than the species found in the cooler, moist forest floor. Fire intensity may be greater because of more abundant fuel in forest areas. In addition, most of the grassland studies were made on small burned areas where recolonization from unburned land could be rapid. Second, with the exception of the mesofaunal species and spiders, the population reductions do not seem to be directly caused by heat of fire. More important in these decreases are postfire changes in the environment.

The transition to xeric conditions is frequently mentioned, along with lack of food and greater temperature fluctuations.

References

Ahlgren, C. E. (1960). Some effects of fire on reproduction and growth of vegetation in northeastern Minnesota. *Ecology* **41**, 431–445.

Ahlgren, C. E. (1970). Some effects of prescribed burning on jack pine reproduction in northeastern Minnesota. *Minn., Agr. Exp. Sta., Misc. Rep.* **94.**

Ahlgren, I. F., and Ahlgren, C. E. (1965). Effects of prescribed burning on soil microorganisms in a Minnesota jack pine forest. *Ecology* **46**, 304–310.

Basham, J. T. (1957). The deterioration by fungi of jack, red, and white pine killed by fire in Ontario. *Can. J. Bot.* **35**, 155–172.

Bollen, W. B., and Wright, E. (1961). Microbes and nitrates in soils from virgin and young-growth forests. *Can. J. Bot.* **7**, 785–792.

Bornebusch, C. H. (1930). The fauna of forest soil. *Proc. Int. Congr. Forest Exp. Sta., 1st, 1929* pp. 541–545.

Buffington, J. D. (1967). Soil arthropod populations of the New Jersey pine barrens as affected by fire. *Ann. Entomol. Soc. Amer.* **60**, 530–535.

Cantlon, J. E., and Buell, M. F. (1952). Controlled burning—its broader ecological aspects. *Bartonia* **26**, 48–52.

Chapman, H. H. (1932). Some further relationships of fire to long leaf pine. *J. Forest.* **30**, 602–604.

Cohen, C. (1950). The occurrence of fungi in soil after different burning and grazing treatments in the veld of Transvaal. *S. Afr. J. Sci.* **46**, 245–246.

Corbet, A. S. (1934). Studies on tropical soil microbiology. II. The bacterial numbers in the soils of the Malay Peninsula. *Soil Sci.* **38**, 407–416.

Coults, J. R. H. (1945). The effect of veld burning on the base exchange capacity of a soil. *S. Afr. J. Sci.* **41**, 218–224.

Davis, K. P., and Klehm, K. A. (1939). Controlled burning in the western white pine type. *J. Forest.* **37**, 399–407.

Demaree, J. B., and Wilcox, M. S. (1947). Fungi pathogenic to blueberries in the eastern United States. *Phytopathology* **37**, 487–506.

Düggeli, M. (1938). Studien über den Einfluss der im Stadtwald Zoflingen angewandten Massnahmen zur Bodenverbesserung auf die Bakterienflora des Waldbodens. *Mitt. Schweiz. Zentralanst. Forstl. Versuchsw.* **20**, 307–444.

Ebert, P. (1958). Das *Geopyxidetum carbonariae*, ein carbophile Pilzassociation. *Z. Pilzk.* **24**, 32–44.

El-Abyad, M. S. H., and Webster, J. (1968). Studies on pyrophilous discomycetes. I. Comparative physiological studies. *Trans. Brit. Mycol. Soc.* **51**, 353–367.

French, J. R. S., and Keirle, R. M. (1969). Studies in fire damaged radiata pine plantations. *Aust. Forest.* **33**, 175–180.

Fritz, E. (1930). The role of fire in the redwood region. *J. Forest.* **29**, 939–950.

Fowells, H. A., and Stephenson, R. S. (1933). Effect of burning on forest soils. *Soil Sci.* **38**, 175–181.

Fuller, W. N., Shannon, S., and Burges, P. S. (1955). Effect of burning on certain forest soils of northern Arizona. *Soil Sci.* **1**, 44–50.

Garren, K. H. (1943). Effects of fire on vegetation of the southeastern United States. *Bot. Rev.* **9**, 617–654.

Gillon, Y. (1971). The effect of bush fire on the principal acridid species of an Ivory Coast Savannah. *Proc. 11th Annu. Tall Timbers Fire Ecol. Conf.* pp. 419–471.

Haig, I. T. (1938). Fire in modern forest management. *J. Forest.* **36**, 1045–1051.

Hall, A. D. (1921). "The Soil," Vol. XV. Murray, London.

Heikinheimo, O. (1915). Der Einfluss der Brandwirtschaft auf die Wälder Finnlands. *Acta Forest. Fenn.* **4**, 1–264.

Hesselman, H. (1917). On the effect of our regeneration measures on the formation of saltpetre in the ground and its importance in the regeneration of coniferous forests. *Medd. Statens Skogsfoersoeksanstalt* **13–14**, 923–1076 (English summary).

Heyward, F., and Tissot, A. N. (1936). Some changes in soil fauna associated with forest fires in the long leaf pine region. *Ecology* **17**, 659–666.

Hintikka, V. (1960). Zur Ökologie einiger an Brandplätzen vorkommender Blätterpilzarten. *Karstenia* **5**, 100–106.

Hurst, G. A. (1970). The effects of controlled burning on arthropod density and biomass in relation to bobwhite quail brood habitat on a right-of-way. *Proc. Tall Timbers Conf. Ecol. Anim. Contr. Habitat Manage.*, Vol. 2, pp. 173–183.

Isaac, L. A., and Hopkins, H. G. (1937). The forest soil of the Douglas fir region, changes wrought upon it by logging and slash burning. *Ecology* **18**, 264–279.

Iwanami, Y. (1969). Temperatures during *Miscanthus* type grassland fires and their effect on regeneration of *Miscanthus sinensis. Sci. Rep. Res. Inst., Tohoku Univ., Ser. D* **20**, 47–88.

Jalaluddin, M. (1968). Fire as an agent in the establishment of a plant disease. *Pak. J. Sci.* **20**, 42–44.

Jalaluddin, M. (1969). Micro-organic colonization of forest soil after burning. *Plant Soil* **30**, 150–152.

Johnson, J. (1919). Influence of heated soils on seed germination and plant growth. *Soil Sci.* **7**, 1–104.

Jorgensen, J. R., and Hodges, C. S. (1970). Microbial characteristics of a forest soil after twenty years of prescribed burning. *Mycologia* **62**, 721–726.

Jorgensen, J. R., and Hodges, C. S. (1971). Effects of prescribed burning on microbial characteristics of soil. *U.S., Forest Serv., Southeast. Forest Exp. Sta., Prescribed Burning Symp.* pp. 107–114.

Kivekäs, J. (1939). Kaskiveljelyksen vaikutus eraisiin maan ominarsuuksiin. *Commun. Inst. Forest. Fenn.* **27**, 1–44.

Komarek, E. V. (1970). Insect control—fire for habitat management. *Proc. Tall Timbers Conf. Ecol. Anim. Contr. Habitat Manage.*, Vol. 2, pp. 157–171.

Laine, L. (1968). Kuplamörsky (*Rhizina undulata* Fr.) uusi metsän tukosieni maassamme. *Folia Forest.* **68**, 44.

Lunt, H. A. (1951). Liming and twenty years of litter raking and burning under red and white pine. *Soil. Sci. Soc. Amer., Proc.* **15**, 381–390.

Lutz, H. J. (1934). Ecological relationships in the pitch pine plains of southern New Jersey. *Yale Sch. Forest. Bull.* **38**.

Lutz, H. J., and Chandler, R. F. (1946). "Forest Soils." Wiley (Interscience), New York.

Markim, F. L. (1943). Blueberry diseases in Maine. *Maine, Agr. Exp. Sta., Bull.* **419**.

Meiklejohn, J. (1953). The effect of bush burning on the microflora of some Kenya soils., *Proc. Int. Conf. Microbiol., 6th, 1953* Vol. 10, pp. 317–319.

Metz, L. J., and Farrier, M. H. (1971). Prescribed burning and soil mesofauna on the Santee Experimental Forest. *U.S., Forest. Serv., Southeast. Forest Exp. Sta., Prescribed Burning Symp.* pp. 100–105.

Moser, M. (1949). Untersuchungen über den Einfluss von Waldbränden auf die Pilzvegetation. I. *Sydowia* 3, 336–383.

Muller, K. M. (1929). "Aufbau, Wuchs and Verjüngung der Südosteurepäischen Urwälder." Schaper, Hanover [reviewed in *J. Forest.* 28, 744 (1930)].

Neal, J. L., Wright, E., and Bollen, W. B. (1965). Burning Douglas fir slash: Physical, chemical, and microbial effects in the soil. *Oreg. State Univ., Forest Res. Lab., Res. Pap.* pp. 1–32.

Norkrans, B., and Hammarstrom, A. (1963). Studies on growth of *Rhizina undulata* Fr. and its production of cellulose and pectin decomposing enzymes. *Physiol. Plant.* 16, 1–10.

Paarlahti, J., and Hanoija, P. (1962). Methodological studies on colony counts of soil microbes. *Metsantutkimuslaitoksen Julkaisuju* 55, 1–7.

Pearse, A. S. (1943). Effects of burning over and raking off litter on certain soil animals in the Duke Forest. *Amer. Midl. Natur.* 29, 406–424.

Petersen, P. M. (1970). Danish fire place fungi: An ecological investigation of fungi on burns. *Dan. Bot. Ark.* 27, 1–96.

Pirk, W. (1950). Pilze im Mossgesellschaften auf Brandflächen. *Mitt. Florist.-Soziol. Arbeits gemeinsch.* [N.S.] 2, 3–5.

Remezov, N. P. (1941). Ammonification and nitrification in forest soils. Trud. Vsesoyuz. nauchnoissledov. *Inst. Lesnojo Khoz.* 24, 89–128 [cited from *Forest. Abstr.* 7, 408 (1946)].

Rice, L. (1932). The effect of fire on prairie animal communities. *Ecology* 13, 392–401.

Rickard, W. H. (1970). Ground dwelling beetles in burned and unburned vegetation. *J. Range Manage.* 23, 293–297.

Romell, L. G. (1935). An example of miriapods as mull formers. *Ecology* 16, 67–71.

Rubzova, Z. I. (1967). Elateridae in soil of pine forest types in West Belorussia. *Oikos* 18, 41–54.

Sarvas, R. (1937). Beobachtungen über die Entwicklung der Vegetation auf den Waldbrandflächen nord-Finnlands. *Silva Fenn.* 39, 1–64.

Schmitz, H., and Jackson, L. W. R. (1927). Heart rot of aspen with special reference to forest management in Minnesota. *Minn., Agr. Exp. Sta., Tech. Bull.* 50.

Seaver, F. J. (1909). Studies on pyrophilous fungi. I. Occurrence and cultivation of *Pyronema*. *Mycologia* 1, 131–139.

Seaver, F. J., and Clark, E. D. (1910). Studies on pyrophilous fungi. II. *Mycologia* 2, 109–124.

Shields, L. M., and Durrell, L. W. (1964). Algae in relation to soil fertility. *Bot. Rev.* 30, 92–128.

Siggers, P. V. (1934). Observations on the influence of fire on the brown spot needle blight of long leaf pine seedbeds. *J. Forest.* 32, 556–562.

Smith, D. W., and Sparling, J. H. (1966). The temperature of surface fires in the jack pine barrens. I. The variation in temperature with time. *Can. J. Bot.* 44, 1285–1292.

Stinson, K. J., and Wright, H. A. (1969). Temperatures of headfires in the southern mixed prairie of Texas. *J. Range Manage.* 22, 169–174.

Sushkina, N. N. (1933). Nitrification of forest soil with reference to the composition of the stands, cutting, and fire. *Izv. Akad. Nauk SSSR, Otd. Mat. Estest. Nauk* **1**, 111–160 [cited from *U.S., Forest Serv. Transl.* **56**].

Taylor, W. P. (1935). Some animal relations to soils. *Ecology* **16**, 127–136.

Tester, J. R., and Marshall, W. H. (1961). A study of certain plant and animal interrelationships on a prairie in northwestern Minnesota. *Univ. Minn., Mus. Natur. Hist., Occas. Pap.* No. 8.

Tresner, H. D., Backus, M. P., and Curtis, J. T. (1954). Soil microfungi in relation to the hardwood continuum in southern Wisconsin. *Mycologia* **46**, 314–444.

Tryon, E. H. (1948). Effects of charcoal on certain physical, chemical, and biological properties of forest soils. *Ecol. Monogr.* **18**, 81–115.

Vandecayve, S. C., and Baker, G. O. (1938). Microbial activities in the soil. III. Activity of specific groups of microbes in different soils. *Soil Sci.* **45**, 315–333.

Veretennikova, A. V. (1963). Algae—pioneers on forest burns. *Priroda (Moscow)* **52**, 105.

Viro, P. J. (1969). Prescribed burning in forestry. *Metsantutkimuslaitoksen Julkaisuja* **67**, 1–49.

Vogl, R. J. (1964). Effects of fire on a muskeg in northern Wisconsin. *J. Wildl. Manage.* **28**, 317–327.

Webster, J., Rifai, M. A., and El-Abyad, M. S. H. (1964). Culture observations on some discomycetes from burnt ground. *Trans. Brit. Mycol. Soc.* **47**, 445–454.

Wright, E., and Bollen, W. B. (1961). Microflora of Douglas fir forest soil. *Ecology* **42**, 825–828.

Wright, E., and Tarrant, R. F. (1957). Microbial soil properties after logging and slash burning. *U.S., Forest Serv., Pac. Northwest Forest Range Exp. Sta., Res. Notes* **157**.

Wright, E., and Tarrant, R. F. (1958). Occurrence of mycorrhizae after logging and slash burning in the Douglas fir forest type. *U.S., Forest Serv., Pac. Northwest Forest Range Exp. Sta., Res. Notes* **160**.

. 4 .

Effects of Fire on Birds and Mammals

J. F. Bendell

I. Introduction

Fire is a tremendous force for change in nature and requires explanation. This chapter begins with a catalog of effects of fire on the life and environment of birds and mammals and provides a starting point for understanding how fire may influence them. During their lifetime, birds and mammals encounter many changes in their environment and easily cope with most of them. What we usually seek to understand are those factors of the animal or its environment that are crucial to its occurrence and abundance. Hence, Sections III and IV attempt an analysis of how kinds and numbers of animals are influenced by fire.

Beyond occurrence and abundance, there is the broader consideration of how species have evolved to live in an environment frequently razed by fire. How have species adapted t conditions caused by fire and what interactions are there between wildlife and their environment that perpetuate the sweep of fire through the forest?

73

A major problem in attempting to develop general statements about fire ecology is the wide variation that occurs in fires and their effects. When a fire occurs, many factors in the environment are changed, and this makes it difficult to pinpoint cause and effect relationships between the action of fire and response of animals. Fires may vary in intensity, duration, frequency, location, shape, and extent. Their effects may differ with season, nature of the fuel, and properties of the site and soil. Added to these sources of variation are the effects of man; a fire may have vastly different results in natural forest as compared to fire over the same area after logging or some other use by man. Thus, it may be wise to consider each fire as a special case and not yet attempt to apply the findings from one burn to another. It may be too early to generalize about the effects of fire on birds and mammals.

Most of the extensive literature on fire deals with its effects on plants and soil. Such studies offer clues to the animal ecologist but may not produce much more than speculation as to the effects of fire on animals. Few studies are quantitative, have adequate controls, and have been carried on long enough to really assess the effect of a particular fire on birds and mammals. There are virtually no data on how varying the properties of fire (e.g., size) will affect wildlife.

II. Fire and the Life of Birds and Mammals

A. Immediate Reactions to Fire

As fire moves through the forest much of what happens to birds and mammals will depend upon their site attachment, mobility, and ability to find refuge. In 1915, a forest fire in western Siberia burned for about 50 days in an area 1,600,000 km² (smoke covered an area the size of Europe). Many mammals, including squirrels (*Sciurus*), bears (*Ursus*), and moose (*Alces*), were observed swimming across large rivers to escape the fire (Udvardy, 1969b). In contrast to this scene of confusion, Hakala *et al.* (1971) described the behavior of birds and mammals during two fires that totaled 86,000 acres on the Kenai National Moose Range in south central Alaska. Despite many observers, there were no reports of large birds and mammals reacting wildly to the fire. A family of swans (*Cygnus*) and a moose moved and fed in a small lake while the forest burned to the shore. A small group of caribou (*Rangifer*) rested on the ground while encircled by fire and later moved away. Vogl (1973) reported similar calm behavior by birds and mammals in a fire in Florida grasslands and hardwoods. Some moved back into the burned area immediately after the fire.

Komarek (1969) observed cotton rats (*Sigmodon*), some carrying young, run before a fire in grassland and shrubland. At times, a cotton rat ran into the flames, became singed, and died. Komarek believed that many rats went into burrows or holes and escaped the fire. When slash was burned from 16 acres of logged Douglas-fir (*Pseudotsuga*), mice (*Peromyscus*), chipmunks (*Eutamias*), shrews (*Sorex*), and wood rats (*Neotoma*) ran from the fire. Some wood rats dashed from one patch of shelter to another until they disappeared into surrounding forest. Most behaved as if reluctant to leave their homes, lingered before running from the clear-cut stand, and a few actually returned to the inferno and perished. The persistence of the wood rats on an area probably relates to their establishment on a home range and shows how such behavior may make an animal vulnerable to fire. Finally, some wood rats appeared to be panic-stricken. They remained immobile until the flames almost reached them or actually singed their fur. Some were set afire with burning resin. Animals that panicked ran wildly from the fire or fell or plunged into the flames (Tevis, 1956).

Sunquist (1967) monitored movements of radio-tagged raccoons (*Procyon*) in a forested area before it was set afire. The raccoons simply moved away, and after the fire neither avoided nor traveled the burn more than usually. However, many birds and mammals are attracted by fires, probably to feed upon prey driven from their homes. Komarek (1969) mentioned species of birds in Australia, Africa, and North America that come to and hunt in front of fires. In another report Komarek (1967) noted that various primates (*Gorilla, Pan, Gibbon*) were attracted to bush fires and abandoned campfires. The Phillippine tarsier (*Tarsier carbonarius*) apparently picks up hot embers from smoldering fires.

Several investigators, including Aldo Leopold, believed wildfire to be extremely destructive of wildlife (Ahlgren and Ahlgren, 1960; Lutz, 1956). Many, however, considered direct mortality caused by fire to be negligible (Hakala *et al.*, 1971; Howard *et al.*, 1959; Komarek, 1969; Vogl, 1973). Dead animals found after a fire are used as evidence that fire was the cause of death. For example, Hakala *et al.* (1971) stated that dead small birds and voles (*Clethrionomys?*) were found in the ashes immediately after a fire. In Africa, Brynard (1971) reported that old long grass may burn and kill large game such as elephant (*Loxodonta*), lion (*Felis*), warthog (*Macrocephalus*), and a variety of antelope. In Alberta, Keith and Surrendi (1971) searched an area of 640 acres of mixed spruce (*Picea*) and aspen (*Populus*) after a severe wildfire in May and found no dead hares (*Lepus*) and only 3 voles (*Clethrionomys?*) apparently killed by the fire. Ruffed grouse (*Bonasa*) populations were under study on the same burn (Doerr *et al.*, 1970) and no dead grouse were found. However, there was a reproductive failure on the burned area and this was attributed to destruction of grouse nests by the fire.

Tevis (1956) tagged 41 *Peromyscus* on 16 acres of logged Douglas-fir just before the slash was burned. A day later he trapped on and around the burn and recovered 13 marked mice, 4 in the burn, on small patches of unburned vegetation, the rest in the surrounding forest. The fire caused the loss of most mice either by death or dispersal, yet a few either stayed and survived or moved away and returned immediately before and after the fire. Most fires burn unevenly and probably refugia are always left for some birds and mammals.

Tevis continued trapping the burn and in $2\frac{1}{2}$ weeks after the fire caught nearly twice the number of mice as before. Similar results were obtained by Sims and Buckner (1973) with *Peromyscus* and Tester (1965) with *Peromyscus* and *Clethrionomys*. Komarek (1963) marked cotton rats before a burn and retrapped after the fire. He found no dead rats on the burn and no change in capture rate to indicate loss caused by burning.

Howard *et al.* (1959) placed caged mammals (*Citellus, Peromyscus, Rattus*) and snakes (*Crotalus*) in rocky crevices and underground before starting a fire in grass and brush (*Ceanothus, Quercus*). Although the animals were not allowed to react freely to the fire, about half of them survived, and it is difficult to believe they could have done worse if they had been free. A similar experiment was performed by Lawrence (1966), who placed *Peromyscus* in cages in burrows under chaparral (*Adenostema*) fire. When the burrow had only one opening, the animals apparently suffocated. When the burrow was a tunnel open at both ends, presumably allowing adequate ventilation, the mice (*Microtus*) survived.

The microclimate, particularly the temperature and humidity of places where small animals might hide to escape fire, appears to be very important to survival. Numerous investigators showed the great decrease in temperature that occurs in the ground only a few centimeters away from the hottest part of a fire (Ahlgren and Ahlgren, 1960; Cooper, 1961a; Martin, 1963; Mcfadyen, 1968; D. W. Smith, 1968). As examples, Kahn (1960) found that during a chaparral fire the soil temperature $\frac{1}{2}$ inch below the surface did not ordinarily rise above 60°C, and Lawrence (1966), also studying chaparral fire, found 4 inches of soil deep enough to provide tolerable temperatures, while temperatures on the surface were 1000°F.

The discussion thus far suggests that burning does not cause much immediate loss of life. However, other wildlife may be more sensitive to fire, and other fires may be more destructive. Chew *et al.* (1958) counted bodies of birds, small mammals, and deer (*Odocoileus*) after a fire in chaparral in a canyon. They believed that the very open nature of the ground gave them a total count. The numbers of dead wood rats and rabbits (*Sylvilagus*) that were found suggested the fire had killed all those in the area. Only a few bodies were charred, so death was from heat or asphyxiation.

How are mammals killed by fire? Howard *et al.* (1959) found that caged mice and rats died when the temperature exceeded about 145°F. Lawrence (1966) speculated that they may suffocate when caught in burrows without an adequate supply of fresh air. He also noted that as the soil warms, the vapor pressure of water increases so that the air within a burrow becomes hot and humid. Birds and mammals hold their body temperatures against a rising ambient temperature partly by evaporative cooling (King and Farner, 1961). Should the temperature in a burrow exceed the upper lethal temperature for the mammal and evaporation cooling become impossible because of high vapor pressure of water, death from heat damage will result. Bearing in mind the relatively low temperature found short distances in the soil from a surface fire, it seems unlikely that mammals are killed by heat. Suffocation is a more probable cause of death. This is also true for humans. According to Zikria *et al.* (1972) out of 105 fire deaths in New York City, 76% were the result of some type of respiratory failure caused by one or a combination of carbon monoxide poisoning, lack of oxygen, and toxic chemicals in the smoke.

B. LONG-TERM EFFECTS OF FIRE

Some effects of fire may be immediate and direct; others are complicated and may not culminate as major influences for a very long time. For example, the African elephant lives as a browser and grazer, on the margin between dry forest and grassland (Darling, 1960; West, 1971). Repeated burning tends to reduce the forest and increase grassland, and the elephant helps by killing trees (Buechner and Dawkins, 1961). Without repeated fire much grassland might revert to forest and cause the elephant population to decline. This series of events may take hundreds of years. Both forest and elephant have many ecological consequences. The elephant helps maintain grassland which in turn supports a great variety of birds and mammals, including famous African game species. Elephants also plough the ground by uprooting trees, make paths on which other animals may penetrate the bush, dig for water which is used by birds and mammals, crop tall trees and shrubs, stimulating production of low lateral shoots—a food source for smaller animals—distribute large amounts of manure that may contain viable seeds of trees, and even make homes for fish that use their footprints in a stream bottom as a place for nesting (Darling, 1960). Clearly, fire, vegetation, and animals may be bound together in many ways and over a long span of time.

1. *Local Climate and Microclimate*

Burns have their own local climate and microclimates that may be important to wildlife. In general, birds and mammals can cope with the direct effects of climate, which probably has its greatest impact through the kind of food and cover it produces on a burn. What would seem most important to birds and mammals in the long term on most burns are extremes in temperature and moisture.

a. Smoke and Blackening. Smoke at the time of a fire is probably too transitory to have much effect on climate and in turn on birds and mammals. Smoke from the huge forest fire described by Udvardy (1969b) decreased sunlight by about two-thirds for approximately 50 days and caused a delay of 10–15 days in ripening of crops. Even these effects would seem of little consequence to wildlife.

The black of charred vegetation and soil might increase the heat input to an area to directly or indirectly influence birds and mammals. Snow and ground temperature are influenced by the color of the soil, and both may affect distribution of wildlife (Klein, 1960; Pruitt, 1959a). Several investigations attributed the early and vigorous growth of plants in the spring to blackened ground after fire (Ahlgren and Ahlgren, 1960; Anderson, 1972; Daubenmire, 1968; Sykes, 1971), and both developments could improve the food supply to birds and mammals at this critical time of the year (Siivonen, 1957).

b. Light. Removal of forest, blackening of the ground, and exposure of light-colored mineral soil and rocks will change the input and reflectance of light, which may affect birds and mammals. Warblers (*Parulidae*) and vireos (*Vireonidae*) that live in the canopy of deciduous forest require shade from direct sunlight (Brewer, 1958; Kendeigh, 1945). Udvardy (1969b) reported that the wood mouse (*Apodemus*) avoided light whereas the ground squirrel (*Spermophilus*) chose bright sunlight. When areas were logged and burned they were invaded by ground squirrels (*Citellus*) (Gashwiler, 1970; Tevis, 1956), which are diurnal and live in open areas. While numerous factors may be involved, the increased light reaching the ground may have played some role in making the opened areas acceptable to them.

c. TEMPERATURE. As discussed elsewhere in this book, fire generally creates higher maximum and lower minimum temperatures on burned than on unburned land. A number of studies, while not directly concerned with fire, suggest how temperature might affect wildlife on a burned area (Brewer, 1958; Cruickshank, 1956; Drent, 1972; Horvath, 1964; Klein, 1960; Pruitt, 1959a; Salt, 1952; Udvardy, 1969b). Several investigators noted the abrupt disappearance of red-backed voles (*Clethrionomys*) after fire (Ahlgren, 1966; Beck and Vogl, 1972; Gashwiler, 1970). Usually, increased temperatures were given as the cause of the decline and this seems probable considering the relatively cold climates in which *Clethrionomys* occurs. Hurst (1971) observed that burned areas in the range of bobwhite quail (*Colinus*) were warmer and drier by exposure to the sun and wind. He believed that both factors improved the habitat for quail, particularly the chicks.

d. HUMIDITY. Humidity is directly related to temperature and wind and is therefore difficult to consider alone. I studied temperature and humidity on a recent burn (Fig. 1) and a nearby older burn with dense

Fig. 1. Open cover measured for temperature and humidity, in summer used by blue grouse and avoided by ruffed grouse. The white length of the stick is 1 ft (30.48 cm).

Fig. 2. Dense cover measured for temperature and humidity, avoided by blue grouse, used somewhat by ruffed grouse.

regrowth (Fig. 2) in cedar (*Thuja*)–hemlock (*Tsuga*)–Douglas-fir forest on Vancouver Island. Both areas were logged and replanted to Douglas-fir. The open or recent burn was burned in 1951, the area of denser regrowth was burned in 1938. One hygrothermograph within a weather screen was placed on the ground in each burn. The instruments were about ½ mile apart at approximately the same elevation. In May through August, weekly averages of daily maximum and minimum temperatures were on the average 4°F hotter and 5°F colder on the recent burn. At times, the open burn was 15°F hotter by day and 12°F colder by night than in the denser vegetation.

The open burn was strikingly drier throughout the day than the dense regrowth (Table I). As with temperatures, the open cover showed the widest fluctuations in humidity and was much drier by day than the dense regrowth. Hence, dense cover was cooler and more humid by day and warmer by night than the younger and more open regrowth.

The amount of moisture in the air and soil may determine the local distribution and range of birds and mammals (Henderson, 1971; Pruitt, 1953; Salt, 1952). We have observed that in summer when it is hot and dry, ruffed grouse live in the parts of burns that are wet and grow

TABLE I

RELATIVE HUMIDITY (%) IN OPEN BURN
AND ADJACENT DENSE REGROWTH

Month	Open burn			Dense regrowth		
	Mean max.[a]	Mean min.	Lowest value	Mean max.	Mean min.	Lowest value
May	96.5	21.6	4	100	60.3	34
June	95.3	23.5	3	100	68.1	38
July	99.4	26.9	9	99.9	68.3	46
August	98.9	38.9	6	99.3	80.7	55

[a] Calculated as the average of daily readings.

thick willow (*Salix*) and alder (*Alnus*). Temperature and humidity in such habitats are comparable to what I have described for dense regrowth of conifers (Table I). Blue grouse (*Dendragapus*) live in the dry, open burn. Both species eat the same foods in summer (King, 1968, 1969), and there is no evidence that one affects the other in a significant way. In winter, when it is cool and wet, ruffed grouse are found on the open burn and in their summer habitat. Blue grouse winter in the relatively dry subalpine forest. In captivity, ruffed grouse drank more water than blue grouse (Bendell and Elliott, 1966). Hence, moisture and humidity probably partly determine the local distribution of blue and ruffed grouse under our conditions.

Cold and wet conditions together or coupled with wind chill may affect birds and mammals. Cold and wet weather has been blamed for killing chicks and causing population decrease in ruffed grouse (Larsen and Lahey, 1958; Ritcey and Edwards, 1963), capercaillie (*Tetrao*) (Marcström, 1960), willow ptarmigan (*Lagopus*) (Höglund, 1970), and sharp-tailed grouse (*Pedioecetes*) (Shelford and Yeatter, 1955). Unfavorable weather is believed to kill chicks either directly or indirectly by making their food supply of insects unavailable. All these grouse live in habitats that may burn and therefore cold and wet conditions as affected by fire may be important influences on their lives.

e. WIND. The layers of a forest greatly modify the speed of winds that blow through them (Spurr, 1964). After logging and fire there may be an increase in the frequency and velocity of winds over the burn. Winds probably have their greatest effect in concert with temperature and humidity. A cold wind may greatly increase heat loss from a bird or mammal, especially if the animal is wet (Hart *et al.*, 1961).

Wind might make burned areas cooler and better for birds (Stoddard, 1962) or too cold, as possibly for some grouse, as noted. Robinson (1960) tested the shelter requirements of penned white-tailed deer. The animals were held in winter in different densities of natural cover but otherwise treated similarly. No differences were found in survival and loss of deer in the range of cover tested. This tends to reduce the importance of wind, at least under the conditions of this experiment. Robinson concluded that even in open cover, deer could find adequate shelter. However, a greater amount of exposure might show an effect. In contrast, Cheatum (1949) observed that deer in exposed places apparently starved to death sooner than those in shelter.

Moose appear to be sensitive to wind and will leave an open area to avoid it (D. H. Pimlott, personal communication). A moose was recently found dead in the barren grounds beyond the tree line and wind chill may have been a factor in its demise (Miller et al., 1972).

The barren-ground caribou migrates from tundra into the boreal forest in winter. Possibly shelter from wind is one advantage gained by the animal in this migration (Kelsall, 1968). Both moose and caribou avoid open, burned areas in winter (Scotter, 1971) and increased winds may partly cause this response.

f. SNOW. Snow depth, duration, and crust may profoundly influence birds and mammals. These factors are the result of the interaction of sunlight, temperature, humidity, and wind which also may directly affect wildlife. Hence, it is difficult to consider the effect of snow alone. Generally, the amount of snow, like rainfall, that reaches the ground is less in a forest than on a burn because of interception by the canopy. Where melting occurs in tree crowns, the dripping water further reduces the depth of snow on the forest floor. Since temperatures fluctuate less in a forest and winds are reduced, any crust that forms on the snow tends to remain. Finally, snow may persist longer in the forest than on the open burn because the forest shields the snow from sunlight and insulates the cold snow and ground. When trees are removed by logging and burning, deeper snow, alternate crusting and thawing, and shorter duration of snow cover may result. All this may be influenced by blackened soil and its resultant heating.

Many northern birds and mammals depend on appropriate conditions on or under the snow for winter survival (Pruitt, 1959a), and when a forest is burned they may find the snow favorable or unfavorable. For example, the abundance of deer in a mountain region of British Columbia was correlated with the amount of snowfall (Edwards, 1956). Apparently deep snow and little crust immobilized deer and excluded them from food supplies. Under these conditions they also were vulnerable to predation by wolves (*Canis*) (Kolenosky, 1972; Pimlott *et al.*, 1969).

Deer and moose leave a burn when snow is soft and deep and live in the surrounding forest where the snow is relatively shallow and retains a firm crust (Kelsall and Prescott, 1971) even when abundant and preferred food may occur on the burn (Gates, 1968). Barren-ground caribou avoid burned areas within their winter range apparently because of the depth and hardness of snow (Pruitt, 1959b). According to Geist (1971), moose tolerate deep snow and find a hard crust a nuisance.

A number of grouse plunge and burrow into the snow for roosting and escape from the cold and predators. These include blue grouse (King, 1971), rock and willow ptarmigan (*Lagopus*) (Weeden, 1965), and ruffed grouse (Gullion, 1967). Gullion (1967) and Dorney and Kabat (1960) give accounts of how shallow snow and heavy crust may deprive ruffed grouse of protection with subsequent heavy loss to cold and raptors.

The early melt of snow and early greening of a burn in spring may explain why some wildlife are attracted to it. The accelerated growth of plants, an effect of increased warmth and increased supply of nutrients from ash, may improve the quality and quantity of foods. Siivonen (1957) correlated fluctuations in abundance of European grouse (*Tetrao, Urogallus, Tetrastes*, and 2 *Lagopus*) with the date of onset of spring growth. An early melt provided early abundance of nutritious food to the hen, which resulted in good production of young and increase in population. Gullion (1967) believes the same effect occurs with ruffed grouse. He also believes that fire benefits ruffed grouse by increasing the numbers of litter-dwelling insects, essential foods of very young chicks. More recently, Watson and Moss (1972) provided evidence that breeding success in red grouse (*Lagopus*) was caused mainly by the nutritive value of food to the hen. Stocks of red grouse have been managed for decades in Britain by controlled burning of heather (*Calluna*), which is a main item in their diet (Lovat, 1911). Whatever the chain of events, it seems possible that burning and blackening may alter the local climate, microclimate, and nutrient supply to influence populations of grouse.

2. Structure of Vegetation

Perhaps the greatest change made by fire to affect birds and mammals is the destruction of trees and large shrubs so that for some time most growth is close to the ground. This is particularly so in logged areas where removal of trunks, branches, stems, twigs, foliage, large debris, and litter on the ground may have profound effects. Again, the action may be complex because structure may affect an animal directly as by providing a song post, or indirectly, as by modifying the local climate to permit operation in an area. At times one factor may be limiting; at other times a number of factors such as structure of the vegetation, temperature, and humidity may interact to determine the fate of an animal (James, 1971; MacArthur, 1958; MacArthur *et al.*, 1962; Mohler *et al.*, 1951; Orians, 1969; Palmgren, 1932; Rosenzweig, 1973; Salt, 1952; Harris, 1952).

Birds and mammals are adapted to their environment in a variety of ways, and adaptations to a forest may be quite inadequate for operation on a burn. One might expect then that an upheaval such as loss of the forest canopy would cause some change in the kind and number of wildlife on a burn. Yet, abundance and species of most birds and mammals change but little after a fire. This remarkable stability within a fluctuating environment is important and needs explanation.

a. Obstruction. A burned forest may either discourage or encourage penetration by wildlife. After some fires, and particularly snag-felling operations, the land may be a jumble of fallen trees that obstruct movement. Kelsall (1968) believed the tangle of dead trees on burns explained the observation of Banfield (1954) that barren-ground caribou avoided burned forest. Gates (1968) showed that deer used burned and debris-free areas more frequently than those that contained unburned logging slash. But he could not rule out the possibility that deer preferred the food on the burned areas. On the other hand, Redfield *et al.* (1970) studied the movements of blue grouse in essentially the same kind of habitat and concluded there was little preference shown by grouse for areas of burned and unburned logging slash. Tevis (1956) thought small mammals (mainly *Peromyscus*) were prevented from moving into a burn until the deep ash from the fire had hardened.

In contrast to the possibility that a burned forest obstructs travel, some investigators reported that wildlife are permitted to move freely and seek their requirements on a burn (Austen, 1971; Lemon, 1968b). For example, bobwhite quail cannot penetrate the "rough" that develops on the ground in unburned forests and so may be excluded as the forest matures. Burning the rough allows them to move about and feed (Hurst, 1971; Stoddard, 1931). Similarly, waterfowl may be kept from wetlands because they are choked with a variety of dense emergent vegetation and terrestrial shrubs and trees. Burning makes open water and encourages seed-bearing plants, which are used by waterfowl as food (Givens, 1962; Vogl, 1967, 1969; Ward, 1968).

The amount of litter on the forest or grassland floor and the hardness of the soil may be modified by fire and incidentally by the action of logging and fire-fighting equipment. Several investigators accounted for disappearance of sparrows (*Passerculus* and *Passerherbulus*), bobolink (*Dolichonyx*), and voles (*Microtus*) after a fire by reduction of the mat of stems and debris that covers the surface of grassland or open forest (Cook, 1959; Gashwiler, 1970; Sims and Buckner, 1973; Tester and Marshall, 1961). However, *Peromyscus* appeared and increased in numbers after the litter was burned (Sims and Buckner, 1973; Tester, 1965). Potter and Moir (1961) observed that mammal burrows were fewer in a burned than in an unburned area. The burned soil had lost some of the A_1 horizon, and the clay mineral soil was baked hard. On the other hand, removal of litter after a fire may make foods that were concealed by it available to birds and mammals (Stoddard, 1931).

b. Requirements for Reproduction. Display posts and nests for reproduction would seem the main factors that affect the response of some wildlife to a burn (Lack, 1937; MacKenzie, 1946; von Haartman, 1956). Some woodpeckers (*Picidae*) rap on large dead snags as part of the process of attracting a mate. The removal of snags by fire or felling may be one factor in the elimination of a woodpecker. Mayfield (1960) believed a large part of the dependence of Kirtland's warbler (*Dendroica*), a ground nester, on early stages of growth of jack pine (*Pinus banksiana*) was the structure required for the nest: porous soils, loose ground vegetation, and low hanging branches. Terrill and Crawford (1946) blamed the lack of nest sites for failure of squirrels (*Sciurus*) to occupy burned areas.

c. Pattern of Cover within a Burn. The pattern of cover is a basic structural feature of wildlife habitat. Fire creates a pattern of cover composed of clumps of vegetation, burned logs and stumps, and the open spaces among them.

Fires rarely burn evenly and result in a larger mosaic of the pattern of old and new cover. The topography of the land also partly determines how a forest burns and revegetates, and wildlife may react to differences in elevation. MacKay (1966) described how fire on the tundra over permafrost resulted in a variety of changes in topography and vegetation: ground heaved by frost, melting and subsidence, and elimination of trees.

i. *Local distribution and density.* The pattern of cover within a burn may determine where birds and mammals live and the density they achieve. For example, blue grouse seek openings and ridges for their territories (Bendell and Elliott, 1966, 1967). Fire frequently clears ridges (Heinselman, 1970); hence, the local distribution of blue grouse may be partly explained by fire-created openings and topography. Martinka (1972) related territories of blue grouse to patches of unburned conifer on burned areas in Montana. The tendency of yellow pine (*Pinus ponder-osa*) to grow in clumps in response to fire (Cooper, 1961b) may well provide hideouts and display posts for males and determine the location of territories.

Watson and Moss (1972) believed that unevenness of the terrain and the pattern of cover provided by heather partly determined the amount of interaction among cocks of red grouse, and, in turn, their density in an area. When males are out of sight of each other, as in thick heather, they fight less and may settle closer to each other. Other work on red grouse suggests that good stocks may be obtained on heather moors that are burned in well-spaced narrow strips or in many small patches on a 10–12-year rotation (Miller, 1964; Picozzi, 1968).

After a fire in Alberta, the density of snowshoe hares decreased and local distribution changed as the hares abandoned the open parts of the burn for remaining cover. The population drop was caused by exodus of young hares, and this seemed to be the result of intraspecific strife caused by reduction of cover. Within a year the population on the burn returned to normal (Keith and Surrendi, 1971).

Biswell *et al.* (1952) recommended spot burns of 5–10 acres in a checkerboard pattern to open up dense chamise (*Adenostema*) brush-land for black-tailed deer, game birds, and mammals in California. Com-pared to a large wildfire burn and dense brushland, the area opened by burning had more deer use per square mile, produced heavier deer, and does had a higher frequency of ovulation, more fawns at heel, and wintered in better condition. The better performance of deer was explained by the enhanced nutritive value of plants growing in the openings. Jack rabbits (*Lepus*) and brush (*Sylvilagus*) rabbits, mourn-ing doves (*Zenaidura*), and valley quail (*Lophortyx*), but not mountain quail (*Oreortryx*), also responded to the opened area. The importance of pattern of cover to behavior and how behavior might explain the responses to spot burning were not considered.

ii. Competition. Interspecific competition is usually avoided by each species seeking a particular resource. Fire may obliterate the specific kind of food or shelter required by two species and throw them against each other for what is left after the burn. Either species alone might persist but the outcome of competition between the two is that one may withdraw. An example of this is provided by ungulates in Banff and Jasper National Parks, Canada (Flook, 1964). Before fire, elk (*Cervus*) were relatively few and mule deer, moose, and bighorn sheep (*Ovis*) were abundant. Each lived more or less separately in appropriate habitats. Following widespread fires which encouraged grassland and shrubland, elk increased in abundance and penetrated areas occupied by the other ungulates. Elk are vigorous competitors for the plant cover that provides food and shelter for mule deer and food for bighorn sheep and moose. As a consequence, all three have declined in abundance. Hence, fire influenced interspecific competition to favor one species of mammal over three others.

iii. Predation. The pattern of cover may greatly influence the relationships between predators and prey and the welfare of both. In Africa, burning thick grassland affects predation and local distribution of prey for game will not enter cover that provides concealment to predators in ambush (Brynard, 1971). On the other hand, a prey moving into an open burn may be exposed to predators that are new to it or operate more effectively in the new habitat. For example, some blue grouse move onto the lowlands of Vancouver Island after logging and burning. Others live in the alpine and subalpine forests. The lowland birds must face a number of predators such as marsh hawk (*Circus*), fox (*Vulpes*), and raccoon that are rare or absent on the uplands. However, they escape from some upland predators [e.g., marten (*Martes*)].

Beyond a direct predator–prey relationship, numerous authors have commented on the importance of cover to the outcome of predation. Rusch and Keith (1971) and Gullion (1972) considered thick clumps of conifers a disadvantage to ruffed grouse because they concealed avian predators. Similarly, Gullion (1967) stated that slash and tangle on the ground gave advantage to mammalian predators of ruffed grouse, an advantage that is removed by burning the concealing cover. Gullion and Marshall (1968) provided impressive data on the length of life of ruffed grouse in different cover types (Fig. 3). The cover types are the result of fire and the loss of life presumably is due to predation. Densities of grouse are highest and males live the longest in stands of 10–30-year-old aspen with few conifers.

iv. Parasites and disease. After a fire, infestations of external and internal parasites may be reduced to the benefit of birds and mammals (Brynard, 1971; Grange, 1949; Isaac, 1963; Lovat, 1911; Stoddard, 1931). Fowle (1944, 1946) sampled parasites in a population of blue grouse 5 years after a wildfire that burned virtually everywhere to mineral soil. Bendell (1955) sampled the same area 12 years later. Differences in kind of parasites and frequency of infection were striking (Table II).

There were differences in the pattern of parasitism shortly after a fire (when the habitat was cleansed) and 12 years later when parasites had time to become established (Table II). Generally, more species of parasites in greater frequency of infection were found with longer time after burning. Not all parasites showed a greater percentage infec-

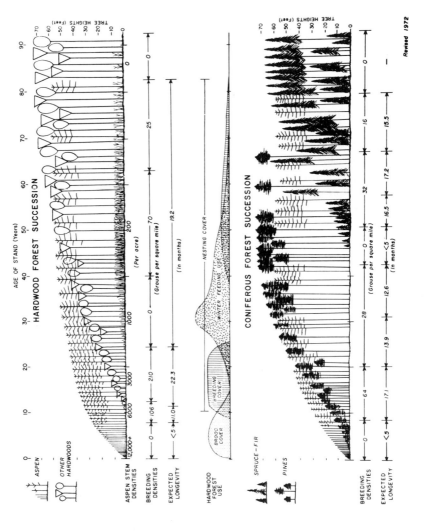

Fig. 3. Forest structure, density, and survival of ruffed grouse. (From Gullion, 1972.)

TABLE II

COMPARISON OF PARASITIC INFECTIONS (%) IN
BLUE GROUSE 5 AND 12 YEARS AFTER WILDFIRE

	Adults		Chicks	
Parasite	5 years	12 years	5 years	12 years
Blood	(12)[a]	(174)	(16)	(89)
Trypanosoma	10	77	6	20
Haemoproteus	83	97	50	66
Leucocytozoon	42	85	19	38
Microfilaria	40	80	0	0
Negative	17	0	—	—
Gut	(20)	(103)	(33)	(107)
Dispharynx	0	4	0	64
Cheilospirura } *Yseria*	0	22	0	10
Rhabdometra	10	39	62	21
Ascaridia	21	14	12	4
Plagiorhynchus	0	0	0	50
External				
Ceratophyllus	0	11	0	0
Lagopoecus	5	38	3	20
Ornithomyia	10	2	6	7

[a] Sample size.

tion. Some remained more or less the same and a few apparently declined. Note that four new parasites occurred at 12 years, and two of
them (*Plagiorhynchus* and *Dispharynx*) caused severe damage to chicks
(Bendell, 1955; Jensen, 1962). The fire apparently reduced infections
of most parasites. A few were even eliminated, but others were unaffected. This suggests that the effect of fire on parasites depends on
species involved, and a blanket statement that fire eliminates or reduces
parasites may need qualification.

Since the population of grouse apparently did not increase greatly
in the 7 years between sampling, the simplest explanation for the change
in the kind and frequency of parasitic infections is alteration in structure
of cover that favored the increase of populations of intermediate hosts.
The appearance and buildup of *Dispharynx* and *Plagiorhynchus* may
be explained by development of a wetter microclimate on the ground
of a revegetating burn. Both worms are transmitted by the sowbug
(*Porcellio*) (Bendell, 1955), which does not tolerate dry air (Paris,
1963).

Casperson (1963) autopsied blue grouse, mostly chicks, shot on an area burned in 1951 (210 sampled) and on an area burned in 1938 (119 sampled). Both areas had similar densities of grouse and similar soils and original forest types. Casperson found *Dispharynx* and *Plagiorhynchus* in approximately twice the frequency of infection on the older burn. He found more ground cover and wood lice on the old than on the young burn, and, like Bendell (1955), explained the frequency of parasites on each burn by development of a vegetation and a microclimate that influenced the intermediate hosts.

Although fire affected the ecology of most grouse parasites, these had very little effect on the population of grouse, which remained about the same for 7 years. Parasites usually cause greatest harm to the young (Jensen, 1962), and, in our case, severe damage was caused to chicks by *Dispharynx* and *Plagiorhynchus*. Twelve years after the fire, there was generally increased parasitemia and two new damaging parasites of young as compared to 5 years after the same burn. However, size of brood in late July and early August was the same in both periods; about three chicks per hen (Bendell, 1955; Fowle, 1960).

d. Size of Burn, Edge, and Interspersion. The size of burn, its edge, and interspersion with cover types beyond the burn are structural features that may be important in the response of birds and mammals to a fire-changed area. Edge between cover types and interspersion of cover types are related in that a number of small fires on an area will create more edge and interspersion than one large burn. As burns become larger, the amount of edge and interspersion becomes less in proportion to the amount of open burn. The relationships also vary with the shape of the burn.

The size and the shape of a burn partly determine how much new and empty habitat there is for animals to exploit, assuming disturbance has made a major change in what was there before. Animals may invade the new habitat and flourish because they have relatively little interference from animals of their own or other species, or they find unlimited resources, or both occur. A small burn might not provide enough change to make a significant impact on wildlife.

i. Edge and interspersion. The edge between a burned and unburned area may be very abrupt and occur within a few feet (West, 1971). Abrupt edges are made when logging and fire open old growth timber (Fig. 4). On the other hand, where burns are close in age the edge between them may show less abrupt change in the structure of the vegetation.

The importance of edge and interspersion is that some wildlife need a variety of resources, and these are best obtained where two or more

Fig. 4. An abrupt edge between an old and new burn. The fire road is about 30 ft (9 m) wide. Blue grouse were in the same density beyond each side of the road and were not concentrated on the edge. (Bendell and Elliott, 1966.)

kinds of cover come together (Biswell *et al.*, 1952; Buckley, 1958; Leopold, 1933; Lovat, 1911). The burning of heather in small patches for red grouse and the spot burning of shrubland for deer, as already mentioned, are examples of producing edge and interspersion on a small scale.

Buckley (1958) stressed the importance of edge in explaining the increase of moose in the Kenai Peninsula of Alaska. McCulloch (1969) counted deer pellets in woodland and on an adjacent burn. On the basis of these counts, deer used the edge of the burn more frequently than they used open burn or forest. Like Buckley, Robinson (1958) believed that small (200–300 acres) cuttings or patch logging and burning in coast forest on Vancouver Island provided large amounts of edge and interspersion and, on this account, would produce more deer, blue grouse, and elk than large (1000 acres and more) clear cuts and burns. On the other hand, Barick (1950) found little use of edge by birds and mammals on his area, and Gullion (1972) also believed ruffed grouse did not make special use of regions between cover types.

We have examined the effect of edge and interspersion on the distribution of blue grouse on two burns on Vancouver Island (Bendell and Elliott, 1966; Zwickel and Bendell, 1972). In both studies the territories of males were mapped in dense, second-growth vegetation and across the edge into very open, recent burns.

Territorial males and other grouse were distributed without concentration on the abrupt edge between cover types (Fig. 5). Moreover, within

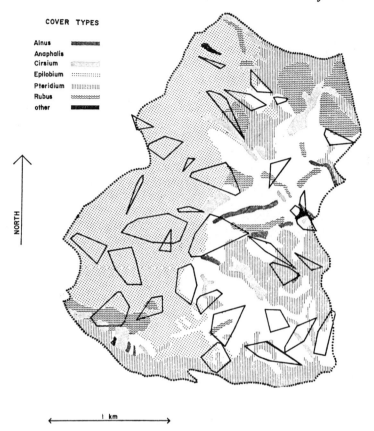

COVER TYPES

Alnus
Anaphalis
Cirsium
Epilobium
Pteridium
Rubus
other

NORTH

⊢____ 1 km ____⟶

Fig. 5. Diagram showing that territories of male blue grouse are spaced over a burn and not located on the edge of a particular type of cover. (From Zwickel and Bendell, 1972; courtesy of E. J. Brill, Leiden.)

a burn, the edges between and the interspersion or pattern of types of herbaceous cover also did not affect distribution of territorial males and other grouse (Fig. 5). Hence, blue grouse found all their requirements within, and without special use of edge between, the open and dense cover and types of herbaceous cover.

On the other hand, Gullion (1972) believes that ruffed grouse require an interspersion of cover types for maximum abundance. Within an area that supports a breeding pair or population, he identified distinctive types of cover essential to the life cycle of the species. The cover types are dense stands of aspen less than 10 years old for broods, more open stands 10–25 years old for wintering and for territories of breeding males, and older aspen forest for food, nesting, and again for wintering. These requirements suggest that ruffed grouse or any wildlife will live only in habitats that provide many needs and habitat relationships are specific and complex. This is not in keeping with the widespread distribution of ruffed grouse (Gullion, 1972) and the fact that it may occur in abundance in regions without aspen (Chapman and Turner, 1956). It also conflicts with the observation that many wildlife do not change in density or trend of abundance after their habitat is devastated by fire.

Finally, compare the abundance of black-tailed deer near Campbell River with that near Northwest Bay. The two areas are about 60 miles apart in more or less the same forest on the east coast of Vancouver Island. The Campbell River area is one of large clear cuts and wildfires, notably 75,000 acres in 1938 and 31,000 acres in 1951. Northwest Bay is an example of patch logging. Since 1954, about 400 acres or more have been cleared each year by cutting and slash burning with most "settings" of about 100 acres (Gates, 1968). Thus, Northwest Bay had more edge and interspersion than Campbell River and might be expected to have a higher population of deer.

Deer populations were measured by checking hunters from each area from 1952 through 1964. The results are tabulated below:

Amount of edge and interspersion	Average No. of hunters checked each year	Average No. of deer/ hunter and range
More	2300	0.19 (0.10–0.36)
Less	2260	0.17 (0.09–0.27)

The number of deer taken per hunter apparently was similar on areas with different amounts of edge and interspersion. The results suggest that edge and interspersion were not important to populations of deer.

ii. Size of burn. Data on the relationship between size of burn and density of wildlife population are few. According to Watson (in Picozzi, 1968) red grouse will not occupy burns in heather moor 200 m wide or larger because of lack of cover. In Australia, Mount (1969) observed that areas burned by bush fires may be so large that forest animals will not move into them. Small burns and their edges are used because animals can readily retreat to nearby unburned forest. In California, wildfire burns of about 1000 acres in brushland were used at times by large numbers of deer, but did not hold populations that moved into adjacent brushland, presumably for food and cover (Biswell *et al.*, 1952; Taber and Dasmann, 1957). For ruffed grouse, Gullion (1972)

TABLE III
Size of Burn and Density of
Breeding Blue Grouse

Size of logged and burned area (acres)	Density (hooting males/100 acres)	Source of data
75,000 ⎤ᵃ	37	Bendell (1955)
20,000 ⎦	5	J. F. Bendell (unpublished)
31,000 ⎤ᵃ	13	Bendell and Elliott (1967)
20,000 ⎦	5	Bendell et al. (1972)
21,000	9	Redfield (1972)
20,000	35	Mossop (1971)
3,000	8	Zwickel (1972)

ᵃ Same area in different years but stable population.

recommends an area of 10 acres to maintain a breeding pair and 30 to 60 acres to manage a population by cutting and burning.

There may be a relationship between size of burn and breeding density of blue grouse. The burns were on the east coast of Vancouver Island within 100 miles of each other and within, more or less, the same kind of forest type. Censuses were made in spring on sample areas of about 300 acres and larger within a period of 20 years. Since the grouse inhabited essentially the same kind of forest and there were apparently no significant changes in climate in the 20-year period, it seems reasonable to compare the populations in relation to size of burn (Table III).

The data are few, and there is uncertainty as to size of burn represented by each sample area. If the data are taken as they stand, there is little relationship between size of burn and density of grouse within the ranges studied. On the other hand, the density reported by Mossop (1971) may be too high. If this is reduced, there may be a positive correlation between density of breeding grouse and size of burn.

When birds and mammals move into a new area they may increase to a spectacular density. Examples of this are blue grouse, as just noted, and deer on the logged and burned areas of Vancouver Island (Bendell, 1955; Robinson, 1958), deer mice on a clear-cut and burned area in Oregon (Gashwiler, 1970), Himalayan tahr (*Hemitragus*) in New Zealand (Caughley, 1970), and pheasants (*Phasianus*) on Protection Island in Washington (Einarsen, 1945). However, the same and other species may not show this response. One explanation for this difference is that extrinsic factors of the environment such as food supply, cover, parasites, and predators favor or disfavor the expansion of population (Klein, 1970; Lack, 1966). Another view is that while extrinsic factors are important to populations, the basic cause of population change is the genetic quality of the stock. Populations expand, remain stable, or decline essentially as a result of intraspecific interaction and the selection of the appropriate genotypes that determine population trend (Chitty, 1967a; Krebs *et al.*, 1973). The theoretical point made here is that size of burn may determine the growth of populations that move into it by providing new and unlimited resources and releasing genotypes that will expand in numbers when free from the interference of established genotypes that tend to keep population in check. The larger the burn the greater the opportunity for establishment of genotypes that cause population expansion.

Redfield (1972) showed that the genotypes of blue grouse that first arrived on new summer range created by logging and burning changed as their populations expanded and leveled off. Thus, quality of stock must be considered in cataloging the factors that affect the response of wildlife to burns.

e. THE MOSAIC OF FORESTS CREATED BY FIRE. The large mosaic of forests of different sizes, shapes, and ages created by fire (Daubenmire, 1968; Heinselman, 1970; Vogl, 1970) provides heterogeneity of environments and wildlife and colonists to new burns when they occur. If forests in different stages of growth occur nearby, then different species of wildlife, each associated with a particular stage of forest growth, may also be found. For example, on Vancouver Island on a burn in herb and shrub stage of succession, white-crowned sparrows (*Zonotrichia*), towhees (*Pipilo*), and robins (*Turdus*) are common residents. Across the edge of this burn in another burn in the young tree stage, western tanagers (*Piranga*), warbling vireos (*Vireo*), and varied thrush (*Ixoreus*) are found. Many studies show a relationship between stage of forest growth and associated birds and mammals (Brewer, 1958; Grange, 1948; Karr, 1968; Martin, 1960; Stelfox and Taber, 1969).

One implication of a variety of wildlife in the landscape is the presence of forms that are adapted to each kind of environment that occurs over a wide area. Thus, there is a native fauna, and it persists without extreme changes in abundance and destructive impact to the places where it lives. Large disturbances in the native biota may permit exotic species to become established and later expand beyond the area of establishment (Udvardy, 1969b). Similarly, native forms may increase to epidemic levels when the usual checks on abundance are removed by some change in their environment. Should a burn cause a species to increase in numbers and spread to damage beyond burn habitats, the different species in adjacent forest may help contain the outbreak and maintain the overall balance between wildlife and their habitats (Elton, 1958).

Wildlife that respond to a burn were there before the fire or move in from adjacent forests. Thus, a variety of nearby forests will offer a diversity of wildlife to exploit the opportunities of a new burn. For example, mature coast forest may have a few blue grouse that occur where tall trees break into open rock face. Should a burn occur, some grouse may move to it and expand in numbers. A number of studies show that birds and mammals may persist in marginal habitats, usually forced there by intraspecific competition for space in preferred habitat (Evans, 1942; Klein, 1970; Krebs, 1971; Tompa, 1964).

It is not only the species of colonists that are important to what happens on a burn, but it may be the genetic kind of the species as well, as noted. The wildlife of genotypes able to rapidly exploit a new environment may live in the marginal habitats offered by a variety of forests. Given the opportunity of a new burn, they may invade and increase greatly in density.

3. *Food*

Food is such an obvious and important need to us we take for granted that it is the pivotal point in the lives of birds and mammals. Perhaps it is, but we are a long way from proving it so, and we need better tests for showing the importance of food to wildlife (Lack, 1954, 1966, 1970; Watson, 1970). Some balance was restored by Chitty (1967b) to the argument that food resources dominate the lives of birds and mammals.

There are difficulties in measuring the importance of food to wildlife. We may not know the proper food or mixture of foods required by a particular species (Dahlberg and Guettinger, 1956; Hill, 1971; Moss *et al.*, 1971; Negus and Pinter, 1966). The growth and condition of animals may be simple indicators of their dependence on food stocks (Klein, 1970). However, while food may be the ultimate factor behind a set of observations, it may not reveal itself in as simple a proximate fashion as poor condition (Watson and Moss, 1972). Animals often compensate for apparently inadequate foods by selective feeding, voluntary reduction in intake, and increased intake (Bissell, 1959; Brown, 1961; Crouch, 1966; Gardarsson and Moss, 1970; Miller, 1968; Moss, 1972; Pendergast and Boag, 1971; Wood *et al.*, 1962). Finally, animals may show a selection or preference for certain foods, but preference may not mean need so that correlations between foods that are simply preferred and the welfare of wildlife may be trivial. For example, Wynne-

Edwards (1970) believes that aspen and birch (*Betula*) are required by beaver (*Castor*) and they adjust their numbers to this food. Actually, populations of beaver may persist without these trees where the beaver feed on a variety of aquatic plants (Hall, 1971).

a. KIND OF FOOD. Birds and mammals are adapted to eat particular kinds of food such as grasses, seeds, and browse, and the abundance and distribution of wildlife may depend on the supply of the appropriate kind of food. The concentration of plants near the ground after a fire should affect the mammals, particularly the large forms, more than the birds, because the former cannot feed more or less beyond standing height. Edwards (1954) described the dramatic change in mammal life, particularly increase of moose and deer, with burning of mature conifer forest and growth of herbs and shrubs in Wells Gray Park, British Columbia. Munro and Cowan (1947) listed some of the birds that disappeared from and those that appeared on logged and burned areas on Vancouver Island. As might be expected, some insectivorous birds that fed in coniferous foliage were lost, while some seed- and fruit-eating, ground-dwelling birds flourished in the new habitat. A number of investigators reported that seed-eating birds and mammals increased dramatically, at times by ingress, when a burn yielded large amounts of this kind of food (Ahlgren, 1966; Cook, 1959; Garman and Orr-Ewing, 1949; Gashwiler, 1970; Hagar, 1960; Lawrence, 1966).

Severe and repeated burning may reduce production of grasses, herbs, and shrubs, and, in turn, grazing and browsing wildlife (Darling, 1960; Daubenmire, 1968; Eddleman and McLean, 1969; Penfound, 1968; Van Wyk, 1971). Leopold and Darling (1953a,b) observed that fire in Alaska did not always produce birch and willow that made good range for moose. Some burns returned almost directly to spruce forest, or through a grass–herb stage, or through aspen. Other authors reported fires that either failed to produce expected species of birds and mammals or actually caused them to decline (Dahlberg and Guettinger, 1956; Scotter, 1971; Vogl, 1967).

Many birds and mammals eat small animals associated with the forest floor. Fire may alter the environment on and in the ground in many ways (Ahlgren and Ahlgren, 1960; Eddleman and McLean, 1969; Isaac and Hopkins, 1937; Potter and Moir, 1961; Scotter, 1971) to profoundly change the microfauna found there (Gill, 1969; Gillon, 1971; Hurst, 1971; Lussenhop, 1971; Pearse, 1943; Vlug, 1972). The next link is a possible effect on wildlife. Woodcock (*Philohela*) eat many earthworms (*Lumbricus*), and the increased alkalinity of the soil that follows most fires may increase the abundance of earthworms and indirectly the number of woodcock (Wing, 1951). Gullion (1967) believes that the burned litter of the forest floor increases production of insects used as food by chicks of ruffed grouse.

b. DEPENDENCE ON A SINGLE KIND OF FOOD. Generally, there is little relationship between a particular kind of plant food and wildlife. The association is usually with some broad group of plants such as grasses, herbs, or shrubs. However, some birds and mammals seem to have relatively narrow preferences for food and may be particularly sensitive to fire. A number of investigators believe that forest fires limit the present abundance and winter distribution of caribou by destruction of their main winter foods: the tree and ground lichens, especially *Cladonia* (Cringan, 1957; Edwards, 1954; Scotter, 1971). In contrast, Bergerud (1971) stated that caribou use a variety of herbs and shrubs as winter forage and *Cladonia* is not an essential food.

Since some animals do seem to specialize on a particular kind of food (Keith, 1965; Kemp and Keith, 1970; Newton, 1970; M. C. Smith, 1968), fluctuations in it may cause parallel changes in distribution and abundance of the dependent animal. At times, woodpeckers move into burns in large numbers, presumably in response to the new food supply provided by the insects that attack dead trees (Blackford, 1955; Koplin, 1969). Grange (1949) and Lauckhart (1957) believe that cycles in numbers of many herbivores are linked to changes that may be caused by fire, in the kind, quantity, and quality of food available. However, a fire through hare range in Alberta had little effect on long-term trend of population, at least in the few years so far studied (Keith and Surrendi, 1971).

On the other hand, forest fires may cause large numbers of birches or conifers to set seed at the same time. Redpolls (*Carduelis*) feed almost exclusively on birch seed and are notorious for their irruptive appearance and disappearance throughout their range (Evans, 1969). Crossbills (*Loxia*) are similar in that they feed on spruce seed and may make mass movements from place to place (Newton, 1970). Perhaps, at times, production of even aged stands of spruce and birch by forest fires is behind the local concentrations of both redpolls and crossbills.

c. QUANTITY. In explaining the response of birds and mammals, particularly ungulates, to conditions after a fire, most investigators argue that quantity and/or quality of food limit the species of wildlife and their abundance (Biswell *et al.*, 1952; Dasmann, 1971; Cringan, 1958; Gullion, 1967; Hagar, 1960; Klein, 1970; Komarek, 1967; Leege and Hickey, 1971; Watson and Moss, 1972). After a fire there may be prolific growth of vegetation near the ground to greatly increase food supply to some birds and mammals. For example, Gates (1968) measured the amount of deer food present on logged and burned areas in the summer with

TABLE IV

ANNUAL PRODUCTION OF DEER FOOD ON LOGGED AND
BURNED AREAS OF DIFFERENT AGES[a]

Food[b]	Years after fire				Mature timber
	4	10	12	14	
Grasses and forbs	151	125	13	97	Trace
Ferns	15	69	140	166	2
Shrubs	782	847	791	744	423
Conifers	Trace	249	28	105	5
Total	948	1290	972	1114	430

[a] Modified from Gates (1968).
[b] Pounds/acre, wet weight.

years after disturbance (Table IV). Many investigators have arrived at similar results (Brown, 1961; Dills, 1970; Leege, 1968; Leege and Hickey, 1971; Lyon, 1971).

Grasses and forbs were important in the early years after disturbance, and shrubs represent major stocks of food throughout all stages of regrowth, even into mature timber (Table IV). The shrubs in mature timber appear to be important winter food, since coniferous vegetation is beyond reach, and food plants in the open may be dead or unavailable under snow. Food is abundant compared to mature timber from 4 years after logging and fire up to at least 14 years.

Gates, on the basis of pellet counts, found that deer use was concentrated on the new areas from immediately after logging and burning until about 6 years. What this selective feeding meant to the numbers of deer was unknown.

d. QUALITY. Numerous investigators believe that quality of food is the main factor that limits abundance of many herbivores (Klein, 1970; Moss, 1967; Schultz, 1969; Watson and Moss, 1972). Many kinds of nutrients are required by birds and mammals, but most ecologists have concentrated on nitrogen as protein, and phosphorus, with lesser interest in calcium and magnesium (Moss, 1967, 1969; Moss and Parkinson, 1972; Ullrey *et al.*, 1973). After most fires there is an impressive change in vigor of plants, which may be caused by their increased uptake of nutrients released in the ash (Ahlgren and Ahlgren, 1960; Hayes, 1970; Humphrey, 1962; Komarek, 1967; Trevett, 1962; Vogl, 1969). Hence, a cogent argument is that the abundance of wildlife on a burn is set by the amount of nutrient release expressed through the quality of food.

Burning is not the only way the nutrient content of plants may be increased. Exposing plants to sunlight, clipping, mowing, or logging old

growth, ploughing the soil, and removing smothering litter may have
the same result (Allen et al., 1969; Blaisdell and Mueggler, 1956; Buell
and Cantlon, 1953; Daubenmire, 1968; Hulbert, 1969; Miller and Miles,
1970; Penfound, 1964; Scott, 1971; Stoddard, 1931).

Like most effects of fire, there is not a simple relationship among
burning, release of nutrients, and what may be taken up by plants that
may be used by animals. The level of nutrients in plants after burning
may be unchanged, increased, or decreased, depending on season, soil,
weather, nature of the fuel and fire, and other factors (Austin and Baisinger, 1955; Beeson, 1941; Daubenmire, 1968; Einarsen, 1946a; Gessell
and Balci, 1963; Hayes, 1970; Isaac, 1963; Leege and Hickey, 1971;
Lemon, 1968b; Mayland, 1967; Smith, 1970; Wagle and Kitchen, 1972).
This complex relationship may help explain why burns may produce quite
different kinds and numbers of wildlife.

The length of time that nutrients are available to animals after fire
is obviously important. An increased nutrient level in food plants may
last from only 1 year to as long as 14 years or more (Cowan et al.,
1950; DeWitt and Derby, 1955; Einarsen, 1946a,b; Gates, 1968; Gimingham, 1970; Trevett, 1962). In Australia, stimulation of plant growth
by ash in burned eucalypt (Eucalyptus) forest may last 20 years (Mount,
1969). If wildlife respond to an increased nutrient supply in their food,
then the response may last from 1–20 years.

Gates (1968) found that after logging and burning levels of protein,
fat, ash, crude fiber, and nitrogen-free extract in foods were adequate
for deer and the same in 4 through 14 year stages of forest growth.
Hence, food was in abundant quality and quantity (Table IV) over
this time. Brown (1961) also found little difference in protein in deer
foods from four stages of forest growth in northwest Washington. The
seral stages ranged from recently logged and burned to dense second
growth. These results are at variance with Cowan et al. (1950), who
reported nutrient levels declined as plant succession advanced, and with
Einarsen (1946a), who worked in a similar forest as Gates and found
a sharp increase of protein in browse the year after a fire, with a decline
to preburn level in about 8 years. Nevertheless, from all three studies,
if quantity and quality of food determine deer populations, we would
expect numbers to expand within a few years after disturbance by logging and fire and then remain stable or decline.

Figure 6 brings together some published data on nutrient content
of deer foods as affected by burning. Individually or in combination
these nutrients are considered critical to wildlife (Dietz, 1965; Klein,
1970; Moss, 1967; Ullrey et al., 1973). The main comparison is between
browse before fire and up to about 5 years after burning.

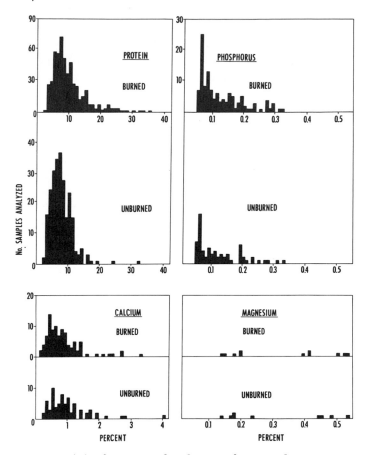

Fig. 6. Amounts (%) of protein, phosphorus, calcium, and magnesium in foods of deer from ranges before fire and up to about 5 years after burning. Source and month of sampling: Brown (1961), January; DeWitt and Derby (1955), May–November; Einarsen (1946a), January, December; Gates (1968), March, December, June; Lawrence and Biswell (1972), July, August; Lay (1957), April, July, October, January; Leege (1969), February; Leege and Hickey (1971), spring, fall; Taber and Dasmann (1958), all year.

Protein appears to increase after a fire while there is little evidence for change in phosphorus or magnesium. Calcium appears to decrease with burning. Even if change in protein occurs, the amount is not very great, perhaps 1–2% on average in the browse. Considering the ability of deer to feed selectively, the increases in protein and decreases in calcium after a fire are not impressive. Again, changes in abundance of deer caused by an increase in protein in their food should occur within approximately 5 years.

Ranges in amounts of each of the four nutrients are surprisingly close, considering that the determinations were made during all seasons of the year in widely separated regions of North America and by different observers. This suggests that generally wide fluctuations do not occur in nutrient contents of plants, perhaps from 5 to 15% in the case of protein. Apparently, many browse plants maintain a fairly constant protein level despite much wider changes in mineral supply. If levels of nutrients are relatively constant, then change in quality of food cannot explain the changes in condition and abundance observed in wildlife.

Wildlife usually obtain nutrients from a variety of foods. There may be cases where the nutrient level of a particular food is changed by fire to affect wildlife. Gullion (1970) and Svoboda and Gullion (1972) observed that ruffed grouse fed selectively in male aspen trees that were old and often injured by fire. They believe these trees were of higher nutritional quality than others and therefore important to reproduction and winter survival. Ungulates frequently use salt licks, places where they paw and lick or eat the ground, presumably to acquire minerals. Komarek (1969) reported that hare, deer, and African elephant eat ash after a fire, perhaps as a form of salt licking.

e. WATER. Burning may alter the supply of water (Adams et al., 1970; Ahlgren and Ahlgren, 1960; Arnold, 1963; Nieland, 1958). Some dry areas of North America are uninhabitable to various birds and mammals because of lack of water (Dasmann, 1964; Dasmann, 1971; Salt, 1952). Hill (1971) observed that African game seriously overgrazed a range if water was provided to them. When water supplies ordinarily disappeared, the game migrated to other parts of their range.

Fire has its greatest impact on stocks of water where water is in short supply. Some fires may improve the supply of free water to wildlife by reducing the loss by vegetation and causing the water table to rise (Arnold, 1963; Dasmann, 1971; Ward, 1968). Large and hot fires may reduce water supplies and eliminate some wildlife (Dasmann, 1971).

Blue grouse in 1000 acres burned by wildfire in the fall on Vancouver Island returned to the area next spring in the same abundance despite a large reduction in food and cover. The area was virtually bare except for burned logs and stumps (Zwickel and Bendell, 1967) (Fig. 7). By midsummer grouse left the burn, a departure about 2 months ahead of normal time of migration to the uplands and before grouse departed from nearby unburned areas. There was abundant food of presumably high quality on the burned lowland range, but much of it was very dry. Most likely a shortage of water caused by the burn affected the birds in the year after the fire. In subsequent years, occupancy and migration were as in the surrounding unburned areas.

Fig. 7. An area burned in the fall and occupied the following spring by blue grouse in density comparable to adjacent unburned forest. (Photo by J. F. Bendell, May.)

f. COMPETITION FOR FOOD. As mentioned, fire may alter the competition for food between wildlife to determine the number of species living on a burn (Flook, 1964). Elk in the western parks feed on grasses and small trees and shrubs. Thus, they deprive mountain sheep and mule deer of food, and stocks of these animals have declined (Flook, 1964). Vesey-Fitzgerald (1971) describes a similar situation in Africa where fire turns mixed grass and shrubland, with a variety of game, into grassland. The buffalo (*Syncercus*) now keep grasses and shrubs grazed to a close sward, and, in the process exclude shrub-feeding game. As a final example, the moose is an acknowledged fire follower and may become abundant on burned areas. As dominant herbivores moose might be expected to compete for food with other wildlife. In central Newfoundland, moose reduced the amounts of white birch and in turn the numbers of hare and beaver (Bergerud and Manuel, 1968).

III. Fire and the Kinds of Birds and Mammals

If species are sensitive to fire-caused changes in environment, as already discussed, then rather large changes in kinds of wildlife should be evident after fire. A comparison between species found in undisturbed forest and those found on land changed by logging and fire offers a way of testing the requirements of a species to get at the fundamental question of what constitutes its habitat.

There are a number of difficulties in evaluating the literature on change in wildlife species after fire. Forest fires and subsequent plant succession may vary. Hence, two burns may produce conflicting results as to the effect of burning. Methods of study also add variability. A count of species may not represent the effect of a burn because census areas and number of animals caught or observed are too small, studies are not long enough to detect a response, and transient species are called residents in burned or unburned habitat. Finally, different species of wildlife may be present in a recent burn than in an old one.

Some birds and mammals are prized as game or for a variety of other reasons. Since many of these forms may increase after a fire, in this respect, burning is beneficial to wildlife. Some of the large mammals that may follow fire are moose, white and black-tailed deer, elk, cougar (*Felis*), coyote (*Canis*), black bear, beaver, and hare (Dahlberg and Guettinger, 1956; Edwards, 1954; Grange, 1949; Hansen *et al.*, 1973; Hayes, 1970; Jonkel and Cowan, 1971; Kelsall, 1972; Lawrence, 1954; Stelfox and Taber, 1969). Mammals dependent on late stages of forest development that may be eliminated or displaced by fire are mountain, woodland, and barren-ground caribou; marten; red squirrel (*Tamiasciurus*); grizzly bear (*Ursus*); wolverine (*Gulo*); and fisher (*Martes*) (Cringan, 1958; Edwards, 1954; Hayes, 1970; Scotter, 1971).

Among birds, those favored by conditions after burning are some of the galliforms: wild turkey (*Galapavo*), ring-necked pheasant, bobwhite quail; the grouse: sharp-tailed, prairie chicken (*Tympanuchus*), ruffed grouse, blue grouse, willow ptarmigan; and some of the waterfowl (Amman, 1963; Gullion, 1967; Lovat, 1911; Stoddard, 1931, 1963; Vogl, 1967; Zwickel and Bendell, 1972). On the other hand, the spruce grouse (*Canachites*) retreats as dense spruce and pine (*Pinus*) forests are removed (Grange, 1948). Thompson and Smith (1970) concluded that control of fire and subsequent loss of open prairie contributed to extinction of the heath hen (*Tympanuchus*).

A. Species Change after Fire

Some data on breeding birds and mammals in unburned and adjacent burned habitat are given in Table V. Where several census plots were used the results were pooled. Neither predators nor mammals larger than a hare were included.

TABLE V

CHANGE IN SPECIES OF BREEDING BIRDS
AND MAMMALS AFTER BURNING

Foraging zone	Before burn	After burn	Gained[b] (%)	Lost[b] (%)
No. of species of birds[a]				
Grassland and shrub	48	62	38 (18)	8 (4)
Tree trunk	25	26	20 (5)	16 (4)
Tree	63	58	10 (6)	17 (11)
Totals	136	146	21 (29)	14 (19)
No. of species of mammals[c]				
Grassland and shrub	42	45	17 (7)	10 (4)
Forest	16	14	13 (2)	25 (4)
Totals	58	59	16 (9)	14 (8)

[a] Sources: Biswell *et al.* (1952); Bock and Lynch (1970); Emlen (1970); Ellis *et al.* (1969); Hagar (1960); Kilgore (1971); Lawrence (1966); Michael and Thornburgh (1971); Tester and Marshall (1961); Vogl (1973).

[b] Numbers of species are in parentheses.

[c] Sources: Ahlgren (1966); Beck and Vogl (1972); Biswell *et al.* (1952); Cook (1959); Gashwiler (1970); Keith and Surrendi (1971); Lawrence (1966); LoBue and Darnell (1959); Sims and Buckner (1973); Tester and Marshall (1961); Tester (1965); Vogl (1973).

Remarkably, most breeding species of birds after a forest fire simply stayed (136 before fire, 146 after), only a few species disappeared, and only a few new species moved in (Table V). The result of fire was a slightly richer avifauna. The greatest loss of species was from foragers of the tree trunk (16%) and tree canopy (17%), and the greatest gain (38%) was among those that fed on or near the ground, as might be predicted from the destruction of trees and the growth of grasses, herbs, and shrubs after burning.

The mammals were like the birds in that there was little change in the total number of species after their habitat was changed, and the few species that were lost were replaced by the few new species that moved in (Table V). Most gain (17%) was in grassland and shrub forms; most loss (25%) was in forest species. A difference between birds and mammals was that fewer new species of mammals (17%) than birds (38%) seemed to appear in grassland and shrub after fire (Table V).

The persistence of many species of birds and mammals on an area after a drastic change in surroundings indicates that they either tolerate a wide range of conditions, or fire burns unevenly to leave some of all habitats. Both are probably involved. The loss of some species and gain of others indicate that some habitats were changed, and the greatest new opportunities were for birds that live on or near the ground. After a fire most of the preburn species of plants may persist although some new forms may appear (Ahlgren, 1960; Mueller-Dombois, 1960; Nieland, 1958; Shafi and Yarranton, 1973; Wein and Bliss, 1973). Change and lack of it in species of wildlife appear similar to that in plants and may be related to them. More detailed studies of the use of plant species by wildlife are needed to elaborate this point.

While species of plants may remain essentially the same after logging and fire, there usually is great change in the local climate and in structure and proportion of each kind of species in the vegetation. The change in habitat but the stability of species composition suggests that most wildlife of burnable forests are broadly adapted and can live under a wide range of conditions. Udvardy (1969a) found that none of the birds of the coniferous forests of North America appeared to have special adaptations, and 20 to 70% utilized other forests. Hence, some birds and mammals apparently are not sensitive to fine details of habitat or they tolerate a wide range of conditions of their environment. Perhaps this is a feature of wildlife that live in fluctuating environments, such as forests that frequently burn.

Data of Hagar (1960) on birds and Ahlgren (1966) on mammals (included in Table V) fit the pattern of species response to fire just described. Note this because both censused on areas that were logged and not burned. Again, disturbance by means other than fire gave the same effect. Thus fire may not produce unique conditions for wildlife.

Why do more new species of birds than mammals appear on a fresh burn? Either mammals have a wider habitat tolerance than birds or there are more new opportunities for birds than for mammals. Probably both are involved. However, birds have a number of adaptations that may permit them to exploit new habitats better than mammals. All relate to flight and the ability to find new places, exploit volume rather than the surface area of a habitat, and make long migrations so that a species may live in two or more places where one place may be inadequate.

The response of birds to burned forest may vary throughout the year. After burning, a pine–hardwood forest attracted different species and numbers in winter, spring, and during spring migration (Michael and Thornburgh, 1971).

B. Energy Flow in a Burned and Unburned Forest

Bock and Lynch (1970) analyzed the biomass or standing crop of birds and their consuming biomass (a measure of food intake) in burned and unburned forest. The ratio between consuming biomass to standing crop biomass was offered as a measure of efficiency of use of food. The smaller the ratio, the more efficiently food was used by a bird or group of birds. Although the numbers of birds on the burned and unburned plots were virtually the same, the biomass (gm/100 acres) and consuming biomass (gm/100 acres) were both greater on the burned plot. The efficiency of food utilization was also greater on the burn. The explanation for these results was that the burned area supported more large birds such as robin (*Turdus*) and red-shafted flicker (*Colaptes*), which utilized the grasses, herbs, and shrubs on the burn. Because of their size, and perhaps lesser activity than in small birds, these species utilized food more efficiently. Moreover, the burn gave greater production (or energy capture) than the unburned reference area as a result of the increase in the plants close to the ground and their avifauna of larger body size.

IV. Fire and the Abundance of Birds and Mammals

Generally, we are interested not only in the species that occur in an area, but also in the density they achieve and how this may be affected by fire. As in the study of species response to fire, changes in abundance with fire may provide natural experiments to test how populations are determined.

After a fire we would expect a change in the abundance of species. This must happen if the numbers of a species are more or less closely adjusted to their environment. The striking thing about our examination of species change after a fire was that most kinds of birds and mammals did not change at all. What then about changes in density? Populations may be determined by extrinsic (Klein, 1970; Watson and Moss, 1972) or intrinsic factors (Krebs *et al.*, 1973). Change in abundance after fire would suggest extrinsic factors were operative; little change would indicate an intrinsic process.

A. Change in Density and Trend after Fire

Some data from several sources on the effect of fire on abundance of birds and mammals are given in Table VI. The hazards in this type of compilation are as noted for the analysis of species change after fire. Major difficulties included small sample size and very short periods of count. The birds did not include raptors; the mammals were the size of a hare or smaller.

TABLE VI

CHANGE IN DENSITY AND TREND OF POPULATIONS OF
BREEDING BIRDS AND MAMMALS AFTER BURNING[a]

	Density			Trend		
Foraging zone	Increase (%)	De- crease (%)	No change (%)	Increase (%)	De- crease (%)	No change (%)
	Birds[b]					
Grassland and shrub	50 (33/66)	9 (6)	41 (27)	24 (10)	10 (4)	66 (27)
Tree trunk	28 (9)	16 (5)	56 (18)	4 (1)	8 (2)	88 (21)
Tree	24 (17)	19 (14)	57 (41)	6 (3)	6 (3)	88 (41)
Totals	35 (59)	15 (25)	50 (86)	12 (14)	8 (9)	80 (89)
	Mammals[c]					
Grassland and shrub	24 (9)	13 (5)	63 (24)	20 (4)	5 (1)	75 (15)
Forest	23 (6)	42 (11)	35 (9)	0 (0)	11 (1)	89 (8)
Totals	23 (15)	25 (16)	52 (33)	14 (4)	7 (2)	80 (23)

[a] Numbers of species are in parentheses.

[b] Sources: Biswell et al. (1952); Bock and Lynch (1970); Emlen (1970); Ellis et al. (1969); Hagar (1960); Kilgore (1971); Lawrence (1966); Michael and Thornburgh (1971); Tester and Marshall (1961); Vogl (1973).

[c] Sources: Ahlgren (1966); Beck and Vogl (1972); Biswell et al. (1952); Cook (1959); Gashwiler (1970); Keith and Surrendi (1971); Lawrence (1966); Sims and Buckner (1973); Tester and Marshall (1961); Tester (1965); Vogl (1973).

Among the birds, most populations showed no change in density after fire (50%) or increased (35%); relatively few (15%) declined. The greatest response to burning in terms of increased abundance was shown by ground-dwelling species (50%); most tree dwellers did not change in density after the burn (56 and 57%) (Table VI).

With respect to trends in abundance shown by birds: A few increased (12%) or decreased (8%), but most (80%) did not change at all after fire and after an immediate adjustment of density on the burn. Actually, most remained at a steady density both on the burned and unburned areas.

The populations of mammals behaved very much like the birds; most (52%) did not change at all in density in response to the fire. Among forest mammals, most decreased in density or showed no change. Unlike the birds where 41% of the populations dwelling in grassland and shrub showed no change, most ground and shrub mammals (63%) showed no change in density after forest fire; they carried on as usual (Table VI).

Trends in most populations of small mammals after fire showed no change (80%) as in the birds (80%) (Table VI). Remarkably, after-fire populations of wildlife remained about the same in density and trend, indicating a high degree of stability in the face of wide fluctuations in the external environment because of forest fire. There was some increase and decrease in level and trend of populations but not the wholesale adjustment one might expect if birds and mammals were closely tuned to their environment. The result is comparable to what we have already seen in the little change of species of wildlife in an area after disturbance. The remarkable stability of populations within a rapidly changing environment strongly suggests that many birds and mammals control their own populations more or less independently of it. If so, intrinsic mechanisms should be sought to explain how populations are regulated.

B. Case Histories

There are some long-term, in-depth studies on responses of wildlife to forest fire that were not included in Table VI. Examination of each of these studies gives some insight into how and why populations may change after a fire.

1. Moose

Increase in abundance of moose after 1920 on the Kenai Peninsula of Alaska is explained by fire (Lutz, 1956; Spencer and Hakala, 1964), but the relationship is not a simple one. There are records of relatively small fires in the Kenai from 1890 to 1910. Fires may produce herbs and grasses which peak at about 5 years and the prime browse species, aspen, willow, and birch, which are most abundant about 15 years after a burn. Some moose may use one or more burns for an entire year, others live there only in the winter and seek spring and summer range elsewhere. Thus, moose populations may have been affected by conditions on burned and other ranges and by movements of moose.

According to Spencer and Hakala (1964), the Kenai herd expanded relatively slowly from 1890 to 1920, and then increased from approximately 2000 to about 5500. The sudden irruption in 1920 was not consistent with the series of fires dating back to 1890 in terms of production of browse. In 1926, there was a large burn, but by 1940, when food should have peaked, the population was back to about 2000 animals. In 1947 there was a very large wildfire and the population expanded from 2000 in 1949 to asymptote again at about 5500 in 1960. Here, peak abundance of moose more or less coincided with peak abundance of food. Thus, the history of moose in the Kenai seems to be a slow response, then an irruption in 1920 after the fires of 1890–1910 that cannot be correlated with the abundance of food, no expansion after the fire of 1926, and another major increase that coincided with peak browse production from the fire of 1947. Apparently, abundance of moose was not related to forest fires in any simple fashion. Some of the causes of variation in the relationship between fire and moose may have been changes in local distribution of moose, the variable response of vegetation to burning, and intrinsic properties of the stock that influenced population growth. It is interesting that moose increased after the very large fire in 1947 despite the fact that Lutz (1956) considered this burn of little value to them because it came back mainly to spruce.

2. Black-Tailed Deer

Taber and Dasmann (1957) made relatively detailed studies of population dynamics of three herds of black-tailed deer in California. Each herd was assumed to reflect about 1000 acres of three nearby kinds of habitat: more or less natural chaparral, newly burned chaparral or wildfire burn, and shrubland where chaparral was spot burned and seeded. The main shortcomings of the study are that deer may have moved from one area to another, and some calculations are based on samples of small size. Pertinent data are as tabulated below:

	Chaparral	Wildfire	Shrubland
Density	Low	High	High
Trend	Stable	Decline	Decline
Fawns/adult female	Low	Low	High
Death rate of females	Low	Low	High
Dispersal rate	None	High	Low

The study made two main points, both the open wildfire and the spot burning of shrubland produced an increase and then decline in density of population. Clearly, burning caused an increase in abundance, but the effect did not last. Second, population responded differently to wildfire than to spot burning. There were high rates of movement in and out of the wildfire and high birth and death rates and low dispersal on the spot burn. Thus the response of deer to burning was complex, involving mainly changes in dispersal or local distribution in one case and changes in birth and death rates in another. These observations can be explained on the basis of changes in quality and quantity of food and cover (Biswell *et al.*, 1952; Taber and Dasmann, 1957, 1958). The possibility of an intrinsic basis for how the stocks performed was not considered.

Brown (1961) reported on a study of black-tailed deer in western Washington where there had been a typical extension of the distribution of deer following logging and burning of old-growth coniferous forest. Data on the density of deer and the quantity and quality of food in January with stages of regrowth and in an area of abundant deer were as tabulated below:

| | Stages of forest growth after logging and burning | | | | Abundant deer |
	I	II	III	IV	
Deer/mile²	35	34	57	36	43
Protein (%) in January in preferred browse					
Huckleberry (*Vaccinium*)	7	8	8	—	7
Trailing blackberry (*Rubus*)	11	10	12	—	7
Western red cedar (*Thuja*)	—	6	8	—	7

Note that the stage of forest at or past the peak of production of food but with food in apparently highest quality (III) contained the most deer when, in relation to food supply, stage of peak production (II) should have had as dense a population or a denser one than any other stage of forest succession. This and the equivalent density of deer in mature timber (IV), fresh logging (I), and the 10–25-year growth after logging (II) where food quantity varied but quality was comparable, suggest that density was related to food quality. However, where deer were at a density of 34/square mile, protein values apparently were greater than on a range with 43 deer/square mile. Differences in density of deer may exist but they are not adequately explained by differences in quantity and quality of food.

The last example concerning black-tailed deer is their failure to respond to a 31,000-acre wildfire on Vancouver Island. Since Taber and Dasmann considered 1000 acres to hold separate populations of the same species, the much larger wildfire discussed here should have been large enough to have had a population of its own.

The Sayward Forest is about 500 square miles of forest and logged and burned land on the east coast of Vancouver Island. The area had produced very large populations of deer and blue and ruffed grouse since about 1900. From approximately 1955, populations of deer and

TABLE VII

NUMBER OF DEER AFTER A LARGE FIRE IN 1951 IN THE
SAYWARD FOREST OF VANCOUVER ISLAND[a]

Year	No. of hunters checked	Deer/hunter	Year	No. of hunters checked	Deer/hunter
1952	641	0.27	1959	—	—
1953	2359	0.25	1960	1973	0.17
1954	1755	0.17	1961	2553	0.15
1955	887	0.20	1962	3902	0.16
1956	768	0.20	1963	4395	0.14
1957	1344	0.12	1964	4517	0.09
1958	1995	0.16			

[a] Data courtesy Dr. P. J. Bandy and the British Columbia Fish and Wildlife Branch.

blue grouse, but not ruffed grouse, have declined. The usual explanation for these declines is that with regrowth of the old clear-cuts by natural regeneration and reforestation, food and cover became inadequate for deer and blue grouse but suitable for ruffed grouse (Robinson, 1958).

Each year since 1951, deer hunters coming from the Sayward Forest have been checked. In August, 1951, there was a wildfire over 31,000 acres of timbered and logged and slash-burned land within the Sayward Forest. If deer, and blue grouse as noted later, respond to conditions after logging and fire, surely a pulse should have occurred in the number of deer taken by hunters each fall. We might expect this increase anytime from the year of the burn to 10–15 years after when nutrient supply and shrub growth were at their peak.

However, the success of deer hunters in the Sayward Forest from 1952 through 1964 apparently changed little or declined (Table VII).

These data lend little support to the idea that deer populations increase in abundance in direct relationship to burning and the increased supply of food and cover (Cringan, 1958; Klein, 1970; Robinson, 1958). The result fits the general conclusion that most birds and mammals change little in abundance after fire, as discussed. Perhaps the burn was not a "good burn," the stock that lived there or moved into it prevented increase, or both occurred.

3. Bobwhite Quail

Stoddard (1931) was one of the few wildlife biologists of his day who favored the use of fire in wildlife management. By burning or ploughing, Stoddard manipulated the habitat to manage quail. The manipulative approach not only produced quail but also provided new insights into their habitat requirements and population dynamics. Manipulation or experimentation is a powerful method to identify suspected limiting factors, and, once identified, they may be altered to manage wildlife.

Ellis *et al.* (1969) describe a recent study in the tradition of Stoddard on the bobwhite quail in Illinois. Areas were burned and sharecropped, sharecropped, or burned only. One area was left under the planting form of management and served as a control. Sharecropping was essentially putting the land back to the farming of grain crops but, like burning, in patches and with a mixture of crops.

The response of the quail to the combination of burning and sharecropping was impressive. The counts of quail were made in November and represent numbers per 100 acres. Some of the results were as in the tabulation below:

| Year | Number of quail in fall | | | |
	Burn and sharecrop	Sharecrop	Burn	Statewide reference
1965	23	28	23	17
1966	[57][a]	18	[23]	32
1967	95	27	52	32
1968	96	[44]	67	37

[a] Bracket indicates first year of manipulation.

Clearly, all manipulations increased numbers of quail. It is unfortunate that there is not a longer run of data, particularly for the response to sharecropping. Burning and sharecropping an area produced the most quail, but sharecropping, and burning alone, produced almost as large an increase in abundance. On the area that was burned only, increase in population did not occur until a year after the fire.

How did the changes in density come about? The similar responses to sharecropping and burning tend to rule out any unique effects of burning such as blackening of soil or quick release of nutrients. Both burning and sharecropping cause many structural changes in the habitat such as clearing the ground and altering the amount and interspersion of cover. They also alter the growth and availability of many foods used by quail. While we cannot separate these and other possible effects on population, we can conclude that fire was probably not an essential cause of them. We must also explain why quail did not respond in the year of burning to this management. It may be that the burn lacked an essential element such as cover. Another possibility is that expansion was prevented by the quality of stock present in that year.

4. *Scottish Red Grouse*

Red grouse may live mainly in and on heather and the proper burning of the heath is important to their occurrence and abundance (Lovat, 1911). Grouse densities may vary within and between areas and territorial behavior in fall is the proximate mechanism of population regulation (Watson and Moss, 1972).

What causes territorial behavior? Miller *et al.* (1966) correlated average density of grouse with the amount of cover by heather, which is interpreted as the quality of food. The correlation coefficient (r) was 0.732, $P < 0.01$, and 54% of the variation (r^2) in breeding density was accounted for by variation in heather cover. Grouse stocks were also manipulated by spreading fertilizer and by burning (Miller *et al.*, 1970). Heather grew better on a fertilized area, contained more nitrogen, and numbers of grouse were doubled. The result fits the theory of Watson and Moss (1972) that the territorial behavior of grouse is determined ultimately by the quality of food.

The number of birds involved in the fertilizer experiment was very small; only seven breeding adults on each area at the start. The differences in amounts of nitrogen in the fertilized and unfertilized heather were also small and perhaps insignificant to grouse:

AMOUNT OF NITROGEN (%) IN
FERTILIZED AND UNFERTILIZED HEATHER

Time	Fertilized	Unfertilized
1966		
April	1.31	1.11
November	0.95	0.86
1967		
November	1.15	1.13

It is difficult to separate the effects of cover from food, and there was no control over the stocks of grouse that occupied the experimental and control areas. Perhaps the birds on the experimental side would have performed the same way if they had colonized the control side of the moor.

In the microtines (*Microtus*, *Lemmus*, *Dicrostonyx*), levels and changes of abundance are explained by the quality of food (Batzli and

Pitelka, 1971), the flow of phosphorus through soil, voles, and plants (Schultz, 1969), and amount of sodium in the soil (Aumann, 1965). However, Krebs and DeLong (1965) reported that a population of voles did not respond significantly to fertilizer, and Krebs *et al.* (1971) found no correlation between sodium levels in the soil and changes in the abundance of voles. Perhaps nutrient supply limits the abundance of some species only at some times, and we should not expect a general process of population control.

Miller *et al.* (1970) also experimented with burning. Heather was burned in patches from 1961 until 1965 when 30% of the study area was in a mosaic of heather of different age and size. Grouse populations showed no response until 1965 when they increased, held steady for 1966 and 1967, and then fell to the original number in 1968. The number of young produced up to fall was the same on both the burned and reference areas. Burning did increase numbers, but the experimental population neither stayed in relatively high density nor increased by better breeding, as predicted by theory.

Picozzi (1968) surveyed 26 moors for correlations among average breeding density of grouse, number, and size of heather fires, and base-richness of rocks underlying each moor. The elemental composition of the rocks was assumed to influence nutritional quality of the heather, and Moss (1969) has shown that the samples he picked from a moor underlain by base-rich rocks were slightly higher in phosphorus than heather from a moor upon granites. The study by Picozzi has already assumed textbook status as an example of a bird population limited by nutrient supply as affected by burning (Odum, 1971). The main conclusions were that the average number of grouse on a moor was positively correlated with the average number of fires and the base status of the underlying rock.

I reanalyzed the data presented by Picozzi and found his main correlation between frequency of burns and mean abundance of grouse rested on two moors, estates 1 and 2, where the birds were very abundant and there were frequent small fires. If these moors are removed, his correlation is not significant ($r = 0.42792$, $P < 0.025$, $r^2 = 0.18311$), which casts doubt on a general relationship between frequency of fires and grouse density. Beyond this, some aspects of the data make analysis difficult. For example, frequent and small fires occur over rich rocks more frequently than over rocks low in nutrients. Thus it is difficult to separate the effects of burning from those of rock in correlations with the abundance of grouse.

In summary, densities of red grouse may change in response to heather burning and seem to be limited by food quality or amount of cover, or both. Territorial behavior is the proximate mechanism that adjusts density, and the quality of food is strongly implicated as the main factor behind this. However, there is no direct evidence that quality of food actually determines territorial behavior and what is responsible is still an open question. Territorial behavior may be caused by extrinsic factors (the quality of food) or the genetic quality of the stock, or both. If genotype is involved, there is still the question as to what it may be adjusted.

5. *Blue Grouse*

We have studied the population dynamics of blue grouse in the Pacific northwest since 1950 (Zwickel and Bendell, 1972). Blue grouse invade new breeding range created by logging and fire and at times reach spectacular densities (approximately 1 pair per acre in spring (Bendell, 1955)). Some of our main findings are (1) grouse populations after logging and fire may remain relatively stable, increase, or decrease; (2) there are large differences in breeding density between areas; (3) population regulation is accomplished by interaction among grouse in spring; (4) within broad limits we cannot correlate any of the ordinary variables of the environment with density and trend of populations; and (5) changes in genetic and behavioral types seem associated with the invasion and density achieved by grouse on new summer range (Bendell *et al.*, 1972; Redfield *et al.*, 1970; Redfield, 1972; Zwickel, 1972; Zwickel and Bendell, 1972).

If the behavior among grouse explains how populations are regulated, how is behavior geared to the environment? We cannot correlate density and change in abundance of grouse with the amount of cover on their range. Within the limits of cover from very open (Fig. 7) to very dense (Fig. 2), populations of breeding grouse may vary from dense to sparse. Consider the densities of grouse observed in open cover (Fig. 1). On four separate areas, 3 low, 1 high, and 3 medium densities occurred in what was apparently similar cover (J. F. Bendell, unpublished; Bendell, 1955; Bendell and Elliott, 1967; Bendell *et al.*, 1972; Mossop, 1971; Redfield, 1972; Zwickel, 1972).

Since the quality of food might relate to density of population, in 1970 we made chemical analyses of soils from separate burns of 1938, 1951, and 1961, where populations had been high, medium, and relatively low in density, respectively. Pot tests with leaf lettuce (*Lactuca*) were also made. No differences were found among soils from under different populations (Zwickel and Bendell, 1972). Perhaps our soils did not reflect the quality of food eaten by grouse so that we cannot rule out the effect of food quality on population. However, two new burns produced vigorous growth of apparently high quality foods and yet the populations remained unchanged on both areas, then expanded on one and declined on the other (Redfield *et al.*, 1970). No relationship was found between population density and change in weather. Similarly, predation and disease do not offer an explanation for what was observed. Both occur but they are not important to population regulation (Zwickel and Bendell, 1972).

The size of the disturbance or burn may be an important factor in how blue grouse respond to new summer range, as noted. The populations on five separate areas were as tabulated below:

DENSITY OF BREEDING GROUSE

Size of burn		
Small	Medium	Large
Low[a]	Medium[f]	High[g]
Low[b]		
Low[c]		
Low[d]		
High[e]		

[a] J. F. Bendell (unpublished).
[b] Bendell *et al.* (1972).
[c] Zwickel (1972).
[d] Redfield (1972).
[e] Mossop (1971)
[f] Bendell and Elliott (1967).
[g] Bendell (1955).

With the exception of a population that reached high density on a small burn (Mossop, 1971) and in this instance high may mean medium density, as noted, the data suggest that densities of breeding grouse were adjusted to the acreage of acceptable summer range. Males did establish larger territories when the size of burn was smaller (Bendell and Elliott, 1967), which supports the idea that density is adjusted to area of acceptable habitat.

Thus far we have looked at how extrinsic factors may explain the response of blue grouse to logging and burning. There is good evidence that this response has an intrinsic basis and some of the speculative arguments involving intrinsic factors may have a place. First, as in the red grouse, intraspecific behavior is critical to the performance of blue grouse populations (Bendell *et al.*, 1972; Zwickel, 1972).

Mossop (1971) tested the aggressive level of territorial males in dense and sparse populations by their reaction to their mirror image. He set up mirrors and a dummy female in the center of territories of males and called the males to the female with the playback of the precopulatory call of the hen. Apparently, territorial males in sparse population were more aggressive and this was how density was set.

Redfield *et al.* (1970) and Redfield (1972) studied colonization of new summer range as it was created by logging and burning. Some cuttings were logged but not burned. The focus of the study was the search for genetic change in stocks as grouse occupied and changed in density on the new range. Birds were classified to a particular genotype by the presence or absence of three alleles that were revealed by electrophoresis of the blood. The main findings were that yearlings rapidly colonized newly logged areas, populations usually increased for about 7 years and then leveled off, and there was a correlation between change in gene frequency and the stage of growth of population. What the change in genotype means in terms of how populations change is unknown, but colonists to new summer range were of a different genotype than established birds, as shown by Krebs *et al.* (1973) for *Microtus*. The change in frequency of genotypes as populations change suggests that genetic selection is part of the process of growth and adjustment of populations of blue grouse to new habitat. Genetic selection may operate through the interaction of grouse for space on the summer range. Thus, an explanation for the response of grouse population to burning involves perhaps the size of the summer range, aggressive behavior, and genetic quality of stock.

How might behavior be adjusted to the amount of acceptable summer range? Genotypes associated with population increase in *Microtus* show a high rate of dispersal (Krebs *et al.*, 1973). As speculation, perhaps the larger the disturbance, the greater the opportunity for genotypes of grouse that expand in density to become established without interference from genotypes that suppress population or cause it to decline. Hence, a large burn may produce a dense population. Eventually, the suppressive genotypes replace the expansionary forms and the rate at which this happens may again be related to size of burn and how rapidly acceptable summer range disappears because of regrowth of vegetation. Populations then stabilize or may enter a decline as summer range shrinks in size and suppressive genotypes become established. Small burns may not show any population increase because the appropriate genotypes cannot escape on them.

Why should grouse in the same habitat except on burns of different size require a large or small amount of space to live in? The answer to this question is again unknown but presumably relates to the persistence and evolution of the species, as discussed later.

Redfield *et al.* (1970) also examined the response of blue grouse to areas that were logged and slash burned, or simply logged with the slash left undisturbed. The two kinds of openings supported essentially the same kinds of population. Thus, as in other species, burning did not seem to add special ingredients to affect population. The factors determining population may be achieved either by mechanical means or by burning.

In summary, wildlife populations may respond in a variety of ways to disturbance by fire. Often there is remarkably little change in populations despite wide fluctuation in the environment. Some factors which determine the abundance of birds and mammals after fire may be quality and availability of food, amount and interspersion of cover, intraspecific interaction, genetic quality of the stock, size of burned area, or some combination of these.

V. Evolution of Birds and Mammals and Burnable Habitat

Wildlife affect their environment, and we should look for ways this may occur in relation to fire. Moreover, an understanding of animal–environment interactions will help explain the properties of wildlife and processes involved in maintaining natural environments.

A. How Wildlife May Affect Fire

Many plant formations are flammable and all or parts of them depend on repeated fire for their existence (Ahlgren and Ahlgren, 1960; Daubenmire, 1968; Flieger, 1970; Heinselman, 1971; Hodgson, 1968; Kayll, 1968; Komarek, 1968; Lutz, 1956; Mount, 1964; Mutch, 1970; Niering *et al.*, 1970; Rowe, 1970; Shafi and Yarranton, 1973; Swan, 1970; Thilenius, 1968). With adaptation of plants to fire it seems possible that some birds and mammals have evolved to exploit burnable forests and grasslands. Moreover, if these habitats are essential to them they may influence the action of fire and perpetuate their habitats, and, in turn, themselves. In Africa, the Nyika Plateau, a rolling grassland, has been frequently burned for a very long time. Nevertheless this grassland is stable and contains a number of endemic vertebrates and plants (Lemon, 1968a,b). Handley (1969) pointed out the broad adaptations of mammals to grasslands. Clearly, some wildlife have evolved to live in flammable habitats. But how might they influence the frequency and action of fire?

The most frequent natural start of fires is by lightning striking the tallest trees and old snags (Komarek, 1968; Kourtz, 1967; Taylor, 1969). Actual ignition of the forest often depends on the quantity of fine combustible material in or near the base of a tree. Birds and mammals, notably the red (J. L. Farrar, personal communication) and Abert squirrel (*Sciurus*) (Keith, 1965), may mutilate the tops of trees and so influence the likelihood of lightning strike. A. R. Taylor (personal communication) remarked that the size and shape of a tree may influence the chance of being struck by lightning and the effect of the strike. Moreover, in feeding, squirrels may make piles of scales from cones close to the base of a tree and provide tinder to convert a lightning discharge into flame (Rowe, 1970). Possibly, birds and mammals, by excavating and placing nests of fine woody material in live and dead trees, may make them more susceptible to lightning and more burnable.

Large herbivores, such as deer, elk, bison (*Bison*), and moose, may alter the fuel in grasslands and forests to influence the way they burn (Bailey and Poulton, 1968; Flook, 1964; Hansen *et al.*, 1973; Hough, 1965; Pimlott, 1963; Ross *et al.*, 1970). For example, Jeffrey (1961) documented the spread of aspen into grassland in Saskatchewan. This began after the disappearance of the bison and suppression of fire because both favored growth of grassland.

Many herbivores feed on fire-sensitive trees and avoid the more flammable species. A selection of this kind may be enough to ensure good stands of the flammable forms, eventual burning of the forest, and maintenance of habitat favorable to herbivores. Fire-tolerant and fire-sensitive conifers are well documented (Flieger, 1970; Kayll, 1968; Knight and Loucks, 1969; Rowe, 1970). The flammable conifers provide the main fuel to burn the forest, and thus initiate extremely rapid regeneration of browse trees and shrubs. When deer and moose browse, they more or less ignore the "torch" trees: the pines, tamaracks (*Larix*), and black spruce. The trees usually eaten include hemlock, cedar, and balsam fir (*Abies*) (Dasmann, 1971; Pimlott, 1963). Hence, ungulates may suppress fire-sensitive species and release the more flammable trees. Unfortunately, the theory does not work for white spruce, which resists fire but is not used as browse by ungulates, and in dense stands is poor habitat for wildlife. Also, Bergerud and Manuel (1968) noted that moose may damage balsam fir and white birch to release white spruce and presumably speed their own demise. However, fire from adjacent stands of more burnable species may spread into white spruce, and there are animals other than ungulates that may discourage expansion of these forests. Hare feed on white spruce (Pimlott, 1963), and some small mammals feed selectively on cones and seed of this species (Brink and Dean, 1966; Radvanyi, 1970).

Ahlen (1968) gives some insight into how young pine (*Pinus silvestris*) may avoid browsing by moose when pines are small and vulnerable by developing sharp needles not found on large pines. Here may be a first step in releasing pine and keeping fire in the forest.

Wildlife may also influence composition of forests by distributing seed, particularly on new burns (Ahlgren, 1960; West, 1968). Much of the seed will be of plants used by wildlife as food so that regrowth will be to the advantage of wildlife. Animal dispersal of seed is enhanced by large size, fleshy parts, and resistant coats, often typical of fire following shrubs and trees.

B. Speciation in Flammable Habitat

An explanation for the great number of species of birds and mammals in the tropics is the many shapes and sizes of plants of the tropical forest (McArthur *et al.*, 1962; Orians, 1969). As noted, birds and mammals of coniferous forest have wide habitat tolerance. There may be a relationship between the fewer species in northern forests and the fact that they frequently burn. Where forests persist there is time and opportunity for many species to evolve. Where forests are frequently changed by fire, selection may be for species that can live under a wide range of conditions; they are broadly adapted. Hence, a few broadly adapted species will occur rather than many species that are specialists. This interpretation stresses the permanency of habitat as a factor in the greater number of species in tropical, less burnable forests. A similar case is made for the relatively few species of plants (Shafi and Yarranton, 1973) and animals (Dunbar, 1968) in northern ecosystems.

Several investigators emphasized that some hardwood forests, such as those of sugar maple (*Acer*) and beech (*Fagus*), have relatively low flammability (Ahlgren and Ahlgren, 1960; Knight and Loucks, 1969; Komarek, 1962, 1968; Swan, 1970). If so, they might be a more stable environment than conifer forest and, as in the stable tropical forests, contain more species of birds. An immediate difficulty in this comparison is that factors other than a relatively long stay of mature trees might affect the composition of the bird fauna, for example, structure of vegetation.

Speirs (1969, 1972) presented counts of breeding birds in balsam fir–black spruce forest at Eaglehead Lake and Black Sturgeon Lake, and sugar maple–beech forest near Dorset, Ontario. Speirs and I classified the birds into those more or less completely associated with mature and successional stages of conifers and hardwoods. In the conifer forest, there was an average of 22 mature and 6 successional stage species. For the hardwood forest there were 22 mature and 1 successional stage species. Clearly, how frequently these forests burn does not appear to relate to the number of species of birds that are adapted to breeding in them.

Martin (1960) surveyed the species of breeding birds in various forest types in Algonquin Park, Ontario. He found 30 species associated with hemlock stands, and only 9 with white pine. The difference seems real and large when one considers that both conifers have relatively similar life forms. White pine depends on repeated fire for its persistence (Heinselman, 1970), while hemlock, when established, may produce relatively fire-resistant (Flieger, 1970) and long-lived stands. The relatively few species of birds in pine forest and the richness of species in hemlock may relate in part to the amount of disturbance by fire in each kind of forest and the length of time opportunities for species exist.

Forest fires and other disturbances may tend to blur the separation of habitats and bring species of wildlife together. The remnants of unburned forest in otherwise open burn may hold both forest and burn species in the same area, as already mentioned. The overlap of species may have many ramifications. As examples, blue grouse in their subalpine habitat do not have the parasite *Dispharynx*. When they move onto the lowland burns, they live closer to ruffed grouse and share this parasite with them (Bendell, 1955; King, 1971). Johnsgard and Wood (1968) reported that the once narrow sympatric zone between the prairie chicken and sharp-tailed grouse has enlarged greatly because of fire and other changes in land use. One consequence has been the increase in hybridization between the two species.

C. Adaptations of Birds and Mammals to Flammable Habitat

Many kinds of wildlife are adapted to living in environments created largely by fire. Some of the features of flammable forests and grasslands that may affect wildlife are much growth of vegetation on and near the ground; growth of trees and shrubs in open stands with thick, often stubby branches and twigs; large fruits, seeds, and nuts that may be retained on the plant and protected in serotinous cones or thick shells; and heavy, slow-rotting litter (Cooper, 1961a; Flieger, 1970; Kayll, 1968; Knight and Loucks, 1969; Mutch, 1970).

Under these conditions we would expect birds and mammals that live as browsers and grazers. There should be a tendency for large size, for this is of advantage where there are wide fluctuations in local climate and for other reasons. Large wildlife are possible where they can feed from the ground and in shrubs and trees with open, strong branches. Thus, in fire forest there may be deer and spruce grouse. In relatively fireproof forest, such as hemlock, there are birds (Martin, 1960) and mammals that are smaller in size.

The birds of the conifer forest have broad adaptations of form and function; there are relatively few species that specialize on conifer trees for living, breeding, and feeding (Udvardy, 1969a). Perhaps this reflects the mixture of deciduous and evergreen vegetation that is often present and of long duration because of recurrent fire. The persistence of many species of birds and mammals after fire, as noted earlier, would support the conclusion by Udvardy that birds of the conifer forest are broadly adapted. Other features of birds and mammals that are probably adaptations to flammable, open grassland, and forest are the ability to run or fly quickly and for long distances, burrowing, storing food, kind of camouflage, pressing flat to avoid detection, and migration (Handley, 1969; Komarek, 1962). Some of the social displays of blue grouse and other birds and mammals can be explained as adaptations to open habitat (Bendell and Elliott, 1966).

Adaptation at the Population Level

Presumably the birth, death, and dispersal rates of a species and how its numbers are regulated reflect the kind of environment to which it is adapted. The sudden and rapid changes in vegetation and other factors that occur after fire must be an important feature of the habitat. Fires presumably offer new opportunities to wildlife but also make problems of existence in a transitory habitat.

Klein (1970) believes that deer populations respond directly to food supply by changes in birth and death rate, and these adaptations fit deer for survival in a rapidly changing environment as created by logging and fire. However, as we have seen, populations of deer may not always respond to fire. Roe deer regulate their numbers by dispersal caused by territorial behavior (Klein and Strandgaard, 1972), and little is known of the role of behavior and dispersal in the population dynamics of deer in North America (Watson and Moss, 1970).

Geist (1971) compared the population properties of a fire-follower, the moose, with those of the bighorn sheep, which he believes lives in a stable, relatively fire-free environment. Some of the relevant features of population of the two ungulates are tabulated below:

	Moose	Bighorn sheep
Birth rate	High and variable	Low and constant
Dispersal rate	High	Low
Abundance	Fluctuating	Steady
Population limited directly by	Food supply	Food supply

Moose are adapted to exploit new habitat by their ability to disperse and quickly expand in numbers on a new area. Populations are then regulated as in deer. For sheep, new habitats rarely appear so dispersal is not an advantage. Populations are held stable by a low birth rate and an appropriate death rate against a steady supply of food. The comparison suggests at least that there are differences in the population parameters of the two ungulates that may relate to their adaptations to a changing or stable environment. Again, moose do not always expand after a forest fire, and Geist cannot rule out the importance of behavior and dispersal in the population regulation of either form.

Geist (1971) considered the unwillingness of sheep to enter timber, unless on traditional routes of travel, an immediate reason for their inability to spread. Perhaps forest fires are the missing factor in dispersal. Repeated fires in the high country may destroy the forest barriers to dispersal and also provide grassland range to sustain dispersing and resident sheep.

If differences in behavior and genotype are important to population dynamics, how might they adapt birds and mammals to fire-created habitats? The blue grouse may serve as a general example. Behavior is the immediate mechanism by which populations are regulated and perhaps determines the genetic quality of the stock. The open forests of the subalpine appear to be stable environment for blue grouse. King (1971) reported that populations of the subalpine appear similar to those on the lowland burns except they persist at low and stable density. Where the environment frequently offers new opportunity, it would seem advantageous to a species to have a form that quickly moves in and reaches high density in the new habitat. If the habitat is transitory, this may lead to extinction. However, if another genotype persists in a more stable habitat, the persistence of the species and a source of colonists are ensured. The price of living in a stable and more or less permanent habitat would seem to be low density or the predominance of suppressive genotypes in the population.

Acknowledgments

Mrs. M. Taitt and Miss W. Ebersberger helped in the search of literature. Miss Ebersberger also assisted with the analysis of data and checked the references cited. Miss Laurie Thompson typed the manuscript. Dr. F. C. Zwickel gave much helpful criticism. I am most grateful for all this assistance; the conclusions and views are my own.

References

Adams, S., Strain, B. R., and Adams, M. S. (1970). Water-repellent soils, fire, and annual plant cover in a desert scrub community of southeastern California. *Ecology* **51**, 696–700.

Ahlen, I. (1968). "Research on Moose-vegetation Interplay in Scandinavia," Proc. 5th Amer. Moose Workshop, pp. 23–33. Kenai Moose Res. Sta. Kenai, Alaska.

Ahlgren, C. E. (1960). Some effects of fire on reproduction and growth of vegetation in northeastern Minnesota. *Ecology* **41**, 431–445.

Ahlgren, C. E. (1966). Small mammals and reforestation following prescribed burning. *J. Forest.* **64**, 614–618.

Ahlgren, I. F., and Ahlgren, C. E. (1960). Ecological effects of forest fires. *Bot. Rev.* **26**, 483–533.

Allen, S. E., Evans, C. C., and Grimshaw, H. M. (1969). The distribution of mineral nutrients in the soil after heather burning. *Oikos* **20**, 16–25.

Amman, G. A. (1963). Status and management of sharp-tailed grouse in Michigan. *J. Wildl. Manage.* **27**, 802–809.

Anderson, R. C. (1972). The use of fire as a management tool on the Curtis prairie. *Arboretum News* **21**, 1–9.

Arnold, J. F. (1963). Uses of fire in the management of Arizona watersheds. *Proc. 2nd Annu. Tall Timbers Fire Ecol. Conf.* pp. 99–111.

Aumann, G. D. (1965). Microtine abundance and soil sodium levels. *J. Mammal.* **46**, 594–604.

Austen, B. (1971). The history of veld burning in the Wankie National Park, Rhodesia. *Proc. 11th Annu. Tall Timbers Fire Ecol. Conf.* pp. 277–296.

Austin, R. C., and Baisinger, D. H. (1955). Some effects of burning on forest soils of western Oregon and Washington. *J. Forest.* **53**, 275–280.

Bailey, A. W., and Poulton, C. E. (1968). Plant communities and environmental interrelationships in a portion of the Tillamook burn, Northwestern Oregon. *Ecology* **49**, 1–12.

Banfield, A. W. F. (1954). Preliminary investigation of the barren-ground caribou. *Can. Wildl. Serv., Wildl. Manage. Bull., Ser. 1* **10A**, 1–79; **10B**, 1–112.

Barick, F. B. (1950). The edge effect of the lesser vegetation of certain Adirondack forest types with particular reference to deer and grouse. *Roosevelt Wildl. Bull.* **9**, 1–146.

Batzli, G. O., and Pitelka, F. A. (1971). Condition and diet of cycling populations of the California vole, *Microtus californicus*. *J. Mammal.* **52**, 141–163.

Beck, A. M., and Vogl, R. J. (1972). The effects of spring burning on rodent populations in a brush prairie savanna. *J. Mammal.* **53**, 336–346.

Beeson, K. C. (1941). The mineral composition of crops with particular reference to soils in which they were grown. *U.S., Dep. Agr., Misc. Publ.* **369**.

Bendell, J. F. (1955). Disease as a control of a population of blue grouse, *Dendragapus obscurus fuliginosus* (Ridgway). *Can. J. Zool.* **33**, 195–223.

Bendell, J. F., and Elliott, P. W. (1966). Habitat selection in blue grouse. *Condor* **68**, 431–446.

Bendell, J. F., and Elliott, P. W. (1967). Behaviour and the regulation of numbers in blue grouse. *Can. Wildl. Serv., Rep. Ser.* No. 4, pp. 1–76.

Bendell, J. F., King, D. G., and Mossop, D. H. (1972). Removal and repopulation of blue grouse in a declining population. *J. Wildl. Manage.* 36, 1153–1165.

Bergerud, A. T. (1971). Abundance of forage on the winter range of Newfoundland caribou. *Can. Field Natur.* 85, 39–52.

Bergerud, A. T., and Manuel, F. (1968). Moose damage to balsam fir-white birch forests in central Newfoundland. *J. Wildl. Manage.* 32, 729–746.

Bissell, H. (1959). Interpreting chemical analyses of browse. *Calif. Fish Game* 45, 57–58.

Biswell, H. H., Taber, R. D., Hedrick, D. W., and Schultz, A. M. (1952). Management of chamise brushlands for game in the north coast region of California. *Calif. Fish Game* 38, 453–484.

Blackford, J. (1955). Woodpecker concentration in a burned forest. *Condor* 57, 28–30.

Blaisdell, J. P., and Mueggler, W. F. (1956). Sprouting of bitterbrush (*Purshia tridentata*) following burning or top removal. *Ecology* 37, 365–370.

Bock, C. E., and Lynch, J. F. (1970). Breeding bird populations of burned and unburned conifer forests in the Sierra Nevada. *Condor* 72, 182–189.

Brewer, R. (1958). Breeding-bird populations of strip-mined land in Perry County, Illinois. *Ecology* 39, 543–545.

Brink, C. H., and Dean, F. C. (1966). Spruce seed as a food of red squirrels and flying squirrels in interior Alaska. *J. Wildl. Manage.* 30, 503–512.

Brown, E. R. (1961). The black-tailed deer of western Washington. *Wash. Dep. Game, Biol. Bull.* 13, 1–124.

Brynard, A. M. (1971). Controlled burning in the Kruger National Park—history and development of a veld burning policy. *Proc. 11th Annu. Tall Timbers Fire Ecol. Conf.* pp. 219–231.

Buckley, J. L. (1958). Effects of fire on Alaskan wildlife. *Proc. Soc. Amer. Forest.* 58, 123–126.

Buechner, H. K., and Dawkins, H. C. (1961). Vegetation change induced by elephants and fire in Murchison Falls National Park, Uganda. *Ecology* 42, 752–766.

Buell, M. F., and Cantlon, J. E. (1953). Effects of prescribed burning on ground cover in the New Jersey pine region. *Ecology* 34, 520–528.

Casperson, K. D. (1963). Visceral parasites in blue grouse, *Dendragapus obscurus fuliginosus* (Ridgway). B. Sc. Thesis, University of British Columbia, Vancouver.

Caughley, G. (1970). Eruption of ungulate populations, with emphasis on Himalayan thar in New Zealand. *Ecology* 51, 53–72.

Chapman, J. D., and Turner, D. B., eds. (1956). "British Columbia Atlas of Resources." B.C. Natur. Resour. Conf., Dept. Recreation and Conserv., Victoria.

Cheatum, E. L. (1949). Bone marrow as an index of malnutrition in deer. *N.Y. State Conserv.* 3, 19–22.

Chew, R. M., Butterworth, B. B., and Grechman, R. (1958). The effects of fire on the small mammal populations of chaparral. *J. Mammal.* 40, 253.

Chitty, D. H. (1967a). The natural selection of self-regulatory behaviour in animal populations. *Proc. Ecol. Soc. Aust.* 2, 51–78.

Chitty, D. H. (1967b). What regulates bird populations? *Ecology* 48, 698–701.

Cook, S. F., Jr. (1959). The effects of fire on a population of small rodents. *Ecology* 40, 102–108.

Cooper, C. F. (1961a). The ecology of fire. *Sci. Amer.* 204, 150–160.

Cooper, C. F. (1961b). Pattern in ponderosa pine forests. *Ecology* 42, 493–499.

Cowan, I. McT., Hoar, W. S., and Hatter, J. (1950). The effect of forest succession upon the quantity and upon the nutritive values of woody plants used as food by moose. *Can. J. Res., Sect. D* **28**, 249–271.

Cringan, A. T. (1957). History, food habits and range requirements of the woodland caribou of continental North America. *Trans. N. Amer. Wildl. Conf.* **22**, 485–501.

Cringan, A. T. (1958). Influence of forest fires and fire protection on wildlife. *Forest. Chron.* **34**, 25–30.

Crouch, G. L. (1966). Preferences of black-tailed deer for native forage and Douglas-fir seedlings. *J. Wildl. Manage.* **30**, 471–475.

Cruickshrank, A. D. (1956). Nesting heights of some woodland warblers in Maine. *Wilson Bull.* **68**, 157.

Dahlberg, B. L., and Guettinger, R. C. (1956). The white-tailed deer in Wisconsin. *Wis. Conserv. Dep., Tech. Wildl. Bull.* No. 14, pp. 1–282.

Darling, F. F. (1960). "Wild Life in an African Territory." Oxford Univ. Press, London and New York.

Dasmann, R. F. (1964). "Wildlife Biology." Wiley, New York.

Dasmann, W. (1971). "If Deer Are to Survive." Stackpole, Harrisburg, Pennsylvania.

Daubenmire, R. (1968). Ecology of fire in grasslands. *Advan. Ecol. Res.* **5**, 209–266.

DeWitt, J. B., and Derby, J. V., Jr. (1955). Changes in nutritive value of browse plants following forest fires. *J. Wildl. Manage.* **19**, 65–70.

Dietz, D. R. (1965). Deer nutrition research in range management. *Trans. N. Amer. Wildl. Nat. Res. Conf.* **30**, 274–285.

Dills, G. G. (1970). Effects of prescribed burning on deer browse. *J. Wildl. Manage.* **34**, 540–545.

Doerr, P. D., Keith, L. B., and Rusch, D. H. (1970). Effects of fire on a ruffed grouse population. *Proc. 10th Annu. Tall Timbers Fire Ecol. Conf.* pp. 25–46.

Dorney, R. S., and Kabat, C. (1960). Relation of weather, parasitic disease and hunting to Wisconsin ruffed grouse populations. *Wis. Conserv. Dep., Tech. Bull.* No. 20, pp. 1–64.

Drent, R. (1972). Adaptive aspects of the physiology of incubation. *Proc. Int. Ornithol. Congr., 15th, 1970* pp. 255–280.

Dunbar, M. J. (1968). "Ecological Development in Polar Regions." Prentice-Hall, New Jersey.

Eddleman, L., and McLean, A. (1969). Herbage—its production and use within the coniferous forest. *In* "Coniferous Forests of the Northern Rocky Mountains" (R. D. Taber, ed.), pp. 179–196. Centre for Natural Resources, Missoula, Montana.

Edwards, R. Y. (1954). Fire and the decline of a mountain caribou herd. *J. Wildl. Manage.* **18**, 521–526.

Edwards, R. Y. (1956). Snow depths and ungulate abundance in the mountains of western Canada. *J. Wildl. Manage.* **20**, 159–168.

Einarsen, A. S. (1945). Some factors affecting ring-necked pheasant population density. *Murrelet* **26**, 39–44.

Einarsen, A. S. (1946a). Crude protein determination of deer food as an applied management technique. *Trans. N. Amer. Wildl. Conf.* **11**, 309–312.

Einarsen, A. S. (1946b). Management of black-tailed deer. *J. Wildl. Manage.* **10**, 54–59.

Ellis, J. A., Edwards, W. R., and Thomas, K. P. (1969). Responses of bobwhites to management in Illinois. *J. Wildl. Manage.* **33**, 749–762.

Elton, C. S. (1958). "The Ecology of Invasions by Animals and Plants." Methuen, London.

Emlen, J. T. (1970). Habitat selection by birds following a forest fire. *Ecology* **51**, 343–345.

Evans, F. C. (1942). Studies of a small mammal population in Bagley Wood, Berkshire. *J. Anim. Ecol.* **11**, 182–197.

Evans, P. R. (1969). Ecological aspects of migration, and pre-migratory fat deposition in the lesser redpoll, *Carduelis flammea cabaret. Condor* **71**, 316–330.

Flieger, B. W. (1970). Forest fire and insects: The relation of fire to insect outbreak. *Proc. 10th Annu. Tall Timbers Fire Ecol. Conf.* pp. 107–114.

Flook, D. R. (1964). Range relationships of some ungulates native to Banff and Jasper National Parks, Alberta. *In* "Grazing in Terrestrial and Marine Environments" (D. J. Crisp, ed.), pp. 119–128. Blackwell, Oxford.

Fowle, C. D. (1944). The sooty grouse (*Drendragapus fuliginosus fuliginosus* Ridgway) on its summer range. M.A. Thesis, University of British Columbia, Vancouver.

Fowle, C. D. (1946). The blood parasites of the blue grouse. *Science* **103**, 708–709.

Fowle, C. D. (1960). A study of the blue grouse (*Dendragapus obscurus* (Say)) on Vancouver Island, British Columbia. *Can. J. Zool.* **38**, 701–713.

Gardarsson, A., and Moss, R. (1970). Selection of food by Icelandic ptarmigan in relation to its availability and nutritive value. *Brit. Ecol. Soc. Symp.* **10**, 47–71.

Garman, E. H., and Orr-Ewing, A. L. (1949). Direct-seeding experiments in the southern coastal region of British Columbia, 1923–1949. *Brit. Columbia, Forest Serv., Tech. Publ.* **T-31**.

Gashwiler, J. S. (1970). Plant and mammal changes on a clearcut in west-central Oregon. *Ecology* **51**, 1018–1026.

Gates, B. R. (1968). Deer food production in certain seral stages of the coast forest. M.Sc. Thesis, University of British Columbia, Vancouver.

Geist, V. (1971). "Mountain Sheep. A Study in Behavior and Evolution." Univ. of Chicago Press, Chicago, Illinois.

Gessel, S. P., and Balci, A. N. (1963). Amount and composition of forest floors under Washington coniferous forests. *In* "Forest–Soil Relationships in North America" (C. T. Youngberg, ed.) pp. 11–23. Oregon State Univ. Press, Corvallis.

Gill, R. W. (1969). Soil microarthropod abundance following old-field litter manipulation. *Ecology* **50**, 805–816.

Gillon, D. (1971). The effect of bush fire on the principal pentatomid bugs (Hemiptera) of an Ivory Coast savanna. *Proc. 11th Annu. Tall Timbers Fire Ecol. Conf.* pp. 377–417.

Gimingham, C. H. (1970). British heathland ecosystems: The outcome of many years of management by fire. *Proc. 10th Annu. Tall Timbers Fire Ecol. Conf.* pp. 293–321.

Givens, L. S. (1962). Use of fire on southeastern wildlife refuges. *Proc. 1st Annu. Tall Timbers Fire Ecol. Conf.* pp. 121–126.

Grange, W. B. (1948). "Wisconsin Grouse Problems." Wis. Conserv. Dept., Madison.

Grange, W. B. (1949). "The Way to Game Abundance." Scribner's, New York.

Gullion, G. W. (1967). Factors affecting ruffed grouse populations in the boreal forests of northern Minnesota, U.S.A. *Proc. 8th Int. Congr. Game Biol. 1967, Finn. Game Res.* **30**, 103–117.

Gullion, G. W. (1970). Factors influencing ruffed grouse populations. *Trans. N. Amer. Wildl. Natur. Resour. Conf.* **35**, 93–105.

Gullion, G. W. (1972). Improving your forested lands for ruffed grouse. *Minn., Agr. Exp. Sta., Misc. J. Ser., Publ.* **1439**, 1–34.

Gullion, G. W., and Marshall, W. H. (1968). Survival of ruffed grouse in a boreal forest. *Living Bird* **7**, 117–167.

Hagar, D. (1960). The interrelationships of logging, birds, and timber regeneration in the Douglas-fir region of northwestern California. *Ecology* **41**, 116–125.

Hakala, J. B., Seemel, R. K., Richey, R. A., and Kurtz, J. E. (1971). Fire effects and rehabilitation methods—Swanson—Russian Rivers fires. *In* "Fire in the Northern Environment" (C. W. Slaughter, R. J. Barney, and G. M. Hansen, eds.), pp. 87–99. U.S. Pac. Northwest Forest Range Exp. Sta., Portland, Oregon.

Hall, A. M. (1971). Ecology of beaver and selection of prey by wolves in central Ontario. M.Sc. Thesis, University of Toronto.

Handley, C. O., Jr. (1969). Fire and mammals, *Proc. 9th Annu. Tall Timbers Fire Ecol. Conf.* pp. 151–159.

Hansen, H. L., Krefting, L. W., and Kurmis, V. (1973). The forest of Isle Royale in relation to fire history and wildlife. *Univ. Minn., Agr. Exp. Sta., Forest. Ser. 13, Tech. Bull.* **294**, 1–43.

Harris, Van T. (1952). An experimental study of habitat selection by prairie and forest races of the deermouse, *Peromyscus maniculatus. Contrib. Lab. Vertebr. Biol. Univ. Mich.* **56**, 1–53.

Hart, J. S., Heroux, O., Cottle, W. H., and Mills, C. A. (1961). The influence of climate on metabolic and thermal responses of infant caribou. *Can. J. Zool.* **39**, 845–856.

Hayes, G. L. (1970). Impacts of fire use on forest ecosystems. *In* "The Role of Fire in the Intermountain West," pp. 99–118. Intermt. Fire Res. Counc., Missoula, Montana.

Heinselman, M. L. (1970). Restoring fire to the ecosystems of the Boundary Water Canoe Area, Minnesota, and to similar wilderness areas. *Proc. 10th Annu. Tall Timbers Fire Ecol. Conf.* pp. 9–23.

Heinselman, M. L. (1971). The natural role of fire in northern conifer forests *In* "Fire in the Northern Environment" (C. W. Slaughter, R. J. Barney, and G. M. Hansen, eds.), pp. 61–72. U.S. Pac. Northwest. Forest Range Exp. Sta., Portland, Oregon.

Henderson, C. W. (1971). Comparative temperature and moisture responses in gambel and scaled quail. *Condor* **73**, 430–436.

Hill, P. (1971). Grass foggage—food for fauna or fuel for fire, or both? *Proc. 11th Annu. Tall Timbers Fire Ecol. Conf.* pp. 337–375.

Hodgson, A. (1968). Control burning in eucalypt forests in Victoria, Australia. *J. Forest.* **66**, 601–605.

Höglund, N. H. (1970). On the ecology of the willow grouse (*Lagopus lagopus*) in a mountainous area in Sweden. *Proc. 8th Int. Congr. Game Biol. 1967, Finn. Game Res.* **30**, 118–120.

Horvath, O. (1964). Seasonal differences in rufous hummingbird nest heights and their relation to nest climate. *Ecology* **45**, 235–241.

Hough, A. F. (1965). A twenty-year record of understory vegetational change in a virgin Pennsylvania forest. *Ecology* **46**, 370–373.

Howard, W. E., Fenner, R. L., and Childs, H. E., Jr. (1959). Wildlife survival on brush burns. *J. Range Manage.* **12**, 230–234.

Hulbert, L. C. (1969). Fire and litter effects in undisturbed bluestem prairie in Kansas. *Ecology* **50**, 874–877.

Humphrey, R. R. (1962). "Range Ecology." Ronald Press, New York.

Hurst, G. A. (1971). The effects of controlled burning on arthropod density and biomass in relation to bobwhite quail brood habitat on a right-of-way. *Proc. 2nd Tall Timbers Conf. Ecol. Anim. Contr. Habitat Manage. 1970* pp. 173–183.

Isaac, L. A., and Hopkins, H. G. (1937). The forest soil of the Douglas-fir region, and changes wrought upon it by logging and slash burning. *Ecology* **18**, 264–279.

Isaac, N. (1963). Fire a tool, not a blanket rule in Douglas-fir ecology. *Proc. 2nd Annu. Tall Timbers Fire Ecol. Conf.* pp. 1–17.

James, F. C. (1971). Ordinations of habitat relationships among breeding birds. *Wilson Bull.* **83**, 215–236.

Jeffrey, W. W. (1961). A prairie to forest succession in wood buffalo park, Alberta. *Ecology* **42**, 442–444.

Jensen, D. (1962). The pathological effects of infections of *Dispharynx nasuta* (Nematoda: Spiruroidea) on the blue grouse, *Dendragapus obscurus* (Say.). Ph.D. Thesis, University of British Columbia, Vancouver.

Johnsgard, P. A., and Wood, R. E. (1968). Distributional changes and interaction between prairie chickens and sharp-tailed grouse in the mid-west. *Wilson Bull.* **80**, 173–188.

Jonkel, C. J., and Cowan, I. McT. (1971). "The Black Bear in the Spruce-fir Forest," Wildl. Monogr. No. 27. Wildl. Soc., Washington, D.C.

Kahn, W. C. (1960). Observations on the effect of a burn on a population of *Sceloporus occidentalis. Ecology* **41**, 358–359.

Karr, J. R. (1968). Habitat and avian diversity on strip-mined land in east-central Illinois. *Condor* **70**, 348–357.

Kayll, A. J. (1968). The role of fire in the boreal forest of Canada. *Petawawa Forest Exp. Sta., Chalk River, Ont., Info. Rep.* **PS-X-7**, 1–15.

Keith, J. O. (1965). The Abert squirrel and its dependence on ponderosa pine. *Ecology* **46**, 150–163.

Keith, L. B., and Surrendi, P. C. (1971). Effects of fire on a snowshoe hare population. *J. Wildl. Manage.* **35**, 16–26.

Kelsall, J. P. (1968). The migratory barren-ground caribou of Canada. *Can. Wildl. Serv., Monogr.* No. 3, pp. 1–340.

Kelsall, J. P. (1972). The northern limits of moose (*Alces alces*) in western Canada. *J. Mammal.* **53**, 129–138.

Kelsall, J. P., and Prescott, W. (1971). Moose and deer behaviour in snow in Fundy National Park, New Brunswick. *Can. Wildl. Serv., Rep. Ser.* No. 15, pp. 1–27.

Kemp, G. A., and Keith, L. B. (1970). Dynamics and regulation of red squirrel (*Tamiasciurus hudsonicus*) populations. *Ecology* **51**, 763–779.

Kendeigh, S. C. (1945). Community selection by birds on the Helderberg Plateau of New York. *Auk* **62**, 418–436.

Kilgore, B. M. (1971). Response of breeding bird populations to habitat changes in a Giant Sequoia forest. *Amer. Midl. Natur.* **85**, 135–152.

King, D. G. (1971). The ecology and population dynamics of blue grouse in the sub-alpine. M.Sc. Thesis, University of British Columbia, Vancouver.

King, J. R., and Farner, D. S. (1961). Energy metabolism, thermoregulation and body temperature. *In* "Biology and Comparative Physiology of Birds" (A. J. Marshall, ed.), vol. 2, pp. 215–288. Academic Press, New York.

King, R. D. (1968). Food habits in relation to the ecology and population dynamics of blue grouse. M.Sc. Thesis, University of British Columbia, Vancouver.

King, R. D. (1969). Spring and summer foods of ruffed grouse on Vancouver Island. *J. Wildl. Manage.* 33, 440–442.

Klein, D. R. (1970). Food selection by North American deer and their response to over-utilization of preferred plant species. *Brit. Ecol. Soc. Symp.* 10, 25–46.

Klein, D. R., and Strandgaard, H. (1972). Factors affecting growth and body size of roe deer. *J. Wildl. Manage.* 36, 64–79.

Klein, H. G. (1960). Ecological relationships of *Peromyscus noveboracensis* and *P. maniculatus gracilis* in central New York. *Ecol. Monogr.* 30, 387–407.

Knight, D. H., and Loucks, O. L. (1969). A quantitative analysis of Wisconsin forest vegetation on the basis of plant function and gross morphology. *Ecology* 50, 219–234.

Kolenosky, G. B. (1972). Wolf predation on wintering deer in east-central Ontario. *J. Wildl. Manage.* 36, 357–369.

Komarek, E. V. (1962). The use of fire: An historical background. *Proc. 1st Annu. Tall Timbers Fire Ecol. Conf.* pp. 7–10.

Komarek, E. V. (1963). Fire, research, and education. *Proc. 2nd Annu. Tall Timbers Fire Ecol. Conf.* pp. 181–187.

Komarek, E. V. (1967). Fire—and the ecology of man. *Proc. 6th Annu. Tall Timbers Fire Ecol. Conf.* pp. 143–170.

Komarek, E. V. (1968). Lightning and lightning fires as ecological forces. *Proc. 8th Annu. Tall Timbers Fire Ecol. Conf.* pp. 169–197.

Komarek, E. V. (1969). Fire and animal behavior. *Proc. 9th Annu. Tall Timbers Fire Ecol. Conf.* pp. 161–207.

Koplin, J. R. (1969). The numerical response of woodpeckers to insect prey in a subalpine forest in Colorado. *Condor* 71, 436–438.

Kourtz, P. (1967). Lightning behaviour and lightning fires in Canadian forests. *Can. Dep. Forest Rural Develop., Forest Branch Publ.* 1179, 1–33.

Krebs, C. J., and DeLong, K. T. (1965). A *Microtus* population with supplemental food. *J. Mammal.* 46, 566–573.

Krebs, C. J., Keller, B. L., and Myers, J. H. (1971). *Microtus* population densities and soil nutrients in southern Indiana grassland. *Ecology* 52, 660–663.

Krebs, C. J., Gaines, M. S., Keller, B. L., Myers, J. H., and Tamarin, R. H. (1973). Population cycles in small rodents. *Science* 179, 35–41.

Krebs, J. R. (1971). Territory and breeding density in the great tit, *Parus major* L. *Ecology* 52, 2–22.

Lack, D. (1937). The psychological factor in bird distribution. *Brit. Birds* 31, 130–136.

Lack, D. (1954). "The Natural Regulation of Animal Numbers." Oxford Univ. Press, London and New York.

Lack, D. (1966). "Population Studies of Birds." Oxford Univ. Press, London and New York.

Lack, D. (1970). Introduction. *Brit. Ecol. Soc. Symp.* 10, xiii–xx.

Larsen, J. A., and Lahey, J. F. (1958). Influence of weather upon a ruffed grouse population. *J. Wildl. Manage.* 22, 63–70.

Lauckhart, J. B. (1957). Animal cycles and food. *J. Wildl. Manage.* 21, 230–234.

Lawrence, G. E. (1966). Ecology of vertebrate animals in relation to chaparral fire in the Sierra Nevada foothills. *Ecology* 47, 278–291.

Lawrence, G., and H. Biswell. (1972). Effect of forest manipulation on deer habitat in giant sequoia. *J. Wildl. Manage.* **36**, 595–605.

Lawrence, W. H. (1954). Michigan beaver populations as influenced by fire and logging. Ph.D. Thesis, University of Michigan, Ann Arbor.

Lay, D. W. (1957). Browse quality and the effect of prescribed burning in southern pine forests. *J. Forest.* **55**, 342–347.

Leege, T. A. (1968). Prescribed burning for elk in northern Idaho. *Proc. 8th Annu. Tall Timbers Fire Ecol. Conf.* pp. 235–253.

Leege, T. A. (1969). Burning seral brush ranges for big game in northern Idaho. *Trans. N. Amer. Wildl. Natur. Resour. Conf.* **34**, 429–438.

Leege, T. A., and Hickey, W. O. (1971). Sprouting of northern Idaho shrubs after prescribed burning. *J. Wildl. Manage.* **35**, 508–515.

Lemon, P. C. (1968a). Effects of fire on an African plateau grassland. *Ecology* **49**, 316–322.

Lemon, P. C. (1968b). Fire and wildlife grazing on an African plateau. *Proc. 8th Annu. Tall Timbers Fire Ecol. Conf.* pp. 71–88.

Leopold, A. (1933). "Game Management." Scribner's, New York.

Leopold, A. S., and Darling, F. F. (1953a). Effects of land use on moose and caribou in Alaska. *Trans. N. Amer. Wildl. Conf.* **18**, 553–562.

Leopold, A. S., and Darling, F. F. (1953b). "Wildlife in Alaska." Ronald Press, New York.

LoBue, J., and Darnell, R. M. (1959). Effect of habitat disturbance on a small mammal population. *J. Mammal.* **40**, 425–436.

Lovat, L. (1911). Heather-burning. *In* "The Grouse in Health and in Disease, Being the Final Report of the Committee of Inquiry on Grouse Disease" (A. S. Leslie, ed.), Vols. 1 and 2, pp. 392–413. Smith, Elder & Co., London.

Lussenhop, J. F. (1971). Response of a prairie soil arthropod population to burning. Ph.D. Thesis, University of Wisconsin, Madison.

Lutz, H. J. (1956). Ecological effects of forest fires in the interior of Alaska. *U.S., Dep. Agr., Tech. Bull.* **1133**, 1–121.

Lyon, L. J. (1971). Vegetal development following prescribed burning of Douglas-fir in south-central Idaho. *U.S., Forest Serv., Intermt. Forest Range Exp. Sta., Res. Pap.* **INT-105**, 1–30.

MacArthur, R. H. (1958). Population ecology of some warblers of northeastern coniferous forests. *Ecology* **39**, 599–619.

MacArthur, R. H., MacArthur, J. W., and Preer, J. (1962). On bird species diversity. II. Prediction of bird census from habitat measurements. *Amer. Natur.* **96**, 167–174.

McCulloch, C. Y. (1969). Some effects of wildfire on deer habitat in pinyon-juniper woodland. *J. Wildl. Manage.* **33**, 778–784.

McFadyen, A. (1968). Measurement of climate in studies of soil and litter animals. *Brit. Ecol. Soc. Symp.* **8**, 59–67.

MacKay, J. R. (1966). Tundra and taiga. *In* "Future Environments of North America, Transformation of a Continent" (F. F. Darling and J. P. Milton, eds.), pp. 156–171. Natur. Hist. Press, Doubleday, New York.

MacKenzie, J. M. D. (1946). Some factors influencing woodland birds. *Quart. J. Forest.* **40**, 82–88.

Marcström, V. (1960). Studies on the physiological and ecological background to the reproduction of the capercaille (*Tetrao urogallus* Lin.). *Viltrevy* **2**, 1–85.

Martin, N. D. (1960). An analysis of bird populations in relation to forest succession in Algonquin Provincial Park, Ontario. *Ecology* 41, 126–140.

Martin, R. E. (1963). A basic approach to fire injury of tree stems. *Proc. 2nd Annu. Tall Timbers Fire Ecol. Conf.* pp. 151–162.

Martinka, R. R. (1972). Structural characteristics of blue grouse territories in south western Montana. *J. Wildl. Manage.* 36, 498–510.

Mayfield, H. (1960). "The Kirtland's Warbler," Bull. No. 40. Cranbrook Inst. Sci., Bloomfield Hills, Michigan.

Mayland, H. F. (1967). Nitrogen availability on fall-burned oak-mountain mahogany chaparral. *J. Range Manage.* 20, 33–35.

Michael, E. D., and Thornburgh, P. (1971). Immediate effects of hardwood removal and prescribed burning on bird populations. *Southwest. Natur.* 15, 359–370.

Miller, F. L., Broughton, E., and Land, E. M. (1972). Moose fatality resulting from overextension of range. *J. Wildl. Dis.* 8, 95–98.

Miller, G. R. (1964). The management of heather moors. *Advan. Sci.* 21, 163–169.

Miller, G. R. (1968). Evidence for selective feeding on fertilized plots by red grouse, hares, and rabbits. *J. Wildl. Manage.* 32, 849–853.

Miller, G. R., and Miles, J. (1970). Regeneration of heather (*Calluna vulgaris* (L.) Hull) at different ages and seasons in northeast scotland. *J. Appl. Ecol.* 7, 51–60.

Miller, G. R., Jenkins, D., and Watson, A. (1966). Heather performance and red grouse populations. I. Visual estimates of heather performance. *J. Appl. Ecol.* 3, 313–326.

Miller, G. R., Watson, A., and Jenkins, D. (1970). Responses of red grouse populations to experimental improvement of their food. *Brit. Ecol. Soc. Symp.* 10, 323–325.

Mohler, L. L., Wampole, J. H., and Fichter, E. (1951). Mule deer in Nebraska National Forest. *J. Wildl. Manage.* 15, 129–157.

Moss, R. (1967). Probable limiting nutrients in the main food of red grouse (*Lagopus lagopus scoticus*). *In* "Secondary Productivity of Terrestrial Ecosystems" (K. Petrusewicz, ed.), Vol. 1, pp. 369–379. Inst. Ecol., Polish Acad. Sci., Warsaw.

Moss, R. (1969). A comparison of red grouse (*Lagopus lagopus scoticus*) stocks with the production and nutritive value of heather (*Calluna vulgaris*). *J. Anim. Ecol.* 38, 103–112.

Moss, R. (1972). Food selection by red grouse (*Lagopus lagopus scoticus* (Lath.)) in relation to chemical composition. *J. Anim. Ecol.* 41, 411–428.

Moss, R., and Parkinson, J. A. (1972). The digestion of heather (*Calluna vulgaris*) by red grouse (*Lagopus lagopus scoticus*). *Brit. J. Nutr.* 27, 285–298.

Moss, R., Watson, A., Parr, R., and Glennie, W. (1971). Effects of dietary supplements of newly growing heather on the breeding of captive red grouse. *Brit. J. Nutr.* 25, 135–143.

Mossop, D. H. (1971). A relation between aggressive behaviour and population dynamics in blue grouse. M.Sc. Thesis, University of British Columbia, Vancouver.

Mount, A. B. (1964). The interdependence of the eucalypts and forest fires in southern Australia. *Aust. Forest.* 28, 166–172.

Mount, A. B. (1969). Eucalypt ecology as related to fire. *Proc. 9th Annu. Tall Timbers Fire Ecol. Conf.* pp. 75–108.

Mueller-Dombois, D. (1960). The Douglas-fir forest associations on Vancouver Island in their initial stages of secondary succession. Ph.D. Thesis. University of British Columbia, Vancouver.

Munro, J. A., and Cowan, I. McT. (1947). A review of the bird fauna of British Columbia. *Brit. Columbia Prov. Mus., Spec. Publ.* No. 2, pp. 1–285.

Mutch, R. W. (1970). Wildland fires and ecosystems—a hypothesis. *Ecology* **51**, 1046–1051.

Negus, N. C., and Pinter, A. J. (1966). Reproductive responses of *Microtus montanus* to plants and plant extracts in the diet. *J. Mammal.* **47**, 596–601.

Newton, I. (1970). Irruptions of crossbills in Europe. *Brit. Ecol. Soc. Symp.* **10**, 337–357.

Nieland, B. J. (1958). Forest and adjacent burn in the Tillamook burn area of northeastern Oregon. *Ecology* **39**, 660–671.

Niering, W. A., Goodwin, R. H., and Taylor, S. (1970). Prescribed burning in southern New England: Introduction to long-range studies. *Proc. 10th Annu. Tall Timbers Fire Ecol. Conf.* pp. 267–286.

Odum, E. P. (1971). "Fundamentals of Ecology," 3rd ed. Saunders, Philadelphia, Pennsylvania.

Orians, G. H. (1969). The number of bird species in some tropical forests. *Ecology* **50**, 783–801.

Palmgren, P. (1932). Zur biologie von *Regulus r. regulus* (I.) und *Parus atricapillus borealis* Selys. *Acta Zool. Fenn.* **14**, 1–113.

Paris, O. H. (1963). The ecology of *Armadillidium vulgare* (Isopoda: Oniscoidea) in California grassland: Food, enemies, and weather. *Ecol. Monogr.* **33**, 1–22.

Pearse, A. S. (1943). Effects of burning-over and raking-off litter on certain soil animals in the Duke Forest. *Amer. Midl. Natur.* **29**, 406–424.

Pendergast, B. A., and Boag, D. A. (1971). Nutritional aspects of the diet of spruce grouse in central Alberta. *Condor* **73**, 437–443.

Penfound, W. T. (1964). Effects of denudation on the productivity of grassland. *Ecology* **45**, 838–845.

Penfound, W. T. (1968). Influence of a wildfire in the Wichita Mountains wildlife refuge, Oklahoma. *Ecology* **49**, 1003–1006.

Picozzi, N. (1968). Grouse bags in relation to the management and geology of heather moors. *J. Appl. Ecol.* **5**, 483–488.

Pimlott, D. H. (1963). Influence of deer and moose on boreal forest vegetation in two areas of eastern Canada. *Trans. Congr. Int. Union Game Biol.* **6**, 105–116.

Pimlott, D. H., Shannon, J. A., and Kolenosky, G. B. (1969). The ecology of the timber wolf in Algonquin Provincial Park. *Ont. Dep. Lands Forest Res. Rep.* (*Wildl.*) **87**, 1–92.

Potter, L. D., and Moir, D. R. (1961). Phytosociological study of burned deciduous woods, Turtle Mountains, North Dakota. *Ecology* **42**, 468–480.

Pruitt, W. O., Jr. (1953). An analysis of some physical factors affecting the local distribution of the shorttail shrew (*Blarina brevicauda*) in the northern part of the lower peninsula of Michigan. *Univ. Mich., Mus. Zool., Misc. Publ.* **79**, 1–39.

Pruitt, W. O., Jr. (1959a). Microclimates and local distribution of small mammals on the George Reserve, Michigan. *Univ. Mich., Mus. Zool., Misc. Publ.* **109**, 1–27.

Pruitt, W. O., Jr. (1959b). Snow as a factor in the winter ecology of the barren ground caribou (*Rangifer arcticus*). *Arctic* **12**, 159–179.

Radvanyi, A. (1970). Small mammals and regeneration of white spruce forests in western Alberta. *Ecology* **51**, 1102–1105.

Redfield, J. A. (1972). Demography and genetics in colonizing populations of blue grouse (*Dendragapus obscurus*). Ph.D. Thesis, University of Alberta.

Redfield, J. A., Zwickel, F. C., and Bendell, J. F. (1970). Effects of fire on numbers of blue grouse. *Proc. 10th Annu. Tall Timbers Fire Ecol. Conf.* pp. 63–83.

Ritcey, R. W., and Edwards, R. Y. (1963). Grouse abundance and June temperatures in Wells Gray Park, British Columbia. *J. Wildl. Manage.* 27, 604–606.

Robinson, D. J. (1958). Forestry and wildlife relationships on Vancouver Island. *Forest. Chron.* 34, 31–36.

Robinson, W. L. (1960). Test of shelter requirements of penned white-tailed deer. *J. Wildl. Manage.* 24, 364–371.

Rosenzweig, M. L. (1973). Habitat selection by rodents. *Ecology* 54, 110–117.

Ross, B. A., Bray, J. R., and Marshall, W. H. (1970). Effects of long-term deer exclusion on a *Pinus resinosa* forest in north central Minnesota. *Ecology* 51, 1088–1093.

Rowe, J. S. (1970). Spruce and fire in Northwest Canada and Alaska. *Proc. 10th Annu. Tall Timbers Fire Ecol. Conf.* pp. 245–254.

Rusch, D. H., and Keith, L. B. (1971). Ruffed grouse–vegetation relationships in central Alberta. *J. Wildl. Manage.* 35, 417–429.

Salt, G. W. (1952). The relation of metabolism to climate and distribution in three finches of the genus *Carpodacus*. *Ecol. Monogr.* 22, 121–152.

Schultz, A. M. (1969). A study of an ecosystem: The Arctic tundra. *In* "The Ecosystem Concept in Natural Resource Management" (G. M. Van Dyne, ed.), pp. 77–93. Academic Press, New York.

Scott, J. D. (1971). Veld burning in Natal. *Proc. 11th Annu. Tall Timbers Fire Ecol. Conf.* pp. 33–51.

Scotter, G. W. (1971). Fire, vegetation, soil, and barren-ground caribou relations in northern Canada. *In* "Fire in the Northern Environment" (C. W. Slaughter, R. J. Barney, and G. M. Hansen, eds.), pp. 209–230. U.S. Pac. Northwest Forest Range Exp. Sta., Portland, Oregon.

Shafi, M. I., and Yarranton, G. A. (1973). Diversity, floristic richness, and species evenness during a secondary (post-fire) succession. *Ecology* 54, 897–902.

Shelford, V. E., and Yeatter, R. E. (1955). Some suggested relations of prairie chicken abundance to physical factors, especially rainfall and solar radiation. *J. Wildl. Manage.* 19, 233–242.

Siivonen, L. (1957). The problem of the short-term fluctuation in numbers of tetraonids in Europe. *Pap. Game Res.* 19, 1–44.

Sims, P. H., and Buckner, C. H. (1973). The effect of clear cutting and burning of *Pinus banksiana* forests on the populations of small mammals in southeastern Manitoba. *Amer. Midl. Natur.* 90, 228–231.

Smith, D. W. (1968). Surface fires in northern Ontario. *Proc. 8th Annu. Tall Timbers Fire Ecol. Conf.* pp. 41–54.

Smith, D. W. (1970). Concentrations of soil nutrients before and after fire. *Can. J. Soil Sci.* 50, 17–29.

Smith, M. C. (1968). Red squirrel responses to spruce cone failure in interior Alaska. *J. Wildl. Manage.* 32, 305–316.

Speirs, J. M. (1969). Birds of Ontario's coniferous forest region. *Can. Audubon* 31, 1–8.

Speirs, J. M. (1972). Birds of Ontario's deciduous forest region. *Ont. Natur.* 10, 27–31.

Spencer, D. L., and Hakala, J. B. (1964). Moose and fire in the Kenai. *Proc. 3rd Annu. Tall Timbers Fire Ecol. Conf.* pp. 11–33.

Spurr, S. H. (1964). "Forest Ecology." Ronald Press, New York.

Stelfox, J. G., and Taber, R. D. (1969). Big game in the northern Rocky Mountain coniferous forest. *In* "Coniferous Forests of the Northern Rocky Mountains" (R. D. Taber, ed.), pp. 197–222. Center for Natural Resources, Missoula, Montana.

Stoddard, H. L. (1931). "The Bobwhite Quail, its Habits, Preservation and Increase." Scribner's, New York.

Stoddard, H. L., Sr. (1962). Use of fire in pine forests and game lands of the deep Southeast. *Proc. 1st Annu. Tall Timbers Fire Ecol. Conf.* pp. 31–42.

Stoddard, H. L., Sr. (1963). Bird habitat and fire. *Proc. 2nd Annu. Tall Timbers Fire Ecol. Conf.* pp. 163–175.

Sunquist, M. E. (1967). Effects of fire on raccoon behavior. *J. Mammal.* 48, 673–674.

Svoboda, F. J., and Gullion, G. W. (1972). Preferential use of aspen by ruffed grouse in northern Minnesota. *J. Wildl. Manage.* 36, 1166–1180.

Swan, F. R., Jr. (1970). Post-fire response of four plant communities in south-central New York State. *Ecology* 51, 1074–1082.

Sykes, D. J. (1971). Effects of fire and fire control on soil and water relations in northern forests—a preliminary review. *In* "Fire in the Northern Environment" (C. W. Slaughter, R. J. Barney, and G. M. Hansen, eds.), pp. 37–44. U.S. Pac. Northwest Forest Range Exp. Sta., Portland, Oregon.

Taber, R. D., and Dasmann, R. F. (1957). The dynamics of three natural populations of the deer *Odocoileus hemionus hemionus*. Ecology 38, 233–246.

Taber, R. D., and Dasmann, R. F. (1958). The black-tailed deer of the chaparral. *Calif. Dep. Fish Game, Bull.* 8, 1–163.

Taylor, A. R. (1969). Lightning effects on the forest complex. *Proc. 9th Annu. Tall Timbers Fire Ecol. Conf.* pp. 127–150.

Terrill, H. V., and Crawford, B. T. (1946). Using den boxes to boost squirrel crop. *Missouri Conserv.* 7, 4–5.

Tester, J. R. (1965). Effects of a controlled burn on small mammals in a Minnesota oak-savanna. *Amer. Midl. Natur.* 74, 240–243.

Tester, J. R., and Marshall, W. H. (1961). A study of certain plant and animal interrelations on a native prairie in northwestern Minnesota. *Univ. Minn., Mus. Natur. Hist., Occas. Pap.* 8, 1–51.

Tevis, L., Jr. (1956). Effect of slash burn on forest mice. *J. Wildl. Manage.* 20, 405–409.

Thilenius, J. F. (1968). The *Quercus garryana* forests of the Willamette Valley, Oregon. *Ecology* 49, 1124–1133.

Thompson, D. Q., and Smith, R. H. (1970). The forest primeval in the Northeast—a great myth? *Proc. 10th Annu. Tall Timbers Fire Ecol. Conf.* pp. 255–265.

Tompa, F. S. (1964). Factors determining the numbers of song sparrows, *Melospiza melodia* (Wilson), on Mandarte Island, B.C., Canada. *Acta Zool. Fenn.* 109, 1–73.

Trevett, M. F. (1962). Nutrition and growth of the lowbush blueberry. *Maine, Agr. Exp. Sta., Bull.* 605, 1–151.

Udvardy, M. D. F. (1969a). Birds of the coniferous forest. *In* "Coniferous Forests of the Northern Rocky Mountains" (R. D. Taber, ed.), pp. 99–129. Center for Natural Resources, Missoula, Montana.

Udvardy, M. D. F. (1969b). "Dynamic Zoogeography with Special Reference to Land Animals." Van Nostrand-Reinhold, Princeton, New Jersey.

Ullrey, D. E., Youatt, W. G., Johnson, H. E., Fay, L. D., Schoepke, B. L., Magee, W. T., and Keahey, K. K. (1973). Calcium requirements of weaned white-tailed deer fawns. *J. Wildl. Manage.* 37, 187–194.

Van Wyk, P. (1971). Veld burning in Krueger National Park. *Proc. 11th Annu. Tall Timbers Fire Ecol. Conf.* pp. 9–31.

Vesey-Fitzgerald, D. (1971). Fire and animal impact on vegetation in Tanzania National Parks. *Proc. 11th Annu. Tall Timbers Fire Ecol. Conf.* pp. 297–317.

Vlug, M. (1972). The effects of logging and slash burning on soil *Acari* and *Collembola* in a coniferous forest near Maple Ridge, B.C. M.Sc. Thesis, Simon Fraser University, Burnaby, British Columbia.

Vogl, R. J. (1967). Controlled burning for wildlife in Wisconsin. *Proc. 6th Annu. Tall Timbers Fire Ecol. Conf.* pp. 47–96.

Vogl, R. J. (1969). One hundred and thirty years of plant succession in a southeastern Wisconsin lowland. *Ecology* 50, 248–255.

Vogl, R. J. (1970). Fire and plant succession. *In* "The Role of Fire in the Intermountain West," pp. 65–75. Intermt. Fire Res. Counc., Missoula, Montana.

Vogl, R. J. (1973). Effects of fire on the plants and animals of a Florida wetland. *Amer. Midl. Natur.* 89, 334–347.

von Haartman, L. (1956). Territory in the pied flycatcher (*Muscicapa hypoleuca*). *Ibis* 98, 460–475.

Wagle, R. F., and Kitchen, J. H., Jr. (1972). Influence of fire on soil nutrients in a ponderosa pine type. *Ecology* 53, 118–125.

Ward, P. (1968). Fire in relation to waterfowl habitat of the delta marshes. *Proc. 8th Annu. Tall Timbers Fire Ecol. Conf.* pp. 255–267.

Watson, A., ed. (1970). "Animal Populations in Relation to Their Food Resources," Brit. Ecol. Soc. Symp. No. 10. Blackwell, Oxford.

Watson, A., and Moss, R. (1970). Dominance, spacing behaviour and aggression in relation to population limitation in vertebrates. *Brit. Ecol. Soc. Symp.* 10, 167–220.

Watson, A., and Moss, R. (1972). A current model of population dynamics in red grouse. *Proc. Int. Ornithol. Congr., 15th, 1970* pp. 134–149.

Weeden, R. B. (1965). "Grouse and Ptarmigan in Alaska, Their Ecology and Management." Alaska Dept. Fish and Game, Juneau.

Wein, R. W., and Bliss, L. C. (1973). Changes in Arctic *Eriophorum* tussock communities following fire. *Ecology* 54, 845–852.

West, N. E. (1968). Rodent-influenced establishment of ponderosa pine and bitterbrush seedlings in central Oregon. *Ecology* 49, 1009–1011.

West, O. (1971). Fire, man and wildlife as interacting factors limiting the development of climax vegetation in Rhodesia. *Proc. 11th Annu. Tall Timbers Fire Ecol. Conf.* pp. 121–145.

Wing, L. (1951). "Practice of Wildlife Conservation." Wiley, New York.

Wood, A. J., Cowan, I. McT., and Nordan, H. C. (1962). Periodicity of growth in ungulates as shown by deer of the genus *Odocoileus*. *Can. J. Zool.* 40, 593–603.

Wynne-Edwards, V. C. (1970). Feedback from food resources to population regulation. *Brit. Ecol. Soc. Symp.* 10, 413–427.

Zikria, B. A., Weston, G. C., Chodoff, M., and Ferrer, J. M. (1972). Smoke and carbon monoxide poisoning in fire victims. *J. Trauma* 12, 641–645.

Zwickel, F. C. (1972). Removal and repopulation of blue grouse in an increasing population. *J. Wildl. Manage.* 36, 1141–1152.

Zwickel, F. C., and Bendell, J. F. (1967). Early mortality and the regulation of numbers in blue grouse. *Can. J. Zool.* 45, 817–851.

Zwickel, F. C., and Bendell, J. F. (1972). Blue grouse, habitat, and populations. *Proc. Int. Ornithol. Congr., 15th, 1970* pp. 150–169.

. 5 .

Effects of Fire on Grasslands

Richard J. Vogl

When the grass has been burnt by the fire of the steppe, it will grow anew in summer.   *Mongolian proverb*

I. Fire in the Grassland Environment

A. FIRE AND THE ORIGIN OF GRASSLAND SPECIES

Early grassland ecologists considered North American grasslands as "climatic climax" formations, ignoring the role of fire except to consider its possible detrimental effects (Daubenmire, 1968; Weaver and Albertson, 1956). Others, such as Harper (1911), Gleason (1913), Aldous (1934), Hanson (1938, 1939), Aikman (1955), Curtis (1959), Cooper (1961), Humphrey (1962), Komarek (1965), and Costello (1969), provided evidence that grassland fires were not unusual occurrences but a natural and integral part of most grassland environments prior to

139

Fig. 1. Grasslands, by their very nature, are conducive to the unimpeded spread of fires. Grasslands often occupy dry sites even in wet regions as does this leeward Hawaiian grassland. This open and level site is swept by descending trade winds, and the contiguous and continuous plant growth dies back to ground levels at least once a year.

the arrival of European man in North America (Fig. 1). Investigators in foreign countries have generally accepted fires as common to most grasslands, but sometimes considered them to conflict with man's interests (Batchelder and Hirt, 1966; Cooper, 1963; Glover, 1971; Phillips, 1965; Van Wyk, 1971; West, 1965; Wheater, 1971). Grasslands are defined here as areas dominated by herbaceous vegetation, particularly grasses or other monocots, and include those areas that support an open overstory of scattered shrubs or trees.

Sauer (1950) and Stewart (1951, 1955, 1956, 1963) considered fire the most important or controlling factor in the origin and development of grasslands, which they called a "fire grass climax." This is perhaps an overstatement, since monocots like grasses and the graminoid sedges and rushes probably evolved as opportunistic pioneers, responding at least as much, and possibly more, to climatic fluctuations than to fire (Stebbins, 1972). Extremes created by variations in rainfall and/or temperatures probably helped to promote the establishment and expansion of grassland herbaceous monocots at the expense of woody angiosperms that survived best under more stable environmental conditions. Grasslands commonly survive and sometimes even thrive under such extremes as strong winds, extensive dust storms, violent summer thunderstorms

Fig. 2. Some grasses occupy hydric sites existing primarily as emergent aquatics. The shallow water along this Florida lakeshore is dominated by *Panicum hemitomon*, which can burn during dry periods, with the fire stimulating seedstalk production and yields.

(often with hail), tornadoes, blizzards, and fires. Grassland precipitation and evaporation fluctuate to produce everything from desertlike conditions to flooded conditions. Some montane, coastal, and lowland grasslands (Figs. 2 and 3) appear to be ever wet, persistently shrouded in chilling mists, or drenched by daily rains.

Extreme and fluctuating temperatures occur commonly, and severe weather conditions are ordinary phenomena. Many temperate and montane grasslands are subject to bitterly cold winters in contrast to very hot summers, and many exhibit wide diurnal temperature fluctuations, with periodic frosts often punctuating growing seasons. Alpine and arctic grasslands, for example, are often underlain with permafrost, and their extremely short growing seasons are interrupted almost nightly by frost.

Growth limitations are also produced by the unusual soil-parent materials characteristic of many grasslands, which are often intensified as they react with climatic conditions. Grasses and graminoid monocots possess broad amplitudes of tolerance which have enabled them to dominate such diverse habitats as tidal flats, brackish marshes, saline sinks, freshwater shallows and marshes (Fig. 2), acidic to alkaline substrates, water-logged to xeric soils, stable bedrock to shifting sands, from the tropics to the tundra (Fig. 3), and from fertile valleys to sterile slopes.

Fig. 3. Grasslands frequently occupy alpine sites with deep soils. The summit regions of Haleakala Crater, East Maui (right) are dominated by *Deschampsia australis*. Heath-scrub occupies the broken terrain and ridges. The grasses (left) grow on these cold, wet, and windswept slopes in bunches or tussocks, with individual clumps often reaching 1 m in height.

Many of these environmental extremes hamper the growth of woody plants, thereby permitting the unrestricted establishment and growth of the better-adapted grassland species.

Individual grasslands are sometimes initiated by fire, which functions as a retrogressive agent setting back succession from shrub- or tree-dominated stages to pioneer conditions, but the grasses and associated graminoid monocots comprising these grasslands originated in the antiquity of the past in response to a number of interrelated environmental factors (Beetle, 1957; Stebbins, 1972). Fires probably did not become an important evolutionary force among grassland species until the establishment of large grasslands conducive to the free spread of repeated fires. Graminoid monocots were joined by dicot herbs from many families, including the Compositae, Leguminosae, Labiatae, Umbelliferae, Euphorbiaceae, and Scrophulariaceae. Those species that survived to reproduce despite repeated fires became dominants, and species possessing growth and/or reproductive fire dependencies emerged. Fire, then, is not considered to be directly and solely responsible for the origin of grassland species, but rather to have been a natural selective force in the subsequent development of most grassland species. Gleason (1913) aptly put this by stating that in order to have a prairie fire there must first be a prairie (the term "prairie" being a North American synonym for open grassland).

Fires may have also influenced natural selection since the heat generated might act as a mutagenic agent (Clements, 1920; Komarek, 1965; Vogl, 1967b), just as high temperatures have been demonstrated to do so experimentally. Grassland fires can occur during the growing season (Moore, 1972) when plants are still in flower. It is conceivable that plants not directly destroyed in such fires and still undergoing meiosis–gametogenesis could be affected by the heat, thereby inducing mutations and providing, along with other factors, genotypic variation for selection. This might have been a contributing factor in creation of the variety of species and ecotypes present in some grassland families (MacMillan, 1959). Fire-induced change and selection might be related, in some degree, to hybridization, polyploidy, apomixis, and other specialization present among grassland species—characteristics which have contributed to their superiority and versatility (Clewell, 1966a; Hanson, 1972; Johnson, 1972; Stebbins, 1972).

B. Grassland Origin and Maintenance

Considerable inquiry has been made as to the origins of various grasslands (Curtis, 1959; Daubenmire, 1968; Vesey-FitzGerald, 1963, 1971; Wedel, 1957, 1961; Wells, 1970). According to their origins, grasslands can be divided into anthropogenic types, derived or secondary grasslands, and true or natural grasslands.

Anthropogenic grasslands are areas dominated by grasses and other low growing herbs that originated as a result of man's activities, including the actions of domesticated and feral livestock. They usually exist in regions that normally support brush or forest without cutting and burning and are most common in the tropics and subtropics. Maintenance is usually provided by grazing and/or burning (Hill, 1971).

Derived grasslands are seral vegetation types that also occur in areas capable of supporting forest. They usually result from, and are maintained by, naturally occurring fires that often combine with climatic, edaphic, and biotic conditions to form more or less permanent grasslands. These and anthropogenic grasslands may be open, that is, free of trees and/or brush thickets or exist as savannas—grasslands with an open overstory of widely spaced trees. Sometimes parklands develop where discrete and sometimes dense groves or galleries of forest and/or brush are surrounded and separated by open grasslands. The arborescent components common to savannas and parklands vary with the region but usually consist of local or widespread fire-tolerant or fire-adapted pioneers. Members of the Leguminosae, a family with strong herbaceous representation in grasslands, comprise the woody overstory of many Old and New World savannas. Monocotyledonous palms are common in African, Central American, South American, Australian, and Asian savannas. Tree ferns and arborescent monocots are common to many areas. Northern Hemisphere savannas support species of *Quercus* and other hardwoods, as well as species of *Pinus* and other gymnosperms (Budowski, 1966).

Natural or true grasslands are open grasslands devoid of woody forest species that have originated primarily because climatic and edaphic conditions favored grassland species and adversely affected woody plants. True grasslands are often extensive and separated from other vegetation formations by grassland–forest, grassland–brush, or grassland–desert ecotones. Curtis (1959) suggested that true grasslands might develop along grassland–forest ecotones from the intense burning of old climax forests. These sites are usually occupied by fire-intolerant woody species incapable of recovering from fire, leaving postfire sites occupied exclusively by herbaceous plants (Vogl, 1969a).

There is general agreement that fire is usually an important factor in the development and continuance of most anthropogenic grasslands, derived grasslands, and savannas. But differences of opinion exist as to the role of fire in the origin and maintenance of true grasslands. These differences are generally related to past attempts to explain their existence by exclusively implicating or emphasizing individual factors (Curtis, 1959). There is little doubt that the origin and maintenance of true grasslands are related to a multiple of environmental factors reacting with the vegetation (Cowles, 1928). Fire is just one of these factors, which is linked to increasing aridity, droughts, warming trends, and thunderstorms. Many other factors which react either singly or collectively include the presence of certain mycorrhizal and other fungi, decomposition rates, the occurrence of heavy rains, frost, snow, hail, or high winds, allelopathic effects of plants, diseases, insect activities and infestations, predator–prey relationships, animal migrations, animal population buildups, grazing, browsing, girdling, trampling, wallowing, digging, and burrowing activities. Although fires are important in the maintenance of most grasslands, some of these and other factors can temporarily, and under the right circumstances, even permanently replace fire as the necessary disturbance, renewal, and maintenance agent.

C. IGNITION SOURCES

Ignition of grassland fires has been caused by early and modern man, lightning, spontaneous combustion, sparks from falling rock, and volcanic eruptions.

A widespread cause of grassland fires was man. In many tropical

grasslands man may have been the only important source of fires (Blydenstein, 1967; Budowski, 1966; Fosberg, 1960). With this easy-to-use
tool, the aborigine was able to create openings in the forest, convert
forest to savanna, and change forest and brush to open grasslands. Fire
became man's first great force with which he asserted widespread influence over a considerable portion of the face of the earth.

An important difference between pre-man and man was that *Homo
sapiens* acquired the control and use of fire (Eiseley, 1954). Man is
largely a grassland animal, having spent most of his existence in grasslands or grassland–forest ecotones, hunting, gathering grassland foods,
domesticating grassland animals, and cultivating grassland plants
(Komarek, 1965, 1967a; Wedel, 1961). Early man apparently found
grassy openings and grasslands, and the organisms they supported, more
productive and fitting to his needs than the forest, and he literally burned
down the forests and cleared brush to create, maintain, and expand
the more exploitable grasslands. Modern man is still largely dependent
upon domesticated grassland animals and crop plants developed in early
times, or the recently "improved" versions of the original progenitors.
Some human diet and nutrition experts even claim that a number of
modern sicknesses and ailments may be related to the recent substitution
of superrefined grains for the coarsely ground grains, whole cereals, bran,
and roughage that man's digestive system evolved with and still needs.

Reasons why primitive groups employed fire have been summarized
by Hough (1926), Sauer (1950), Stewart (1951), Day (1953), Bartlett
(1955), and Komarek (1967a). A commonly mentioned use of fire was
to drive game (Moore, 1972), but this seems questionable since many
grassland animals show little fear of fire (Beck and Vogl, 1972; Komarek,
1965, 1969; Vogl, 1967a, 1973) and are reluctant to be driven by it.
Big game is particularly difficult to be directionally herded by fire unless
a large number of beaters are present or burning is related to special
land features or traps. Game that is not harassed will just as often
pass back through a fire or remain until the fire passes, as they will
flee before oncoming flames (Vogl, 1967a). Hunting fires were more
often used to expose or flush out animals seeking cover in impenetrable
coverts or to destroy cover, thereby making the game more accessible
to the hunter (West, 1971). The aborigine probably used fire as do
some predators (Gillon, 1971), taking advantage of the increased animal
movements triggered by the fire, the smoke which helps to screen the
hunter and his scent, and the momentary confusion of animals confronted
with coverless terrain as the fire passes. Grassland rodents can be easily
captured, for example, by walking immediately behind a headfire and
collecting the animals as they scurry through the fire only to find their

old haunts and protective cover reduced to ash. Heat- and smoke-dazed insects such as grasshoppers, moving and concentrating before a fire, are also easy to gather (Gillon, 1971; Komarek, 1969). Fire was also used by the hunter to concentrate game by burning small areas that became more desirable for herbivores than the surrounding unburned grasslands (West, 1971). Burning widespread areas of grassland was conducted to maintain maximum productivity of hunting stocks and perpetuate the best hunting conditions.

Agricultural objectives were accomplished by burning to improve grazing for domestic stock and to clear areas for cultivation (Bartlett, 1955, 1956; Harris, 1972; West, 1971). Burning was also conducted to create fertilizing ash on fields, to selectively favor certain plant species over others, to stimulate flower, fruit, and seed production for subsequent harvest, to facilitate harvesting of certain crops (Heady, 1972), and to eliminate crop residues and weeds.

Additional reasons for grassland burning were for communication, to increase visibility and mobility, to reduce and repel insect pests, to minimize attack by enemies or predators, and to minimize the threat and intensity of uncontrolled and warfare fires (Biswell, 1972; Moore, 1972). In some regions, grassland fires were set, particularly where heavier fuels were unavailable, to warm transient natives during cold periods. Accidental fires also resulted from unattended and careless campfires (Moore, 1972).

Regardless of the reasons why man might have used fire and the role that fire and grasslands might have had in the origin and development of *Homo sapiens,* most ecologists agree that man-caused fires were a common grassland phenomenon that covered widespread areas (Curtis, 1956). These fires were often superimposed on natural fire occurrences, resulting in everything from subtle to profound vegetational changes because of these additional and often unusual and untimely fires.

Another source of fires in many grasslands is lightning. Lightning ignition is produced when cloud-to-ground discharges occur without precipitation, when lightning precedes precipitation, or when it produces fires beyond the range of the usually local thunderstorm showers (Budowski, 1966; Komarek, 1964, 1966, 1967b, 1968, 1971; Steenbergh, 1972; Uman, 1969; West, 1971). Lightning fires often persist despite rains, and grasslands thoroughly soaked by rain can often burn after just a few hours of drying winds and/or sun (Budowski, 1966). Rains followed by clearing conditions are often produced by the passage of cold fronts in the southeastern United States, for example, and are considered ideal for the controlled burning of pine savannas (H. L. Stoddart, personal communication). At these times, the seeds, rootstocks, and soils

Fig. 4. Prior to man's intervention, grassland fires could spread in an unimpeded manner, often traveling long distances before reaching natural firebreaks or being extinquished. These eastern Colorado sand hills are currently dominated by *Artemisia* shrubs as a result of sustained heavy grazing by livestock, which, in turn, has affected the natural spread of fires.

are wet and protected, and the fuel consumption is incomplete, leaving a protective covering of organic matter on the soils. Lower fire temperatures are maintained and the strong steady winds associated with the front tend to dissipate the heat horizontally thereby minimizing scorch to the overstory trees. Even in grasslands with high rainfall, fires can occur after just a few rainless hours or days (Vogl, 1969b).

A little lightning could go a long way in the past (Harper, 1911; West, 1971), since lightning fires could spread in an unimpeded manner. Then, most grasslands were uninterrupted by modern man's activities and structures, and fuels were continuous, contiguous, and conducive to freely spreading fires and the development of wide fire fronts (Fig. 4) (Curtis, 1959; Drummond, 1855; Gregg, 1954; Jackson, 1965; Rose Innes, 1971). Sometimes these fires gained enough force and size to generate fire–storm systems that ultimately produced violent rainstorms sometimes accompanied by lightning (Drummond, 1855; Gregg, 1954). Natives have been known to use fires for rainmaking in Rhodesia in this connection (West, 1971). Once large fires were started they often covered vast areas, even burning around fuel-free pockets (Wells, 1965) and noncombustible vegetation (Curtis, 1959; Gleason, 1913; Harper, 1911; Vogl, 1969a). These fires were not stopped until extinguished by weather changes, reaching bare ground created by a previous fire or intense grazing, or until reaching extensive natural firebreaks.

The importance of lightning-caused grassland fires has been largely ignored in the past (Komarek, 1968, 1971). Most lightning fires occur in grasslands without leaving physical evidence of their causes and were and still are often erroneously recorded as having been started by other or unknown causes (Vogl, 1967b, 1969b). Areas burned by lightning fires were reduced to insignificant amounts as widespread overgrazing, mowing, and plowing eliminated and interrupted fuels, while man-caused fires that accompanied the pioneer settlement of grasslands became abundant. Even when lightning was recognized as a source of grassland fires, the significance of these fires was generally ignored, or they were considered to be intrinsically detrimental (Weaver and Albertson, 1956).

Numerous studies have documented that lightning has been and still is a major cause of fires in most grasslands of the world (Batchelder and Hirt, 1966; Curtis, 1959; Ehrenfried, 1965; Granfelt, 1965; Hind, 1859; Komarek, 1971). These fires are infrequent, as in some arctic, montane, or tropical grasslands, or are common, as when rapidly moving frontal systems sweep across midcontinental plains leaving rashes of grassland fires in their swaths (Blevins and Marwitz, 1968; Wolfe, 1972). Regions subject to prolonged drought, drying winds, high temperatures, as well as to dry lightning storms burn more frequently and completely than those subject to wet weather, high humidities, cool temperatures, and thunderstorms normally accompanied by heavy rains. Regardless of their frequencies, lightning-caused fires are natural environmental factors that must be recognized to fully understand grassland ecology.

Another possible source of grassland fires is spontaneous combustion. This cause, like sparks from falling rocks (Henniker-Gotley, 1936), is seldom witnessed or documented, and therefore is rarely considered. Viosca (1931) witnessed fires occurring spontaneously in a Louisiana marsh under conditions that are undoubtedly duplicated in grasslands. Marshes, wet prairies, humid tropical grasslands, grasslands during wet periods, or other vegetation types contiguous with grasslands are characterized by wet soils, humid microclimates, and heavy vegetal accumulations which are often compacted, rotted, and fermented. These conditions are conducive to the biological and chemical heating necessary to produce spontaneous combustion. Until research is conducted on the conditions necessary to produce spontaneous ignition, using native plant materials in natural environments, this source of fires cannot be eliminated. Although conditions for spontaneous combustion may not occur often, it may be that when conditions are right, ignition occurs simultaneously in widely scattered locations within the vegetation type (Viosca, 1931). Investigators tend to consider environments only in terms of short time spans, so that seemingly exceptional happenings and rare events, like spontaneous combustion or lightning fires, appear to be so incidental that they do not warrant attention. In reality, they may be important in terms of the life spans of grasslands, or may be controlling or selective forces because they often occur during the growing season.

D. The Inseparables—Grasslands, Droughts, and Fires

Recorded accounts of early explorers, travelers, settlers, scientists, and historians include descriptions of encounters with lightning ignition, man's use of fire, grassland fires, burned grasslands, and related phenomena (Batchelder and Hirt, 1966; Costello, 1969; Curtis, 1959; Daubenmire, 1968; Drummond, 1855; Ehrenfried, 1965; Gleason, 1913; Gregg, 1954; Harper, 1911; Hind, 1859; Humphrey, 1962; Jackson, 1965; Lehmann, 1965; Malin, 1967; Moore, 1972; Vogl, 1964b; Weaver and Albertson, 1956; West, 1971). Most early witnesses to grassland fires or their effects appeared to accept them as rather commonplace, some even being aware of the ecological roles that these fires played.

The occurrence of fires is not considered to be any more coincidental than are the inherent features of the grassland environment (Odum, 1969). Fire has been an integral part of some grassland environments long enough to have resulted in the development of interesting plant–animal interactions (Hocking, 1964; Janzen, 1967). The elements of the fire triangle, ignition, oxygen, and fuel, relate to the grassland climate, topography, and plant materials in a way that makes fires almost inevitable (Wells, 1970).

Borchert (1950) stated that ". . . grassland climates favor fire, just as they favor grass whether there are fires or not." Grassland climates facilitate fires by the occurrence of dormant periods, dry seasons, and periodic droughts (Coupland, 1958; Lapham, 1965; Malin, 1967). True grasslands often dominate areas with semiarid climates or dry seasons, and large grasslands often exist in rain shadows subject to descending and drying winds which produce low humidities, abundant sunshine, high temperatures, transpiration loss, and evaporative stress. Dry periods are often interrupted or terminated by lightning storms, with ignition times coinciding with the driest conditions (Lehmann, 1965). This occurs consistently enough in some grasslands so that lightning is the most common cause of fires (Gartner, 1972; Granfelt, 1965; Komarek, 1964). Even in those tropical and subtropical grasslands where lightning storms and droughts are rare, it appears that when meteorological conditions are finally conducive to droughts, conditions also tend to favor lightning storms (Vogl, 1969b).

Grasslands usually occupy broad and unbroken level plains or rolling hills that lend themselves well to the free spread of surface fires (Figs. 1 and 4) (Wells, 1970). The low growth forms of the plants contribute to the open nature of grasslands, and with the large areas of continuous vegetation, allow for the unrestricted flow of winds. Winds are caused by differences in temperatures and/or pressures, which are often generated or intensified by the juxtaposition of mountains on plains and of forests on grasslands. Even small grassland pockets have greater air circulation than surrounding forest. Prairies are windy places, then, and the winds not only provide oxygen for grass fires but drive headfires at alarming speeds, even with heavy fuel moisture or after nightfall when fires in other vegetation types die down or go out. Great Plains grasslands are the windiest in the spring when the vegetation is dormant and usually very dry (Coupland, 1958; Vogl, 1964b). Wind movement has also been found to be higher in dry years (Albertson and Weaver, 1942). Winds also tend to dry grassland fuels. Grasslands occurring on light soil types with poor water-retaining capacities are particularly susceptible to drying by transpiration and evaporation. Drying is often aided by sunlight penetration, which can reach ground levels as the winds bend, whip, and separate the plant cover.

Grassland fuels burn readily. Most grassland plants are surface deciduous hemicryptophytes, with the aboveground portions dying back at least once a year (Blydenstein, 1968), even in regions with seasonless climates. As a result, grasslands are particularly vulnerable to fires as standing plants dry and cure to ground level (Broido and Nelson, 1964; Dawkins, 1939; Schroeder and Buck, 1970; Wells, 1970). Most grassland species are xerophytically adapted, often with stiff, scabrous leaves and rigid stems whose structure may be aided by high silica content (Komarek, 1965). Many grasses and associated herbs also orient their leaves, minimizing exposure to sun and air and thus reducing transpiration. Coriaceous xerophytic species usually possess low moisture content even before drying. Shoots produced after a fire have also been found to be stiffer and more erect than ordinary shoots (McCalla, 1943; O'Connor and Powell, 1963). Their rigid and erect nature and behavioral adaptations not only help to keep stem and leaves upright even after growth terminates but also expose the grassland understory and soils to sun and wind. This results in ideal combustion conditions (very dry, loosely arranged,

Fig. 5. Seedstalks of grasses on postburn sites are more persistent and rigid than those of unburned plants. *Andropogon virginicus* (left) produced in Georgia during the summer growing season still persists the following May despite heavy winter rains. *Andropogon gerardi* (right) still stands in April after a northwestern Wisconsin winter with snow. Cured, dried, standing, and loosely arranged grasses are highly flammable fuels, with one fire helping to create conditions that lead to another.

and finely divided fuels with numerous air spaces) that permit further drying, as well as abundant oxygen for burning. Compaction of grassland fuels seldom reaches the degree attained by heavier forest fuels, even after heavy snows, rains, or inundation (Fig. 5) (Aikman, 1955; Vogl, 1973).

Because of these characteristics, conditions are not usually conducive to the rapid decomposition of plant materials by bacteria, fungi, and soil invertebrates. Therefore, grassland plant debris often accumulates faster than it decomposes, with variations in decomposition rates being largely determined by temperatures, amount of rainfall, and the moisture present in the litter. Accumulation not only results from slow breakdown of plant materials but from the rapid and prodigious growth characteristic of many grassland plants, with entire plant tops being added to the litter layer at the end of each growing season rather than becoming a functioning increment as in woody vegetation types.

The chemical structure of many species (Philpot, 1968) also contributes to this decay resistance. The persistent nature or decay resistance of grasses and other monocots has long been recognized in their desirability as materials for building shelters (Vogl, 1969b), mats (Vogl, 1964b), boats, baskets, and other items. Interestingly, straw, grasses, and sedges grown on burned sites are more resistant to decay when used as thatching for European roofs than comparable material collected from unburned sites (E. V. Komarek, personal communication). Burning may actually produce decay resistance in the stems and leaves of some plants, or it might be that the resistance is a result of the increased proportion of seedstalks produced with burning, which are more resistant to decay than stems and leaves (McCalla, 1943). Flower stalks of a number of grassland monocots and dicots are not only slow to decay, but after fires often remain unburned and standing erect, or only burned off at the base with intense fires. The relative fire resistance of seedstalks of some grassland grasses and forbs may have evolved as a fire adaptation that ensures survival of seeds and may be related to mineral, ash (Broido and Nelson, 1964; Philpot, 1970, 1971), and/or silica content (Komarek, 1965).

The rapid growth and accumulation, the usually slow decomposition rates, the chemical and physical composition of grassland plants, and the highly flammable nature of the plant debris point to a vegetation type that can readily burn. Grasslands that can be readily and repeatedly burned have apparently evolved with fire (Mutch, 1970), becoming dependent upon it as the primary decomposition agent and key nutrient recycler. At the same time, the grassland plants create conditions that make the necessary fires almost inevitable.

II. Physical Characteristics of Grassland Fires

The physical nature of grassland fires is simple in comparison to fires in more stratified vegetation types. Rapidly moving headfires consume most of the vegetation and often develop broad fronts because of the extensive and unbroken terrain and the continuous nature of the fuels of many grasslands. These fire fronts tend to become irregular in outline as topography, fuel loads, winds, and developing fire storms speed up or retard fire movements (Daubenmire, 1968). Headfires in dense fuels and tall grasslands often generate large flames, but sparks or embers that cause fires to spot ahead of the main front are generally not produced because most grassland fuels are consumed too quickly and thoroughly. Fires spotting ahead of fronts are more characteristic of savannas where volatile woody plant elements contribute long-lasting embers that are readily transported by winds (Wright, 1972). Fire fronts often become so uneven that flank fires burning at right angles to the wind develop and spread laterally until meeting other portions of the fire. Sometimes sections of a fire move so far ahead that backfires develop as they eat their way against the wind into unburned fuels. Because of the uncertain path that a fire may take and the capricious nature of winds and large fires, sections of vegetation often escape burning. Patches of burned and unburned grassland are particularly characteristic of areas burned under waning climatic conditions that often lead to the fire's suppression.

Grassland fire behavior is not only affected by the prevailing winds at the time of burning (Byram, 1948) but is also influenced by the direction and angle of slope, local winds, drafts, and convectional movements created as fires build, near each other, and merge. These factors often result in the temporary intensification of fires, the development of firewhirls or whirlwinds, and sometimes in the production of fire storms (Schroeder and Buck, 1970). Large grassland fires have been known to generate extensive convection columns that can result in the development of cumulus clouds that sometimes build to thunderstorm dimensions, generating lightning, thunder, and rain.

Despite the extensive fronts, roaring speeds, and ominous nature that grassland fires can assume, extremely high temperature at the plant and ground levels, complete consumption of all fuels, and damage to basal portions of plants and rootstocks are uncommon (Fig. 6). Dry grassland fuels tend to be flashy, igniting readily and burning quickly (Fig. 7). In addition, grassland fires generally produce a narrow belt of flames and pass rapidly because of the low growth, open terrain, level to gently rolling topography, and the general presence of winds. As a result, fires seldom tarry long enough to build high surface temperatures. The highest temperatures are usually produced well above the ground at the apex of the flames and above, and are rapidly dissipated by winds. Grassland fuels are normally converted to ash so rapidly that one can follow immediately behind most fires without experiencing any discomfort from heat, smoke, prolonged burning, or hot ashes, and the soils are generally cool to the touch (McArthur, 1966; Morton, 1964). Higher temperatures are usually produced at ground levels by slow moving fires (Davis and Martin, 1960). Thus, the slower moving backfires tend to generate more heat at these levels than headfires (Byram, 1958; Heyward, 1938; Iwanami and Iizumi, 1966; Lindenmuth and Byram, 1948). Woody fuels and large accumulations of grassland litter take longer to consume and therefore produce higher temperatures (Conrad and Poulton, 1966; Hare, 1961; Hopkins *et al.*, 1948; Stinson and

Fig. 6. Headfires in grassland fuels, as in this savanna, burn rapidly and pass quickly usually leaving behind uninjured seeds and rootstocks and stimulated plant meristems, as well as a protective layer of unburned litter and organic matter on the soil surface.

Wright, 1969; Wright, 1971). Hot fires in brush and tree fuels in savannas often operate to the detriment of the woody species that produced them. Hot spots created by persistent fires are uncommon except where animals have amassed plant materials, or unusual accumulations of fuel have resulted from wind, water, fire protection, or other actions. A review of temperatures recorded in grass fires was presented by Daubenmire (1968). Considerable time has been spent developing techniques and instrumentation to record fire temperatures. Efforts have also been made to determine the amounts of heat needed to damage and kill plants (Daubenmire, 1968; Hare, 1961). Many of the recorded temperatures are not comparable, however, and are, at best, relative temperatures that usually have not been positively related to the organisms in question. More meaningful results might be obtained by using organisms directly to evaluate the effects of heat, or by using the ash or other biological indicators to determine the amounts of heat generated and the damage produced.

    Ground fires that burn the roots and/or substrate occasionally occur, particularly in wet and mesic grasslands during prolonged droughts that lower water tables (Vogl, 1969a), thereby exposing dried layers of or-

Fig. 7. Grassland fuels (upper) tend to be flashy, igniting readily and burning quickly and thoroughly, even when fuel moisture levels are high or when fires in other fuel types would go out. Shoot meristems in this grass (lower) exist at ground level and thereby survive fire damage. The new postburn growth is usually more vigorous and productive.

ganic matter and peat to fires. Ground fires serve as retrogressive agents as they consume soils, sometimes down to mineral substrates, creating depressions that often fill with water and become, or revert to, prairie potholes or grassland ponds. Light soil types that are dry conduct less heat then heavy soils although surface temperatures may be higher.

Heat conduction tends to increase with the soil moisture content (Heyward, 1938). Under dry conditions the rhizomes of ferns (Vogl, 1969b), and possibly the root systems of other grassland plants, occasionally facilitate the subterranean spread of fires. Ground fires usually generate extremely high temperatures and not only destroy those plants contacted but tend to be persistent and serve as reignition sources when conditions are again conducive to the spread of surface fires. Under normal conditions, however, grassland fires usually do not create temperatures that are destructive or lethal to living plants.

The kinds and amounts of smoke generated vary with general moisture conditions at the time of burning, the stage of growth of the plants, and the species present. Arid grasslands burned during the dry season will often produce little observable smoke, dormant grasslands with relatively high moisture content produce billowy white smoke, and green grasslands tend to burn with roily gray to black smoke and usually require higher winds to sustain fires than do those with drier fuels (McArthur, 1966). The smoke of large fires is known to reach great heights and long distances and will often carry fly ash with it.

The ash produced in grassland fires is usually gray to black, light in weight, and fine textured, and occurs in negligible quantities when compared to that produced in brush and forest fires (Fig. 8) (Daubenmire, 1968). The finer fuels are often feathery white immediately after burning but quickly disintegrate. The light ash is easily airborne, a feature that annoyed early grassland travelers (Gregg, 1954; Malin, 1967). The burned surfaces warm readily and generate whirlwinds that lift ash into the air. Dry and windy conditions following large fires produce dust storms that sometimes carry ash for considerable distances. Despite occasional aeolian erosion, most ash quickly settles, joining charred and unburned bits of vegetation on the soil to create a general darkened surface (Fig. 8). The blackened soil surface is known to absorb solar heat, thereby causing higher temperatures than are produced on comparable unburned soils (Daubenmire, 1968). Some postburn soil surfaces may actually reach higher temperatures from direct insolation on the exposed and darkened surfaces than they did during the fire. As the new plant cover begins to develop, the blackened surfaces deteriorate. These changes might create favorable growing conditions by first promoting warm soils that stimulate speedy seed germination, sprouting, and growth, particularly in cold seasons or climates (Aikman, 1955; Curtis, 1959; Ehrenreich and Aikman, 1963; Hadley and Kieckhefer, 1963; Kelting, 1957). Once growth has been initiated and the growing season progresses, the breakdown of the dark ash layer reduces temperatures that could cause evaporation–transpiration stress to plants and soils.

Fig. 8. The soil surface (upper) immediately after the passage of a grassland fire is cool to the touch and covered with a thin layer of feathery white ash. The basal portions of the grasses are still intact. The fine ash soon settles (lower) and combines with charcoal and charred plant remains to produce a darkened soil surface. The black surface absorbs more insolation than unburned surfaces and promotes earlier growth.

III. Effects of Fire on Productivity

Burning has been found generally to increase the production of most grassland vegetation, but occasionally it is ineffective and sometimes even deleterious to individual species. Reactions to fire vary with the grassland type, fuels, soils, moisture conditions, fire frequencies, and burning times. Grasslands with optimum growing conditions are more consistently stimulated by fire than those occupying marginal sites, particularly areas with critical moisture conditions (Reynolds and Bohning, 1956; Staples, 1945; Trlica and Schuster, 1969; West, 1965). Yields have been found to be reduced by burning marginal grasslands too frequently, at the wrong season, after severe abuse, during periods of critical soil moisture (Jackson, 1965), or when exceptionally hot fires damage plants (Hopkins et al., 1948; Launchbaugh, 1964).

The higher yields of the aboveground portions of plants, which can be several times greater than from comparable unburned areas and which are related to corresponding increases in root systems (Kucera, 1970), are associated with a number of factors. The removal by fire of the plant tops triggers latent primordial regions to initiate new growth. Growth is produced, sometimes very rapidly (Hopkins, 1963; Lewis, 1964), often regardless of the soil moisture content and the occurrence of precipitation (Figs. 7 and 8). Increased numbers of grass and forb flowers and seeds are usually stimulated by burning (Fig. 5) (Biswell and Lemon, 1943; Curtis and Partch, 1950; Ehrenreich and Aikman, 1963; Lemon, 1949; Lloyd, 1972; Parrott, 1967). A summary of species that produced increased and decreased numbers of inflorescences with burning is given by Daubenmire (1968). Flower, fruit, and/or seed production are also stimulated in some woody plants present, although this response may not occur until after the first growing season. The vegetative reproduction of perennial species on most postburn sites occurs more rapidly and vigorously than growth on unburned sites (Duvall, 1962; Hadley, 1970; Old, 1969; Ralston and Dix, 1966; Vogl, 1965; Wright, 1969).

The fire removal of the litter permits the development of denser growth. Otherwise these accumulations, which are often abundant on optimum sites, produce a dominating mantle that stifles growth or physically impairs vigorous growth by depriving plants of space and light (Fig. 9). Chemical substances leached from undecomposed plant remains may further inhibit growth. Complete consumption of all litter sometimes occurs, however, and may lead to impaired growth, particularly during periods of critical moisture and when soils are exposed to excessive weathering. The conversion of the litter and standing crop to a blackened layer of ash and charcoal is considered to stimulate earlier growth by creating warmer surface temperatures (Fig. 8), particularly following winter or spring burning in temperate regions (Ehrenreich and Aikman, 1963; Lloyd, 1972; Penfound and Kelting, 1950). The earlier growth increases annual yields by effectively extending

Fig. 9. Tall-grass species, whether they grow in temperate (left) or in subtropical grasslands (right), produce growth that usually accumulates faster than it decomposes. Litter buildups tend to physically impair vigorous and abundant growth. Fire removes these accumulations and releases the nutrients.

growing seasons. Contrarily, the standing plant remains create shade and the unburned litter serves as an insulation which keeps the ground cooler until later in the growing season, thus shortening it (Weaver and Rowland, 1952). The activities of soil organisms, as well as chemical reaction rates, are also increased with the higher soil temperatures.

The blanket of dead and decaying vegetation contains nutrients that are largely unavailable for growth until released slowly and incompletely by decay, or rapidly and more completely by fire. There is little evidence as to the fertilizing effects of the ash increment on increased growth (Daubenmire, 1968; Lloyd, 1972), but nutrients in this form are considered to be more readily available. The effects of ash can be assessed by examining the effects of fire on grassland soils or by comparing the nutrient contents of plants growing on burned and unburned sites.

Soils in burned grasslands usually have slightly higher pH values due to the release of alkaline earth metals (Baldanzi, 1961; Cook, 1939; Daubenmire, 1968; Ehrenreich and Aikman, 1963; Garren, 1943; Moore,

1961). In neutral or alkaline soils these pH increases appear to have little effect, except that they may alter microbial activities. In acidic grasslands, pH increases often temporarily improve germination and growth conditions and may accelerate succession (Vogl, 1969a). There is usually an increase in fertility and/or organic matter in the soils of burned grasslands (Baldanzi, 1961; Daubenmire, 1968; Heyward and Barnette, 1934), except on steep sites subject to heavy runoff, sites subject to postburn adversities, or areas swept by severe fires.

There are generally no direct losses of soil nutrients except for the volatilization of nitrogen and sulfur. The nitrogen losses are often recovered through precipitation and the increased actions of nitrogen-fixing plants, particularly legumes (Burton, 1972), soil algae, bacteria, and certain fungi (Cohen, 1950; Orpurt and Curtis, 1957; Stewart, 1967). The activities of these organisms on burned sites often result in more available nitrogen than comparable unburned areas (Daubenmire, 1968). Mineral salts of Ca, P, K, and Mg often increase with burning (Metz et al., 1961). These salts are vulnerable to lateral movements by wind and water, and since they are water soluble, they are readily taken up by plants and soil organisms or leached to lower levels. The temporal and transient nature of the nutrient elements makes them difficult to measure with conventional methods, and along with inherent soil variability, might account for some of the soil study discrepancies discussed by Daubenmire (1968). Under normal circumstances, fires do not appear to affect grassland soils adversely (Lloyd, 1971) and generally appear to improve them (Hole and Watterston, 1972).

Soil erosion is not a major concern in most grasslands unless adverse environmental conditions follow fire. Even then, the remaining basal crowns, fibrous and extensive root systems, ash, charcoal, and unconsumed litter usually protect the soils from severe and massive erosion. Fortunately, grasslands commonly occupy level sites that minimize water-caused erosion. Erosion is most evident in bunch or tussock grasslands occupying steep slopes in regions subject to heavy rains (Dawkins, 1939). Most severe erosion in grasslands can be traced to causes other than fire, such as excessive rodent activities, heavy grazing, trampling, rooting, and disturbances and compaction by machinery. More detailed information of the effects of fire on soils is to be found in Chapter 2 of this work.

Plants recovering from fire often reflect improved soil conditions by their healthy green colors, larger sizes, and higher water content (Aldous, 1934; Cook, 1939; Halls, 1952; Mes, 1958; Vogl, 1965). The nutrient content of grassland plants on burned sites tends to be higher (Daubenmire, 1968) or relatively unchanged (Lloyd, 1971), and Nielsen and

Hole (1963) even noted symptoms of nitrogen and phosphorus deficiencies in unburned grassland plants. Postburn plants are preferred by herbivores (Vogl and Beck, 1970), who seek out burned sites and unerringly select the more palatable, and apparently more nutritious forage (Smith *et al.*, 1960; Vogl, 1973). Animals grazing burned grasslands have been found to gain weight more rapidly (Anderson, 1964; Greene, 1935; Hilmon and Hughes, 1965; Southwell and Hughes, 1965; Wahlenberg *et al.*, 1939) than those grazing unburned grasslands, and burned grasslands can generate higher animal densities (Vogl, 1973; Vogl and Beck, 1970). Big game in Africa may prefer the emerging growth of postburn sites, not only because it is more palatable but because the regrowth offers improved visibility and a better chance of detection and escape from predators than the tall grasses of unburned sites. In addition, there are reduced numbers of ticks, mites, and flies in the open burned sites (Hill, 1971; J. Phillips, personal communication; Van Rensburg, 1971).

The time it takes the standing crop and litter on a burned site to return to preburn conditions has been found to vary from 3 to 6 years on tall grass prairies (Dix, 1960; Ehrenreich and Aikman, 1963; Hadley and Kieckhefer, 1963; Tester and Marshall, 1961; Vogl, 1965). Recovery times are probably longer in marginal site grasslands and in areas swept by severe fires. Other fire effects, such as altered vegetational composition, particularly in savannas (Vogl, 1964b), and animal utilization (Vogl and Beck, 1970) are longer lasting. Studies that emphasize the lengths of time it takes grasslands to revert to preburn conditions often consider unburned grasslands as normal, thereby implying that burning is an unusual event without considering the natural fire frequencies. It may be that in many regions burned grasslands are closer to normal, or at least they were until modern man entered the scene.

IV. Effects of Fire on Vegetational Composition

The presence of recurring disturbances such as fire favors grasslands, savannas, or parklands in regions with climates capable of supporting brush or forest. Repeated fires generally promote grasses at the expense of woody species (Glover, 1971; Hare, 1961; Robertson and Cords, 1957; Scott, 1971; Trollope, 1971; Van Rensburg, 1971), although a number of woody plants are extremely fire tolerant and even fire dependent. If an area is subject to high fire frequencies, the chances of it being occupied by grassland as opposed to brushland, chaparral, or forest are influenced by burning times, fire intensities, climate, soils, biotic factors, and other factors. With few exceptions, fire-adapted woody species that can withstand fire or recover from it cannot continue to survive in large numbers in grassland areas that support intense fires on a very frequent, annual, or repeated basis (Glover, 1971).

Woody plants have difficulty invading established grasslands, particularly if the grassland is healthy and subject to recurring fires (Lemon, 1970; Pearson, 1936; Wilde, 1958). The fires maintain vigorous herbaceous growth which successfully competes with the woody species for space, moisture, and light (Blydenstein, 1967). Fire also injures or kills most living woody plant tops while generally leaving the living portions of grassland species undamaged. In regions capable of supporting both grassland and forest, the number of woody elements is often related to fire frequencies and intensities, with the most frequent and intense fires resulting in the fewest woody species (Olindo, 1971).

In forest communities, climax or subclimax stages are considered to be more diverse and stable than the pioneer stages of plant succession (Loucks, 1970; McIntosh, 1967). Fires in forests generally act as retrogressive agents, returning succession to earlier and less stable conditions (Vogl, 1970a). As a result, repeated burning tends to simplify species composition and stand structure, often producing monotypes of a fire-resistant tree or shrub. Grassland burning may reduce the number of woody species present, but this is usually offset by a corresponding increase in herbs. Repeated burning in native grassland communities generally does not reduce the species diversity and may even increase it by promoting growth of additional grasses, legumes, and other forbs, including annual plants. Although studies have not concentrated on vegetational changes in the same grassland with repeated burning, a quasi-equilibrium is probably reached after a certain number of fires whereby the species composition remains fairly constant (McMurphy, 1963), with fire primarily affecting the number of individuals per species. The increases in the number of grass stems per plant and the number of grass plants per area with burning have been well documented for many species (Biswell and Lemon, 1943; Burton, 1944; Curtis and Partch, 1950; Czuhai and Cushwa, 1968; Dix and Butler, 1954; Kucera, 1970; Old, 1969; Ralston and Dix, 1966; Vogl, 1965). Increases in leguminous species and densities also often occur with burning (Clewell, 1966b; Cushwa *et al.*, 1966, 1968, 1970; Hilmon and Hughes, 1965; Hodgkins, 1958; Lemon, 1967, 1970; Martin and Cushwa, 1966). Sometimes fires favor forbs over grasses but often promote the reverse (Daubenmire, 1968; Kucera and Koelling, 1964; Wright, 1969).

Fires may create monotypes in grasslands or marshes that are already very low in species because of extreme conditions. Fires under these circumstances may stimulate vegetative reproduction of the prevailing dominants to such an extent that they physically compete with and ultimately eliminate any incidental species present. Dense stands of grasses, particularly sod-forming and rhizomatous tall grasses in wet habitats and species belonging to such monocotyledonous families as the Cyperaceae, Typhaceae, and Juncaceae, may be aided in forming pure stands by burning.

Grassland fires sometimes create disturbed sites or pioneer conditions that permit invasion by certain opportunistic species such as annuals or short-lived perennials and "weedy" natives or aggressive alien species. Because of disturbance by fire (Stewart, 1956) and other agents, grassland diversity is commonly assured by heterogeneous mixtures of invaders, opportunistic pioneers, annuals, short-lived, long-lived, and stable perennials (Lemon, 1949; Quinnild and Cosby, 1958; Ramsay and Rose Innes 1963). Conversely, grasslands free from disturbance soon decline in species numbers.

Although the effects of fire on annual grassland species have been complicated and confused by studies on nonnative species in overgrazed and abused native perennial grasslands (Daubenmire, 1968), a few generalizations can be made. Native annuals are usually encouraged by burning, provided the fires occur at the appropriate times. Seed production, germination, and seedling establishment of annuals as well as perennial species are generally promoted by fire (Curtis and Partch, 1948; Cushwa *et al.*, 1968; Ehrenreich and Aikman, 1957; Lloyd, 1972; Mark, 1965; Shaw, 1957; Van Rensburg, 1971). Heat treatment of seeds has been found to increase the germination rates of some species (Capon and Van Asdall, 1967; Martin and Cushwa, 1966; West, 1965). Included among the annuals are a number of "phoenix" plants, species that usually appear after a fire, since germination and/or establishment is restricted to postburn sites. Most annual grasses and forbs are pioneers requiring open soils and full sunlight, conditions common on postburn sites.

If an area is burned after the annual plants have started growth, burning is detrimental and if repeated can eliminate the annuals. Some annuals like the California poppy (*Eschscholtzia* spp.) can withstand repeated top removal by fire or other agents up to the time of floral initiation, surviving for several years with vegetative regrowth until the plants can terminate with flower and seed production. But most annuals cannot survive fire once growth is initiated, particularly those whose germination is triggered by factors other than fire. Fire-stimulated annuals are seldom threatened by fires before setting seeds because of the reduced fuels. H. Wright (personal communication) maintained that spring burning is detrimental to Texas annual grassland species triggered by winter rains, since these plants are destroyed by fire before setting seed, whereas the perennial species survive as they are still dormant when burned or resprout after being burned. Burning, then, does not necessarily favor perennial over annual species, unless the fires occur after the annuals commence growth. Seeds of native annuals are probably seldom destroyed by the heat or flame of grassland fires, and conditions for germination and seedling establishment are often created or enhanced by fire.

Many perennial grasses and forbs are capable of vegetative reproduction. This trait aids in the colonization of new or open areas, as do pollination and seed dissemination by wind which are also typical of grassland species (Vogl, 1969b). Vegetative reproduction gives competitive advantage and helps species to survive catastrophes, including fire. Aggregated species patterns typify many grasslands: clones, clumps, and stands of individual species formed by vegetative spread (Fig. 10). Vegetative spread may be, in turn, aided allelopathically (Muller, 1966; Rice, 1967; Wilson, 1970), and inhibitory effects of plants and litter may also be regulated by fire. The environmental extremes common to many grasslands, including those created by fires, were probably selective forces in favoring perennials with vegetative habits. Additional selection was applied as fires caused repeated defoliation and top removal, leaving behind mainly grassland species with vulnerable meristems or perennating buds at ground level or below that could survive fires and produce new shoots (Fig. 7).

These generalizations were not presented to minimize the different results often obtained under a variety of grassland conditions. Real differences do exist in the reactions of various grassland types and grassland species to fire. For example, Palouse grasslands react differently than short-grass prairies (Daubenmire, 1968), sandveld reacts differently than waterlogged "vleis" (Kennan, 1971), and sourveld reacts differently than thornveld (Scott, 1971). Bunch grass reactions contrast with those of sod grasses, with upland sod-forming species varying from rhizomatous swamp species (Van Rensburg, 1971), and cool-season grasses respond differently than warm-season grasses. The time of burning and the frequencies of fire can be so critical in some grasslands that the results can be either beneficial or detrimental.

V. Grassland Succession

Pioneer American ecologists focused their attention on grassland ecotones (Clements, 1916; Gleason, 1913, 1923; Vestal, 1914; Weaver, 1954;

Fig. 10. Clonal growth as a result of asexual reproduction or vegetative spread, as illustrated by the composite *Rudbeckia hirta*, is characteristic of many grassland grasses and forbs. This habit has undoubtedly helped many species to become dominants and to survive fire.

Weaver and Albertson, 1956; Weaver and Clements, 1938). These early studies coincided with the general cessation of natural and widespread prairie fires, as the sweeping grasslands were interrupted by plowed fields, roads, fence lines, and settlements, or as grassland fuels were reduced by grazing and haying. The elimination of these fires permitted dramatic and dynamic vegetational changes, as trees and shrubs previously held in check began to invade and grow along prairie–forest ecotones. Ecologists observed herbaceous vegetation being replaced by woody vegetation and grasslands giving way to forests, occurrences that became so commonplace they seemed to relate to some universal property of the vegetation. These observations influenced the concept of plant succession; that is, of one species replacing another in an unidirectional series until an end point is reached. Few of the originators of this concept apparently considered these vegetational changes as atypical or unnatural (Gleason, 1913, 1923; Harper, 1911, 1913; Vogl, 1967a, 1970a).

Embodied in the theory of plant succession is the premise that each species has a distinct amplitude of tolerance, or a specific range of environmental factors that it can withstand. Species are thus classified as pioneer, intermediate, or climax types (Burgess, 1965). Problems have resulted when attempts were made to classify open grassland species according to these types and to apply the concept of plant succession to treeless or natural grasslands. Plant succession was considered universal, when in reality succession in forest and in natural grasslands have little in common.

Light requirements or shade tolerances, used to classify plants successionally, play a minor role in grassland as compared to forest. In grasslands composed of mixtures of short and tall species, there is usually a vertical succession of growth and flowering from the shortest to the tallest throughout the growing season. Each species grows primarily while it receives maximum sunlight and therefore cannot be considered shade tolerant or intolerant. Only rarely are grassland species shade tolerant in a fashion similar to that of forest understory species. The degree of shading produced in forests, however, is considered to exist seldom in grasslands because of their open canopies (Curtis, 1959).

If climax forest species are defined as mesic plants that maintain themselves by reproducing in their own shade, this definition is meaningless when applied to grasslands. Climax grassland species have been determined in ways that largely ignore shade and reproduction tolerances. Dominant grassland species are sometimes considered the climax species. Long-lived perennials are often classified as climax (Weaver and Clements, 1938), but under certain circumstances may also act as aggressive pioneers. Species that do not readily reinvade disturbed grasslands or seldom contribute to secondary successional seres are also considered as climax. Certain legumes (Curtis, 1959), along with plants like wiregrass (*Aristida stricta*) and *Lilium* spp., are not known to reinvade disturbed areas and could also be called climax species (Wells and Shunk, 1931). None of these classification criteria are completely satisfactory, however.

It is somewhat easier to classify pioneer grassland species, since they are often annuals or short-lived perennials that readily invade disturbed sites, grow on exposed soils, and thrive in full sunlight. Not all grassland annuals act as pioneers. For example, the winter and summer annuals of Southwest grasslands, including grasses, respond more to the seasonal rains than to disturbances or pioneer conditions. A more reasonable approach is to dismiss successional classifications of grassland species and classify them as increasers, decreasers, neutrals, invaders, or retreaters (Vogl, 1964a), depending upon their responses to such factors as grazing, droughts, or burning.

Any succession that takes place in a grassland deviates from the unidirectional and stepwise replacement series, or is at least complicated by the asexual habits and clonal development of many species, by antibiotic effects, and by species aggregations (Fig. 10). In other words, strong vegetative reproductive habits and allelopathic properties (Curtis and Cottam, 1950; Muller, 1966), along with prolific seed production, may be more important to plants becoming the dominants of a grassland, or replacing other species, than moisture or shade tolerances.

Curtis (1959) suggested that internal succession is more important than typical plant succession in grasslands. Internal succession occurs continuously as fossorial vertebrates, invertebrates, and their predators disturb and mix the grassland surface and subsurface soils and destroy plants, thus producing pioneer sites or gaps available for reinvasion. Any grassland is constantly being turned over, a little at a time, with the result that "pioneer" species occur adjacent to and mixed with "climax" species.

A more reasonable approach to grassland succession is to abandon the traditional straight-lined and unidirectional approach and consider it as a cyclic or circular phenomenon (Vogl, 1970a). Instead of progressing through a replacement series, occasionally checked and set back by catastrophes, grasslands are maintained as vegetative cycles, the driving forces of these cycles often being fires. Most grassland climates fluctuate from wet to dry and back again on a seasonal, yearly, cyclic, or irregular basis, with a growth cycle or response superimposed on these fluctuations (Cowles, 1911, 1928; Jackson, 1965; Malin, 1967). Grasses and associated herbaceous species die back annually or during dry periods, and with the usual slow decomposition rates and resultant fuel accumulations, lead to inevitable fires.

In grassland–forest ecotones and other seral situations, recurring fires retard the encroachment of woody trees into the grasslands. But more important than the retrogressive checking of woody plant invasion, particularly in natural grasslands, is the requirement of fire to sustain the vegetational composition, vitality, and productivity and to serve as the agent necessary for cycling nutrients back through the grassland system, thereby maintaining it. In some grasslands with high rainfall, or during wet periods in arid regions, the usual decomposition agents are active in recycling grassland nutrients, thereby reducing or temporarily eliminating the need for fires to assume this essential function. Outbreaks of termites and other insects (Bourlière and Hadley, 1970) and grazing by herbivores may also reduce the need for fire. But under normal conditions, fires are usually the key to maintaining grasslands and sustaining grassland vigor.

Grasslands are generally noted for their dark rich soils and their inherent soil-building properties that are related to the highly productive grasses and legumes, including their extensive root systems. These properties may be related to fire as the primary decomposition agent and related to the increased production of both above- and belowground portions of burned plants. With normal decomposition, nutrients are gradually moved through a series of steps and cycles, with quantities of matter being shunted off into organisms and other systems. Burning, however, rapidly and directly breaks down the vegetable matter, thereby providing more complete and immediately usable nutrients for new growth. The products of incomplete combustion, including charcoal (Pritchett, 1971; Tryon, 1948), organic carbon, and other compounds, contribute significantly to the formation and enrichment of soils as do the ash and minerals. Grassland soil profiles almost universally contain the persistent products of past fires that contributed to soil development. Postburn conditions may also favor the activities of some bacteria and fungi. Thus grassland soil genesis is not only related to the postfire productive growth of the vegetation but also to the more rapid and efficient recycling of nutrients. In summary, succession can be better understood if the traditional concepts and terms are discarded and if grassland succession is considered as a cyclic or circular phenomenon in which fire, or its ecological equivalent, is essential.

VI. Discrepancies in Evaluating the Effects of Fire

A number of investigations and reviews of the effects of fire on grasslands (Ahlgren and Ahlgren, 1960; Daubenmire, 1968; Garren, 1943) present contrasting differences that are sometimes completely contradictory, even from the same area. These discrepancies have been presented objectively or have been evaluated by accepting certain findings while dismissing others. Few attempts, however, have been made to explain these differences.

Some studies have been affected by a conscious or unconscious bias, with investigators setting out to demonstrate that fires produce ill effects, or inherently assuming that fires are always detrimental (Stewart, 1963). Some agencies actually promoted such one-sidedness, favoring research that showed fires to be detrimental while suppressing studies that pointed to beneficial aspects of burning (Schiff, 1962). Perhaps of greater consequence was the unconscious prejudice toward fire that usually starts in early childhood (Vogl, 1967a,b) and was, and sometimes still is, present among scientists. In either case, a lack of objectivity resulted which was responsible for the development of some of these contradictory research results. Conversely, some investigators went to the other extreme by overstating the case for fire and excluding all other factors or explanations.

An important factor affecting results is the variable nature of fires and the environments in which they occur. No two fires, or the conditions under which they occur, are alike. When these variable fires are superimposed upon heterogeneous landscapes that have been almost certainly exposed to differing histories, divergent research findings result. The pre- and postburning conditions are also unique and tend to produce variances that make comparisons difficult at best. Sometimes these variables are intensified by studies that lack good experimental design and methods that minimize sampling variability (Vogl, 1969c), or that measure parameters that are unresponsive to fire.

Differences are sometimes inferred when comparisons of incompatible results are made. Areas burned by severe wildfires are not comparable to those swept by light, controlled burns, a fertile site to an infertile site, dormant season fires to those occurring during the growing season, and the effects of natural fires are not directly comparable to artificial burning done with added or altered fuels, butane torches, fire chambers, and the like. The effects of uncontrolled wildfires, for example, which often occur under the worst climatic conditions and after abnormal fuel buildup, cannot begin to approximate the effects of frequent lightning-caused fires that commonly spread when rain-soaked grasslands dry.

A number of objective studies have positively demonstrated that fires produce damaging effects to grassland vegetation. If most grasslands evolved with fires and became adjusted to them through time, it is difficult to understand why these recent fires should produce adverse results, particularly since these same grasslands have withstood countless trials by fire in the past. This, of course, does not deny that even fire-adapted species have times and conditions when they are affected adversely by fire. Negative findings have been used to support hypotheses that fires had not previously been a part of those grassland environments.

Whether stated or not, most studies that have found fires damaging to grasslands were conducted in arid or semiarid areas or on marginal sites, or on sites which had been severely degraded by prolonged heavy grazing or other abuses. In North America, for example, fires have seldom been found to adversely affect the vegetation of tall-grass prairies where climatic conditions are usually more than adequate for growth and where grazing is presently absent, as opposed to short-grass prairies with restricted growing conditions and almost universal heavy grazing (Daubenmire, 1968; Penfound, 1968).

The presence of livestock or concentrations of native herbivores before, and particularly after burning, can completely alter the vegetational responses to fire (Arnold, 1950, 1955; Vesey-FitzGerald, 1971). Few investigators have taken this into account and often simply ignore the influences of grazing, despite the fact that the western Great Plains, southwestern United States, and most other extensive grasslands in the world have been so completely exploited by livestock or big game grazing that it is nearly impossible to find even small ungrazed areas for comparative studies. This sustained heavy grazing, along with such things as the introduction of weeds, promiscuous burning (Pickford, 1932), and burning at unnatural times and intervals, has resulted in species compositions and gene pools that little resemble the original presettlement formations. In addition, livestock and game will concentrate on fresh burns, closely grazing tender shoots exposed by the removal of the accumulated dead herbage and the new shoots as they emerge after burning. This type of heavy grazing (Anderson, 1964; Daubenmire, 1968; Jackson, 1965), coupled with trampling and soil compaction, usually alters or negates the normal responses of grassland plants to fire. Responses are often further complicated on marginal sites by droughts, unfavorable temperatures, and nutrient problems. It is somewhat unfair, then, to use the results of such studies alone to evaluate the effects of fire on an entire grassland and to imply that it always was this way.

Another factor that can produce extraordinary effects with burning is the abnormal accumulation of fuel that results from the exclusion of fire beyond natural frequencies. The effects of fire on bunch grasses (Fig. 3), for example, are particularly variable between studies and from species to species. The nature of many bunch or tussock-forming grasses is such that as the aboveground growth dies back it tends to become self-lodging. This results in plants impacted with litter, which causes decline in growth that ultimately leads to decadence. When fires finally occur, they are often detrimental since lethal temperatures are attained or the weakened plants are slow to recover. Shoot meristems

Fig. 11. Many grasslands are presently represented by remnant areas, as this Midwest railroad right-of-way, and may be atypical in species composition, genetic makeup, and topography. Such areas may react differently to fire than would the original grasslands from which they were derived.

are particularly vulnerable to fire damage when they become severely pediceled as a result of prolonged fire protection and/or erosion. High temperatures for long periods are also attained when experimental burns occur with little or no wind that would otherwise dissipate the heat and hasten the burning. Some bunch grass species may have evolved under a regime of frequent fires normally spread by strong winds and, therefore, cannot respond favorably to conditions that deviate from these.

The results of studies conducted on remnant or relict grasslands are atypical. In many regions, the remaining unplowed areas do not represent the original grassland, occupying sites with poor soils and exceptional topography, and supporting atypical assemblages of plants. Studies of the effects of fire on old fields (Curtis and Partch, 1948; Robocker and Miller, 1955; Swan, 1970; Zedler and Loucks, 1969), cemetery lots, railroad rights-of-way (Fig. 11) (Ralston and Dix, 1966), remnant prairies (Dix and Butler, 1954), and sand hills are necessary since they are often all that are left to study, but they should be used cautiously with their limitations in mind when extrapolating or generalizing about the role or fire in an entire grassland. Species at the edges of their ranges and on marginal sites also react differently because of peripheral selectivity and the presence of ecotypes (MacMillan, 1959). Such sites are often created by extreme environmental conditions and are delicately balanced, fragile systems that are readily upset by man's uses and abuses. Adverse reactions to fire occur, not so much because the species present are not adapted to fires but because the grasslands are already precarious systems at the time of burning.

VII. Changes Resulting from Cessation of Fires

Many grasslands have received protection from fires, particularly where fires have been considered incompatible with current land uses (Daubenmire, 1968). The elimination of fires that had been an intricate part of the environment has various effects, depending upon numerous factors including land use, grassland type and condition, and climate. Light to moderate grazing or repeated mowing, for example, have replaced fires in some grasslands since these uses tend to produce many of the same results (Daubenmire, 1968). These practices are not entirely equivalent, because of their incomplete recycling of nutrients and growth stimulation, but have helped to maintain grasslands.

Grasslands that produce prodigious growth but possess slow decomposition rates change most dramatically when fires are eliminated. Fires appear necessary to remove the litter accumulations before they become excessive and eventually suppress growth in productive areas (Hulbert, 1969; Kucera and Ehrenreich, 1962; West, 1965). Conversely, grasslands exhibiting slow or little annual growth generally remain unchanged for the longest time after fires have been excluded.

Aside from the losses of the positive effects of fire previously mentioned, changes resulting from cessation of fires are most apparent in grassland ecotones and savannas capable of supporting woody vegetation where livestock grazing often removed the fuels or destroyed their continuous nature. The presence of livestock also complicated the resulting changes by exerting selective pressures on certain species, making it difficult to separate the changes produced by fire elimination from those caused by grazing.

Grassland borders appear to be unstable in many regions, shifting back and forth through time as competitive battles are won or lost between the grassland plants and the woody species, along with members of the Palmae, Cactaceae, and Agavaceae (Gleason, 1912, 1913; Lapham, 1965). Periods of wet climatic conditions promote the range expansion of woody plants. The spread of grassland is assisted by fires that are usually more frequent and intense during droughts, and that help to eliminate or reduce the woody elements thereby providing additional ground for grassland expansion. Although grassland expansion or retraction is controlled primarily by climatic conditions, the elimination of fires that normally accompany dry conditions is enough to tip the balance so that woody species almost universally begin to hold their own or replace the grasslands. Without fires to maintain healthy grasslands and check woody encroachment, many savannas have reverted to forests; and forest, brush, and scrub species have replaced grassland plants.

In North America, the aspen (*Populus* spp.) parkland of Canada exists between Great Plains grassland and boreal forest. This formation consisted of groves of trees growing in depressions and on north-facing slopes with grassland vegetation occurring on the uplands and south-facing slopes. Since settlement, the aspen groves have been advancing on the grasslands (Bird, 1961; Coupland, 1950; Jeffrey, 1961; Moss, 1932, 1952) with fire being the most important factor checking this advance. Since the removal of fire, the retarding effects of snowshoe hares, deer, insects, snow, hail, and frost have been insufficient to stem the woody invasion.

The open brush prairie savannas of Manitoba, Minnesota, and Wisconsin quickly changed to closed forest with the advent of settlement and the establishment of fire protection (Buell and Buell, 1959; Ewing, 1924; Vogl, 1964b). These savannas contained forest elements which were kept in reduced or suppressed forms by repeated fires, so that the conversion to forest was very rapid once the woody species were released from their flaming bonds.

Even in the Lake States where soil-parent material exerts control of such vegetation types as the northern Wisconsin pine barrens, removal of fires has resulted in numerous changes (Vogl, 1970b). Fires every 20–40 years help to maintain the open nature of the barrens by retarding woody plant growth. Fires every 10 years or less promote vigorous herbaceous growth. Fire frequencies determine the overstory conifer composition. Fire may also play a role in recycling nutrients from the ground-layer vegetation and litter to the overstory trees, thereby counteracting the infertile substrates and arrested decay (Vogl, 1970b).

The bracken–grasslands of the Lake States are usually treeless and dominated by bracken fern (*Pteridium aquilinum*), along with grasses and other herbaceous plants (Curtis, 1959). Most bracken–grasslands are considered to have originated after intense forest fires that resulted in increased surface water that converted forests to sites suited to sedges, grasses, and bracken fern. Although additional fires are considered to have little effect on their vegetational composition, woody plants have begun to reinvade a number of these bracken–grasslands since fire protection began (Vogl, 1964a).

The oak openings of Minnesota, Iowa, Wisconsin, and Illinois occurred between eastern deciduous forest and tall-grass prairie and consisted of groves of fire-resistant oaks (*Quercus* spp.) scattered across the grasslands. Stands of more mesic and fire-sensitive trees were confined to the leeward sides of natural firebreaks such as lakes, rivers (Gleason, 1912, 1913), or swamps (Vogl, 1969a), where they were afforded protection from prairie fires pushed by the prevailing southwest winds. Most of these oak openings sprang up into forest with fire exclusion, surrounding the original open-grown oaks with even-aged forests that date back to the last fire or to the start of fire exclusion (Cottam, 1949).

The ecotones between grassland and forest in Missouri, Nebraska, Kansas, and Oklahoma are occupied by oak (*Quercus* spp.) and oak–hickory (*Carya* spp.) savannas (Rice and Penfound, 1959). The trees are confined primarily to watercourses, valleys, broken terrain, or specific substrates, with open grasslands occupying the level or undulating uplands (Aikman and Gilly, 1948). Since fire elimination, tree densities have increased and tree compositions have changed within the woodland zones. In many areas, the fire-sensitive *Juniperus virginiana* has become abundant, and the more mesic tree species have increased (Fitch and McGregor, 1956). A general expansion of shrubs has also occurred. Invasion of the open grasslands by woody species, however, has been generally slow to practically nonexistent, being hampered by such factors as semiarid climatic conditions, available moisture, soil differences, grassland competition, and livestock grazing. Conditions for expansion and growth of woody species have not been favorable, and fires to check such invasions are less critical than in those areas conducive to growth of woody plants.

In the southeastern United States, marked changes in vegetation have occurred with the elimination of fires, particularly in the piney-woods region of the Southeast, along with the Big Thicket of Texas and pine regions of Arkansas. Historical accounts attest that most uplands were once occupied by pine savannas (Vogl, 1972). The fire-stimulated understory growth provided abundant, continuous, contiguous, and highly flammable fuels necessary for freely spreading and recurring fires in a region with a high lightning incidence (Fig. 12). Repeated fires maintain low tree densities and open canopies by selectively favoring pines (*Pinus* spp.) over numerous hardwood species, thereby allowing adequate sunlight and principally pine litter to reach the ground. The absence of fire results in hardwood growth and/or invasion, closing canopies, and abundant hardwood litter accumulations which suppress grassland growth and spreading fires (Vogl, 1972). Frequent fire was an integral part of southeastern pine savannas (Lemon, 1970), and without its continued presence the grassland understories cease to exist, and pine savannas are converted to dense hardwood forests. Interestingly, in these grassland–deciduous forest transitions there is no conclusive

Fig. 12. Southeastern United States pine savannas were swept by frequent surface fires that maintained a vigorous grassland understory. The thick-barked and fire-adapted pines survived the grassland fires while the fire-tender hardwoods were checked and prevented from invading.

evidence that repeated burning alone will eliminate the more xeric hardwoods once they have become established, since even annual burning does little more than kill the aboveground portions of the trees, while simultaneously stimulating new sprouts year after year. In these vegetation types, the changes produced by fire suppression must be considered nearly irreversible, for the reinstatement of fire in these types will not completely return the vegetation to preprotection compositions.

Grasslands also form various transitional types as they come in contact with deserts, chaparral, and western coniferous forests. As in other grassland ecotones, the elimination of fires has generally resulted in the expansion of the trees and shrubs at the expense of grasslands (Blackburn and Tueller, 1970; Box, 1967; Box *et al.*, 1967; Brown, 1950; Dwyer and Pieper, 1967; Humphrey, 1962; Jameson, 1962; Leopold, 1924). Although fire exclusion has favored a general expansion of woody plants in these types, it has not necessarily benefited them (Christensen and Hutchinson, 1965; McIlvain and Armstrong, 1966). Many of these areas are now crowded with trees that are economically undesirable. In addition, excessive densities of woody plants have led to stand stagnation, weakened resistance, and decline, and to establishment of species whose

Fig. 13. Short-grass prairies in the southwestern United States once covered the valleys, plains, and plateaus, with trees confined to the steep slopes, canyon and mesa walls, broken terrain, and rocky outcrops as illustrated in the background. With overgrazing (foreground), fires could no longer sweep the open grasslands, the grasslands were weakened, and conifer encroachment resulted.

life cycles and life history requirements are out of adjustment with their environments. Ponderosa pine (*Pinus ponderosa*) savannas that have been denied fires in central Arizona, for example, were previously productive for cattle and quality lumber. These areas are now occupied by dog-haired thickets of stunted and weak trees that have excluded the grassland understory and have essentially ceased to grow and reproduce, currently posing serious wildfire threats (Arnold, 1950; Hare, 1961; Marshall, 1963). Some conifer species originally existed on scarps and other broken terrain that provided natural fire protection since these rugged sites were free of continuous grassland to carry fires (Fig. 13). The effects of fire on western desert, forest, and chaparral ecotones are discussed in other chapters.

Although the examples presented above are in North America, comparable situations can be found along grassland edges and savannas, particularly derived savannas, in other parts of the world wherever man or animals have eliminated the occurrence of natural fires (Blydenstein, 1968; Bourlière and Hadley, 1970; Glover, 1971; Skovlin, 1971). With the realization that there are exceptions, it can be stated that grassland and woody species tend to be competitive and even mutually exclusive, seldom coexisting without the periodic assistance of fires. In addition, the grassland component in transitional areas seldom continues to exist in regions capable of vigorous woody growth without the repeated occurrence of fires or comparable forces.

VIII. Current and Future Uses of Fire in Grassland Management

A. RANGE MANAGEMENT

The most extensive use of fire for range purposes is in brush control to improve livestock grazing (Love, 1970). It is most often used as a clean-up measure with chemical treatment, brush cutting or crushing, bulldozing and brush piling, or root raking. Conversion of brush sites to grasslands, or the temporary eradication of invading brush from grasslands, has been somewhat successful using these methods, although costly at times. Ranchers are being forced to use marginal areas for grazing as many of the better lands, such as fertile valleys, are relinquished to farming and crop production, particularly where land values, produce prices, and taxes have risen, and where irrigation waters have become available or economically feasible. Marginal sites are occupied by trees, brush, or other woody plants, have little or no grass cover, and often consist of infertile soil types in rugged terrain. Most of the present site conversions apply fire only once or utilize a wildfire, but it is conceivable that in the future controls over the use of herbicides, the need to utilize more marginal areas, economic pressures, and other ecological considerations will dictate working more closely with nature by using repeated fires alone. Burning is most effective in areas with native grassland species still present that can expand under this type of management. Pure brush types such as steep mountainsides of California chaparral, which evolved as shrub-dominated sites, are the most difficult and perhaps foolhardy to convert to grassland, and these artificial grasslands tend to be tenuous and temporary in nature. The use of alien grasses instead of native species contributes further to their instability.

In North America, the general use of fire in grazing management has declined. In the West and Southwest, most ranchers cannot afford the losses of forage needed for fuels, which must often be equivalent to a standing crop to support an effective fire (Daubenmire, 1968). Other factors discouraging the use of fire are the difficulties of executing controlled burns, the liabilities of escape fires, the need to remove livestock before and after burning, fear of fires, lack of burning experience, and economic pressures to run ranches to their fullest capacities. The poor results often obtained because of ineffective burns, burn-deteriorated ranges, adverse postburn climatic conditions, and the postburn presence of alien and often unpalatable plants as well as livestock (Christensen, 1964; Countryman and Cornelius, 1957; Hervey, 1949; McKell *et al.*, 1962; Pechanec and Hull, 1945) have also convinced many to dismiss burning as a management practice. The use of fire in open grasslands presently holds its greatest promise in noxious weed control, particularly against plants that are toxic or damaging to livestock. In these situations grazing animals must usually be temporarily eliminated anyway, and the application of fire at the time when the pest plants are most vulnerable can be effective and yet the least costly and disturbing method (Furbush, 1953; Major *et al.*, 1960; Sharp *et al.*, 1957). Grassland burning can also be used to improve the palatability of coarse species which, in turn, can result in increased livestock production.

Even in the Southeast, where burning maintains forage palatability and maximum livestock production and checks hardwood invasion, the use of fire is declining. Dual purpose lumber and cattle production are giving way to single objective and intense pasture management. This usually involves replacement of the native grasses (Vogl, 1972) with imported pasture grasses. At present there is little use for fire in this type of management, since the emphasis has been on selection of suitable grasses and management of cattle by manipulating densities and rotating pastures. Most eastern Great Plains and Lake States native ranges have been similarly replaced by exotic species with fire management seldom considered. A current attitude toward fire in managing grasses for forage, turf, and range is perhaps illustrated by the Youngner and McKell (1972) symposium on grass biology and utilization where only one out of thirty papers considered fire as a possible management tool, and that one paper dealt with the negative aspects of fire.

The widespread use of fire in open grassland management is currently confined to portions of those countries that still have extensive open ranges that are not being intensively utilized or managed. These areas are often under the control of native herdsmen or tribes and are located in remote or undeveloped regions.

Interest in the effects of fire on North American and many other grasslands is becoming more and more academic as it relates to range management. This is apparent from the general decline in studies of the effects of fire and from the low priority fire has received in the North American Grassland Biome studies. Each year increasing acreages of grassland are lost as farms expand, cattle are put in feeder lots, cities grow, and higher priorities for rangelands develop (Burgess, 1964; Moir, 1972). Given time, available water, increasing human populations, food shortages, and economic incentives, many, and in some regions all, the remaining grasslands will be turned "wrong side up."

Most of the world's grasslands have been altered and weakened by overgrazing and other general abuses, including misuse of fire, and faintly resemble the original native grasslands (Fig. 13). As a result, many grasslands face more imminent, basic, and remedial problems than whether to burn or not to burn. There is little hope that future range management will often use fire unless grasslands are allowed to recover and long-range objectives and sound ecological management are considered to be more important than short-term economic gains.

The prospects of using fire in turf management, in growing grasses, sods, and grass seed for lawns and other ornamental uses, are also dim since most turf production areas are adjacent to urban centers. People in these areas are demanding that the agricultural burning be stopped because of the smoke produced, reacting to the nuisance, and being unable or unwilling to differentiate between vegetation smoke, a natural product, and the smog created by internal combustion. Those seeking to prohibit burning also tend to dismiss the benefits derived from turf burning, including high seed production, effective disease control, and an absence of complex and synergistic side effects that chemical substitutes might create (Jenkins, 1970).

B. WILDLIFE MANAGEMENT

If the present trends prevail, the greatest potential use of fire will be in wildlife management since many wildlife preserves, refuges, and hunting grounds still support native grasslands. Most of the national parks of Africa, for example, contain various grasslands and/or savannas and have management programs that include fire as an important and often essential tool (Boughey, 1963; Brynard, 1964; Hill, 1971; Lemon, 1968; Van Rensburg, 1971).

The majority of grassland mammals and birds respond favorably to the changes created by the judicial use of controlled fire (Komarek, 1969), a response that extends to the nongame species (Beck and Vogl, 1972; Vogl, 1973). A number of grassland animals are currently mistaken as forest inhabitants, but in reality were confined to forests more by necessity than by choice as adjacent grasslands and grassland borders were destroyed or invaded by forest. When the preferred grasslands and grassland edges were eliminated, some of the more versatile species continued to survive by retreating to the forest (Marshall, 1963). As a result, spectacular increases in wildlife reproduction often occur as forest types are reconverted to grasslands or savannas with burning.

Although instances of animal mortality have been reported from grassland fires (Brynard, 1971; Moore, 1972; Vogl, 1967a), the general benefits derived from improved habitats, increased productivity, growth stimulation, and other changes usually offset any direct mortality (Cancelado and Yonke, 1970; Hurst, 1970; Leopold, 1933; Riechert and Reeder, 1972; Vogl, 1967a). Wildlife losses are considered to be unusual events that have been overemphasized. Decisions to exclude fire from grassland types because of potential mortality may actually increase wildlife losses indirectly by permitting the habitat to deteriorate, change, or succeed to less productive vegetational stages (Chamrad and Dodd, 1972; Lehmann, 1965). Fire exclusion programs often result in exceptionally destructive fires by permitting abnormal fuel buildups and vegetational deterioration.

Although controlled burning of grassland types is a sound ecological practice and an effective wildfire management tool, most wildlife managers seldom use fire. Numerous reasons are given for their reluctance or refusal to use fire, or why they are prohibited from burning, but most are excuses founded on prejudice toward fire or ignorance of the essential role fire plays in most grassland environments. Managers often appear to be more receptive to new management techniques, looking for substitutes that are easier to administer and "safer" to use. In the October, 1969, *Journal of Wildlife Management*, for example, only 3 of 49 articles mentioned fire, and that issue included a special management section. As more wildlife biologists realize that vegetation management is the key to most wildlife problems and that fire can be easy to use, safe, natural, and an inexpensive tool that can be helpful and not harmful, the uses of fire in the management of grassland animals should increase. As interest in the use of fire increases, burning techniques and methods should develop that will make it easier to prescribe and apply fire.

C. Natural Area Management

Because of the rapid, widespread, and often thorough destruction of grasslands in the United States and elsewhere, governments, universities, conservation groups, and other agencies have been attempting to preserve the last remnants or representative portions of these vanishing grasslands. Attempts have also been made to restore or recreate grasslands where they have been completely eliminated (Anderson, 1972; Cottam and Wilson, 1966; Greene and Curtis, 1953). As acquisition or restoration of grasslands has become a reality, the question of management and maintenance has arisen, particularly in those grasslands where protection alone has not guaranteed their continued or healthy existence. Bray (1957) and others were aware of the necessity of controlled disturbances in the preservation of natural area grasslands and recommended burning over other forms of disturbance. Mowing, and mowing and raking, have been the most common substitutes for burning, but these practices have some shortcomings compared to fire (Christiansen, 1972; Richards, 1972).

The use of fire has been advocated in management of grassland preserves, particularly in tall-grass prairies where woody plant invasion is a problem and where heavy growth accumulations become self-defeating (Anderson, 1972; Burt, 1971; Jenkins, 1971; Lindsey *et al.*, 1970; Thompson, 1972). Hanson (1938, 1939), Stone (1965), Boardman (1967), Vogl (1967a), Butts (1968), Loucks (1968), and Odum (1969) have presented arguments for the use of fire in management of all fire-type communities, including grasslands, that are part of national parks, refuges, and preserves. They reason that if fire was a natural part of such communities as grasslands prior to man's interventions, then natural area management will be incomplete and the grassland environments deficient until controlled burns or wildfires are allowed to again take their place among the natural order of things.

IX. Epilogue

Because of limited space, discussions of, and references to, those topics already adequately surveyed in the review of fire and grasslands by Daubenmire (1968) were kept to a minimum or updated. Emphasis was placed on those subjects that have been neglected or on themes that were considered to be of particular importance. A general approach was taken in most sections because past discussions have been frequently confused by "the facts"—observations and data that are often contradictory, extremely variable, or unfounded.

Elucidation of some points with examples from the literature was avoided because of their limited and questionable nature. Controversy over the effects of fire on grasslands will only be resolved with an awareness of bias and recognition of the limitations and conditions under which studies are conducted. A renewed interest in fire and grasslands is needed. Although grasslands are one of the most studied plant communities (Curtis, 1959), general knowledge of fire and grasslands is fragmentary and basic issues are still to be resolved. Vigorous research programs, small and large, are needed that will objectively seek basic understanding of the interrelationships between fire and the various components of the grassland community. Grassland managers, from private ranchers to agency directors, should be encouraged to "experiment" with fire and learn to use it as an effective tool. It is time that we realize that "playing with fire" will not necessarily lead to getting burned and just might help us to relearn a lost art and gain a powerful and natural tool.

The generalizations presented were made with the realization that there are exceptions to each but, nevertheless, were attempts to synthesize basic concepts and focus attention on fundamental issues. The majority of the grasslands of the world have coexisted with fire to some degree through time. Once man understands the natural relationships between fire and individual grasslands and grassland organisms, he can again begin to control and utilize the great fire force to his advantage and to the vigorous continued existence and dynamic maintenance of grassland ecosystems.

Acknowledgments

My interest in grasslands originated with my mother's fascination with prairie flowers and the countless days that my brother Mike and I roamed Wisconsin prairies. I acknowledge the exposure to grasslands and fire that I received from Professors R. L. Dix, H. C. Hanson, G. Cottam, J. T. Curtis, and H. C. Green. I thank Drs. G. Cottam, J. D. Sauer, R. A. Schlising, and H. A. Wright for critically reading this chapter, and Carol Vogl for helping in its preparation.

References

Ahlgren, I. F., and Ahlgren, C. E. (1960). Ecological effects of forest fires. *Bot. Rev.* **26**, 483–533.

Aikman, J. M. (1955). Burning in the management of prairie in Iowa. *Proc. Iowa Acad. Sci.* **62**, 53–62.

Aikman, J. M., and Gilly, C. L. (1948). A comparison of the forest floras along the Des Moines and Missouri Rivers. *Proc. Iowa Acad. Sci.* **55**, 63–73.

Albertson, F. W., and Weaver, J. E. (1942). History of the native vegetation of western Kansas during seven years of continuous drought. *Ecol. Monogr.* **12**, 23–51.

Aldous, A. E. (1934). Effects of burning on Kansas bluestem pastures. *Kans., Agr. Exp. Sta., Bull.* **38**, 65.

Anderson, K. L. (1964). Burning Flint Hills bluestem ranges. *Proc. 3rd Annu. Tall Timbers Fire Ecol. Conf.* pp. 88–103.

Anderson, R. C. (1972). Prairie history, management and restoration in southern Illinois. *Midwest Prairie Conf., 2nd, 1970,* pp. 15–21.

Arnold, J. F. (1950). Changes in ponderosa pine bunchgrass ranges in northern Arizona resulting from pine regeneration and grazing. *J. Forest.* **48**, 118–126.

Arnold, J. F. (1955). Plant life-form classification and its use in evaluating range condition and trend. *J. Range Manage.* **8**, 176–181.

Baldanzi, G. (1961). Burning and soil fertility. *Trans. Int. Congr. Soil Sci., 7th, 1960* Vol. 2, pp. 523–530.

Bartlett, H. H. (1955). "Fire in Relation to Primitive Agriculture and Grazing in the Tropics." Univ. of Michigan Bot. Gard., Ann Arbor.

Bartlett, H. H. (1956). Fire, primitive agriculture and grazing in the tropics. *In* "Man's Role in Changing the Face of the Earth" (W. L. Thomas, ed.), pp. 692–714. Univ. of Chicago Press, Chicago, Illinois.

Batchelder, R. B., and Hirt, H. F. (1966). Fire in tropical forest and grasslands. *U.S. Army Natick Lab., Tech. Rep.* **67-41-ES**, 380.

Beck, A. M., and Vogl, R. J. (1972). The effects of spring burning on rodent populations in a brush prairie savanna. *J. Mammal.* **53**, 336–346.

Beetle, A. A. (1957). Grassland climax, evolution, and Wyoming. *Univ. Wyo., Publ.* **21**, 64–70.

Bird, R. D. (1961). Ecology of the aspen parkland of western Canada in relation to land use. *Can., Dep. Agr., Publ.* **1066**, 155.

Biswell, H. H. (1972). Fire ecology of the ponderosa grasslands. *Proc. 12th Annu. Tall Timbers Fire Ecol. Conf.* pp. 69–96.

Biswell, H. H., and Lemon, P. C. (1943). Effect of fire upon seedstalk production of range grasses. *J. Forest.* **41**, 844.

Blackburn, W. H., and Tueller, P. T. (1970). Pinyon and juniper invasion in black sagebrush communities in east-central Nevada. *Ecology* **51**, 841–848.

Blevins, L. L., and Marwitz, J. D. (1968). Visual observations of lightning in some Great Plains hailstorms. *Weather* **23**, 192–194.

Blydenstein, J. (1967). Tropical savanna vegetation of the Llanos of Colombia. *Ecology* **48**, 1–15.

Blydenstein, J. (1968). Burning and tropical American savannas. *Proc. 8th Annu. Tall Timbers Fire Ecol. Conf.* pp. 1–14.

Boardman, W. S. (1967). Wildfire and natural area preservation. *Proc. 6th Annu. Tall Timbers Fire Ecol. Conf.* pp. 134–142.

Borchert, J. R. (1950). The climate of the central North American grassland. *Ann. Ass. Amer. Geogr.* **40**, 1–39.

Boughey, A. S. (1963). Interaction between animals, vegetation and fire in Southern Rhodesia. *Ohio J. Sci.* **63**, 193–209.

Bourlière, F., and Hadley, M. (1970). The ecology of tropical savannas. *Annu. Rev. Ecol. Syst.* **1**, 125–152.

Box, T. W. (1967). Brush, fire, and west Texas rangeland. *Proc. 6th Annu. Tall Timbers Fire Ecol. Conf.* pp. 7–19.

Box, T. W., Powell, J., and Drawe, D. L. (1967). Influence of fire on south Texas chaparral communities. *Ecology* **48**, 955–961.

Bray, J. R. (1957). Preservation of natural areas. *Minn. Natur.* **8**, 117–119.

Broido, A., and Nelson, M. A. (1964). Ash content: Its effect on combustion of corn plants. *Science* **146**, 652–653.

Brown, A. L. (1950). Shrub invasion of southern Arizona desert grassland. *J. Range Manage.* **3**, 172–177.

Brynard, A. M. (1964). The influence of veld burning on the vegetation and game of the Kruger National Park. *Mongr. Biol.* **14**, 371–393.

Brynard, A. M. (1971). Controlled burning in the Kruger National Park—history and development of a veld burning policy. *Proc. 11th Annu. Tall Timbers Fire Ecol Conf.* pp. 219–231.

Budowski, G. (1966). Fire in tropical American lowland areas. *Proc. 5th Annu. Tall Timbers Fire Ecol. Conf.* pp. 5–22.

Buell, M. F., and Buell, H. F. (1959). Aspen invasion of prairie. *Bull. Torrey Bot. Club* 86, 264–265.

Burgess, R. L. (1964). Ninety years of vegetational change in a township in southeastern North Dakota. *Proc. N. Dak. Acad. Sci.* 18, 84–94.

Burgess, R. L. (1965). A study of plant succession in the sandhills of southeastern North Dakota. *Proc. N. Dak. Acad. Sci.* 19, 62–80.

Burt, D. (1971). The imperiled prairie. *Nat. Conservancy News* 21, 6–9.

Burton, G. W. (1944). Seed production of several southern grasses as influenced by burning and fertilization. *J. Amer. Soc. Agron.* 36, 523–529.

Burton, J. C. (1972). Nodulation and symbiotic nitrogen fixation by prairie legumes. *Midwest Prairie Conf., 2nd, 1970,* pp. 116–121.

Butts, D. B. (1968). Fire for management in national parks. *Rocky Mt.—High Plains Parks Recreation J.* 3, 1–105.

Byram, G. M. (1948). Vegetation temperature and fire damage in the southern pines. *Fire Control Notes* 9, 34–36.

Byram, G. M. (1958). Some basic thermal processes controlling the effects of fire on living vegetation. *U.S., Forest Serv., Southeast. Forest Ext. Sta., Res. Notes* 114, 2.

Cancelado, R., and Yonke, T. R. (1970). Effects of prairie burning on insect populations. *J. Kans. Entomol. Soc.* 43, 274–281.

Capon, B., and Van Asdall, W. (1967). Heat pre-treatment as a means of increasing germination of desert annual seeds. *Ecology* 48, 303–306.

Chamrad, A. D., and Dodd, J. D. (1972). Prescribed burning and grazing for prairie chicken habitat manipulation in the Texas coastal prairie. *Proc. 12th Annu. Tall Timbers Fire Ecol. Conf.* pp. 257–276.

Christensen, E. M. (1964). Changes in composition of a *Bromus tectorum–Sporobolus cryptandrus–Aristida longiseta* community following fire. *Utah Acad. Sci., Arts Lett., Proc.* 41, 53–57.

Christensen, E. M., and Hutchinson, M. A. (1965). Historical observations on the ecology of Rush and Tooele Valleys, Utah. *Utah Acad. Sci., Arts Lett., Proc.* 42, 90–105.

Christiansen, P. A. (1972). Management of Hayden Prairie: Past, present and future. *Midwest Prairie Conf., 2nd, 1970,* pp. 25–29.

Clements, F. E. (1916). Plant succession. *Carnegie Inst. Wash. Publ.* 242, 512.

Clements, F. E. (1920). Adaptation and mutation as a result of fire. *Carnegie Inst. Wash., Yearb.* 19, 348–349.

Clewell, A. F. (1966a). Native North American species of *Lespedeza* (Leguminosae). *Rhodora* 68, 359–405.

Clewell, A. F. (1966b). Natural history, cytology, and isolating mechanisms of the native American *Lespedezas*. *Bull. Tall Timbers Res. Sta.* No. 6.

Cohen, C. (1950). The occurrence of fungi in the soil, and different burning and grazing treatments of the veld in the Transvaal. *S. Afr. J. Sci.* 46, 245–246.

Conrad, E., and Poulton, C. E. (1966). Effect of a wildfire on Idaho fescue and bluebunch wheatgrass. *J. Range Manage.* 19, 138–141.

Cook, L. (1939). A contribution to our information on grass burning. *S. Afr. J. Sci.* 36, 270–282.

Cooper, C. F. (1961). The ecology of fire. *Sci. Amer.* 204, 150–160.

Cooper, C. F. (1963). "An Annotated Bibliography of the Effects of Fire on Australian Vegetation." Mimeo. Soil Conserv. Auth., Victoria, Australia.

Costello, D. (1969). "The Prairie World." Crowell-Collier, New York.

Cottam, G. (1949). The phytosociology of an oak woods in southwestern Wisconsin. *Ecology* 30, 271–287.

Cottam, G., and Wilson, H. C. (1966). Community dynamics on an artificial prairie. *Ecology* 47, 88–96.

Countryman, C. M., and Cornelius, D. R. (1957). Some effects of fire on a perennial range type. *J. Range Manage.* 10, 39–41.

Coupland, R. T. (1950). Ecology of mixed prairie in Canada. *Ecol. Monogr.* 20, 271–316.

Coupland, R. T. (1958). The effects of fluctuations in weather upon the grasslands of the Great Plains. *Bot. Rev.* 24, 273–317.

Cowles, H. C. (1911). The causes of vegetative cycles. *Bot. Gaz. (Chicago)* 51, 161–183.

Cowles, H. C. (1928). Persistence of prairies. *Ecology* 9, 380–382.

Curtis, J. T. (1956). The modification of mid-latitude grasslands and forests by man. *In* "Man's Role in Changing the Face of the Earth" (W. L. Thomas, ed.), pp. 721–736. Univ. of Chicago Press, Chicago, Illinois.

Curtis, J. T. (1959). "The Vegetation of Wisconsin." Univ. of Wisconsin Press, Madison.

Curtis, J. T., and Cottam, G. (1950). Antibiotic and autotoxic effects in prairie sunflower. *Bull. Torrey Bot. Club* 77, 187–191.

Curtis, J. T., and Partch, M. L. (1948). Effect of fire on the competition between blue grass and certain prairie plants. *Amer. Midl. Natur.* 39, 437–443.

Curtis, J. T., and Partch, M. L. (1950). Some factors affecting flower production in *Andropogon gerardi*. *Ecology* 31, 488–489.

Cushwa, C. T., Brender, E. V., and Cooper, R. W. (1966). The response of herbaceous vegetation to prescribed burning. *U.S., Forest Serv., Res. Note* SE-53, 2.

Cushwa, C. T., Martin, R. E., and Miller, R. L. (1968). The effects of fire on seed germination. *J. Range Manage.* 21, 250–254.

Cushwa, C. T., Hopkins, M., and McGinnes, B. S. (1970). Response of legumes to prescribed burns in loblolly pine stands of the South Carolina piedmont. *U.S., Forest Serv., Res. Note* SE-140, 6.

Czuhai, E., and Cushwa, C. T. (1968). A resume of prescribed burnings on the Piedmont National Wildlife Refuge. *U.S., Forest Serv., Res. Note* SE-86, 4.

Daubenmire, R. (1968). Ecology of fire in grasslands. *Advan. Ecol. Res.* 5, 209–266.

Davis, L. S., and Martin, R. E. (1960). Time–temperature relationships of test headfires and backfires. *U.S., Forest Serv., Southeast. Forest Exp. Sta., Res. Notes* 148, 2.

Dawkins, C. J. (1939). Tussock formation by *Schoenus nigricans:* The action of fire and water erosion. *J. Ecol.* 27, 78–88.

Day, G. M. (1953). The indian as an ecological factor in the northeastern forest. *Ecology* 34, 329–346.

Dix, R. L. (1960). The effects of burning on the mulch structure and species composition of grasslands in western North Dakota. *Ecology* 41, 49–56.

Dix, R. L., and Butler, J. E. (1954). The effects of fire on a dry, thin-soil prairie in Wisconsin. *J. Range Manage.* 7, 265–268.

Drummond, A. T. (1855). Our north-west prairies, their origin and their forests. *Can. Rec. Sci.* 2, 145–153.

Duvall, V. L. (1962). Burning and grazing increase herbage on slender bluestem range. *J. Range Manage.* 15, 14–16.

Dwyer, D. D., and Pieper, R. D. (1967). Fire effects on blue grama–pinyon–juniper rangeland in New Mexico. *J. Range Manage.* **20**, 359–362.

Ehrenfried, G. (1965). Grassland vegetation: Historical note. *Science* **148**, 1173.

Ehrenreich, J. H., and Aikman, J. M. (1957). Effect of burning on seedstalk production of native prairie grasses. *Proc. Iowa Acad. Sci.* **64**, 205–212.

Ehrenreich, J. H., and Aikman, J. M. (1963). An ecological study of certain management practices on native prairie in Iowa. *Ecol. Monogr.* **33**, 113–130.

Eiseley, L. C. (1954). Man, the fire-maker. *Sci. Amer.* **191**, 52–57.

Ewing, J. (1924). Plant succession of the brush prairie in northwestern Minnesota. *J. Ecol.* **12**, 238–266.

Fitch, H. S., and McGregor, R. L. (1956). The forest habitat of the University of Kansas Natural History Reservation. *Publ. Mus. Natur. Hist. Univ. Kans.* **10**, 77–127.

Fosberg, F. R. (1960). Nature and detection of plant communities resulting from activities of early man. *In* "Impact of Man on Humid Tropics Vegetation, Goroka, Territory Papua and New Guinea," pp. 251–262.

Furbush, P. B. (1953). Control of medusa-head on northern California ranges *J. Forest.* **51**, 118–121.

Garren, K. H. (1943). Effects of fire on vegetation of the southeastern United States. *Bot. Rev.* **9**, 617–654.

Gartner, F. R. (1972). Fire and the Black Hills forest–grass ecotone. *Proc. 12th Annu. Tall Timbers Fire Ecol. Conf.* pp. 37–68.

Gillon, D. (1971). The effect of bush fire on the principal pentatomid bugs (Hemiptera) of an Ivory Coast savanna. *Proc. 11th Annu. Tall Timbers Fire Ecol. Conf.* pp. 377–417.

Gleason, H. A. (1912). An isolated prairie grove and its phytogeographical significance. *Bot. Gaz. (Chicago)* **53**, 38–49.

Gleason, H. A. (1913). The relation of forest distribution and prairie fires in the middle west. *Torreya* **13**, 173–181.

Gleason, H. A. (1923). The vegetational history of the middle west. *Ann. Ass. Amer. Geogr.* **12**, 39–85.

Glover, P. E. (1971). Nature, wildlife, and the habitat with a discussion on fire and other influences. *Proc. 11th Annu. Tall Timbers Fire Ecol. Conf.* pp. 319–336.

Granfelt, C. (1965). Grassland fires. *Science* **149**, 816.

Greene, H. C., and Curtis, J. T. (1953). The re-establishment of prairie in the University of Wisconsin Arboretum. *Wild Flower Mag.* **29**, 77–88.

Greene, S. W. (1935). Relation between winter grass fires and cattle grazing in the longleaf pine belt. *J. Forest.* **33**, 338–341.

Gregg, J. (1954). "Commerce of the Prairies (1844)." Univ. of Oklahoma Press, Norman.

Hadley, E. B. (1970). Net productivity and burning responses of native eastern North Dakota prairie communities. *Amer. Midl. Natur.* **84**, 121–135.

Hadley, E. B., and Kieckhefer, B. J. (1963). Productivity of two prairie grasses in relation to fire frequency. *Ecology* **44**, 389–395.

Halls, L. K. (1952). Burning and grazing in coastal plain forests. *Univ. Ga. Coastal Plain Exp. Sta. Bull.* **51**, 33.

Hanson, A. A. (1972). Breeding of grasses. *In* "The Biology and Utilization of Grasses" (V. B. Youngner and C. M. McKell, eds.), pp. 36–52. Academic Press, New York.

Hanson, H. C. (1938). Ecology of the grassland. *Bot. Rev.* 4, 51–82.

Hanson, H. C. (1939). Fire in land use and management. *Amer. Midl. Natur.* 21, 415–434.

Hare, R. C. (1961). Heat effects on living plants. *U.S., Forest Serv., S. Forest Exp. Sta., Occas. Pap.* 183, 32.

Harper, R. M. (1911). The relation of climax vegetation to islands and peninsulas. *Bull. Torrey Bot. Club* 38, 515–525.

Harper, R. M. (1913). Forests of Alabama. *Ala., Geol. Surv., Monogr.* 8, 228.

Harris, D. R. (1972). The origins of agriculture in the tropics. *Amer. Sci.* 60, 180–193.

Heady, H. F. (1972). Burning and the grasslands in California. *Proc. 12th Annu. Tall Timbers Fire Ecol. Conf.* pp. 97–107.

Henniker-Gotley, G. R. (1936). A forest fire caused by falling stones. *Indian Forest.* 62, 422–423.

Hervey, D. F. (1949). Reaction of a California annual-plant community to fire. *J. Range Manage.* 2, 116–121.

Heyward, F. (1938). Soil temperatures during forest fires in the longleaf pine region. *J. Forest.* 36, 478–491.

Heyward, F., and Barnette, R. M. (1934). Effect of frequent fires on chemical composition of forest soils in the longleaf pine region. *Fla. Agr. Exp. Sta., Bull.* 265, 39.

Hill, P. (1971). Grass forage, food for fauna or fuel for fire or both? *Proc. 11th Annu. Tall Timbers Fire Ecol. Conf.* pp. 337–375.

Hilmon, J. B., and Hughes, R. H. (1965). Fire and forage in the wiregrass type. *J. Range Manage.* 18, 251–254.

Hind, H. Y. (1859). "Report on the Assiniboine and Saskatchewan Exploring Expedition of 1859." Publ. by authority of the Legislative Assembly, Toronto.

Hocking, B. (1964). Fire melanism in some African grasshoppers. *Evolution* 18, 332–335.

Hodgkins, E. J. (1958). Effects of fire on undergrowth vegetation in upland southern pine forests. *Ecology* 39, 36–46.

Hole, F. D., and Watterston, K. G. (1972). Some soil water phenomena as related to manipulation of cover in the Curtis Prairie, The University of Wisconsin Arboretum. A preliminary report. *Midwest Prairie Conf., 2nd, 1970* pp. 49–57.

Hopkins, B. (1963). The role of fire in promoting the sprouting of some savanna species. *J. West Afr. Sci. Ass.* 7, 154–162.

Hopkins, H. F., Albertson, F. W., and Riegel, A. (1948). Some effects of burning upon a prairie in west-central Kansas. *Trans. Kans. Acad. Sci.* 51, 131–141.

Hough, W. (1926). Fire as an agent in human culture. *Smithson. Inst., Bull.* 139, 270.

Hulbert, L. C. (1969). Fire and litter effects in undisturbed bluestem prairie in Kansas. *Ecology* 50, 874–877.

Humphrey, R. R. (1962). "Range Ecology." Ronald Press, New York.

Hurst, G. A. (1970). The effects of controlled burning on arthropod density and biomass in relation to bobwhite quail (*Colinus virginianus*) brood habitat. Ph.D. Thesis, Mississippi State University, University, Mississippi.

Iwanami, Y., and Iizumi, S. (1966). On the relations between the burning temperature and the amount of fuels in natural grassland. I. Measurements of burning temperatures with varying amounts of *Miscanthus sinensis*. *Tohoku Univ. Sci. Rep. Res. Inst.* 17, 27–33.

Jackson, A. S. (1965). Wildfires in the Great Plains grasslands. *Proc. 4th Annu. Tall Timbers Fire Ecol. Conf.* pp. 241–259.

Jameson, D. A. (1962). Effects of burning on a galleta–black grama range invaded by juniper. *Ecology* 43, 760–763.

Janzen, D. H. (1967). Fire, vegetation structure, and the ant × acacia interaction in Central America. *Ecology* 48, 26–35.

Jeffrey, W. W. (1961). A prairie to forest succession in Wood Buffalo Park, Alberta. *Ecology* 42, 442–444.

Jenkins, J. (1970). The $31 million smoke problem. *Farm Quart.* pp. 53–56.

Jenkins, R. E. (1971). Prairie management. *Nat. Conservancy News* 21, 12–14.

Johnson, B. L. (1972). Polyploidy as a factor in the evolution and distribution of grasses. *In* "The Biology and Utilization of Grasses" (V. B. Youngner and C. M. McKell, eds.), pp. 18–35. Academic Press, New York.

Kelting, R. W. (1957). Winter burning in central Oklahoma grassland. *Ecology* 38, 520–522.

Kennan, T. C. D. (1971). The effects of fire on two vegetation types at Matopos, Rhodesia. *Proc. 11th Annu. Tall Timbers Fire Ecol. Conf.* pp. 53–98.

Komarek, E. V. (1964). The natural history of lightning. *Proc. 3rd Annu. Tall Timbers Fire Ecol. Conf.* pp. 139–183.

Komarek, E. V. (1965). Fire ecology—grasslands and man. *Proc. 4th Annu. Tall Timbers Fire Ecol. Conf.* pp. 169–220.

Komarek, E. V. (1966). The meteorological basis for fire ecology. *Proc. 5th Annu. Tall Timbers Fire Ecol. Conf.* pp. 85–125.

Komarek, E. V. (1967a). Fire—and the ecology of man. *Proc. 6th Annu. Tall Timbers Fire Ecol. Conf.* pp. 143–170.

Komarek, E. V. (1967b). The nature of lightning fires. *Proc. 7th Annu. Tall Timbers Fire Ecol. Conf.* pp. 5–41.

Komarek, E. V. (1968). Lightning and lightning fires as ecological forces. *Proc. 8th Annu. Tall Timbers Fire Ecol. Conf.* pp. 169–197.

Komarek, E. V. (1969). Fire and animal behavior. *Proc. 9th Annu. Tall Timbers Fire Ecol. Conf.* pp. 160–207.

Komarek, E. V. (1971). Lightning and fire ecology in Africa. *Proc. 11th Annu. Tall Timbers Fire Ecol. Conf.* pp. 473–511.

Kucera, C. L. (1970). Ecological effects of fire on tallgrass prairie. *Proc. Symp. Prairie Prairie Restoration, 1968,* p. 12.

Kucera, C. L., and Ehrenreich, J. H. (1962). Some effects of annual burning on central Missouri prairie. *Ecology* 43, 334–336.

Kucera, C. L., and Koelling, M. (1964). The influence of fire on composition of central Missouri prairie. *Amer. Midl. Natur.* 72, 142–147.

Lapham, I. A. (1965). The great fires of 1871 in the Northwest (reprinted). *Wis. Acad. Rev.* 12, 6–9.

Launchbaugh, J. L. (1964). Effects of early spring burning on yields of native vegetation. *J. Range Manage.* 17, 5–7.

Lehmann, V. W. (1965). Fire in the range of Attwater's prairie chicken. *Proc. 4th Annu. Tall Timbers Fire Ecol. Conf.* pp. 127–143.

Lemon, P. C. (1949). Successional responses of herbs in the longleaf-slash pine forest after fire. *Ecology* 30, 135–145.

Lemon, P. C. (1967). Effects of fire on herbs of the southeastern United States and central Africa. *Proc. 6th Annu. Tall Timbers Fire Ecol. Conf.* pp. 112–127.

Lemon, P. C. (1968). Effects of fire on an African plateau grassland. *Ecology* **49**, 316–322.

Lemon, P. C. (1970). Prairie ecosystem boundaries in North America. *Proc. Symp. Prairie Prarie Restoration, 1968,* pp. 13–18.

Leopold, A. (1924). Grass, brush, timber and fire in southern Arizona. *J. Forest.* **22**, 1–10.

Leopold, A. (1933). "Game Management." Scribner's, New York.

Lewis, C. E. (1964). Forage response to month of burning. *U.S., Forest Serv., Exp. Sta., Res. Note* **SE-35**, 4.

Lindenmuth, A. W., and Byram, G. M. (1948). Headfires are colder near the ground than backfires. *Fire Control Notes* **9**, 8–9.

Lindsey, A. A., Schmelz, D. V., and Nichols, S. A. (1970). "Natural Areas in Indiana and Their Preservation." Amer. Midl. Natur., Notre Dame, Indiana.

Lloyd, P. S. (1971). Effects of fire on the chemical status of herbaceous communities of the Derbyshire Dales. *J. Ecol.* **59**, 261–273.

Lloyd, P. S. (1972). Effects of fire on a Derbyshire grassland community. *Ecology* **53**, 915–920.

Loucks, O. L. (1968). Scientific areas in Wisconsin: Fifteen years in review. *BioScience* **18**, 396–398.

Loucks, O. L. (1970). Evolution of diversity, efficiency, and community stability. *Amer. Zool.* **10**, 17–25.

Love, R. M. (1970). The rangelands of the western U.S. *Sci. Amer.* **222**, 88–96.

McArthur, A. G. (1966). Weather and grassland fire behavior. *Aust. Forest. Res. Inst. Leafl.* **100**, 23.

McCalla, T. M. (1943). Microbiological studies of the effect of straw used as a mulch. *Trans. Kans. Acad. Sci.* **43**, 2–56.

McIlvain, E. H., and Armstrong, C. G. (1966). A summary of fire and forage research on shinnery oak rangelands. *Proc. 5th Annu. Tall Timbers Fire Ecol. Conf.* pp. 127–129.

McIntosh, R. P. (1967). An index of diversity and the relation of certain concepts to diversity. *Ecology* **48**, 392–404.

McKell, C. M., Wilson, A. M., and Kay, B. L. (1962). Effective burning of rangelands infested with medusahead. *Weeds* **10**, 125–131.

MacMillan, C. (1959). The role of ecotypic variation in the distribution of the central grassland of North America. *Ecol. Monogr.* **29**, 287–308.

McMurphy, W. E. (1963). Burning Flint Hills grassland: Effects on range condition, forage production, and soil moisture. Ph.D. Thesis, Kansas State University, Lawrence.

Major, J., McKell, C. M., and Berry, L. J. (1960). Improvement of medusahead infested rangelands. *Calif., Agr. Exp. Sta., Ext. Serv. Leafl.* **123**, 8.

Malin, J. C. (1967). "The Grassland of North America. Prolegomena to its History with Addenda and Postscript." P. Smith, Gloucester, Massachusetts.

Mark, A. F. (1965). Flowering behavior of New Zealand mountain grass. *Naturwissenschaften* **7**, 180–193.

Marshall, J. T. (1963). Fire and birds in the mountains of southern Arizona. *Proc. 2nd Annu. Tall Timbers Fire Ecol. Conf.,* pp. 135–142.

Martin, R. E., and Cushwa, C. T. (1966). Effects of heat and moisture on leguminous seed. *Proc. 5th Annu. Tall Timbers Fire Ecol. Conf.* pp. 159–175.

Mes, M. G. (1958). The influence of veld burning or mowing on the water, nitrogen and ash content of grasses. *S. Afr. J. Sci.* **54**, 83–86.

Metz, L. J., Lotti, T., and Klawitter, R. A. (1961). Some effects of prescribed burning on coastal plain forest soil. *U.S., Forest Serv., Southeast. Forest Exp. Sta., Pap.* 133, 10.

Moir, W. H. (1972). Tall grass prairie in Colorado and its aesthetic value. *Midwest Prarie Conf., 2nd, 1970,* pp. 40–46.

Moore, A. W. (1961). The influence of annual burning on a soil in the derived savanna zone of Nigeria. *Trans. Int. Congr. Soil Sci., 7th, 1960* Vol. 4, pp. 257–264.

Moore, C. T. (1972). Man and fire in the central North American grassland 1535–1890: A documentary historical geography. Ph.D. Thesis, University of California, Los Angeles.

Morton, B. R. (1964). Fire and wind. *Sci. Progr. (London)* 52, 249–258.

Moss, E. H. (1932). The poplar association and related vegetation of central Alberta. *J. Ecol.* 20, 380–415.

Moss, E. H. (1952). Grassland of the Peace River region, western Canada. *Can. J. Bot.* 30, 98–124.

Muller, C. H. (1966). The role of chemical inhibition (allelopathy) in vegetational composition. *Bull. Torrey Bot. Club* 93, 332–351.

Mutch, R. W. (1970). Wildland fires and ecosystems—a hypothesis. *Ecology* 51, 1046–1051.

Nielsen, G. A., and Hole, F. D. (1963). A study of the natural processes of incorporation of organic matter into soil in the University of Wisconsin Arboretum. *Trans. Wis. Acad. Sci., Arts Lett.* 52, 213–227.

O'Connor, K. F., and Powell, A. J. (1963). Studies in the management of snow-tussock grassland. I. The effects of burning, cutting and fertilizer on narrow-leaved snow-tussock at a mid-altitude site in Canterbury, New Zealand. *N. Z. J. Agr. Res.* 6, 354–367.

Odum, E. P. (1969). The strategy of ecosystem development. *Science* 164, 262–270.

Old, S. M. (1969). Microclimates, fire, and plant production in an Illinois prairie. *Ecol. Monogr.* 39, 355–384.

Olindo, P. M. (1971). Fire and conservation of the habitat in Kenya. *Proc. 11th Annu. Tall Timbers Fire Ecol. Conf.* pp. 243–257.

Orpurt, P. A., and Curtis, J. T. (1957). Soil microfungi in relation to the prairie continuum in Wisconsin, *Ecology* 38, 628–637.

Parrott, R. T. (1967). A study of wiregrass (*Aristida stricta* Michx.) with particular reference to fire. M.A. Thesis, Duke University, Durham, North Carolina.

Pearson, G. A. (1936). Why the prairies are treeless. *J. Forest.* 34, 405–408.

Pechanec, J. F., and Hull, A. C., Jr. (1945). Spring forage lost through cheatgrass fires. *Nat. Wool Grower* 35, 13.

Penfound, W. T. (1968). Influence of a wildfire in the Wichita Mountains Wildlife Refuge, Oklahoma. *Ecology* 49, 1003–1006.

Penfound, W. T., and Kelting, R. W. (1950). Some effects of winter burning on a moderately grazed pasture. *Ecology* 31, 554–560.

Phillips, J. (1965). Fire as a master and servant: Its influence in the bioclimatic regions of Trans-Sahara Africa. *Proc. 5th Annu. Tall Timbers Fire Ecol. Conf.* pp. 7–109.

Philpot, C. W. (1968). Mineral content and pyrolysis of selected plant materials. *U.S., Forest Serv., Res. Note* INT-84, 4.

Philpot, C. W. (1970). Influence of mineral content on the pyrolysis of plant materials. *Forest Sci.* 16, 461–471.

Philpot, C. W. (1971). The pyrolytic effect of treating cottonwood with plant ash. *U.S., Forest Serv., Res. Note* INT-139, 5.

Pickford, G. D. (1932). The influence of continued heavy grazing and of promiscuous burning on spring–fall ranges in Utah. *Ecology* 12, 159–171.

Pritchett, W. L. (1971). Comments: Effects of prescribed burning on soils. *Prescribed Burning Symp. Proc., 1971,* pp. 97–99.

Quinnild, C. L., and Cosby, H. E. (1958). Relicts of climax vegetation on two mesas in western North Dakota. *Ecology* 39, 29–32.

Ralston, R. D., and Dix, R. L. (1966). Green herbage production of native grasslands in the Red River Valley—1965. *Proc. N. Dak. Acad. Sci.* 20, 57–66.

Ramsay, J. M., and Rose Innes, R. (1963). Some quantitative observations on the effects of fire on the Guinea Savanna vegetation on northern Ghana over a period of eleven years. *Sols Afr.* 8, 41–85.

Reynolds, H. G., and Bohning, J. W. (1956). Effects of burning on a desert grass-shrub range in southern Arizona. *Ecology* 37, 769–777.

Rice, E. L. (1967). Chemical warfare between plants. *BioScience* 38, 67–74.

Rice, E. L., and Penfound, W. T. (1959). The upland forests of Oklahoma. *Ecology* 40, 593–607.

Richards, M. S. (1972). Management of Kalsow Prairie. *Midwest Prairie Conf., 2nd, 1970,* pp. 30–33.

Riechert, S. E., and Reeder, W. G. (1972). Effects of fire on spider distribution in southwestern Wisconsin prairie. *Midwest Prairie Conf., 2nd, 1970,* pp. 73–90.

Robertson, J. H., and Cords, H. P. (1957). Survival of rabbitbrush, *Chrysothamnus* spp., following chemical, burning, and mechanical treatments. *J. Range Manage.* 10, 83–89.

Robocker, W. C., and Miller, B. J. (1955). Effects of clipping, burning and competition on establishment and survival of some native grasses in Wisconsin. *J. Range Manage.* 8, 117–120.

Rose Innes, R. (1971). Fire in West African vegetation. *Proc. 11th Annu. Tall Timbers Fire Ecol. Conf.* pp. 147–173.

Sauer, C. O. (1950). Grassland climax, fire, and man. *J. Range Manage.* 3, 16–21.

Schiff, A. L. (1962). "Fire and Water." Harvard Univ. Press, Cambridge, Massachusetts.

Schroeder, M. J., and Buck, C. C. (1970). Fire weather, *U.S., Forest Serv., Agr. Handb.* 360, 229.

Scott, J. D. (1971). Veld burning in Natal. *Proc. 11th Annu. Tall Timbers Fire Ecol. Conf.* pp. 33–51.

Sharp, L. A., Hironaka, M., and Tisdale, E. W. (1957). Viability of medusa-head (*Elymus caput-medusae* L.) seed collected in Idaho. *J. Range Manage.* 10, 123–126.

Shaw, N. H. (1957). Bunch spear grass dominance in burnt pastures in southeastern Queensland. *Aust. J. Agr. Res.* 8, 325–334.

Skovlin, J. M. (1971). The influence of fire on important range grasses of East Africa. *Proc. 11th Annu. Tall Timbers Fire Ecol. Conf.* pp. 201–217.

Smith, E. F., Young, V. A., Anderson, K. L., Ruliffson, W. S., and Rogers, S. N. (1960). The digestibility of forage on burned and non-burned bluestem pasture as determined with grazing animals. *J. Anim. Sci.* 19, 388–391.

Southwell, B. L., and Hughes, R. H. (1965). Beef cattle management practices for wiregrass-pine ranges of Georgia. *Ga., Agr. Exp. Sta., Bull.* 129, 26.

Staples, R. R. (1945). Veld burning. *Rhodesia Agr. J.* 42, 44–52.

Stebbins, G. L. (1972). The evolution of the grass family. *In* "The Biology and Utilization of Grasses" (V. B. Youngner and C. M. McKell, eds.), pp. 1–17. Academic Press, New York.

Steenbergh, W. F. (1972). Lightning-caused destruction in a desert plant community. *Southwest. Natur.* **16**, 419–429.

Stewart, O. C. (1951). Burning and natural vegetation in the United States. *Geogr. Rev.* **41**, 317–320.

Stewart, O. C. (1955). Forest and grass burning in the mountain West. *Southwest. Lore* **21**, 5–8.

Stewart, O. C. (1956). Fire as the first great force employed by man. *In* "Man's Role in Changing the Face of the Earth" (W. L. Thomas, ed.), pp. 115–133. Univ. of Chicago Press, Chicago, Illinois.

Stewart, O. C. (1963). Barriers to understanding the influence of use of fire by aborigines on vegetation. *Proc. 2nd Annu. Tall Timbers Fire Ecol. Conf.* pp. 117–126.

Stewart, W. D. P. (1967). Nitrogen-fixing plants. *Science* **158**, 1426–1432.

Stinson, K. J., and Wright, H. A. (1969). Temperatures of headfires in the southern mixed prairie of Texas. *J. Range Manage.* **22**, 169–174.

Stone, E. C. (1965). Preserving vegetation in parks and wilderness. *Science* **150**, 1261–1266.

Swan, F. R., Jr. (1970). Post-fire response of four plant communities in south-central New York state. *Ecology* **51**, 1074–1082.

Tester, J. R., and Marshall, W. H. (1961). A study of certain plant and animal interrelations on a native prairie in northwestern Minnesota. *Occas. Pap. Minn. Mus. Natur. Hist.* No. 8.

Thompson, P. W. (1972). The preservation of prairie stands in Michigan. *Midwest Prairie Conf., 2nd, 1970,* pp. 13–14.

Trlica, M. J., Jr., and Schuster, J. L. (1969). Effects of fire on grasses of the Texas High Plains. *J. Range Manage.* **22**, 329–333.

Trollope, W. S. W. (1971). Fire as a method of eradicating macchia vegetation in the Amatole Mountains of South Africa—experimental and field scale results. *Proc. 11th Annu. Tall Timbers Fire Ecol. Conf.* pp. 99–120.

Tryon, E. H. (1948). Effect of charcoal on certain physical, chemical, and biological properties of forest soils. *Ecol. Monogr.* **18**, 81–115.

Uman, M. A. (1969). "Lightning." McGraw-Hill, New York.

Van Rensburg, H. J. (1971). Fire: Its effect on grasslands, including swamps— southern, central and eastern Africa. *Proc. 11th Annu. Tall Timbers Fire Ecol. Conf.* pp. 175–199.

Van Wyk, P. (1971). Veld burnings in the Kruger National Park. An interim report of some aspects of research. *Proc. 11th Annu. Tall Timbers Fire Ecol. Conf.* pp. 9–31.

Vesey-FitzGerald, D. (1963). Central African grasslands. *J. Ecol.* **51**, 243–274.

Vesey-FitzGerald, D. (1971). Fire and animal impact on vegetation in Tanzania National Parks. *Proc. 11th Annu. Tall Timbers Fire Ecol. Conf.* pp. 297–317.

Vestal, A. G. (1914). A black-soil prairie station in northeastern Illinois. *Bull. Torrey Bot. Club* **41**, 351–364.

Viosca, P., Jr. (1931). Spontaneous combustion on marshes of southern Louisiana. *Ecology* **12**, 439–442.

Vogl, R. J. (1964a). The effects of fire on the vegetational composition of bracken-grasslands. *Trans. Wis. Acad. Sci., Arts Lett.* **53**, 67–82.

Vogl, R. J. (1964b). Vegetational history of Crex Meadows, a prairie savanna in northwestern Wisconsin. *Amer. Midl. Natur.* **72**, 157–175.

Vogl, R. J. (1965). Effects of spring burning on yields of brush prairie savanna. *J. Range Manage.* **18**, 202–205.

Vogl, R. J. (1967a). Controlled burning for wildlife in Wisconsin. *Proc. 6th Annu. Tall Timbers Fire Ecol. Conf.* pp. 47–96.

Vogl, R. J. (1967b). Fire adaptations of some southern California plants. *Proc. 7th Annu. Tall Timbers Fire Ecol. Conf.* pp. 79–109.

Vogl, R. J. (1969a). One-hundred and thirty years of plant succession in a southeastern Wisconsin lowland. *Ecology* **50**, 248–255.

Vogl, R. J. (1969b). The role of fire in the evolution of the Hawaiian flora and vegetation. *Proc. 9th Annu. Tall Timbers Fire Ecol. Conf.* pp. 5–60.

Vogl, R. J. (1969c). Quantitative ecology: Comments and criticism. *Biologist* **51**, 85–90.

Vogl, R. J. (1970a). Fire and plant succession. *Symp. Role Fire Interm. West, 1970,* pp. 65–75.

Vogl, R. J. (1970b). Fire and the northern Wisconsin pine barrens. *Proc. 10th Annu. Tall Timbers Fire Ecol. Conf.* pp. 175–209.

Vogl, R. J. (1972). Effects of fire on southeastern grasslands. *Proc. 12th Annu. Tall Timbers Fire Ecol. Conf.* pp. 175–198.

Vogl, R. J. (1973). Effects of fire on the plants and animals of a Florida wetland. *Amer. Midl. Natur.* **89**, 334–347.

Vogl, R. J., and Beck, A. M. (1970). Response of white-tailed deer to a Wisconsin wildfire. *Amer. Midl. Natur.* **84**, 269–272.

Wahlenberg, W. G., Greene, S. W., and Reed, H. R. (1939). Effects of fire and cattle grazing on longleaf pine lands, as studied at McNeill, Mississippi. *U.S., Dep. Agr., Tech. Bull.* **683**, 52.

Weaver, J. E. (1954). "North American Prairie." Johnson Publ. Co., Lincoln, Nebraska.

Weaver, J. E., and Albertson, F. W. (1956). "Grasslands of the Great Plains." Johnson Publ. Co., Lincoln, Nebraska.

Weaver, J. E., and Clements, F. E. (1938). "Plant Ecology." McGraw-Hill, New York.

Weaver, J. E., and Rowland, N. W. (1952). Effects of excessive natural mulch on development, yield and structure of native grassland. *Bot. Gaz. (Chicago)* **114**, 1–19.

Wedel, W. R. (1957). The central North American grassland: Man-made or natural? *Soc. Sci. Monogr.* **3**, 36–46.

Wedel, W. R. (1961). "Prehistoric Man on the Great Plains." Univ. of Oklahoma Press, Norman.

Wells, B. W., and Shunk, I. V. (1931). The vegetation and habitat factors of the coarser sands of the North Carolina coastal plain: An ecological study. *Ecol. Monogr.* **1**, 465–520.

Wells, P. V. (1965). Scarp woodlands, transported grassland soils, and concept of grassland climate in the Great Plains region. *Science* **148**, 246–249.

Wells, P. V. (1970). Postglacial vegetational history of the Great Plains, new evidence reopens the question of the origin of treeless grasslands. *Science* **167**, 1574–1582.

West, O. (1965). "Fire in Vegetation and its Use in Pasture Management, with Special Reference to Tropical and Sub-tropical Africa" (Mimeo.). Commonwealth Bureau of Pastures and Field Crops, Hurley, Berkshire, England.

West, O. (1971). Fire, man, and wildlife as interacting factors limiting the development of climax vegetation in Rhodesia. *Proc. 11th Annu. Tall Timbers Fire Ecol. Conf.* pp. 121–145.

Wheater, R. J. (1971). Problems of controlling fires in Uganda national parks. *Proc. 11th Annu. Tall Timbers Fire Ecol. Conf.* pp. 259–275.

Wilde, S. A. (1958). "Forest Soils." Ronald Press, New York.

Wilson, R. E. (1970). The role of allelopathy in old-field succession on grassland areas of central Oklahoma. *Proc. Symp. Prairie Prairie Restoration, 1968,* pp. 24–25.

Wolfe, C. W. (1972). Effects of fire on a Sand Hills grassland environment. *Proc. 12th Annu. Tall Timbers Fire Ecol. Conf.* pp. 241–255.

Wright, H. A. (1969). Effect of spring burning on tobosa grass. *J. Range Manage.* **22,** 425–427.

Wright, H. A. (1971). Why squirreltail is more tolerant to burning than needle-and-thread. *J. Range Manage.* **24,** 277–284.

Wright, H. A. (1972). A progress report on techniques to burn dozed juniper. *Proc. 12th Annu. Tall Timbers Fire Ecol. Conf.* pp. 169–174.

Youngner, V. B., and McKell, C. M., eds. (1972). "The Biology and Utilization of Grasses." Academic Press, New York.

Zedler, J., and Loucks, O. L. (1969). Differential burning response of *Poa pratensis* fields and *Andropogon scoparius* prairies in central Wisconsin. *Amer. Midl. Natur.* **81,** 341–352.

. 6 .

Effects of Fires on Temperate Forests:

North Central United States

C. E. Ahlgren

I. Introduction

As a dramatic agent of change, wildfire has played a major role in shaping the vegetation of north central United States and adjacent central Canada for many centuries. Telltale layers of charcoal found in soil profiles reveal that at least 95% of the forests of the area burned at one time or another (Maissurow, 1935) (Fig. 1). Charcoal in different

Fig. 1. Charcoal layer between humus and mineral soil, indicating fire origin of existing 80-year-old jack pine forest, northeastern Minnesota.

strata of peat in northern Minnesota shows that fire occurred periodically in the area long before white settlement (Soper, 1919). Nearly all forest stands in northeastern Minnesota have burned from one to several times in the past three or four centuries (Heinselman, 1969). Most of the natural, even-aged pine stands in the area developed after fire. Uneven-aged stands of mixed tree species usually have developed on land that has long escaped extensive burning. Much of the grassland—the "stump prairies" of the area—also developed after repeated fires following logging (Vogl, 1964b).

II. Physical Factors in Ecological Change

Fires in this region usually destroy the aboveground portions of all existing vegetation, in contrast to fires in some parts of southern and western portions of North America where fire-tolerant tree species frequently survive repeated burning. In most cases, fire occurs only once in the rotation period of the tree species involved. Sudden, drastic changes in the physical environment are closely associated with the vegetation removal. The most obvious of these changes is the increase in light intensity, the direct result of removal of vegetation. If this were the only change, however, the postfire vegetation would be similar to the vegetation which follows logging. There are, however, more fundamental changes in soil moisture, nutrients, and texture discussed in other chapters, plus changes in temperature and humidity.

A. Postfire Soil Temperature Extremes

The alterations in postfire soil temperature are most evident on the soil surface, where the average maximum temperatures at midday may be from 5° to 30°F higher than on comparable, unburned lands (Harper, 1944; Shirley, 1932; Wahlenberg *et al.*, 1939; Beaufait, 1960a; Kittredge, 1938; Maini and Horton, 1966; Ehrenreich, 1959; Beaton, 1959; Old, 1969; Lloyd, 1965; Scotter, 1963, 1964; Isaac, 1930; Hensel, 1923; Lipas and Mäki-Petäys, 1961). In northern Minnesota jack pine (*Pinus banksiana*) stands, maximum temperatures 2 inches below the surface averaged 1°F warmer on burned sites. Minimum temperatures 2 inches below the surface on both burned and unburned sites were the same. In other areas, differences detectable from 1 to 6 inches below the surface have been reported (Maini and Horton, 1966; Scotter, 1963, 1964; Smith, 1951). Such differences continue for 2–20 years, depending on vegetation and soil conditions.

Increases in soil temperature extremes following fire have been found in various regions (Pearse, 1943; Ahlgren, 1970). In prescribed burning experiments in logged, northeastern Minnesota jack pine forests in which the slash was scattered, the writer found that both maximum and minimum temperature extremes were greater on the soil surface than on comparable uncut, unburned lands. Maxima averaged 3.6° higher and minima 5.4°F lower during June, July, and August for 7 years after burning. During years 8–11, differences were not as evident. The only exception to this condition was the month of June, the year of fire, when maximum peaks were frequently higher on unburned land. This difference may be related to the complete freedom of air movement on the denuded, burned surface prior to recovery of vegetation. Air temperatures at 20 cm and 3 m above the soil surface were similar on burned and unburned land, but somewhat lower minima were recorded on burned sites, undoubtedly also the result of lack of insulation during cool night hours (Ahlgren, 1970).

B. POSTFIRE SOIL TEMPERATURE DURATION

Of perhaps more significance to recovering vegetation are differences in duration of warmer temperatures noted on the northern Minnesota jack pine prescribed burn areas (Fig. 2). Soil surface temperatures exceeded 140°F for 2 hr and 122°F for 2 hr up to 3 and 8 consecutive

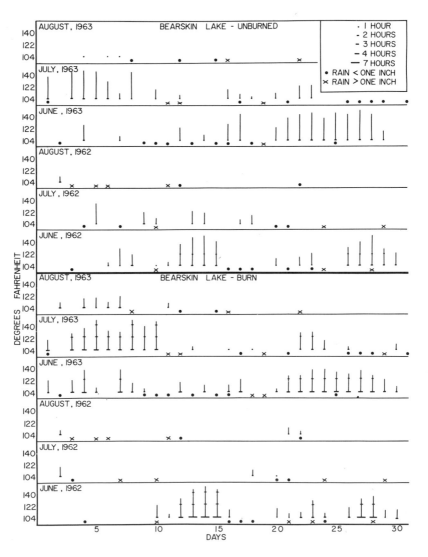

Fig. 2. High temperature duration at soil surface, burned and unburned jack pine forest, northeastern Minnesota.

days, respectively, during June and July of the first two postfire growing seasons. On comparable, unburned land, maximum temperatures were similar or slightly lower, but of shorter duration, usually less than $\frac{1}{2}$ hr, on fewer consecutive days.

The greatest increase in soil temperature on burned land may occur in the spring, before the surface is covered with vegetation (Hensel, 1923; Old, 1969). This heat can hasten spring development of plants on burned areas and increase flowering of some species, especially grasses (Ehrenreich, 1959; Curtis and Partch, 1950). In other cases, the increase in heat can be sufficient to kill young tree seedlings and impede forest development (Boyce, 1925; Isaac, 1929; McArdle and Isaac, 1934; Tryon, 1948). LeBarron (1944) reported jack pine seedling mortality of 5% and black spruce (*Picea mariana*) mortality of 25% as a result of heat damage on burned land. Since these species and others survive well on burned land throughout the area, however, it would appear that postfire heat is not usually a determining factor in overall seedling mortality. Increase in temperature has been related to increased light absorption by blackened soil surface and to the presence of charcoal in the soil (Isaac, 1929; Lutz, 1956; Shirley, 1932; Tryon, 1948). However, reports of higher day and lower night temperatures would seem to relate such temperature changes more closely to the lack of insulating vegetation and duff.

C. RELATIVE HUMIDITY

In northern Minnesota, the minimum relative humidity 20 cm above ground is consistently lower on burned than on unburned lands. This difference is probably related to the drying effect of the greater air movement on burned land, the result of sparse cover of newly established postfire vegetation. Maximum relative humidty at this level was similar on both burned and unburned land.

Postfire vegetation, then, develops on land which has been altered in soil chemistry, texture, and moisture. In addition, the surface 2 inches of soil, where the new roots first become established, are subject to longer periods of warm temperature and wide fluctuations in both temperature and humidity.

III. Fire Adaptation in Pine Species

Among the native tree species, white (*Pinus strobus*), red (*P. resinosa*), and jack pines are frequently cited as typical postfire species. Many workers have concluded that these species do not reestablish themselves extensively without the aid of fire in either the natural forest or after logging (Maissurow, 1941; Van Wagner, 1970; Ahlgren, 1960). There is a close relationship between the silvicultural characteristics of these species and the postfire environment.

A. FLAMMABILITY AND RESISTANCE TO FIRE DAMAGE

The high flammability of needles and duff in a pine forest enhances the probability that fire will occur. Pure stands of red pine are considered the most flammable because the crown structure is ideal for the spread of the crown fires, and the well-aerated, noncompacting litter burns readily (Van Wagner, 1970). Since fire hazard is high in large acreages of pure red pine, the current practice of planting red pine in pure stands may be increasing future fire danger. Red and white pine occur naturally in mixed stands, often associated with shade-tolerant hardwoods and brush understory. Fire hazard is usually less in such mixed stands. Naturally occurring pure stands of jack pine are more common and more flammable than these stands of mixed red pine, white pine, and deciduous species.

As mentioned earlier, most extensive fires in this area have been characterized by high mortality of standing timber. In mature, mixed red and white pine stands, fire is less likely to "crown" than in younger stands of these species or in jack pine stands where crowns are closer to the ground. Bark of older red pine is corky, thicker, and more resistant to fire than that of jack or white pine. Consequently, stems of older red pine trees are more likely to survive ground fires in areas where no crown damage occurs. Van Wagner (1970) found that more red pines are destroyed in fire by crown damage than by stem damage. Lower temperatures are required to destroy the crown than to destroy the stem cambium. Fire scars are not unusual on trunks of large red pines, indicating survival in past fires. Spurr (1954) reported a 220-year-old red pine with fire scars at six different ages in its life. Such survival is a negligible part of fire ecology in this region. Fire scars are rare in old jack and white pine stands.

B. FIRE ADAPTATION OF JACK PINE CONES

Nonserotinous cones of red and white pine are easily destroyed by crown fire or scorch, whereas the serotinous cones of most jack pines remain closed on live trees for many years (Fig. 3). Although seed

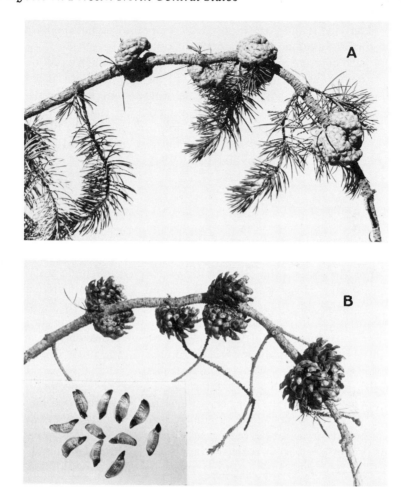

Fig. 3. Jack pine cones. A, Closed; B, open after fire. Inset shows seeds shed by cones in B.

viability decreases with age of cone, viable seeds have been obtained from 20-year-old cones (Roe, 1963; Eyre and LeBarron, 1944; Ahlgren, 1959). The resinous material closing the cones is destroyed by heat, allowing cone scales to open readily and release seed quickly after fire. Cones open best after exposure to temperatures between 120° and 140°F (Beaufait, 1960b; Reitz, 1937; Cameron, 1953; Eyre, 1938; Cayford, 1963). In prescribed burn areas in northern Minnesota, temperatures reached between 122° and 212°F for 7 min at a height of 13 m in the tree crowns (Ahlgren, 1970). Seed within the cones can remain viable even when cones are exposed to temperatures between 700° and 1000°F for brief periods (Beaufait, 1960b).

Some jack pine do produce nonserotinous cones which open naturally without heat. These are more prevalent in the southern part of the region. It has been suggested that the serotinous cone is a fire-adaptation which has increased through natural selection in the northern conifer areas where fire is more prevalent (Rudolph *et al.*, 1957). Black spruce is frequently associated with northern jack pine forests, and its semiserotinous cones could also be fire adaptations.

C. SEED PRODUCTION

The three major pine species of the area differ in their rates of seed production. This difference has a marked effect upon their reestablishment after fire. Jack pine cones with viable seeds remain on the tree for many years. Hence seeds of this species are accumulated and are readily available for immediate germination whenever fire occurs. Red and white pines do not accumulate seeds but shed them during the early autumn of each seed year. Bumper seed crops occur every 3–5 years in white pine and about every 5–10 years in red pine. The probability that fire will occur when abundant seed is available on surviving or nearby trees is much reduced for these latter two species. In areas where spring and summer fires occur, invading vegetation becomes established and may provide serious competition before pine seed is available from adjacent areas. It is only where fires occur in late summer or autumn before seed is shed that a favorable environment for the establishment of red and white pine is created. Maissurow (1935) found that the disappearance of pure white pine stands was not solely the direct effect of either forest fires or logging but was caused by a disturbed balance between seed bearing capacity of the forest and frequency or destructiveness of forest fires. Seed of jack pine continues to be shed for 3 years after fire (Ahlgren, 1970).

These fire-oriented characteristics of jack pine have been responsible for the postfire conversion of many mixed red and white pine stands to jack pine. Prior to the use of jack pine for pulp production, red and white pines were harvested for timber, and the jack pine was left standing. Slash fires were common, advance reproduction and the few remaining seed trees of red and white pine were destroyed, and seeds of scattered jack pine were released (Wackerman *et al.*, 1929; Eyre, 1938; Eyre and LeBarron, 1944; Kilburn, 1960, Cooper, 1961; Cayford, 1963).

Fig. 4. Upper, unburned humus on soil surface before fire. Lower, humus after prescribed burning. Note reduction in depth.

D. SEEDLINGS

Because of their short, slow growing vertical roots, pine seedlings survive best in mineral soil or on reduced, burned humus 1 inch or less in depth (Fig. 4). Without adequate moisture, this type of root is unable to penetrate the thick, dry, fibrous humus layers to the moisture in the mineral soil below, before the shoots develop water deficits as a result of transpiration. White pine is more adaptable than the other pines to a thicker humus layer. Smith (1968) reported that roots of white pine seedlings could penetrate as much as 2 inches of pine needle litter within 2 weeks, but survival was good only if there were heavy and frequent rains during the germination period. The fleshy cotyledons or endosperm of some hardwood seeds permit development of 5- to 6-inch seedling root systems before the first leaves begin to function. Thus, the roots of such hardwood species are able to penetrate unburned, deep litter to mineral soil before the leaves have created an increased water demand.

Ooyama (1954) reported that extracts of humus contain substances that inhibit pine germination. Others report no evidence of this (Farrar and Fraser, 1953; Jarvis, 1966) and point to limited moisture as the cause of low germination on humus seedbeds. Jarvis found that on humus, initial jack pine germination is good, but survival to the second year is much reduced. Litter removal in nature is most frequently accomplished by burning. Consequently, all three of the native pine species are associated with postfire vegetation. Without the fire-caused alteration of duff-humus, it is only where trees are uprooted or mineral soil exposed in other ways that pine seedlings become established. Logging operations often expose patches of mineral soil, and if viable seed is available soon after logging and competition of other vegetation not severe, pines can become established.

The first year after fire, moisture content immediately below the soil surface may be 5% higher on burned land than on comparable, unburned land, the result of increased evaporation and capillarity of the burned surface (Ahlgren, 1970). Although species typical of the postfire vegetation, including pines, have relatively low water demands, they undoubtedly benefit from this first-year increase.

In the early first-year stages of growth, pine seedlings thrive best in partial shade (Fraser and Farrar, 1953), a condition which may be related to the tenuous hold these young seedlings have on moisture supply. On burned land, shade is usually provided by the vigorous herbaceous growth that covers the area during the first growing season after fire (Ahlgren, 1959) and by the standing and fallen dead trees (Eyre and LeBarron, 1944). In later years, seedlings of all three species become more shade intolerant, although there is a marked difference among species. Jack pine is the least shade tolerant; red pine is referred to as "low mid-tolerant," and white pine is classified as "shade high mid-tolerant" (Graham, 1954; Bakuzis, 1959). Five years after fire, light reaching the forest floor was reduced by 23%, 10 years after fire, light intensity was reduced by 47%, and there was an 88% reduction in light intensity on the forest floor in a mature jack pine stand. Light reduction at 4-ft and 6-ft levels occurs later, as the shrub and tree overstory close in, so that in the mature (80-year-old) jack pine forest, light intensity at the 6-ft level was reduced by 80% (C. E. Ahlgren, unpublished).

IV. Postfire Development of Other Plant Species

A. SPRUCE

Spruce has been associated with postfire vegetation in many parts of the United States. Mineral soil or reduced humus layer provides an ideal seedbed for the characteristically short, slow-growing spruce seedling root as it does for the pines. White spruce (*Picea glauca*) and black spruce, the two native species, may vary in their response to fire, however.

Reports of the overall effect of fire on the abundance of white spruce vary. Although fire improves the seedbed, it can eliminate all seed sources, since the seed and cone are easily destroyed by fire. Seeds are shed in the autumn, with good seed years every 3–4 years. Roe (1952) reported 11,900 cones with 271,000 viable seeds on an open grown 75-year-old white spruce tree in Minnesota. A late summer or early autumn fire during a good seed year in an area where such trees survive will result in the establishment of white spruce. Fires in the spring or early summer, however, usually favor the establishment of other tree species. Consequently, there are reports of favorable postfire white spruce reproduction (Rowe, 1970; Holman and Parker, 1940; Bedell, 1948; Millar, 1939), as well as reports of slow reestablishment of this species after fire (Lutz, 1956; Saari, 1923; Heikinheimo, 1915; Breitung, 1954; Minckler, 1945). In northern Minnesota, several mature seed trees adjacent to a late summer prescribed burn resulted in abundant spruce seedlings 2–3 years later. First-year seedling mortality of white spruce is often high, the result of frost heaving and smothering by leaf litter. Survival improves after the second growing season. Since seedlings of this species are shade tolerant, successful seedling establishment can continue for a number of years on burned land, as seed is available.

Black spruce, in contrast, has persistent, semiserotinous cones that retain seed for several years and are opened by heat. The cones are located near the center of the upper crown where they are least likely to be damaged by fire, especially in taller trees. Seed is shed gradually during the first postfire year (Wilton, 1963). When black spruce is present in stands which burn, sufficient seed is usually available for reestablishment (Bloomberg, 1950; Holman and Parker, 1940; LeBarron, 1940; MacLean, 1957; Millar, 1939). Further, since fire frequently destroys the seed crop in the shorter trees, a natural selection of genetic strains with most vigorous height growth may take place in areas where fires occur every century (Ahlgren, 1959). In northern Minnesota, most black spruce germinates on burned land the second and third seasons after fire, 1–2 years later than jack pine. This delay may be related to the susceptibility of young spruce seedlings to ash damage (Heikinheimo, 1915). After a season of rainfall, surface ash concentrations may be reduced to nontoxic levels by leaching. The formation of other toxic substances in soil when heated to high temperatures has been postulated (Pickering, 1910; Wilson, 1914). Since spruces are slower growing and more shade tolerant than pines, they become obvious on burned land more slowly and are frequently obscured by surrounding vegetation during their early growth. On organic soils in northern Minnesota, burning has favored black spruce reproduction in the preparation of seedbed and eradication of dwarf mistletoe (Johnston, 1971).

B.   OTHER CONIFERS

Balsam fir (*Abies balsamea*) cones are produced annually and disintegrate rapidly in the fall, leaving seeds on the forest floor where they are destroyed if fire occurs (MacLean, 1957). Some seed is distributed in caches by squirrels and chipmunks. Large quantities of seed are necessary for establishment on organic soil (Van Nostrand, 1965). Balsam fir has a high degree of shade tolerance (Bakuzis and Hansen, 1965). Although it appears sporadically a few years after fire, more pronounced reproduction occurs later after a new seed source has been established. In the Gaspé region, MacArthur (1964) reported peak reproduction 9 years after fire, declining to zero 13–18 years after fire. In other parts of Canada, balsam fir is known to increase in later years as the postfire forest matures (MacLean, 1957; Jarvis, 1966). In northern Minnesota, balsam fir reproduction continues and becomes prominent 30–50 years after fire. Reduction of fires by better fire control is a major reason given for the increase in balsam fir in the northern part of this area (Grant, 1934). Although balsam fir is not an immediate postfire species, it can often play a dramatic role in the fire sequence. The abundant, highly flammable fuel left following spruce budworm epidemics often sets the stage for future fires (Flieger, 1970).

In logged, burned-over, northern Minnesota swamps, tamarack (*Larix laricina*) reproduction becomes established successfully with black spruce, if adequate seed is available in adjoining areas (Johnston, 1973). The writer has observed occasional tamarack seedlings thriving on upland jack pine prescribed burn sites.

C.   DECIDUOUS TREE SPECIES

Both paper birch (*Betula papyrifera*) and trembling aspen (*Populus tremuloides*), the two deciduous species most frequently associated with early postfire vegetation in the region, sprout abundantly after fire and achieve height growth more rapidly than the seed-reproducing conifers during the first few years. By the tenth year, aspen and birch may attain a height of up to 12 ft, but an average height of 4–5 ft is common because of browsing. Although increased light intensity, the result of logging, stimulates aspen sprouting, burning accelerates the process (Lake States Forest Experiment Station, 1931; Shirley, 1932; Breitung, 1954; Lutz, 1956; Rowe, 1955; Horton and Hopkins, 1965), especially after spring fires. Shirley (1932) reported that increased heat in the soil around rhizomes during the first postfire year is a factor in this sprouting.

On moist sites, most successful birch reproduction is by the annually produced seed from nearby trees (Scotter, 1964; Lutz, 1956; Ahlgren, 1960). Birch seedlings are susceptible to desiccation, however, and high mortality is common. On drier sites, birch is reestablished from stump sprouts, where the previous stand contained abundant, well-distributed birch. Seed or vegetative origin of mature birch stands can be readily distinguished by the single stem growth habit of seed origin trees and the clumped growth of stump sprouts.

Most other species occur sporadically and are not directly associated with early postfire succession. One of the prime factors governing their behavior is lack of seed source. Those species which do not provide abundant seed at the time the new vegetation is becoming established develop more slowly later on. Most of these are characteristically shade tolerant and become established under the shade of the postfire forest. They become part of the mixed-species forest which develops when an area has been free of fire for many years. In North Dakota, a comparison of a 72-year-old burn on deciduous woods and a similar, unburned forest revealed the same species present in both areas, but in different frequencies after fire. Bur oak (*Quercus macrocarpa*), ash (*Fraxinus* sp.), and elm (*Ulmus americana*) were less frequent, while aspen increased (Potter and Moir, 1961). Rogers (1959) reported that white oak (*Quercus alba*) increased after fire. In northwestern Minnesota, it is believed that the paper birch–white pine–bur oak forests are being replaced by basswood (*Tilia americana*), fir, and black ash because of the decrease in fire frequency (Buell and Bormann, 1955).

D. Shrubs

The majority of the significant shrubs appearing after fire are those which were present in the preburn forest. Although they sprout vegetatively soon after fire and may change in frequency, few if any can be classified as "fire followers." No shrub species are completely eliminated by fire. Those which have a significant amount of seed reproduction may also appear soon after burning but are slower in reaching any size or abundance to be of ecological significance. Many shrubs respond favorably to increases in light intensity, sprouting rapidly on cutover as well as lightly burned land. When fires are moderate or severe, recovery is somewhat delayed.

Hazel (*Corylus cornuta*) is frequently a serious forest competitor. Buckman (1962, 1964) found hazel sprouting to be stimulated by light fires and retarded by severe fires. In northern Minnesota, fire frequently damages the root collar of hazel, reducing height growth and sprouting for 5 years after fire (Ahlgren, 1960). Early spring fires are not as damaging to growth of hazel as fires in late May, June, and July. On unburned, cutover land, recovery of hazel is rapid and vigorous.

Various species of blueberry, especially *Vaccinium angustifolium*, can be seriously retarded by severe fires (Uggla, 1950; Vogl, 1964a; Smith, 1968; Hall, 1955), although there are reports of induced sprouting and revitalized growth after burning (Sharp, 1970; Trevett, 1962) with complete recovery to abundant fruiting in 3 years. Swan (1970) reported blueberry to be relatively unaffected by fire in New York state. In northern Minnesota, complete recovery and abundant fruiting occurred 3–4 years after fire.

Bush honeysuckle (*Diervilla lonicera*) sometimes decreases in abundance after fire (Swan, 1970). In northern Minnesota, its abundance was relatively unaffected by fire, although variation in fire intensity affected survival of some less frequent shrubs. Canada honeysuckle (*Lonicera canadensis*), for example, is more frequent on lightly burned areas than on severely burned land.

The alders (*Alnus crispa* on well-drained sites and *A. rugosa* on moist sites) reach peak abundance 10 years after fire (Scotter, 1964).

Red raspberry (*Rubus idaeus*) reproduces both by seed and vegetative means after fire (Ahlgren, 1960; Yli-Vakkuri, 1961; Uggla, 1950) and is most frequent within 5–10 years. Seed of this species survives fire in the soil and may be stimulated to germinate by heat (Uggla, 1950). Early growth is enhanced by nitrates provided in the ash. A very vigorous reponse of raspberry occurs on unburned, cleared, scarified soil in northern Minnesota. This vegetative sprouting is stimulated by the disturbance and cutting of underground plant parts during mechanical scarification.

Seeds of pin cherry (*Prunus pensylvanica*) generally survive fire in the moist lower level of the organic soil which seldom burns completely. Numerous seedlings are often found the first year after fire. Although mortality is high, occasional plants become established as part of the high shrub or low tree canopy in the developing forest.

Sweet fern (*Comptonia peregrina*), which also reproduces by seed, is seldom a strong component of the dominant shrubs until the fifth year after fire. It reaches maximum abundance at about 10 years and then declines as shade-tolerant species compete.

Other shrubs, *Cornus, Rhus,* and *Salix* spp., are found sporadically and are usually not frequent until the fifth year following fire.

E. Herbs

Herbaceous vegetation usually develops first on recently burned land. Herbs become active earlier in the growing season than trees and shrubs, grow rapidly, and quickly invade open, burned land. This lush, rapid growth may be related to the fertilizing action of the mineral nutrients released by the ash (Ahlgren, 1959, 1960). Behavior of individual herb species within an area varies. Vegetative sprouting of some species may be enhanced. Seeds of other species survive fire in the lower layers of the organic soil and may be stimulated to germinate by heat (Stone and Juhren, 1951; Wright, 1931; Uggla, 1950). Some species thrive in increased light conditions; others may grow in shade after other vegetation has become established.

Postfire herbs are usually classified by their frequency after fire. In Wisconsin, Vogl (1964a) classified postfire plants as increasers, decreasers, neutrals, invaders, and retreaters. He found the largest portion of invading or increasing herbs in burned muskeg areas to be grasses and sedges.

In northern Minnesota on newly burned forest land, the majority of the first, temporary herbs reproduce by seed (Ahlgren, 1959, 1960, 1970). About one-third of these are wind disseminated and may be blown in after fire. Another one-third are disseminated by birds and small mammals which frequent newly burned lands (Ahlgren, 1966). Other species are undoubtedly brought in by deer and bear.

On recently burned areas, distribution of herb species is often a key to their means of dissemination. Plants of vegetative or seed origin which survived the fire in the soil are the first to appear abundantly over the entire burned area. The wind-borne seeds germinate later and are fairly evenly distributed. Plants brought in by animals and birds during the first 3 years after fire are scattered irregularly over the area. The vegetatively reproducing herbs are part of the more stable vegetation and are restored to preburn levels within a few years. Most of the species which occur on both burned and unburned land reproduce vegetatively, and their distribution is determined largely by their presence in the area before fire and their ability to survive in increased light.

F. Ferns

Among the ferns and fern allies, only bracken (*Pteridium aquilinum*) survives and becomes prominent soon after fire. Most reproduction is vegetative from surviving underground rhizomes. In northern Finland, Oinonen (1967) found that spores do not mature in time to germinate and become established before winter. Spores can be introduced earlier in the summer from southern areas, presumably in guano. Thus, sporelings of bracken are common in burned areas. Oinonen traced the origin of large clones to development from sporelings on battle-charred areas after wars. The same relationship of clone size to fire disturbance occurred in *Lycopodium clavatum* and *L. annotinum* (Oinonen, 1968). Rhizomes of other species are nearer the surface and are destroyed by fire. These species usually do not appear until a number of years after fire and are established by wind-disseminated spores. Many sporelings cannot survive in direct sunlight. In northern Minnesota, for example, *Polypodium virginianum* occurred only in older stands free from burning for many years.

G. Mosses and Lichens

The mosses and lichens of the forest floor are usually destroyed by fire, although Lutz (1956) found evidence that some moss species on burned areas in Alaska survived as fragments in the soil. Most research workers agree that there is a postfire succession of moss species which varies with stand type. For example, mosses and lichens are found in fewer numbers under pure aspen stands than under coniferous forests. Among coniferous stands, lichen development is greater under spruce than under jack pine stands (Lutz, 1956).

Many of the common moss species are rather universally distributed within climate zones and can be found in many countries. Among the frequently mentioned mosses and liverworts which colonize soon after fire and often disappear 5–10 years later are *Ceratodon purpureus, Polytrichum juniperinum*, and *P. piliferum, Marchantia polymorpha, Funaria hygrometrica*, and *Pohlia nutans* (Lutz, 1956; Sarvas, 1937; Scotter, 1964; Uggla, 1950; Aaltonen, 1919; Hustich, 1951; Graff, 1936; Lilienstern, 1929; Skutch, 1929; Summerhayes and Williams, 1926). Daubenmire (1949) suggested that the early postfire appearance of some of these species was related to their low nitrogen requirements which allow them to become established during a brief period of freedom from competition, and to their high pH tolerance.

Ten to 30 years later, these species are replaced by others, including

Aulacomnium palustre, Hedwigia ciliata, Pleurozium schreberi, Polytrichum commune, and *Hylacomnium splendens.* In Finland, Viro (1969) found that mosses were first evident 3 years after fire and were dominated by *Polytrichum* species, especially *P. juniperinum.* Thirty years after fire, 40% of the mosses were *Polytrichum* species. Fifty years after fire, *Polytrichum* still comprised 4% of the mosses on burned land as compared with 1% on unburned land. Feather mosses, *Pleurozium schreberi, Dicranum* spp., and *Hylacomnium splendens,* began appearing 10 years after fire. Recovery of feather mosses was slow, and 50 years after fire, only *Dicranum* spp. had recovered to preburn levels.

Although *Sphagnum* can survive burning and reestablish itself, frequently it may not be found in burned areas until more than 50 years after fire.

The succession of bryophyte species on northern Minnesota burned land is noted in Table I. *Marchantia polymorpha* was the most noticeable early fire follower. *Funaria hygrometrica* and *Polytrichum* spp. reached a peak during the 4- to 14-year period. *Pleurozium schreberi* and *Dicranum* spp. were more typical of the mature forest.

Lichens have rarely been identified in studies of postfire vegetation. Where investigated, however, they have been found also to follow a definite succession (Scotter, 1964; Uggla, 1950; Lutz, 1956). Hustich (1951) and Uggla (1950) found that reindeer moss (*Cladonia* spp.) in northern Europe frequently takes more than 40 years to recover completely from burning. Cup lichens (also *Cladonia* spp.), however, may recover completely in 20 to 30 years. These investigators estimated that complete recovery of typical forest lichen cover takes 30 to 40 years. In Finland, lichens appeared 7 years after fire. Lichen abundance was 5 times greater on 50-year-old burns than on unburned forest. This would

TABLE I

PERCENT COVER BY REPRESENTATIVE BRYOPHYTE SPECIES IN BURNED-OVER JACK PINE–BLACK SPRUCE FORESTS IN NORTHEASTERN MINNESOTA AT DIFFERENT INTERVALS AFTER FIRE

Species	Years after fire										
	1	2	3	4	5	10	15	20	30	50	80
Marchantia polymorpha	1	3	14	9	3	0	0	0	0	0	0
Polytrichum spp.	0	1	9	50	40	27	33	15	2	2	1
Funaria hygrometrica	0	0	0	1	3	3	4	0	0	0	0
Pleurosium schreberi	0	0	0	0	1	2	3	3	3	9	5
Dicranum spp.	0	0	0	0	1	1	1	1	1	1	6

suggest that stabilization of vegetation after burning, especially at this lowest level of the forest floor, takes many years (Viro, 1969). In northern Canada, Scotter (1964) found crustose lichens such as *Lecidea cuprea* and *Peltigera* sp. were the first to appear after burning. Frequently, *Peltigera* replaced the liverwort, *Marchantia*, about 5 years after fire. Cup lichens appeared before reindeer moss. Lichen growth was slower under recovering jack pine than under black spruce.

V. Interrelationships

A. HEIGHT LEVELS

Changes in height of plants in different strata of recovering vegetation in burned jack pine forests in northern Minnesota are shown in Fig. 5. Throughout the first 10 years, vegetative sprouting of aspen maintained a height dominance over other levels of vegetation and did not seriously affect the recovering herb, low shrub, or major shrub layers. Serious competition occurred only where the aspen was especially dense and interfered with developing jack pine seedlings below it. On burned land, aspen sprouting usually is confined to areas where aspen previously existed. On unburned, cut lands where the land is mechanically treated with rock raking, the underground rhizomes of aspen are broken up and distributed more extensively, creating a vigorous aspen forest over most of the area.

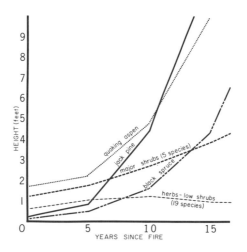

Fig. 5. Height of major vegetation, 0–15 years following fire in northeastern Minnesota jack pine–black spruce stand.

Fig. 6. Typical herb and low shrub vegetation layer under 30-year-old jack pine stand, northeastern Minnesota.

Jack pine becomes established during the first and second postfire years and is below the herb–low shrub layer for the first 5 years. By the seventh or eighth postfire year, it grows above the major shrub stratum. Black spruce, a more shade-tolerant species, usually appears the second and third postfire years. It grows more slowly, does not rise above the herb–low shrub layer until the eighth year, and above the major shrub layer until the thirteenth year after fire.

Height of the herb–low shrub layer does not change appreciably and becomes completely stabilized by the fifteenth year after fire (Fig. 6). Major shrubs, however, continue to increase in height beyond this period, especially in a mature forest where they often grow tall in response to shading of the overstory (Fig. 7).

B. Percent Cover

Changes in postfire percent cover of trees are presented in Table II. There is a close correlation between percent of space occupied by reproduction of each tree species and its height, as seen in Fig. 5. Aspen, because of vigorous vegetative sprouting, has a prominent percent cover the first postfire year and remains high for about 20 years. The slower growing conifers increase in percent cover as they rise above the herb–low shrub stratum between the fifth and tenth postfire years.

Fig. 7. Typical high shrub vegetation layer in mature jack pine–black spruce stand, northeastern Minnesota.

Hazel, usually the most prominent major shrub in jack pine–black spruce forests, increases in cover steadily as it increases in height. From the fifteenth postfire year to the mature forest, it covers from 30 to 50% of the area where it occurs. If moderately shade-intolerant tree

TABLE II

AVERAGE PERCENT COVER OF REPRODUCTION BY COMMON TREE
SPECIES IN BURNED-OVER JACK PINE–BLACK SPRUCE FORESTS IN
NORTHEASTERN MINNESOTA AT DIFFERENT INTERVALS AFTER FIRE[a]

Species	Years after fire										
	1	2	3	4	5	10	15	20	30	50	80
Pinus banksiana	1	1	1	1	3	14	48	55	1	1	1
Picea mariana	1	1	1	1	1	8	16	8	8	7	9
Populus tremuloides	5	12	16	19	13	10	6	8	1	1	1
Betula papyrifera	4	4	5	5	7	8	18	8	1	1	1
Picea glauca	0	0	0	0	1	1	1	3	2	2	2
Abies balsamea	0	0	0	0	1	1	1	2	2	7	2

[a] Figures for 30 years and later represent second-generation reproduction.

seedlings are to survive, they must become established soon after fire and grow above the shrub layer at the 2-ft level before the shrubs have attained a dense cover.

Cones begin to appear on postfire jack pines when the trees are 7 years old. Seeds from the few nonserotinous cones are shed, germinate, and many die. This process is repeated regularly and is reflected in the continued seedling count for this species shown in Table III. The mortality rate of these seedlings is reflected in the low count of larger trees in the maturing forest. Seedlings of the more shade-tolerant balsam fir and spruce, however, can become established as understory in mature pine stands. The reproductive activity of these species is indicated by the high seedling count found 15, 20, and 50 years after fire.

In mature white pine forests, bumper crops of seed are produced at 3- to 5-year intervals, and numerous seedlings cover the forest floor for a few years. Most of these die because of the inability of their roots to penetrate the thick, dry humus to the moist mineral soil below and because of their inability to survive in heavy shade. Abortive waves of white pine reproduction recur after every abundant seed year.

Most reproduction of the common shrubs—hazel, blueberry, raspberry, and bush honeysuckle—in postfire forests is vegetative. Overall abundance of these species is dependent on their prefire density and response to light and competition (Table IV). Hazel increases steadily in both percent cover and height until about the twentieth year after fire and continues as a dominant shrub in mature, old forests. Blueberry, which is less shade tolerant, reaches maximum abundance 10 years after fire, continues in abundance to the twentieth postfire year and declines to low quantities thereafter. Bush honeysuckle is sparsely distributed in the preburn forest and is relatively insensitive to differences in light intensity. It remains constant for about 50 years, declining only in the very old forest.

Raspberry, a shade-intolerant species, reproduces vegetatively and by seed and reaches its peak 5 to 10 years after fire. In Finland, Viro (1969) found that raspberry was prominent on burned land for about 6 years and declined rapidly thereafter. Raspberry seed may be spread by birds, and the heat of fire may stimulate germination (Uggla, 1950).

Sweet fern (*Comptonia peregrina*), which reproduces by seed, returns slowly in the early years and declines after the fifteenth postfire year, as shade-tolerant species compete. It is infrequent in the mature forest.

As a group, the shrubs increase temporarily in percent cover from the fifth to the twentieth years. This increase produces relatively dense shade at the 2- to 4-ft level which undoubtedly influences temperature, humidity, and vegetation below. Consequently, during this period shrub

TABLE III

Numbers of Seedlings and Advanced Growth per Acre by Major Species in Burned-Over Jack Pine–Black Spruce Forests in Northeastern Minnesota at Different Intervals after Fire[a]

Species	\\ Years after fire															
	1		5		10		15		20		30		50		80	
	<1	≥1	<1	≥1	<1	≥1	<1	≥1	<1	≥1	<1	≥1	<1	≥1	<1	≥1
Pinus banksiana	5872	0	4913	0	4212	580	3119	1938	1829	1457	0	1200	0	275	0	185
Picea mariana	607	0	3334	0	3610	0	2835	29	1066	281	485	200	453	104	233	94
Populus tremuloides	2308	0	1468	0	972	9	843	145	115	150	101	172	70	104	59	30
Betula papyrifera	1717	0	992	0	729	51	582	101	341	117	148	197	170	108	145	22
Picea glauca	0	0	152	0	187	0	253	0	486	4	749	47	162	56	22	21
Abies balsamea	0	0	13	0	15	0	33	0	50	2	384	24	1144	31	52	27

[a] <1 and ≥1 represent diameter at breast height in inches.

TABLE IV

AVERAGE PERCENT COVER BY COMMON SHRUB SPECIES IN BURNED-OVER
JACK PINE–BLACK SPRUCE FORESTS IN NORTHEASTERN MINNESOTA AT
DIFFERENT INTERVALS AFTER FIRE

Species	Years after fire							
	1	5	10	15	20	30	50	80
Corylus cornuta	7	17	23	32	51	45	47	42
Vaccinium angustifolium	8	17	30	25	27	17	10	8
Rubus idaeus	2	8	9	1	1	2	3	2
Diervilla lonicera	10	9	10	10	9	8	8	3
Comptonia peregrina	0	12	35	23	17	2	0	0

competition may be a determining factor in the growth of young tree
species.

Herbs exhibit several postfire recovery patterns which are related to
their modes of reproduction and shade tolerance. The vegetatively reproducing species which cover most of the forest floor are relatively unaffected by fire (Table V). *Aster* and *Pteridium* show a slight decrease

TABLE V

AVERAGE PERCENT COVER BY TYPICAL HERB SPECIES IN BURNED-OVER
JACK PINE–BLACK SPRUCE FORESTS IN NORTHEASTERN MINNESOTA AT
DIFFERENT INTERVALS AFTER FIRE

Species	Years after fire										
	1	2	3	4	5	10	15	20	30	50	80
Aster macrophyllus	30	31	44	42	35	22	25	24	35	41	74
Cornus canadensis	5	12	9	12	14	16	12	12	8	9	10
Pteridium aquilinum	24	27	22	25	19	14	12	14	13	17	35
Maianthemum canadense	1	1	1	2	3	3	6	2	2	3	3
Aralia nudicaulis	1	1	2	2	2	2	8	9	10	12	10
Geranium Bicknellii	8	7	1	1	1	1	0	0	0	0	0
Polygonum cilinode	1	4	3	2	2	2	0	0	0	0	0
Epilobium angustifolium	1	2	8	12	10	2	1	1	1	1	0
Fragaria vesca	1	2	2	3	4	5	1	1	1	1	1
Oryzopsis asperifolia	2	2	4	3	6	10	2	3	3	5	3
Calamagrostis canadensis	4	19	23	22	28	32	16	3	1	1	0
Trientalis borealis	1	1	1	1	1	1	2	1	1	2	1
Lycopodium spp.	1	2	2	1	3	2	3	6	5	6	6
Habenaria spp.	0	0	0	0	0	1	1	1	1	1	1
Goodyera spp.	0	0	0	0	0	1	1	1	1	1	1
Cypripedium acaule	0	0	0	0	0	1	1	1	1	1	1
Corallorhiza maculata	0	0	0	0	0	0	0	1	1	1	1

during the 10- to 20-year postfire period, during which time shrub recovery is at its peak. Other plants, *Maianthemum* and *Aralia*, seem to increase under this shrub cover.

Some seed-reproducing species, *Geranium, Polygonum,* and *Epilobium,* as temporary fire followers, are often abundant on newly burned land the first 5 years, becoming infrequent by the fifteenth year. *Fragaria vesca,* which reproduces vegetatively, behaves similarly and is most abundant during the 5- to 10-year postfire period. Behavior of grasses varies with the species. *Oryzopsis asperifolia* was relatively unchanged by fire. *Calamagrostis canadensis,* a less shade-tolerant grass, reached its maximum abundance by the tenth postfire year and then its abundance declined.

Many of the common herbs of the forest floor, for example *Trientalis* and *Lycopodium,* occur in small numbers soon after fire and remain in small numbers, increasing slightly in the mature forest. A few herbs, notably those in the Orchidaceae (*Habenaria, Corallorhiza,* etc.) are seldom found until 20 years after fire.

C. Number of Species

In northeastern Minnesota, the average number of plant species is reduced about 25% the first year after fire, the largest reduction occurring among the herbs and mosses (Table VI). With the appearance of the temporary fire followers the second year, the preburn number of herb species is regained quickly. As the temporary seeders decline in later years, the more shade-tolerant species replace them. Recovery of bryophytes is slow, and the major mosses of the forest floor frequently appear in appreciable numbers between the fifth and fifteenth years after fire.

TABLE VI

AVERAGE NUMBER OF SPECIES OF BRYOPHYTES, HERBS, SHRUBS, AND TREES IN BURNED-OVER JACK PINE–BLACK SPRUCE FORESTS IN NORTHEASTERN MINNESOTA AT DIFFERENT INTERVALS AFTER FIRE

Species	Years after fire										
	1	2	3	4	5	10	15	20	30	50	80
Bryophytes	1	2	2	2	4	5	6	6	5	6	5
Herbs	30	38	37	40	36	37	36	37	36	36	37
Shrubs	12	13	14	14	14	15	13	13	14	14	14
Trees	3	4	5	5	5	5	6	6	5	5	5
All species	46	57	58	61	59	63	59	62	60	61	61

Numbers of shrub species are unaffected by fire, except for some seed-reproducing species which may be temporarily absent during the first postfire year. Similarly, the number of tree species is not appreciably altered, except for the somewhat later development of black spruce and balsam fir.

References

Aaltonen, V. T. (1919). Über die Natürliche Verjüngung der Heidewälder im Finnischen Lappland. *Inst. Quaest. Forest. Finland, Commun.* 1, 1–319 (1–56, German summary).

Ahlgren, C. E. (1959). Some effects of fire on forest reproduction in northeastern Minnesota. *J. Forest.* 57, 194–200.

Ahlgren, C. E. (1960). Some effects of fire on reproduction and growth of vegetation on northeastern Minnesota. *Ecology* 41, 431–445.

Ahlgren, C. E. (1966). Small mammals and reforestation following prescribed burning. *J. Forest.* 64, 614–618.

Ahlgren, C. E. (1970). Some effects of prescribed burning on jack pine reproduction in northeastern Minnesota. *Minn., Agr. Exp. Sta., Misc. Rep.* 94, 1–14.

Bakuzis, E. V. (1959). Synecological coordinates in forest classification and in reproduction studies. Ph.D. Thesis, University of Minnesota Micro Xerox Public. no. 242, Univ. Microfilms, Ann Arbor, Michigan.

Bakuzis, E. V., and Hansen, H. L. (1965). "Balsam Fir, a Monographic Review." Univ. of Minnesota Press, Minneapolis.

Beaton, J. P. (1959). The influence of burning on the soil in the timber range of Lac le Jeune, British Columbia. I. Physical properties. *Can. J. Soil Sci.* 39, 1–11.

Beaufait, W. R. (1960a). "Influence of Shade Level and Site Treatment, Including Fire, on Germination and Early Survival of *Pinus banksiana*." Forestry Div., Michigan Dept. Conserv., Lansing, Michigan.

Beaufait, W. R. (1960b). Some effects of high temperatures on the cones and seeds of jack pine. *Forest Sci.* 6, 194–199.

Bedell, G. H. D. (1948). White spruce reproduction in Manitoba. *Can. Forest Serv., Silvicult. Res. Note* 87.

Bloomberg, W. J. (1950). Fire and spruce. *Forest. Chron.* 26, 157–161.

Boyce, J. S. (1925). "Report on Forest Tree Disease Observations in Great Britain and Denmark," Reprint. Off. Invest. Forest Pathol., Bur. Plant Ind., Washington, D.C.

Breitung, A. J. (1954). A botanical survey of the cypress hills. *Can. Field Natur.* 68, 55–92.

Buckman, R. E. (1962). Two prescribed summer fires reduce abundance and vigor of hazel brush regrowth. *U.S., Forest Serv., Lake States Exp. Sta., Tech. Note* 620.

Buckman, R. E. (1964). Effects of a prescribed burn on hazel in Minnesota. *Ecology* 45, 626–629.

Buell, M. F., and Bormann, F. H. (1955). Deciduous forests of Ponemah Point, Red Lake Indian Reservation, Minn. *Ecology* 36, 646–658.

Cameron, H. (1953). Melting point of the bonding material in lodgepole and jack pine cones. *Can. Dep. Res. Develop. Forest. Branch, Forest Res. Div., Silvicult. Leafl.* **86.**

Cayford, J. H. (1963). Some factors influencing jack pine regeneration after fire in southeastern Manitoba. *Can. Dep. Res. Develop., Forest. Branch, Forest Res. Div., Publ.* **1016.**

Cooper, C. F. (1961). The ecology of fire. *Sci. Amer.* **204,** 150–156, 160.

Curtis, J. T., and Partch, M. L. (1950). Effect of removal of litter on flower production in certain grasses. *Ecology* **31,** 448–489.

Daubenmire, R. F. (1949). "Plants and the Environment." Wiley, New York.

Ehrenreich, J. H. (1959). Effects of burning and clipping on growth of native prairie in Iowa. *J. Range Manage.* **12,** 133–138.

Eyre, F. H. (1938). Can jack pine be regenerated without fire? *J. Forest.* **36,** 1067–1072.

Eyre, F. H., and LeBarron, R. K. (1944). The management of jack pine stands in the Lake States. *U.S. Dep. Agr., Tech. Bull.* **863.**

Farrar, J. L., and Fraser, J. W. (1953). Germination of jack pine seeds on humus. *Can. Dep. Res. Develop., Forest. Branch, Forest Res. Div., Silvicult. Leafl.* **91.**

Flieger, B. W. (1970). Forest fire and insects: The relation of fire to insect outbreak. *Proc. 10th Annu. Tall Timbers Fire Ecol. Conf.* pp. 107–114.

Fraser, J. W., and Farrar, J. L. (1953). Effect of shade on jack pine germination. *Can. Dep. Res. Develop. Forest. Branch, Forest Res. Div., Leafl.* **88.**

Graff, P. W. (1936). Invasion of *Marchantia polymorpha* following forest fires. *Bull. Torrey Bot. Club* **63,** 67–74.

Graham, S. (1954). Scoring tolerance of forest trees. *Mich. Forest.* **4,** 1–2.

Grant, M. L. (1934). The climax forest community in Itasca County, Minn. and its bearing on the successional status of the pine community. *Ecology* **15,** 243–257.

Hall, I. V. (1955). Floristic changes following the cutting and burning of a woodlot for blueberry production. *Can. J. Agr. Sci.* **35,** 143–152.

Harper, V. L. (1944). Effects of fire on gum yields of longleaf and slash pine. *U.S., Dep. Agr., Circ.* **170.**

Heikinheimo, O. (1915). Der Einfluss der Brandwirtschaft auf die Wälder Finnlands. *Acta Forest. Fenn.* **4,** 1–264.

Heinselman, M. L. (1969). Diary of the Canoe Country's landscape. *Naturalist* **20,** 2–13.

Hensel, R. L. (1923). Recent studies on the effect of burning on grassland vegetation. *Ecology* **4,** 183–188.

Holman, H. L., and Parker, H. A. (1940). Spruce regeneration. The Prairie Provinces. *Forest. Chron.* **16,** 79–83.

Horton, K. W., and Hopkins, E. J. (1965). Influence of fire on aspen suckering. *Can., Dep. Forest., Publ.* **1095,** 1–19.

Hustich, I. (1951). The lichen woodlands in Laborador and their importance as winter pastures for domesticated reindeer. *Acta Geogr., Paris* **12,** 1–48.

Isaac, L. A. (1929). Seedling survival on burned and unburned surfaces. *U.S., Forest Serv., Pac. Northwest Forest Exp. Sta., Forest Res. Note* **3.**

Isaac, L. A. (1930). Seedling survival on burned and unburned surfaces. *J. Forest.* **28,** 569–571.

Jarvis, J. M. (1966). Seeding white spruce, black spruce, and jack pine on burned seedbeds in Manitoba. *Can., Dep. Forest., Publ.* **1166,** 1–6.

Johnston, W. F. (1971). Broadcast burning slash favors black spruce reproduction on organic soil in Minnesota. *Forest. Chron.* **47**, 33–35.

Johnston, W. F. (1973). Tamarack reproduction well established on broadcast burns in Minnesota peatland. *U.S., Forest Serv., Res. Note* N.C.-153, pp. 1–3.

Kilburn, P. D. (1960). Effects of logging and fire on xerophytic forests in northern Michigan. *Bull. Torrey Bot. Club* **87**, 402–405.

Kittredge, J. (1938). Comparative infiltration in the forest and open. *J. Forest.* **36**, 1156–1157.

Lake States Forest Experiment Station. (1931). Fire hinders conversion of aspen to pine and spruces. *U.S., Forest Serv., Lake States Forest Exp. Sta., Tech. Note* **34**.

LeBarron, R. K. (1940). The role of forest fires in the reproduction of black spruce. *Proc. Minn. Acad. Sci.* **7**, 10–14.

LeBarron, R. K. (1944). Influence of controllable environmental conditions on regeneration of jack pine and black spruce. *J. Agr. Res.* **68**, 97–119.

Lilienstern, M. (1929). Physiologische Untersuchung über die Ursachen des Vorkommens von *Marchantia polymorpha* L. auf Feuerstätten *Ber. Deut. Bot. Ges.* **47**, 341–347.

Lipas, E., and Mäki-Petäys, E. (1961). Kutotusen vaidutus metsämaan lämpö-ja kosteusoloihin. Master's Thesis, Dept. of Forestry, University of Helsinki (cited from Viro, 1969).

Lloyd, P. S. (1965). The ecological significance of fire in limestone grassland communities of the Derbyshire dales. *J. Ecol.* **56**, 811–826.

Lutz, H. J. (1956). The ecological effects of forest fires in the interior of Alaska. *U.S., Dep. Agr., Tech. Bull.* **1133**.

McArdle, R. E., and Isaac, L. A. (1934). The ecological aspects of natural regeneration of Douglas fir in the Pacific Northwest. *Proc. Pac. Sci. Congr., 5th, 1933* pp. 4009–4051.

MacArthur, J. D. (1964). A study of regeneration after fire in the Gaspé region. *Can., Dep. Forest., Publ.* **1974**.

MacLean, D. W. (1957). "The Aspen-Birch-Spruce-Fir Type in the Boreal Forest Region of Ontario," 1st Progr. Rep. on Proj. H-103. Canada Dept. of Northern Affairs and Natural Resources, Forestry Branch.

Maini, J. S., and Horton, K. W. (1966). Reproductive response of *Populus* and associated *Pteridium* to cutting, burning, and scarification. *Can., Dep. Forest., Publ.* **1155**, 1–20.

Maissurow, D. K. (1935). Fire as a necessary factor in the perpetuation of white pine. *J. Forest.* **33**, 373–378.

Millar, J. B. (1939). Spruce regeneration in northern Ontario. *Forest. Chron.* **15**, 93–96.

Minckler, L. W. (1945). Reforestation in the spruce type in northern Appalachians. *J. Forest.* **43**, 349–356.

Oinonen, E. (1967). Sporal regeneration of ground pine (*Lycopodium complanatum*) in southern Finland in the light of the dimensions and the age of its clones. *Acta Forest. Fenn.* **83**, 1–85.

Oinonen, E. (1968). The size of *Lycopodium clavatum* and *L. annotinum* stands as compared to that of *L. complanatum* and *Pteridium aquilinum* stands, the age of the tree stand, and the dates of fire on the site. *Acta Forest. Fenn.* **87**, 1–53.

Old, S. M. (1969). Microclimate, fire and plant production in an Illinois prairie. *Ecol. Mongr.* **39**, 355–384.

Ooyama, N. (1954). The growth inhibiting substances contained in the leaf litter of trees. *J. Jap. Forest. Soc.* **36**, 38–41.

Pearse, A. S. (1943). Effects of burning-over and raking-off litter on certain soil animals in the Duke Forest. *Amer. Midl. Natur.* **29**, 406–424.

Pickering, S. U. (1910). Studies of the changes occurring in heated soils. *J. Agr. Sci.* **3**, 258–276.

Potter, L. D., and Moir, D. R. (1961). Phytosociological study of burned deciduous woods, Turtle Mountain, North Dakota. *Ecology* **42**, 468–480.

Reitz, R. C. (1937). Kiln temperatures for jack pine cones. *J. Forest.* **35**, 1163–1164.

Roe, E. I. (1952). Seed production of a white spruce tree. *U.S., Forest Serv., Lake States Forest Exp. Sta., Tech. Note* **373**, 1.

Roe, E. I. (1963). Seed stored in cones of some pine stands, northern Minnesota. *U.S., Forest Serv., Lake States Forest Exp. Sta., Res. Pap.* **LS-1**, 1–14.

Rogers, D. J. (1959). Some effects of fire in southern Wisconsin woodlots. *Wis. Coll. Agr., Forest Res. Note* No. 51, pp. 1–2.

Rowe, J. S. (1955). Factors influencing white spruce reproduction in Manitoba and Saskatchewan. *Can., Dep. North. Affairs Natur. Resour., Forest Res. Div., Tech. Note* **3**.

Rowe, J. S. (1970). Spruce and fire in northwestern Canada and Alaska. *Proc. 10th Annu. Tall Timbers Fire Ecol. Conf.* pp. 245–254.

Rudolph, T. D., Libby, W. J., and Pauley, S. S. (1957). Jack pine variation and distribution in Minnesota. *Univ. Minn. Forest. Note* **58**, 1–2.

Saari, E. (1923). Forest fires in Finland with special reference to state fires. *Acta Forest. Fenn.* **26**, 1–255.

Sarvas, R. (1937). Beobachtungen über die Entwicklung der Vegetation auf den Waldbrandflächen nord-Finnlands. *Silva Fenn.* **39**, 1–64.

Scotter, G. W. (1963). Effects of forest fires on soil properties in northern Saskatchewan. *Forest. Chron.* **39**, 41–56.

Scotter, G. W. (1964). Effects of forest fires on the winter range of barren ground caribou in northern Saskatchewan. *Can. Wildl. Serv., Wildl. Mgt. Bull., Ser.* 1 No. 18.

Sharp, W. M. (1970). The role of fire in ruffed grouse habitat management. *Proc. 10th Annu. Tall Timbers Fire Ecol. Conf.* pp. 47–61.

Shirley, H. L. (1932). Does light stimulate aspen suckers? *J. Forest.* **29**, 524–525; **30**, 414–420.

Skutch, A. F. (1929). Early stages of plant succession following forest fires. *Ecology* **10**, 177–190.

Smith, D. M. (1951). The influence of seedbed conditions on the regeneration of eastern white pine. *Conn., Agr. Exp. Sta., New Haven Bull.* **545**.

Smith, D. W. (1968). Surface fires in northern Ontario. *Proc. 8th Annu. Tall Timbers Fire Ecol. Conf.* pp. 41–54.

Soper, E. K. (1919). The peat deposits of Minnesota. *Minn., Univ., Geol. Surv., Bull.* **16**.

Spurr, S. H. (1954). The forests of Itasca in the 19th century as related to fire. *Ecology* **35**, 21–25.

Stone, E. C., and Juhren, G. (1951). Effect of fire on germination of seed of *Rhus ovata. Amer. J. Bot.* **38**, 368–372.

Summerhayes, V. W., and Williams, P. H. (1926). Studies on the ecology of English heaths. II. *J. Ecol.* **14**, 203–243.

Swan, F. R., Jr. (1970). Postfire response of four plant communities in south central New York State. *Ecology* **51**, 1074–1082.

Trevett, M. F. (1962). Nutrition and growth of low bush blueberries. *Maine, Agr. Exp. Sta., Bull.* **605**, 1–151.

Tryon, E. H. (1948). Effect of charcoal on certain physical, chemical, and biological properties of forest soils. *Ecol. Monogr.* **18**, 81–115.

Uggla, E. (1950). "Ecological Effects of Fire in North Swedish Forests." Almqvist & Wiksell, Stockholm.

Van Nostrand, R. S. (1965). Results of experimental seeding of balsam fir on a recent burn. *Can. Dept. Forest. Publ.* **1103**, 1–11.

Van Wagner, C. E. (1970). Fire and red pine. *Proc. 10th Annu. Tall Timbers Fire Ecol. Conf.* pp. 211–214.

Viro, P. J. (1969). Prescribed burning in forestry. *Commun. Inst. Forest. Fenn.* **67.7**, 1–48.

Vogl, R. J. (1964a). The effects of fire on a muskeg in northern Wisconsin. *J. Wildl. Manage.* **28**, 317–327.

Vogl, R. J. (1964b). Effects of fire on bracken-grasslands. *Wis. Acad. Sci., Arts Lett.* **53**, 67–82.

Wackerman, A. E., Zon, R., and Wilson, F. G. (1929). Yield of jack pine in the Lake States. *Wis. Agr. Exp. Sta., Res. Bull.* **90**, 1–23.

Wahlenberg, W. G., Greene, S. W., and Reed, H. R. (1939). Effect of fire and cattle on grazing and longleaf pine lands as studied at McNeill, Miss. *U.S., Dep. Agr., Tech. Bull.* **683**.

Wilson, G. W. (1914). Studies of plant growth in heated soil. *Biochem. Bull.* **3**, 202–209.

Wilton, W. C. (1963). Black spruce seedfall immediately following fire. *Forest. Chron.* **39**, 477–478.

Wright, E. (1931). The effect of high temperatures on seed germination. *J. Forest.* **29**, 679–687.

Yli-Vakkuri, P. (1961). Emergence and initial development of tree seedlings on burnt over forest land. *Acta Forest. Fenn.* **74**, 1–52.

. 7 .

Effects of Fire on Temperate Forests: Northeastern United States

Silas Little

I. Frequency and Type of Presettlement Fires

Fires were common in many sections of the Northeast before the land was settled by Europeans because Indians used fire for land-clearing, for keeping woods in an open condition, and for driving game. Burning was not deliberate in the Adirondacks, in much of northern New England, nor in the hillier sections of southern New England and Pennsylvania (Day, 1953). But elsewhere frequent fires may have been the rule. These fires shaped the parklike forests noted by early explorers in coastal Maine, southern New England, New Jersey, Maryland, central New York, and some other parts of the Northeast (Day, 1953). These forests were so open that travel on foot or by horseback was easy, and in some places deer (*Odocoileus virginianus*) or wild turkeys (*Meleagris gallopavo*) a mile away could be seen easily from the top of a hill.

Fires were partly or wholly responsible for extensive clearings, too. More than 1000 square miles of burned and treeless land in the Shenandoah Valley extended northward into Pennsylvania. And there were other areas of various sizes: for example, open plains 25 or 30 leagues (apparently 60 miles or more) in extent near Narragansett Bay (Day, 1953).

Of course, abandonment of Indian fields also created open meadows until trees became established. According to Day, Indians practiced little agriculture in northern New England or the Adirondacks but had some cultivated clearings in coastal Maine and extensive ones in many other sections, such as coastal areas and along the Taunton River in Massachusetts, in the Connecticut River valley north to New Hampshire, in the Hudson River valley to above Albany, in the Finger Lakes area, on Staten Island, Long Island, and Manhattan Island, and in New Jersey, especially along the Delaware River. Some of the cultivated clearings extended 2, 4, or 6 miles on all sides from an Indian village. Village locations were shifted every 10–20 years. The abandoned fields either grew back to forest or remained grasslands, depending on the fire regime.

The Indians' influence on the vegetation extended over much of the Northeast. Evidence of deliberate use of fire by Indians or of extensive clearing for agriculture seems lacking in northern New England and the Adirondacks—places where travelers stressed the thickness of the woods (Day, 1953). However, even there an occasional fire was caused by an escaped campfire or lightning strike. Elsewhere the frequency of fire varied, depending at least in part on the buildup and dryness of fuels. Wet spots, such as swamps and coves, undoubtedly burned less frequently than the drier uplands (Hawes, 1923). And on uplands the sandy sites dried out faster and were probably burned more frequently than sites with heavier soils.

If the Indians were responsible for maintaining a mosaic of seral stages, this explains the abundance of upland wildlife that was reported by early observers. Deer, turkey, ruffed grouse (*Bonasa umbellus*), quail (*Colinus virginianus*) as in southern New England, bison (*Bison bison*) in western Pennsylvania and New York, and heath hen (*Tympanuchus cupido*) were all favored (Thompson and Smith, 1970). The control of wildfires was considered highly instrumental in the decline and extirpation of the heath hen.

How intense and frequent were the presettlement fires? Day (1953) implied that Lenâpé Indians burned the woods deliberately in the spring and fall and accidentally at other times. One observer said that the Indians of coastal New England burned the woods in November (Hawes, 1923). Many of their early-spring or fall fires might have been set at times when burning conditions were not severe, so the fires would have been of low intensity. Even under more severe burning conditions, fire intensity was usually not great on frequently burned sites because of small amounts of fuel. In many areas where the Indians used fire deliberately, fires may have burned at intervals of 3 years or less on the drier forested sites, at 1-year intervals in the grasslands, and at long intervals—100 years or so—on the wetter forested sites, at which time severe fires could have developed during drought periods in certain types. In an oak stand of the New Jersey piedmont, scars on one stem indicated that fires had occurred at 10- to 15-year intervals between 1640 and 1680 (Buell *et al.*, 1954).

Fire frequency was probably irregular. In certain sections and in certain forest types, there may have been periods without fire long enough to permit seedlings of certain susceptible tree species to start and develop into relatively resistant saplings, whereas with more frequent fires this would have been impossible.

Varying fire history, coupled with varying site and climatic conditions, produced great variation in the composition of the northeastern forests. At some times and in some places it favored spruce fir forest; in others northern hardwoods, white pine (*Pinus strobus*), pitch pine (*P. rigida*) or other species of pine (Fig. 1), Atlantic white cedar (*Chamaecyparis thyoides*), yellow poplar (*Liriodendron tulipifera*), oaks (*Quercus* spp.) or other hardwoods. Varying fire history alone can cause great changes in composition of forests on specific sites within a climatic area (Figs. 1, 2, and 4).

The Indian influence on forests of the Northeast has been traced back for more than 1000 years (Day, 1953), but before that, for an even longer period, lightning fires had occurred in forests throughout the Northeast. Their occurrence was favored by the presence of many large and partly decayed trees containing punky wood that could retain a fire despite the usual accompanying rainfall. Even though in recent years lightning has not been an important cause of forest fires in the Northeast, enough lightning fires probably occurred even before the Indian influence so they had great effect on forest composition.

II. Frequency and Type of Postsettlement Fires

As a general rule, the early European settlers used fire to clear land and to favor quick growth of grass in the spring. However—probably as the result of European influences—the prevailing attitude changed: to discourage burning, especially in areas where the inhabitants prospered.

Fig. 1. An old stand of shortleaf and pitch pines on an upland site in the New Jersey pine barrens, probably similar to many of the original stands that occurred on such sites as the result of fires set by Indians. Compare with Figs. 2 and 4.

There were sectional differences. In the New Jersey pine barrens some landowners held onto the idea of burning woodland through years and generations, usually to reduce fuels and prevent damage from wildfires, especially to improved property. For many years in some mountainous sections and up to the present time in Maine, fire has been employed as a rough tool of management in the production of low-bush blueberries (*Vaccinium* spp.). In Maine the period between fires is 2 or 3 years, allowing only one or two crops of blueberries between burns (Trevett, 1962).

Even though attempts were made, at least since 1750, to restrict the use of fire and to control wildfires, effective fire-control organizations in the Northeast are largely a product of the twentieth century. Some industries and states attempted to control forest fires before passage of the Weeks Law of 1911, which authorized federal cooperation in fire protection on watersheds of navigable streams. However, funds appropriated under the Weeks Law were still not sufficient, so broadened cooperation in fire protection was authorized in the Clarke-McNary Law of 1924. Fire control was further stimulated by the Civilian Conservation Corps, which between 1933 and 1942 built many roads, water holes, and firebreaks and aided in suppression efforts. In subsequent years available manpower declined, but the use of machinery—tractors and plows, bulldozers, and tank trucks on the ground and airplanes and helicopters for aerial drops of retardants—has made fire control still more effective.

Consequently, the frequency and type of fires have changed through the years. Before 1900 wildfires were more common in many sections than in recent years, and their average size was far greater. Under some conditions, as in the northern hardwood type, many of the wildfires occurred in slash-covered areas after heavy cuttings. With the improved protection of recent years, hardly any wildfires occur in wooded areas of northern hardwoods, though appreciably more wildfires occur in the white pine–hardwood and oak types, and proportionally many more in the flammable pitch pine type. Even in this last type most present-day fires are small.

Between 1926 and 1930 the area burned yearly in the Northeast— Maryland northward—was 423,699 acres (Hastings, 1933). Between 1961 and 1970 the yearly average area burned was only 79,998 acres in the same states.

For the Northeast as a whole, fire has become relatively rare in woodlands. In 1971, a fairly typical year, there were 13,026 fires, which burned 116,251 acres, only 0.15% of the 79 million acres in the protected forest areas of the Northeast, including West Virginia (U.S. Forest Service, 1972).

However, during abnormal years both the number of fires and the area burned increase greatly. For example, in 1963 there were 25,514 fires, which burned 512,041 acres in the Northeast, including West Virginia. And the number of large fires, often less than 1% of the total, usually account for half to 95% of the burned area.

These very large fires occur during periods of extremely high hazard, resulting usually from high winds during periods of extended drought. Under such conditions, in 1947 Maine had several blowup fires that burned 240,000 acres (Hall, 1947). On one weekend in 1963 six fires burned about 185,000 acres in southern New Jersey. Similar fires have occurred periodically since settlement, especially in certain sections such as the New Jersey pine barrens. For example, Benjamin Franklin recorded one fire in 1755 that burned for 30 miles in southern New Jersey, and Vermeule (1900) stated that pine forest fires in New Jersey sometimes burn over 100,000 acres in a single season.

In the future, large fires can be expected whenever suitable weather and fuel conditions occur. These weather conditions develop at very rare intervals in Maine, more frequently in the more hazardous fuel types of Cape Cod, the New Jersey pine barrens, and similar areas.

III. Fire Effects in Northeastern Types

The general forest types of the Northeast have been well described by Lull (1968). In the following discussion, white pine is discussed rather than the general white pine–hemlock–hardwood forest type because it is the most important species. Oaks are listed separately because effects of fire differ between oaks and the associated yellow poplar and sweetgum (*Liquidambar styraciflua*). For the same reason pitch pine, Atlantic white cedar, Virginia pine (*P. virginiana*), loblolly pine (*P. taeda*), and pond pine (*P. serotina*) are discussed separately and not as a part of the broad yellow pine–hardwood type in which these species are most commonly found.

A. SPRUCE-FIR

In the spruce-fir type of the Northeast, fire has occurred rarely—most often in slash-covered areas within a few years after a heavy cutting. In uncut stands, fires are uncommon. However, in 1947 about 7000 acres burned in northern Maine (Hall, 1947).

Fire tends to eliminate existing stems of spruce, balsam fir (*Abies balsamea*), and their associates, although McIntosh and Hurley (1964) implied that fire is necessary to perpetuate red and white spruces (*Picea rubens* and *P. glauca*) in the Catskills. Many burns are restocked initially by aspen (*Populus* spp.) or birch seedlings (*Betula* spp.), which start from wind-carried seeds. White or red spruce seedlings invade the burns, developing as an understory to the aspen–birch complex and eventually replacing it. On some of the better sites, northern hardwoods such as sugar maple (*Acer saccharum*) and beech (*Fagus grandifolia*) eventually replace the spruces. Or in some areas the spruces may be replaced by balsam fir (Roy, 1940; McIntosh and Hurley, 1964; Fowells, 1965). Succession to northern hardwoods or fir might perhaps be checked by fire near or at the end of a rotation. Place (1955) thought that the role of fire in management of spruce may be limited to deep mineral soil, to severe burns that consume most of the humus, and to areas where the subsequent supply of seed is plentiful. In his opinion, burning could improve the seedbed, though during dry years charred surfaces are too hot for much seed germination.

In spruce types of the Northeast, on certain mountaintops (such as Spruce Mountain in West Virginia), fire after a heavy cutting destroyed the organic soil. Such sites may have practically no mineral soil, and trees grow in an organic mat over rock. Where most of the organic mat is destroyed by fire or where subsequent erosion on a steep slope removes the small amount of mineral soil, restocking by spruce and fir is extremely slow, depending on gradual accumulation of organic matter so that suitable niches are provided (Korstian, 1937).

Light fires may, of course, only cause wounds on some spruce trees, especially those several inches in diameter; but these wounds serve as invasion sites for insects and fungi. In one study, within 3 years of a fire, 53% of the surviving trees in the red spruce type had been attacked by fungi, insects, or both (Stickel and Marco, 1936).

B. NORTHERN HARDWOODS

In the northern hardwood type, fires have had several different effects on stand composition, largely because of differences in fire intensity and frequency, in soil and aspect, and in available seed sources. A single slash fire after a heavy cutting may favor black cherry (*Prunus serotina*), the birches, red maple (*Acer rubrum*), and to some extent aspen; or it may favor white pine, especially on the drier south and west aspects (Hough and Forbes, 1943). Repeated fires after logging may result in stands of aspen and pin cherry (*Prunus pensylvanica*); still more fires may create open areas dominated by bracken fern (*Pteridium aquilinum*), goldenrod (*Solidago* spp.), or grasses and sedges (Hough and Forbes, 1943). Swan (1970) reported that repeated fires on sites in central New York favored redtop (*Agrostis alba*) and little bluestem (*Andropogon scoparius*) and to some extent goldenrod but adversely affected poverty grass (*Danthonia spicata*). Less frequent but repeated fires favor oak forests over northern hardwoods (Swan, 1970).

The northern hardwoods are less well adapted to fire than oaks for two reasons. Fewer northern hardwoods sprout after fire than do oaks. In Swan's (1970) study 43% of the northern hardwood saplings sprouted, compared to 87% of the oak saplings. The thinner bark of the maples, birches, beech, and aspen makes them more susceptible to complete or partial basal wounding than the oaks.

An associated conifer, hemlock (*Tsuga canadensis*), is particularly vulnerable to fire. In Swan's study 93% of the hemlock saplings died and did not resprout. Older hemlocks, having more dead outer bark, are somewhat more resistant so that a hemlock 9 inches in diameter may be twice as resistant to fire as a balsam fir 15 inches in diameter (Stickel, 1941).

Under certain conditions fire might be used to favor the less tolerant light-seeded species, such as the birches, over sugar maple and beech. However, few attempts have been made—mostly in Canada or the Lake States—to use fire for this purpose. An autumn fire reduced advance seedling density, mainly maple, from 160,000 to 18,000 per acre, but spring fires were more effective in killing advance growth of sapling size: three consecutive spring fires killed up to 55% of all trees 0.6–4.5 inches in diameter (Burton *et al.*, 1969). Because burned seedbeds favor birch establishment and reduced competition from advance growth of maples would favor dominance of birches in the next stand, the use of fire at the end of the rotation might shape composition of the subsequent stand in favor of birches. However, this hypothesis has not been tested.

Fire during the rotation usually affects timber yields adversely because of basal wounds and subsequent decay. However, because fires are rare in the northern hardwood type, such decay is seldom important.

C. EASTERN WHITE PINE

Fire played a major role in establishing even-aged stands of white pine in the original forest, although wind and other destructive forces also played a role in some areas (Cline and Spurr, 1942; Hough and Forbes, 1943). Many of the white pine stands originally found in Maine started following fire, and in the absence of fire this species has not reproduced itself on many sites because of competition from more tolerant species (Cary, 1936). However, too frequent or too severe fires do eliminate white pine (Horton and Bedell, 1960).

The role of fire in favoring white pine has been twofold. First, fire provides suitable seedbeds. Pine litter, lichens, or very dense grass cover are among the unfavorable seedbeds, and all can be made more favorable by burning (Smith, 1940; Smith, 1951). Second, fire checks succession to more shade-tolerant hardwoods such as sugar maple.

White pine is more tolerant of shade than pitch pine, aspen, or gray birch (*Betula populifolia*)—species that may also invade areas after severe fire disturbances. Consequently, some original stands of white pine may have started under a canopy of those species or a partial canopy of other hardwoods. Lutz and McComb (1935) found that the scattered white pines at Heart's Content, Pennsylvania, started under a partial canopy, possibly one that was opened up by fire or windthrow.

On sandy sites in many parts of the Northeast, either white pine or pitch pine can predominate, and fire history often determines which species prevails. Frequent, relatively hot fires favor pitch pine, a species that has thicker bark than white pine and also the ability (1) to sprout at the base or along the bole from dormant buds and (2) to form new crowns after heat kills existing foliage (Little, 1959). Pitch pine also has an advantage in that it produces viable seed at a younger age than does white pine (Fowells, 1965).

In contrast, exclusion of fires for several years seems essential to establishment and growth of white pine. In studies of prescribed fire in southern Connecticut, Niering *et al.* (1970) eliminated 60% of the white pines in the smaller size classes, mostly less than 1 inch in diameter. Until rough bark develops on the lower bole, white pines are susceptible to killing by relatively light fires. However, some may survive, as in the Connecticut study. Once rough bark has developed, white pines may be resistant to damage by light fires that still are effective in killing species with thinner bark, such as maples and beech.

After the pines are resistant, repeated light fires may favor a grassy or herbaceous understory, as Bromley (1935) observed in one area. However, he also noted that a subsequent 18-year period of fire exclusion was sufficient for development of a heavy understory of oaks that crowded out the grasses and forbs formerly present.

Fires may damage roots, boles, or crowns of white pine trees. McConkey and Gedney (1951) observed that root injuries caused by wildfires during a drought were more serious than crown injuries. Where 75% or more of the major surface roots had been killed or severely damaged by fire—but only a third or less of the crown was scorched—mortality during 3 years after the fire was as follows: small trees (2–6 inches in diameter), 100%; medium trees (7–11 inches), 60%; and large trees (12 inches and larger), 40%. But where no more than 25% of the roots were killed or injured—even though two-thirds or more of the crown was scorched—mortality was lower: small trees, 80%; medium trees, 46%; and large trees, 14%.

Total consumption of foliage by fire kills white pine, and scorching of all foliage usually does too. Cambial layers on the bole may be killed by heat. Where the cambium on only part of the bole is killed, a wound occurs and provides entrance for such fungi as *Fomes pini,* cause of red ring rot, the most important heartrot of white pine.

D. OAKS

Oak stands in the Northeast have often been favored by certain fire regimes (Fig. 2). In many areas, hemlock and northern hardwood species that are more shade tolerant—sugar maple, red maple, and sweet birch (*Betula lenta*)—are today tending to capture the oak sites, for example, in Connecticut (Stephens and Hill, 1971), northern New Jersey (Buell *et al.*, 1966), and central New York (Swan, 1970). The trend to northern hardwoods is particularly noticeable on high quality sites. It has apparently been favored by (1) a great reduction in the frequency of fire and (2) longer rotations since the decline of fuel-wood production for domestic and industrial use. This decline started about 1860 but became widespread and was still increasing in the early part of this century. In the meantime, the fire exclusion policy was becoming ever more effective.

So long as the northern hardwoods or hemlocks remain as small understory stems, relatively light fires might eliminate many of them. Repeated fires significantly reduced the amounts of seedlings and saplings of red maple and other northern hardwoods in the oak stands that Swan studied

Fig. 2. The climax forest on upland sites in the New Jersey pine barrens, a mixture of oaks and hickories with an understory dominated by huckleberries (*Gaylussacia* spp.) and blueberries. Such stands develop under fire exclusion on these sites and form a physiographic climax.

(1970). In southern Connecticut prescribed fires eliminated sweet birch in the smaller size classes, 100% of the stems 1–2 inches in diameter and a few stems up to 6 inches in diameter (Niering *et al.*, 1970).

Single fires, particularly light ones, may have relatively little effect on the composition of trees or associated plants. For example, Wood (1971) tried clear-cutting in the fall and a light spring fire in an oak forest on the eastern Allegheny Plateau of Pennsylvania. Clear-cutting alone or clear-cutting followed by fire favored such invading species as pin cherry, fireweed (*Erechtites hieracifolia*), blackberry (*Rubus* sp.), and whorled loosestrife (*Lysimachia quadrifolia*), or increased substantially such species as sedges (*Carex* spp.), hay-scented fern (*Dennstaedtia punctilobula*), and blueberries. Differences were due to cutting, and no apparent difference in community composition was due to extremely light fire.

Annual fires in grasslands stimulated the production of more floriferous, vigorous, and taller stands of *Andropogon* than on unburned areas during a 3-year period (Niering *et al.*, 1970). In an oak woodland of southern Connecticut, spotted wintergreen (*Chimaphila maculata*) increased sixfold after a single prescribed burn (Niering *et al.*, 1970).

Over a long period, repeated fires would have adverse effects on oak regeneration, chiefly by destroying seed and killing seedlings of low vigor. However, at least in some stands, use of fire might eliminate most of the reproduction of tolerant species, such as sugar maple, red maple, and sweet birch, after which a period of no fire might favor establishment of oak seedlings. In one study area in New Jersey, three light fires in an 8-year period caused a reduction of 225 stems per acre of oaks larger than 0.5 inch d.b.h. 10 years after the last fire, but the total number of oaks at that time also increased by 1200 seedlings or sprout clumps per acre (Little, 1973).

In West Virginia, stands that had been lightly burned or otherwise disturbed during the previous two decades had more oak reproduction than did undisturbed stands (Carvell and Tryon, 1961). Even though there is evidence that fire has played an important role in oak ecology, it has not yet been adequately tested so one could prescribe it as an effective and economical tool for obtaining oak regeneration (Clark and Watt, 1971).

Fires can cause basal wounding of oaks, but even severe wounding of the stems does not reduce the rate of diameter growth of surviving trees. However, if the crowns are also appreciably damaged or reduced, then diameter growth is significantly slowed (Jemison, 1944).

Basal-wounded oaks are subject to discoloration and decay of the stem wood, and basal wounding by fire rates high among the causes of cull in hardwoods (Hepting and Hedgcock, 1937). Berry (1969) reported that fire scars were the most important avenue of entry for decay in Kentucky's upland oak stands. However, though he found that decay volume increased with tree age and diameter, the decay losses in trees 30–90 years old were minor. Decay losses are more important in the South than in the Northeast: for example, in Maine the decay rate is slow, and decay is often confined to the wood behind the wound, although discoloration of heartwood extends well beyond the wound (Hepting and Shigo, 1972).

In areas where oak sprouts from stumps will predominate, a single fire just after cutting will reduce decay in the subsequent stand. Roth and Hepting (1943) have shown that burning after cutting ensures low origin sprouts, which are less susceptible to decay than high union sprouts. The latter are more likely to have their heartwood infected by decay that can be traced back to the parent stump.

Shrub oaks, bear oak (*Quercus ilicifolia*), blackjack oak (*Q. marilandica*), and dwarf chinkapin oak (*Q. prinoides*) differ from the tree oaks in being adapted to relatively hot fires at short intervals. All three species produce mature acorns on very young sprouts, 3 years or so old. All three species thrive best in open conditions, and overtopping shade reduces their vigor. Blackjack and dwarf chinkapin oaks are less widespread than bear oak, which occurs with pitch pine on many of the most frequently burned sites of the Northeast. Frequent fires that do not permit fruiting of tree oaks and shortleaf pine (*Pinus echinata*) have been considered the cause of pitch pine-scrub (shrub) oak areas in the New Jersey pine barrens (Little, 1964).

E. YELLOW POPLAR AND SWEETGUM

Yellow poplar, which grows from southern New England southward, is particularly important on the better hardwood sites at low elevations in the Middle Atlantic States. Sweetgum has a somewhat more restricted range, being especially common on the inner Coastal Plain of the Middle Atlantic States.

Relatively poor conditions for regeneration of both species can be improved through burning. In one study in New Jersey and Maryland, burned seedbeds had 2–28 times as many yellow poplar seedlings as unburned areas, but on most sites unburned seedbeds had adequate amounts of reproduction (Little, 1967). In the piedmont of South Carolina a similar increase in number of yellow poplars also occurred after burning. There seedlings starting on burned areas were somewhat taller after 3 years than those on unburned sites, apparently because of earlier germination of yellow poplar seeds on the burns (Shearin *et al.*, 1972). On the whole, however, burning seems desirable only (1) on the dry borderline sites where oak or beech litter is deep enough to hamper establishment of seedlings, or (2) where the supply of seed is low (Little, 1967). My observations indicate that sweetgum requirements are similar.

Both yellow poplar and sweetgum are intolerant and yield to more tolerant species in the absence of severe disturbances. Hence, though both are subject to wounding, particularly before thick bark develops on the lower boles, fire during the latter part of the rotation might be one way of reducing competition and favoring yellow poplar and sweetgum. In this respect, the reactions might be similar to those already described for oaks, but compressed into a shorter period because both sweetgum and yellow poplar seedlings need release sooner than oaks.

Yellow poplar does possess one adaptation to fire not found in oaks and sweetgum. Its seeds remain viable in the forest floor for up to 8 years (Sander and Clark, 1971), while acorns and sweetgum seeds do not remain viable for more than a year.

F. PITCH PINE

Pitch pine is the most fire-adapted species of trees throughout much of the Northeast. It grows on a wide range of sites: thin soil ridges, dry sand plains, coastal dunes, poorly drained sands, and swamps. But its occurrence is limited mainly to poor sites, especially to areas that have a history of frequent and severe fires. In New England and upstate New York, white pine frequently replaces pitch pine in the absence of severe fires. The most extensive areas of pitch pine are found on Cape Cod and in the New Jersey pine barrens. In the latter region the species sometimes grows with, or is replaced by, shortleaf pine (Fig. 1).

Pitch pine has several characteristics that give it greater resistance to fire than most species have. Most of its seedlings, like those of shortleaf pine and of pond pine but unlike those of other northeastern conifers, develop basal crooks that bring dormant buds into or against mineral soil on upland sites (Little and Somes, 1956; Little and Mergen, 1966). Consequently, pitch pines sprout at the base after stem-killing fires (Fig. 3), and retain this ability to a much greater age than does shortleaf pine (Little and Somes, 1956).

Along with shortleaf and pond pines, pitch pine is able to form a new terminal shoot from a protected bud in the bole bark and to develop

Fig. 3. Basal crooks and sprouts on pitch pine seedlings. Most pitch, shortleaf, and pond pines develop basal crooks similar to that of the seedling on the right. Such crooks bring basal dormant buds on upland sites into or against mineral soil, so the buds are protected from damage by fires. The stem of the seedling on the left was killed by fire, and sprouts have started from dormant buds. One or two of these are capable of becoming trees of normal size.

new branches from similar buds along the bole or main branches, provided fire created just enough heat to kill the terminal shoot and all fine branches but not enough to injure protected buds. Even if all the foliage of a pitch pine (or pond pine or often shortleaf pine) is scorched or killed by the heat of a fire, still the tree will refoliate.

In sections where wildfires have been common, most of the pitch pines may have serotinous cones that do not open normally in the fall but remain closed for a long period, often until the heat of a fire strikes them. Pond pine, but not shortleaf pine, is also noted for its serotinous cones. Where the closed-cone characteristic prevails in stands of pitch pine, a large amount of seed may be released on fresh burns, and this characteristic alone helps greatly in maintaining the dominance of pitch pines in some areas (Little, 1959).

In sections where most of the pitch pines produce nonserotinous cones, light winter fires can be used to prepare suitable seedbeds for this species as well as for shortleaf pine. Where pitch and shortleaf pines grow in mixture with black oak (*Quercus velutina*), white oak (*Q. alba*), chestnut oak (*Q. prinus*), or other tree oaks, both advance and subsequent reproduction of the pines are favored by winter fires, and the total number of pine seedlings present after harvest cutting often increases with the number of burns made before the cutting. In plots of one study where the stand was harvested in a seed-tree cutting, the number of pines present 16 years later increased from 40 stems per acre where no burns were made before cutting to more than 2000 stems per acre in areas burned five or six times (Little, 1964).

Although light winter fires are adequate for preparing seedbeds for pine seedlings in oak–pine stands, they are inadequate for stopping the development of hardwood understories, mostly of oak and hickory (*Carya* spp.) seedlings, especially in old-field pine stands (Little and Moore, 1949; Little, 1973). Such fires kill back the smaller hardwood seedlings and eliminate some of them. On the basis of work in the southern United States, far better control of succession would be obtained by the use of some summer fires under the stands after winter fires had reduced the amount of fuel. Chaiken (1952), Grano (1970), and Langdon (1971) reported that in the South repeated summer fires eliminated many understory hardwoods, even stems 2 or 3 inches in diameter near the ground. For example, Grano stated that one winter fire killed 94% of the hardwood stems having diameters at the base of less than 3.5 inches, and eleven annual summer burns eliminated sprouting on 85% of the hardwood rootstocks.

However, on lowland sites where basal dormant buds of shrubs and hardwoods are located in the deep organic mat, a single deep-burning fire before or after a harvest cutting can eliminate nearly all the competing vegetation. In one study in New Jersey the forest floor on poorly drained soils was 6 inches to a foot thick. Shrubs alone on such sites form a dense layer of stems 3–12 ft tall, are often composed as in the study area of 15 species or more, and compete strongly with pine seedlings. However, during droughts in late summer or early fall, a fire can burn deep enough to kill the basal buds of both hardwood trees and shrubs. Where the fire occurs shortly before pine seeds fall or where the overstory pines bear serotinous cones, a large amount of pine seed falls on the burn. This often provides many pine seedlings: in the New Jersey study 6700 to 6900 seedlings per acre after a deep-burning fire but only 200 to 1300 seedlings per acre where a moister forest floor prevented the development of a deep-burning fire (Little and Moore, 1953). On the latter site hardwood shrubs and trees sprouted, and their regrowth competed strongly with the smaller number of pine seedlings. However, where the fire burned deep, most of the hardwood trees and shrubs did not sprout; so the deep-burning fire not only prepared a better seedbed for the establishment of pine seedlings but also eliminated much of their competition.

A deep-burning fire and subsequent exclusion of wildfires for many years apparently provided the proper regime for the relatively high quality stands of pitch pines found on some poorly drained soils in the New Jersey pine barrens. These trees may be relatively large (up to 19 inches d.b.h.), straight, naturally pruned to heights of 50 or more ft, and 75 to 100 ft tall.

The appearance of such pitch pines is in marked contrast to those found on upland areas in association with scrub oaks (bear, blackjack, and dwarf chinkapin oaks). Such a mixture in the New Jersey pine barrens occupies areas most frequently burned by wildfires. Repeated wildfires at frequent intervals, and with intensities sufficient to kill back existing trees, serve to explain the near absence of shortleaf pine and tree oak species in the pine–scrub oak association. Black, white, chestnut, and scarlet (*Quercus coccinea*) oaks and shortleaf pine do not bear viable seed, even on sprouts, until most stems are 20 years or older. In contrast, sprouts of pitch pine and bear oak often start to produce viable seed when 3 years old (Little and Somes, 1964). Hence, wildfires at 8- or 12-year intervals could easily eliminate shortleaf pine and tree oaks when continued over a century or two.

Within the New Jersey pine barrens, stands of pitch pine and scrub oaks vary from an overstory of pitch pines of pulpwood size to multiple sprout stands where dominant stems are less than 11 ft tall. Differences seem related to fire history. Lutz (1934) stated that the former type of stand had been subjected to fires at about 16-year intervals, and the low-sprout stands known as "plains" (Fig. 4) had been burned at about 8-year intervals. In addition, in my opinion, most of the fires in the plains community have been of killing intensity, while at least some of the fires in the taller stands have not been so damaging.

The cause of plains vegetation in the New Jersey pine barrens has long been the subject of discussion. Various hypotheses have been ad-

Fig. 4. A "plains" stand in the New Jersey pine barrens, a low sprout growth of pitch pine, bear oak, and blackjack oak. These sprouts of pitch pine are slow growing, in part because they started from old stools (50 to 80 years old at time of sprouting). Such stands are the result of killing fires at short intervals, 8 years or so.

vanced in attempts to explain the dwarfing of trees on the basis of microclimate, geology, soils, toxic amounts of soluble aluminum, or other factors [see reviews by Little and Somes (1964) and by McCormick and Buell (1968)]. Conclusions from the two most exhaustive studies (Lutz, 1934; Andresen, 1959) agreed that killing fires at short intervals were the primary cause of dwarfed trees.

The most important factor slowing recovery is probably the intense competition between and within sprout clumps. As many as 249 sprouts have been counted on a single stool of pitch pine a year after a killing fire. In one portion of the plains where most stems were about 27 years old, there was still an extremely high number of stems per acre: about 16,000 in 5400 sprout clumps per acre. About 1500 sprout clumps were pitch pines, and in these were 1280 stems per acre larger than 0.5 inch d.b.h. (Little and Somes, 1964).

Other secondary factors slowing recovery of the vegetation after wildfires include (1) old age of many stools when the last sprouts started and (2) damage to pine stems by browsing deer, insects, and cone collectors (for the florist trade).

Recent studies in the plains indicate that (1) because planted white pines have attained heights up to 21 ft in 13 years, indicating potential heights of 60–80 ft at 50 years, soil is not the cause of dwarfing of the natural vegetation, but (2) the fire history may have favored a special race of pitch pine in which trees of seedling origin mature early and have relatively short and crooked stems (Little, 1972).

On the whole, fire has been essential in perpetuating pitch pine stands in the Northeast, but it has also often created a scrub forest appearance in these stands. In part, this appearance is due to confining pitch pines to the poorest sites: deep, sandy plains or the thin soils of rocky outcrops.

But even more important has been the effect of fire on the trees themselves. Many are sprouts that started after the last wildfire, and all too often from relatively old stools, resulting in severe competition within sprout clumps, rapid decline in growth rate of the sprouts, and early maturity. Many of the present-day trees have lived through wild-fires that defoliated the crowns, killed back terminal shoots, and created either crooks (when new terminal shoots developed from dormant buds) or flat-topped crowns (when no new terminal shoot developed). Conse-quently, the best-looking pitch pines occur usually in areas where one or more wildfires favored their establishment as seedlings but where no subsequent wildfires have damaged them. In some of these areas pitch pines form pure stands, but in New England and New York the pitch pine often grows in mixture with white pine.

G. ATLANTIC WHITE CEDAR

Atlantic white cedar occupies swamp sites, predominantly of peat soils, and especially in sections of sandy soils, as in the New Jersey pine barrens and on Cape Cod. These same sections are usually noted, too, for their frequent and large wildfires (Little, 1950).

Fires seldom start and spread on the wet sites of true swamps, except within a few years after heavy cuttings that left dense slash covering the ground. Most of the fires that have affected white cedar stands started on drier sites. When driven by a strong wind, a large, hot fire may cross a wet swamp by crowning in the white cedars. More fre-quently, the large swamps have served as firebreaks, and only white cedar trees along the edge of the stand were killed by the fire (Little, 1950).

White cedars of all ages are very susceptible to injury by fire because of their thin bark and flammable foliage. Fire wounds on living trees are rare because most affected trees are killed.

Effects of fires vary with several factors, including the type of fire. Surface, crown, and ground fires have all occurred in white cedar swamps. Relatively dry, usually sandy swamps do burn more readily than wetter swamps, and fire has prevented white cedars from occupying some of the drier swamps or poorly or very poorly drained soils, where in the New Jersey pine barrens a pitch pine subclimax now prevails. In wet swamps the effects of fire vary with such factors as (1) composition of the affected stand, (2) amount of viable seed by species in the forest floor, (3) amount of viable seed dispersed on the burn from adjoining stands, (4) depth to which fire burns in the forest floor, and (5) position of the water table after the burn.

Lowering the forest floor or peat by burning is, of course, equivalent to raising the water table. During unusually dry periods it is possible for fires to burn into the peat so that for a long period thereafter water normally stands on the area. This may throw the stage of succession back at least to leatherleaf (*Chamaedaphne calyculata*), which in southern New England and New Jersey is the first shrub to invade quaking bogs (Rigg, 1940; Wright, 1941). Once a stand of leatherleaf is established, recurrent fires in its crowns or on the surface may impede establishment of trees. In southern New Jersey white cedar, red maple, or pitch pine may eventually become established, particularly after sphagnum moss builds up the forest floor so that water no longer stands on the surface. But tree growth, especially of white cedars, may be hindered, too, through severe browsing by deer (Little, 1950). The length of time during which leatherleaf dominates sites in the absence of recurring fires is not exactly known but appears to be about 50 years.

Where fire kills tree and shrub stems but does not burn into the forest floor, the area may be restocked by hardwood sprouts, if there were enough hardwoods present before the fire (Little, 1950).

Species occurring outside the burn may restock the area under certain conditions. These conditions include (1) an absence of stored viable seeds in the forest floor or the destruction of these seeds or resulting seedlings (as through prolonged flooding after germination) and (2) an absence of hardwoods or the killing of dormant basal buds on hardwoods by a deep-burning fire (Little, 1950). Under such conditions, and provided of course the site is still suitable for tree growth, the resulting stand may be composed of one or more of several species: white cedar, red maple, gray birch, or pitch pine (Little, 1950).

Under certain conditions pure white cedar stands may restock a burn, chiefly from seeds stored in the forest floor or peat at the time of the fire. Mature white cedar stands often have 1–3.6 million viable white cedar seeds per acre in the top inch of forest floor (Korstian, 1924; Little, 1950). Even after two growing seasons without addition of appreciable amounts of fresh seed, or 1 year of absolute exclusion of additional seeds, the surface inch of forest floor may contain 260,000 to 1,100,000 viable white cedar seeds per acre, and the underlying 2 inches of peat may have 260,000 to 950,000 viable seeds per acre. Pure white cedar stands develop on burns from such stored seed following either a crown fire in uncut pure stands or a slash fire after cutting similar stands; in both cases the swamp being so wet that the peat will not burn (Little, 1950).

In recent years, fire, along with cutting and deer browsing, has tended to reduce the proportion of white cedar in favor of associated species. The conversion has been most complete in narrow swamps and along the edges of wide swamps. In the centers of large swamps some recent fires have created conditions favorable for white cedar, while other fires have hastened the succession to more tolerant hardwoods (Little, 1950).

In the presettlement forests, fire may have favored the occurrence of white cedar in both wet and dry swamps more than it has in recent years. First, there were far greater supplies of white cedar seed. In wet swamps many stands were then relatively old and composed of large trees, hence good producers of seed. The larger production meant more seed stored in the peat and more distributed to adjoining areas and for greater distances than from the younger, shorter trees of today.

Second, if fires on upland sites were frequent and of low intensity, they often would not have spread into the very poorly drained parts of relatively dry swamps. In contrast, in recent years most dry swamps undergo periods of fire exclusion, broken only by severe fires that burn the very poorly drained spots too.

Third, the type and frequency of fires in wet swamps also probably favored white cedar more in the presettlement forest than today. There would have been relatively few fires in young stands of white cedar and, hence, less chance that white cedar would be replaced by other species. Of the infrequent fires that did burn in wet swamps, many probably occurred during droughts when they may have consumed enough peat to prevent sprouting by any associated hardwoods. White cedar with its large amounts of wind-distributed seeds would have had an excellent chance of restocking the burn when moisture conditions became suitable.

At rare intervals fires may have occurred after an extensive blow-down, producing results similar to recent slash fires except that the greater supplies of white cedar seed would have increased the chances of another relatively pure stand. Other factors, such as beaver damage to associated hardwoods, undoubtedly helped in perpetuating relatively pure stands of white cedar in the presettlement forest, even though this relatively long-lived species is successional to a climax of swamp hardwoods (Little, 1950).

H. VIRGINIA PINE

Virginia pine occurs in the Middle Atlantic States chiefly from southern New Jersey and south central Pennsylvania southward through Delaware and Maryland. It is especially common in southern Maryland, the five counties south of Baltimore. The species characteristically occupies poor soils: dry ridges or Coastal Plain soils that either are sandy or have cemented or compacted subsoils (Fenton and Bond, 1964).

Having relatively thin bark and lacking ability to sprout from dormant buds, either at the base or along the bole, Virginia pines are far more susceptible to fire than are pitch and shortleaf pines. Consequently, Virginia pine is not an important component of the forest in most sections of the New Jersey pine barrens.

On the other hand, Virginia pines bear short needles and in pure stands produce a scanty, relatively compact forest floor. Unless hardwood leaves are also present, the forest floor under Virginia pine stands dries slowly, and the available fuel often permits only a light fire with low flames. Older Virginia pines that have rough bark can survive such fires.

Virginia pine is also a prolific producer of seeds. Under mature stands the annual fall of seeds may vary from 48,000 to 996,000 seeds per acre. In openings, as on clear-cut strips 130 to 200 ft wide, the fall of seeds from adjoining mature stands may be at the rate of 8000 to 70,000 seeds per acre (Fenton and Bond, 1964).

In the original forest, infrequent killing fires probably favored establishment of Virginia pine stands. These fires created the needed openings, and the abundant seed supply from Virginia pine may have restocked such burns with dense stands of seedlings. For example, a surface fire in November 1952 burned a 30-year-old pine–hardwood stand near Beltsville, killing many trees but resulting in more than 10,000 pine seedlings per acre (Church, 1955).

Fires that did not kill most of the overstory trees probably encouraged subsequent establishment of Virginia pine reproduction but did not provide the open conditions necessary for its vigor and dominance. In such areas, sprouts of the more tolerant hardwoods would have been favored (Fenton, 1960). Presumably in some areas relatively light, but repeated, fires after Virginia pines had developed rough bark on the basal portions of their boles may have checked development of hardwood understories, permitting Virginia pines to recapture the site after a severe fire or extensive windthrow.

In recent years Virginia pine has been reproduced by clear-cutting strips and then broadcast burning the slash prior to seedfall. Seedbed preparation is essential for adequate reproduction, and on burns the number of established seedlings may be 4 times that on undisturbed seedbeds (Fenton and Bond, 1964).

I. LOBLOLLY AND POND PINES

North of the Virginia–Maryland boundary, loblolly and pond pines occur mostly in southern Delaware and on the eastern shore of Maryland There two of their associates are Virginia and shortleaf pines.

In that section pond pine is most common in areas where there have been more wildfires than usual, probably because it is more resistant than loblolly pine. Like pitch pine, pond pine can sprout from dormant buds at the base or along the bole, and pond pine is noted for its closed, serotinous cones (Little *et al.*, 1967).

Fires in loblolly pine stands can have several different effects on stand composition. Where few hardwood sprouts develop and where there is an ample source of pine seeds nearby, a burn by a killing wildfire is restocked by pines. Where there is a hardwood understory and the burn is not deep enough to kill basal buds, a fire that kills only part of the pine overstory may just hasten the succession to hardwoods. Because many sites on the eastern shore are poorly drained and have thick forest floors, fires on such sites can burn deep enough during droughts to kill the existing trees and provide open areas for establishment of pure pine stands where seed supplies are adequate.

In recent years fire has been one of the tools used by foresters in providing conditions necessary for the establishment of loblolly pine stands. Mainly, severe single fires have been used in late summer after seed-tree cuttings (Little and Mohr, 1954) or after the use of a Marden brush cutter in predominantly hardwood stands. However, light winter fires are also of value in preparing favorable seedbeds for establishment of pine seedlings and in partially controlling associated hardwoods (Little and Mohr, 1963).

In the original forest certain fire regimes may have favored pines over the climax hardwoods. Repeated killing fires, possibly at intervals of 25 years or less, favored pond pine over loblolly pine. Another fire regime—one based on exclusion of fire until pines were of sapling size and then repeated light fires to check the development of hardwood understories—might have favored loblolly pine. Light fires that occurred during the summer were probably more effective in eliminating hardwoods than winter fires. Whenever appreciable holes developed in the overstory canopy as a result of windthrow, damage by bark beetles, or other factors, the rapid growth of lobolly pine seedlings would have enabled them to compete, often successfully, with any small hardwoods. As in more recent times, deep-burning fires on poorly drained sites during droughts may have eliminated the existing stand—pines, hardwoods, and shrubs—and provided open areas for the invasion of pines and subsequent formation of relatively pure pine stands.

IV. Conclusion

Throughout the Northeast the effects of fires on natural vegetation have been and are highly variable. Differences in fire intensity, fire frequency, depth of the forest floor, and season of fire occurrence are only some of the factors that produce great variations in effects. Other important factors include soil and climate differences, and the varying characteristics of the many different species in sprouting ability, in viability of seed stored in the forest floor, in bark thickness at various tree ages, or in other ways that make plants resistant or susceptible to fire. Consequently, a certain fire regime results in covers of grasses, sedges, or forbs, while another regime in the same general region favors one of several compositions of forest stands.

Wildfires have often been destructive to buildings in forested areas and damaging to existing trees. Certain wildfires have killed whole stands of trees, covered large areas, and forced the evacuation of people living in some villages. Monetary losses in destroyed buildings, killed standing trees, and other effects have been great. In this chapter no attempt was made to portray the extent of these damages.

Emphasis has been placed on describing the ecological role that fires have had in shaping the composition of plant communities in the northeastern United States. This role is highly important in certain sections for providing and maintaining a diversity of habitats, both grassland and forest. At the same time, reduction of fuel through prescribed burns can help to prevent catastrophic wildfires in such fire-prone sections as the New Jersey pine barrens.

References

Andresen, J. W. (1959). A study of pseudo-nanism in *Pinus rigida* Mill. *Ecol. Monogr.* **29**, 309–332.

Berry, F. H. (1969). Decay in the upland oak stands of Kentucky. *U.S., Forest Serv., Res. Pap.* **NE-126**, 1–16.

Bromley, S. W. (1935). The original forest types of southern New England. *Ecol. Mongr.* **5**, 61–89.

Buell, M. F., Buell, H. F., and Small, J. A. (1954). Fire in the history of Mettler's woods. *Bull. Torrey Bot. Club* **81**, 253–255.

Buell, M. F., Langford, A. N., Davidson, D. W., and Ohmann, L. F. (1966). The upland forest continuum in northern New Jersey. *Ecology* **47**, 416–432.

Burton, D. H., Anderson, H. W., and Riley, L. F. (1969). Natural regeneration of yellow birch in Canada. *U.S., Forest Serv., Northeast. Forest Exp. Sta., Birch Symp. Proc.* pp. 55–73.

Carvell, K. L., and Tryon, E. H. (1961). The effect of environmental factors on the abundance of oak regeneration beneath mature oak stands. *Forest Sci.* **7**, 98–105.

Cary, A. (1936). White pine and fire. *J. Forest.* **34**, 62–65.

Chaiken, L. E. (1952). Annual summer fires kill hardwood root stocks. *U.S., Forest Serv., Southeast. Forest Exp. Sta., Res. Note* **19**, 1.

Church, T. W. (1955). Observations following wildfire in a young stand of Virginia pine and hardwoods. *U.S., Forest Serv., Northeast. Forest Exp. Sta., Forest Res. Note* **49**, 1–2.

Clark, F. B., and Watt, R. F. (1971). Silvicultural methods for regenerating oaks. *U.S., Forest Serv., Northeast. Forest Exp. Sta., Oak Symp. Proc.* pp. 37–43.

Cline, A. C., and Spurr, S. H. (1942). The virgin upland forest of central New England. *Harvard Forest Bull.* **21**, 1–58.

Day, G. M. (1953). The Indian as an ecological factor in the Northeastern forest. *Ecology* **34**, 329–346.

Fenton, R. H. (1960). Seven years later: Effects of wildfire in a young stand of Virginia pine and hardwoods. *U.S., Forest Serv., Northeast. Forest Exp. Sta., Forest Res. Note* **100**, 1–4.

Fenton, R. H., and Bond, A. R. (1964). The silvics and silviculture of Virginia pine in southern Maryland. *U.S., Forest Serv., Res. Pap.* **NE-27**, 1–37.

Fowells, H. A. (1965). Silvics of forest trees of the United States. *U.S., Dep. Agr., Agr. Handb.* **271**, 1–762.

Grano, C. X. (1970). Eradicating understory hardwoods by repeated prescribed burning. *U.S., Forest Serv., Res. Pap.* **SO-56**, 1–11.

Hall, A. G. (1947). Four flaming days. *Amer. Forests* **53**, 540–542 and 569–570.

Hastings, A. B. (1933). Federal financial and other direct aid to States. 73rd U.S. Congress, 1st Sess., Doc. 12, pp. 1053–1093.

Hawes, A. F. (1923). New England forests in retrospect. *J. Forest.* **21**, 209–224.

Hepting, G. H., and Hedgcock, G. G. (1937). Decay in merchantable oak, yellow poplar, and basswood in the Appalachian Region. *U.S., Dep. Agr., Tech. Bull.* **570**, 1–29.

Hepting, G. H., and Shigo, A. L. (1972). Difference in decay rate following fire between oaks in North Carolina and Maine. *Plant Dis. Rep.* **56**, 406–407.

Horton, K. W., and Bedell, G. H. D. (1960). White and red pine: Ecology, silviculture, and management. *Can., Forest. Branch, Bull.* **124**, 1–185.

Hough, A. F., and Forbes, R. D. (1943). The ecology and silvics of forests in the high plateaus of Pennsylvania. *Ecol. Monogr.* **13**, 299–320.

Jemison, G. M. (1944). The effect of basal wounding by forest fires on the diameter growth of some southern Appalachian hardwoods. *Duke [Univ.] Sch. Forest. Bull.* **9**, 1–63.

Korstian, C. F. (1924). Natural regeneration of southern white cedar. *Ecology* **5**, 188–191.

Korstian, C. F. (1937). Perpetuation of spruce on cut-over and burned land in the higher southern Appalachian Mountains. *Ecol. Monogr.* **7**, 125–167.

Langdon, O. G. (1971). Effects of prescribed burning on timber species in the southeastern Coastal Plain. *U.S., Forest Serv., Southeast. Forest Exp. Sta., Prescribed Burning Symp. Proc.* pp. 34–44.

Little, E. L., Little, S., and Doolittle, W. T. (1967). Natural hybrids among pond, loblolly, and pitch pines. *U.S., Forest Serv., Res. Pap.* NE-67, 1–22.

Little, S. (1950). Ecology and silviculture of white cedar and associated hardwoods in southern New Jersey. *Yale [Univ.] Sch. Forest. Bull.* **56**, 1–103.

Little, S. (1959). Silvical characteristics of pitch pine (*Pinus rigida*). *U.S., Forest Serv., Northeast. Forest Exp. Sta., Sta. Pap.* **119**, 1–22.

Little, S. (1964). Fire ecology and forest management in the New Jersey pine region. *Proc. 3rd Annu. Tall Timbers Fire Ecol. Conf.* pp. 34–59.

Little, S. (1967). Treatments needed to regenerate yellow-poplar in New Jersey and Maryland. *U.S., Forest Serv., Res. Note* NE-58, 1–8.

Little, S. (1972). Growth of planted white pines and pitch seedlings in a south Jersey plains area. *Bull. N.J. Acad. Sci.* **17**, 18–23.

Little, S. (1973). Eighteen-year changes in the composition of a stand of *Pinus echinata* and *P. rigida* in southern New Jersey. *Bull. Torrey Bot. Club* **100**, 94–102.

Little, S., and Mergen, F. (1966). External and internal changes associated with basal-crook formation in pitch and shortleaf pines. *Forest Sci.* **12**, 268–275.

Little, S., and Mohr, J. J. (1954). Reproducing pine stands on the eastern shore of Maryland. *U.S., Forest Serv., Northeast. Forest Exp. Sta., Sta. Pap.* **67**, 1–11.

Little, S., and Mohr, J. J. (1963). Conditioning loblolly pine stands in eastern Maryland for regeneration. *U.S., Forest Serv., Res. Pap.* NE-9, 1–21.

Little, S., and Moore, E. B. (1949). The ecological role of prescribed burns in the pine–oak forests of southern New Jersey. *Ecology* **30**, 223–233.

Little, S., and Moore, E. B. (1953). Severe burning treatment tested on lowland pine sites. *U.S., Forest Serv., Northeast. Forest Exp. Sta., Sta. Pap.* **64**, 1–11.

Little, S., and Somes, H. A. (1956). Buds enable pitch and shortleaf pines to recover from injury. *U.S., Forest Serv., Northeast. Forest Exp. Sta., Sta. Pap.* **81**, 1–14.

Little, S., and Somes, H. A. (1964). Releasing pitch pine sprouts from old stools ineffective. *J. Forest.* **62**, 23–26.

Lull, H. W. (1968). A forest atlas of the Northeast. *U.S., Forest Serv., Northeast. Forest Exp. Sta.*, pp. 1–46.

Lutz, H. J. (1934). Ecological relations in the pitch pine plains of southern New Jersey. *Yale [Univ.] Sch. Forest. Bull.* **38**, 1–80.

Lutz, H. J., and McComb, A. L. (1935). Origin of white pine in virgin forest stands of northwestern Pennsylvania as indicated by stem and basal branch features. *Ecology* **16**, 252–256.

McConkey, T. W., and Gedney, D. R. (1951). A guide for salvaging white pine injured by forest fires. *U.S., Forest Serv., Northeast. Forest Exp. Sta., Res. Note* **11**, 1–4.

McCormick, J., and Buell, M. F. (1968). The plains: Pygmy forests of the New Jersey pine barrens, a review and annotated bibliography. *Bull. N.J. Acad. Sci.* **13**, 20–34.

McIntosh, R. P., and Hurley, R. T. (1964). The spruce–fir forests of the Catskill Mountains. *Ecology* **45**, 314–326.

Niering, W. A., Goodwin, R. H., and Taylor, S. (1970). Prescribed burning in southern New England: Introduction to long-range studies. *Proc. 10th Annu. Tall Timbers Fire Ecol. Conf.* pp. 267–286.

Place, I. C. M. (1955). The influence of seed-bed conditions on the regeneration of spruce and balsam fir. *Can., Forest. Branch, Bull.* **117**, 1–87.

Rigg, G. B. (1940). Comparisons of the development of some sphagnum bogs of the Atlantic coast, the interior, and the Pacific coast. *Amer. J. Bot.* **27**, 1–14.

Roth, E. R., and Hepting, G. H. (1943). Origin and development of oak stump sprouts as affecting their likelihood to decay. *J. Forest.* **41**, 27–36.

Roy, H. (1940). Spruce regeneration—Quebec. *Forest. Chron.* **16**, 10–20.

Sander, I. L., and Clark, F. B. (1971). Reproduction of upland hardwood forests in the Central States. *U.S., Dep. Agr., Agr. Handb.* **405**, 1–25.

Shearin, A. T., Bruner, M. H., and Goebel, N. B. (1972). Prescribed burning stimulates natural regeneration of yellow-poplar. *J. Forest.* **70**, 482–484.

Smith, D. M. (1951). The influence of seedbed conditions on the regeneration of eastern white pine. *Conn., Agr. Exp. Sta., New Haven, Bull.* **545**, 1–61.

Smith, L. F. (1940). Factors controlling the early development and survival of eastern white pine (*Pinus strobus* L.) in central New England. *Ecol. Monogr.* **10**, 373–420.

Stephens, G. R., and Hill, D. E. (1971). Drainage, drought, defoliation and death in unmanaged Connecticut forests. *Conn., Agr. Exp. Sta., New Haven, Bull.* **718**, 1–50.

Stickel, P. W. (1941). On the relation between bark character and resistance to fire. *U.S., Forest Serv., Northeast. Forest. Exp. Sta., Tech. Note* **39**, 1–2.

Stickel, P. W., and Marco, H. F. (1936). Forest fire damage studies in the Northeast. III. Relation between fire injury and fungal infection. *J. Forest.* **34**, 420–423.

Swan, F. R., Jr. (1970). Post-fire response of four plant communities in south-central New York State. *Ecology* **51**, 1074–1082.

Thompson, D. Q., and Smith, R. H. (1970). The forest primeval in the Northeast—a great myth? *Proc. 10th Annu. Tall Timbers Fire Ecol. Conf.* pp. 255–265.

Trevett, M. F. (1962). Nutrition and growth of the lowbush blueberry. *Maine, Agr. Exp. Sta., Bull.* **605**, 1–151.

U.S. Forest Service. (1972). "1971 Wildfire Statistics," GPO 933–009. USFS, Washington, D.C.

Vermeule, C. C. (1900). The forests of New Jersey. *Annu. Rep., N.J. State Geol., 1899* pp. 13–101.

Wood, G. W. (1971). Biomass, production and nutrient distribution in mixed-oak stands following clearcutting and fire. Ph.D. Thesis, Pennsylvania State University, University Park.

Wright, K. E. (1941). The great swamp. *Torreya* **41**, 145–150.

. 8 .

Effects of Fire on Temperate Forests and Related Ecosystems: Southeastern United States

E. V. Komarek

I. Introduction

Studies of effects of fire on the flora and fauna of the southeastern United States have had a long and controversial history. Most of the research effort has had relevance to forest management of longleaf pine (*Pinus palustris*), range management in such forests, and management of wildlife, particularly bobwhite quail. The ecological relationships of fire and most species of plants and animals have yet to be studied in detail.

The southeastern region of the United States (hereafter called the Southeast) ranges from the Appalachians to the sea. Its lands are some of the oldest on the continent, and yet new land is being built. It was long separated from its western plant and animal relatives and yet retained cretaceous plant "relicts." It was the refugium when glaciers reached their southern border. Subsequently man disturbed its plants and animals and turned its soil. Throughout this long period fire has been a part of its environment. Thus, in essence, many aspects of fire ecology can be studied in the Southeast for it has accumulated species from a large portion of the North American continent.

II. Natural Fires

On a continually changing pattern, the kaleidoscope of time, space, mountains and plains, and living things, the added effects of natural fires must be superimposed. Coal beds with their attendant fossil charcoal, fusain, attest that fire was part of ancient ecosystem (Komarek, 1972). Bryson *et al.* (1970) reconstructed the frontal weather systems and associated weather patterns for the late Pleistocene for the North American continent. These in many respects are quite similar to those weather systems that today create thunderstorms whose lightning discharges ignite not only forests and grasslands but buildings in farm and city as well.

Although natural fires can be started by volcanic action, spontaneous ignition, and friction, lightning appears to be the major cause of such fires. The action of this gigantic spark on forest, grassland, and field has been well documented (Arnold, 1964; Requa, 1964; Taylor, 1971; Komarek, 1964, 1965, 1966, 1967a,b, 1968, 1970, 1972).

There are indications that our present weather patterns originated during the Tertiary by the various geological events that changed the weather systems and thus the climate of North America. Earlier the continent was covered by a rather equable and uniform climate so that there was little difference in the vegetation from Alaska to Mexico. However, with geological changes, particularly the development of the Rocky Mountain chain that extends nearly the entire length on the western side of North America, frontal systems developed that swept the continent in summer east of the Rockies from Alaska southeasterly to the southern Atlantic coast and to the tip of Florida. These systems now occur at about 7- to 14-day intervals during the summer months and literally bathe the continent east of the Rockies in lightning, setting fires as they travel southeastward (Komarek, 1968). These frontal systems are greatly affected by the Appalachian and associated mountains so there is a much different distribution pattern of precipitation in this mountain region and the area to the west and northwest as contrasted with that of the Piedmont Plateau and Coastal Plain. Braun (1950) divided this area on the basis of forest vegetation: the "Oak–Chestnut Forest region" on the highlands and the "Mixed and Western Mesophytic Forest regions" on the plateau to the west. These are predominately hardwood forest regions.

To the east of the mountain region in what might be called the "rain shadow" of the Applachians is the Piedmont Plateau. However, because of the proximity of this region to the Atlantic Ocean and the low pressure systems that develop there, the possible lack of precipitation is alleviated to some degree. This is the "Oak–Pine Forest region" of Braun (1950).

To the east and south, somewhat like a concentric band, is the Coastal Plain that reaches from New Jersey southward and then along the Gulf of Mexico to eastern Texas. This is the "Evergreen Forest region" of Braun (1950). In addition, Braun classified the Ozark and Ouachita Mountains, the Interior Highlands, as the "Oak–Hickory Forest region" which is the "drier western part of the deciduous forest."

The precipitation patterns of these forest regions are quite different, and, therefore, the number, size, intensity, and effect of lightning fires vary. For these reasons, Komarek (1968) divided the southeastern ecosystem into two "Provisional Lightning Fire Bioclimatic regions."

III. Man and Fire

When man became a member of the southeastern ecosystems some 20,000 years ago he became another "fire agent" in the region. Man as a fire agent was not limited to specific and short periods of ignition. He could and did set fires during long droughts and during winter months, often when lightning could not, or did not, set fires. Thus, man may have set in motion other ecological changes of which we are not even aware. The litter of hardwood forests in particular is difficult to burn in the summer months, but shortly after leaf fall in autumn and after frost this fuel can be easily burned during the winter months.

IV. Fire and Fire Characteristics

For present purposes fire can be considered as the combustion of recent organic matter which may vary from such compressed substances as peat to very diffuse materials such as pine "needle-drape" that envelops bushes and other understory plants. The fuel can be weathered, partly decomposed (as peat, litter, or "duff"), or recently fallen leaves, needles, grasses, and various kinds of mixtures or blends of all these and other materials. In other words, the fuel for forest or grassland fires is as diversified as the plant kingdom itself.

To this diversity must also be added the complexity of the moisture relationships of the fuel. The moisture content of the fuel along with temperature and humidity of the air determine the characteristics of the fire and its effect on the ecosystem. When the litter and underlying "duff" are wet, they cannot burn. However, sunshine, high temperatures, and brisk winds can dry out the uppermost layer quickly so that a "cool" or "feeble" fire can literally crawl or sweep, depending on the wind velocity, over the underlying wet organic matter. The temperatures of such fire is relatively low. However, after several days this same fuel will burn much hotter under relatively the same temperature, humidity, etc. and after several weeks of drying weather, it will burn with such intensity that a "fire storm" can develop. To the diversity of the fuel then we must add the variations of the moisture content of that fuel as well as the vagaries of the weather, sunshine, humidity, wind velocity, and transpiration rates of plants.

Many studies have shown wide variations in effects among backing fires (backing into the wind), head fires (traveling with the wind), and flank fires (moving sideways to the wind) on level ground. If to this are added the slope and exposure of the land, the ecological effects are compounded. The four types of fire mentioned above occur only under conditions of steady wind movement. However, even though wind velocity and direction are steady, eddy currents and even whirlwinds can be created by the action of the fire itself and its varying convection columns and properties. Thus, fires are dynamic and constantly changing.

There is a large body of literature dealing with fire behavior, meteorological conditions, fuels, and many other aspects of fire in forest and range in the southeastern ecosystems (Ahlgren and Ahlgren, 1960; Cushwa, 1968; Duvall et al., 1968; Davis, 1959; Komarek, 1965, 1967a, 1968; Wahlenberg, 1946, 1960; Williams, 1938).

V. The Southeastern Ecosystem "Fire Regimes"

As mentioned earlier, I have divided the Southeast into two Provisional Lightning Fire Bioclimatic regions, the southern pine forest and the eastern deciduous forest, on the basis of certain prominent forest types as well as "fire weather" and other meteorological phenomena.

Spurr (1964) stated that:

> Fire is the dominant fact of forest history. The great majority of the forests of the world—excepting only the perpetually wet rain forests . . . and the wettest belts of the tropics—have been burned over at more or less frequent intervals for many thousands of years.

And that:

> Around the world, the dominance of pine and oak forests of virtually all species and in virtually all regions is due predominantly to fire.

Although these statements are sweeping and perhaps overstated, as Spurr acknowledged, they appear to apply to the southeastern forest ecosystem. The pines and oaks that have covered most of the forested region in the Southeast in the past certainly have a preponderance of "fire-adapted" pine and oak species. They, along with most of the grasses, legumes and other forbs, and shrubs, particularly of the Coastal Plain and Piedmont Plateau have been selected by fire for certain accumulations of genes not only to live but to multiply and increase in such environments.

A. The Southern Pine Forest Lightning Fire Bioclimatic Region

This region is physiographically divided into two sections, the Coastal Plain and the Piedmont Plateau. Braun has called the former the "Southern Evergreen Forests" because of the vast acreage in pine forest and the latter "Oak–Pine Forests" because of the mixture of these two species. There are also recognizable differences in the fire regimes for the two areas. The oak–pine forest is characterized by a mixture of oak trees (many *Quercus* species) and relatively short-needle pines, loblolly (*Pinus taeda*) and shortleaf (*P. echinata*), whose cast leaves and needles lie relatively flat on the surface of the ground when packed by rains. These fuels are not as flammable as those of the southern evergreen forest but much more so than those of the more mesic eastern deciduous forests. The southern evergreen forest consists of long-needle pines (*Pinus palustris, P. Elliotii,* and *P. elliotii* var. *densa*). The understory in these forests is made up of prairie vegetation along with many deciduous trees and shrubs that are kept in a very low suckering stage by recurring fires. Where fires are frequent enough, the shrub layer in the forests of the Piedmont Plateau is also kept pruned back and a grass understory is maintained.

Braun (1950) pointed out that although the pine forests of the South may dominate the landscape and may be ". . . the most prominent vegetational feature . . . there is scarcely a forest community, upland or lowland, without some admixture of coniferous or broad-leaved evergreen trees and shrubs." However, in spite of this the Coastal Plain pine forests are essentially prairies or savannas and, when burned regularly, maintain a very rich, varied, and endemic prairie flora and fauna. In the absence of fire the pine savannas develop into forests ecologically comparable with the so-called beech–maple forest of the northeastern United States.

B. Eastern Deciduous Forest Lightning Fire Bioclimatic Region

This region consists of the Appalachian Highlands and the "Mesophytic Forest regions" of Braun (1950). It is characterized by relatively low fuel flammability, less critical fire weather which occurs at infrequent intervals. The yearly rainfall is well distributed which prevents widespread fires except under very unusual conditions. Only the southern part is found in the Southeast.

VI. Coastal Plain Forests

Early explorers, visitors, settlers, and botanists remarked on the broad expanse of parklike pine forests that extended from southern Virginia to east Texas. Because of the uniformity of open-spaced pine trees and a grassland understory many remarked about its monotony:

> A report written in 1810 stated that the "Sandhills yield no other timber than yellow pine which from the sameness and uniformity of appearance affords too little variety to be pleasing to the view" [Wells, 1932].

The extent of this forest has been much reduced and the "virgin" forest has been virtually destroyed within a hundred years. It consisted primarily of longleaf pine on the more or less rolling uplands and slash pine on the lower wetter lands, or "flat-woods." In valleys, bottom lands, and along rivers, a large mixture of deciduous hardwoods reminiscent of the eastern deciduous forest occurred. Undergrowth of several bush or shrub species as well as sucker growth of the hardwoods were kept suppressed by regularly occurring fires caused by lightning or Indians.

Thus, there were essentially three main components of this extensive plant association: the pines, the grassland with some shrub overstory, and the hardwoods of the more mesic naturally fire-protected areas.

Before any real botanical or ecological studies were undertaken, most of this region had been drastically affected by settlement and other activities by the white man. However, there are some remnants even though they are few and far between.

A. Longleaf Pine Forest

The longleaf pine forest and its fire ecology have been well studied, perhaps more than any other forest. Wahlenberg (1946) and Croker (1968) cited more than 1300 titles on this one species. About one-sixth of the references are related to fire ecology and fire management. Today controlled or prescribed burning is widely used in the forest, range, and wildlife management of this and associated pine forests.

Wahlenberg (1946) who devoted an extensive discussion to the fire ecology of this pine forest stated that:

> Longleaf pine is a temporary type in that without the influences of fire and man it would eventually be succeeded by a forest of mixed hardwoods. Any management system, therefore, that maintains longleaf as a forest must arrest the trend of natural succession. Among the natural factors controlling the regereration of longleaf, fire is undoubtedly most influential.

Longleaf pine together with all the other southeastern pines, with the possible exception of spruce pine (*P. glabra*) and to some extent slash pine, require a mineral seedbed. After germination longleaf develops a taproot which continues to grow for several years while in the "grass" stage. In these early years the needles may become affected by "brown spot" disease caused by a fungus (*Scirrhia acicola*). The disease can be controlled by burning the infected needles before new growth in spring. The bud is protected from fire by long needles and scales. The bark of young longleaf pine resists heat and, for young trees, is quite thick. After the tree reaches the sapling stage it is virtually fire resistant, except to hot fires when the new shoot axis or "candles" are expanding in late spring and summer.

B. Slash Pine Forest

Next to longleaf pine, slash pine was the dominant tree of the Coastal Plain. As the primary tree of the wet "flat-woods," slash pine was not exposed to frequent fires. Although it is well adjusted to burning, it must be protected from fire until it is in the sapling stage. From then on the tree can survive defoliation, apparently without harm as long as the upper third of the crown is not scorched and the fires do not occur in summer. Its needles apparently are less flammable than those of longleaf pine. Slash pine reproduces best on open mineral soil but will germinate and grow under damp conditions where there is some litter. In this respect it is not as particular in its regeneration requirements as the longleaf pine.

This species also occurs in extensive parklike forests with a grassland understory which, however, contain highly flammable shrubs as gallberry (*Ilex glabra*) and saw-palmetto (*Serenoa repens*).

A variety of slash pine (*P. elliotii* var. *densa*) which occurs in southern Florida has many more fire protective mechanisms than the typical species. In the northern part of the range of *P. elliotii* var. *densa* the forest has a grassland understory with palmetto. In wetter areas cabbage palms (*Sabal palmetto*) give the forest a tropical aspect. At the very southern tip of Florida on the limestone ridges, however, there are many broad-leaved subtropical species that quickly invade the forest unless burned regularly. If regularly burned, here too, a grassland ground cover prevails.

C. LOBLOLLY PINE FOREST

The loblolly pine originally was an inhabitant of the more mesic areas in the Coastal Plain. However, with settlement, logging, and fire exclusion this forest has replaced many areas that were once longleaf or slash pine forests. In some parts of the Coastal Plain and in much of the Piedmont Plateau, loblolly has become the dominant pine forest species. Wahlenberg (1960) discussed the fire ecology and fire management of this species and cited over 1400 references.

Prescribed burning is widely used for regeneration of this pioneer species which requires a mineral seedbed. Until the tree is in the sapling stage, 12 ft or taller, it is relatively fire intolerant. Thereafter, because of its thick bark, it is quite fire tolerant. Controlled burning is used regularly in pole-size or larger forests which are very fire resistant. In the "Red Hill" region of north Florida and southern Georgia, forests of loblolly pine have been burned regularly for management of wildlife for over three-quarters of a century.

Lotti *et al.* (1960) pointed up the primary silvicultural problem in loblolly pine forests.

> If one were to select the particular problem giving pine foresters the most concern, it probably is the natural replacement of pine by the more tolerant but inferior hardwoods.

Today prescribed burning is used regularly to control the invasion of this hardwood understory.

D. POND PINE FOREST

Pond pine (*Pinus serotina*) is found locally and primarily in poorly drained flats from southeastern Virginia to central and southeastern Alabama and in a rather isolated area on Cape May, New Jersey. In eastern North Carolina it is the characteristic tree of extensive poorly drained evergreen bogs or savannas (depending on the severity of fire) called "pocosins." In other areas it is found scattered along low flat and poorly drained stream beds (Wells, 1932; Wenger, 1958).

The cones of pond pine are serotinous and require heat from relatively hot fires to open. Wenger (1958) found 59% viability of seeds from badly charred cones within 30 days of planting. The seeds germinate and grow well on mineral soil exposed by burning.

After it reaches seedling size, pond pine sprouts readily and continues to do so to an advanced age. The seedlings have basal crooks as a fire adaptation (Little and Somes, 1956). Epicormic shoots along stems and branches are produced after defoliation by fire.

E. SAND PINE FOREST

Sand pine (*Pinus clausa*) is of limited distribution and occurs in central Florida and a small section along the Gulf coast of northwestern Florida. Its understory includes such shrubs as rosemary (*Ceratiola ericoides*), saw-palmetto (*Serenoa repens*), scrub palmetto (*Sabal etonia*), several scrub oaks (*Quercus chapmanii, Q. geminata, Q. laevis, Q. myrtifolia*), and some live oak (*Quercus virginiana*). The pine trees are sensitive to fire and have serotinous cones, particularly the Ocala sand pine in the central Florida area. The Choctawahatchee sand pine of northwestern Florida has fewer serotinous cones, and open cones are frequent. The tree and its associated shrubs, etc., are adapted to severe fires that destroy the forest to regenerate it. However, if man-caused or lightning fires occur at the wrong season or if the forest is reburned too quickly, it comes back slowly.

VII. Coastal Plain Pine Savannas and Grasslands

The Coastal Plain grasslands can be divided into five major types as follows.

1. Pine–bluestem savanna (Hilmon and Hughes, 1965)
2. Pine–wiregrass savanna (Hilmon and Hughes, 1965)
3. Pine–cane savanna (Wells, 1932; Hilmon and Hughes, 1965)
4. The Georgia–Florida–Alabama "Red Hills" savanna (Stoddard, 1931; Kurz, 1944)
5. Coastal marshes and prairies

A. Pine–Bluestem Savanna

This savanna is characterized by an overstory of longleaf pine and some slash pine in the lower or wetter areas. The grassland understory is predominately little bluestem (*Andropogon scoparius*) and slender bluestem (*A. tener*). However, there is a great diversity of plants all adjusted to frequent burning. Nearly 200 species of grasses have been collected from this one type. There probably are as many other species of plants as well. Grelen and Duval (1966) listed 155 species of plants from this type. More than 25 species of legumes have been noted in one experimental area in Mississippi.

This grass and forb understory, essentially relict prairie (Grelen and Duvall, 1966; Vogl, 1972), characterized the parklike longleaf pine forests of the Coastal Plain when first observed by early visitors. Like the forest, it is a fire subclimax, a successional arrestment induced by burning. The pine–bluestem savanna, along with its open pine forest which filters the sunlight instead of shading the forest floor, is in relative equilibrium only as long as fires recur at relatively frequent intervals. This equilibrium was established with both climate (lightning and precipitation patterns) and fire long before the region was invaded by man. The Indian probably extended this vegetation type by more frequent burning than was natural, whereas the white man upset this balance largely by fire exclusion. In the latter case, loblolly pine, slash pine, and broad-leaved hardwood trees and shrubs invaded much of the original area even where it was not destroyed by farming and logging.

Studies in the pine–bluestem region have shown that *burning increases protein, phosphorus, and calcium contents of the grasses, enhances the palatibility,* and *improves the composition of the rangeland by increasing the desirable grasses, forbs, and legumes.* No significant effects of burning on soil properties were noted, while mineral elements, nitrogen, and the organic matter improved with burning (Greene, 1931, 1935a,b; Fraps and Fudge, 1940; Campbell, 1946; Campbell and Cassady, 1951; Cassady, 1953; Cassady and Mann, 1954; Campbell *et al.*, 1954, 1963; Smith *et al.*, 1955, 1958; Lay, 1956; Duvall *et al.*, 1968; Langdon *et al.*, 1952).

The above studies have generally shown that the increased nutritive qualities do not last more than a few months after spring burning, but this is usually the time when livestock as well as wildlife need extra protein and minerals. Pioneer Florida cattlemen used a system of progressive burning to take advantage of this increased protein and palatibility (Stoddard, 1962). They would burn adjacent areas at various times so as to have a series of new burns upon which the cattle would graze progressively. In fact, fire was used, one might say, as a sort of a fence or management procedure to assist in handling their cattle. I have witnessed this system being used by the Masai in certain parts of East Africa in recent years (1970 and 1972). From discussions with old cattlemen and farmers in the western United States, I believe it was a generally used range management practice there as well. This was probably adapted from the generally recognized custom of burning for game by many native tribes in many parts of the world.

Studies on the effect of burning on 25 species of shrubs browsed by deer (*Dama virginiana*) showed a temporary 2-year increase in protein and phosphoric acid (Lay, 1957) so that in such vegetation the nutritive effects of burning may last longer than in grass and forbs.

B. Pine–Wiregrass Savanna

This savanna originally extended from southern Virginia to Alabama along the Coastal Plain of the Atlantic Ocean and the Gulf of Mexico. It mingles in southern Alabama with more western bluestem savanna. Its primary grass constituents are pineland threeawn (*Aristida stricta*) commonly called "wiregrass" and Curtis dropseed (*Sporobolus curtissi*) also called "wiregrass." There are many secondary grasses such as toothache grass (*Ctenium aromaticum*), Florida dropseed (*Sporobulus floridanus*), and cutover muhly (*Muhlenbergia expansa*) as well as several species of bluestem (*Andropogon* spp.), Panic grasses (*Panicum* spp.), and many legumes and other forbs.

Lemon (1949) listed 38 species of grasses, forbs, and sedges as part of postfire succession responses in a savanna of this type at Alapha, Georgia. However, many of the experimental areas that have been used for research on grazing effects lack the leguminous flora that was originally present. Stoddard and Komarek (1941) found that many of the most valuable wildlife legumes cannot adjust to even moderate grazing pressure by livestock. Under any kind of grazing they disappear. This is why cattle and quail have never mixed well on this savanna type, except in central Florida where conditions are different (Stoddard, 1931; Stoddard *et al.*, 1961; E. V. Komarek, unpublished).

Several groups of striking and interesting plants occur characteristically in the low acid swales, depressions, steep hillsides, and bogs in the wiregrass–pine savanna. Among these the most attractive are certain species of pitcher plants and orchids which are not only adapted to low, wet, acid areas but to regular and recurring fires for they cannot compete with either heavy accumulations of dead grasses or the shade and competition from brush species. McDaniel (1971) has written that "the genus *Sarracenia* is well adapted to moderate fires, at least in the South, which remove old growth, destroy competition, and help induce flowering" and that "along the Gulf Coast plants of the genus are often associated in bogs with *Sphagnum, Drosera, Pinguicula, Eriocaulon, Calopogon, Bermannia* and other genera characteristic of highly acid sites."

Several species of wild orchids (*Clestes divaricata, Pogonia ophioglossoides, Calopogon barbatus, C. pallidus, C. multiflorus, Habenaria cilliaris, H. cristata, H. integra, H. nivea, H. blephariglottis,* and *Spiranthes* sp.) have been maintained in considerable numbers over a long period of time by late spring annual burning in an original growth longleaf pine and slash pine forest at Greenwood Plantation (R. Komarek, personal communication). This area, a low swale or valley, had been more or less protected from burning for more than several decades and had become a broad-leaved hardwood jungle of several species (*Nyssa biflora, Persea borbonia, P. palustris, Quercus nigra, Liriodendron tulipifera,* etc.)

By repeated burning and some mechanical clearing this brush or hardwood jungle has in 28 years of annual burning developed into a characteristic pine–wiregrass savanna rich in many species of plant genera common to such areas. During this "reclamation" period, the grass orchids mentioned above have greatly increased in numbers. The pitcher plant, *Sarracenia minor,* and other characteristic bog plants as mentioned by McDaniel (1971) have also increased. Previously these only occurred in small fringe areas along the edge of the hardwood jungle.

The nutritional effects of fire on the forage of the pine–wiregrass savanna have been found to be similar to those of the pine–bluestem savanna. Several studies (Leukel and Stokes, 1939; Biswell *et al.,* 1942; Lemon, 1949; Halls *et al.,* 1952) have shown that burning significantly increases levels of protein (or nitrogen), phosphorus, potassium content of the native forage as well as making the grasses more vegetative, succulent, and palatable. Significant increases in cattle weights were also recorded on burned ranges as against unburned ones in all the above studies.

Cultivated pastures of improved grasses have replaced pine–wiregrass savanna in many parts of its former range. Fire is now used on coastal bermuda (*Cynodon*) and bahia (*Paspalum* spp.) grasses regularly to remove old dead mulching grass, to control insects, and to fertilize pastures. Burning not only increases the protein, phosphorus, and potassium content but increases the yield of grass and hay considerably. Recently farmers have been advised to spread nitrogenous fertilizer directly on the ash after burning for the ash inhibits an enzyme, urease, so that the nitrogen from the fertilizer is not released into the air (Manson *et al.*, 1974; Jackson and Burton, 1962).

Several studies have also shown that burning increases seed production of many grasses even up to 20% in the South as well as elsewhere (Burton, 1944; Komarek, 1965).

C. Pine–Cane Savanna

Two species of native bamboos (*Arundinaria tecta, A. gigantea*) once common and locally abundant, but now largely gone, formed various types of "canebreaks" throughout the southeastern United States (Mc-Clure, 1959). Unfortunately, little is known of the fire relationships of *A. gigantea*, locally called "maiden-cane," except that it persists where regular burning is conducted and livestock grazing is excluded. By 1950 most of the "breaks" had been eliminated by a combination of overgrazing, uncontrolled fire and clearing with the exception of switch cane (*A. tecta*). At the present time over 2 million acres of the latter persist in the Carolinas and Virginia, associated primarily with pond pine (*Pinus serotina*) in areas that have had periodic burning (Shepard *et al.*, 1951; Hilmon and Hughes, 1965; Hughes, 1966; Wells, 1932, 1942).

Hilmon and Hughes (1965) and Biswell *et al.* (1943) stated that prior to the 1950's the burning of canebreaks (*A. tecta*) was looked upon with scorn. Since that period, however, research studies have shown that "cane is one of the most nutritious of the native forage plants growing in the eastern United States" (Hilmon and Hughes, 1965) and that where cane is burned properly and regularly it is more nutritious (higher in protein), more palatable and succulent, and more productive (Biswell *et al.*, 1943, and above authors). In the absence of fire the cane becomes unpalatable and "under continuous protection from fire, cane stands lose vigor, thin out, and die" (Hughes, 1966). Under such fire exclusion, heavy accumulations of very flammable fuel develop and some of the most destructive fires in the Coastal Plain occur.

In some regions, in the absence of fire, both species of *Arundinaria* are invaded by several broad-leaved species of trees (*Quercus nigra, Q. prinus, Liriodendron tulipifera, Persea borbonia, P. palustris, Nyssa biflora,* and *N. sylvatica*) and shaded out of existence so that only isolated plants occur. A hot fire will rejuvenate the cane in such situations (E. V. Komarek, unpublished).

Wells (1932, 1942) has reported on several other types of local savannas including those found in pocosins or poorly drained low area and which under certain fire regimes become "evergreen shrub bogs." The grass stage of succession in these contains a wide variety of grasses, sedges, and showy flowers, a plant community in many respects resembling those in the bogs of the wiregrass savanna. Like the latter, these are fire-maintained plant associations because without fire they disappear.

D. GEORGIA–FLORIDA–ALABAMA "RED HILLS" SAVANNA

Originally this tristate red soil region was covered with a mixture of the bluestem– and wiregrass–pine savannas intermingled with hardwoods in the branches or where fire did not occur regularly. However, because of such activities as farming and logging this soil, suitable for growth of many annual and perennial legumes, e.g., *Lespedeza* spp., *Desmodium* spp., and *Strophostyles* spp., was converted to a "derived savanna" that developed because of the regular burning of the region for various reasons. Most of the Red Hill country is now covered with a mixture of pines, largely loblolly, longleaf, slash, and shortleaf pines, intermingled in many areas with various broad-leaved tree species. This region was extensively cultivated prior to 1860–1880 and then reverted, under sporadic burning, to the present-day forests, which have been burned annually for many years. With this treatment the pine forest, mixed in places with widely spaced hardwood trees, has developed into an interesting grassland, extremely valuable to many forms of wildlife. It is rich in grasses, forbs, sedges, and showy flowering plants. In many areas of the region, such management has maintained an extremely high population of quail and also some of the largest populations of the red-cockaded woodpecker (*Dendrocopos borealis*), an endangered species.

Fortunately, also within the Red Hill region, are some remnants of the original longleaf pine forest (trees 300 years and older) and savanna. One of these, on Greenwood Plantation, Thomasville, Georgia, has been burned annually for the last 27 years. The "virgin" forest does not have the abundance or variety of leguminous plants found on lands that were once farmed and then allowed to revert to pine forest. Some aspects of this derived grassland were described by Harper (1914). However, Stoddard (1931) first described it in some detail with reference to the role of fire in its maintenance. He found that the bobwhite quail ate seeds from 93 species of forbs of 47 families. Of these, over half were legumes, most of which occur in these *fire-maintained* derived savannas. He also found 59 species of grasses (17 panicums, 12 paspalums) most of which also occur in these grasslands. Because the bobwhite quail is a *grassland* inhabitant, not of a brush or dense forest, fire or some other kind of disturbance is required for its perpetuation. Stoddard (1931) proved experimentally, as well as on a practical basis on about 100 hunting plantations throughout the Southeast from North Carolina to Arkansas, that the savanna or prairie vegetations were absolutely necsesary for high production and high quality hunting of the bobwhite throughout the Coastal Plain on many soil and forest types.

Stoddard's studies showed that burning was necessary to maintain the grassland understory because in that way competing vegetation (shrubs, hardwoods, mulches of dead grass, pine needles, etc.) was held in check. Furthermore such burning stimulated the grasses and legumes which comprised the basic diet of the quail in these pine savannas.

Stoddard (1931) stated:

> . . . the immediate and direct effect of burning an area, is of course, always apparent, the general effect of long continued annual, or irregular but frequent, burning upon the vegetation of an area, and its indirect effect upon the animal life, *present a complex problem, one that would require years of careful research on the part of the personnel of a well-equipped experiment station to work out. Such research is greatly needed, and should be carried on, for fire may well be the most important single factor in determining what animal and vegetable life will thrive in many areas* [italics Komarek].

In 1958, Stoddard and his associates established a series of plots to be burned at different intervals and some to remain unburned.

Clewell (1966) examined these plots in connection with his studies of *Lespedeza* spp. He found that:

> The lespedezas are well adapted to frequent surface fires, as can be seen at Tall Timbers Research station in Leon County, Florida. . . . Grazing by livestock has been excluded for at least 50 years. The upland of this property has been burned annually since the 1890's to maintain the habitat for bobwhite quail. Such burning arrests plant succession, killing or pruning back invading hardwoods but favoring pines and a great assortment of fire-adapted grasses and forbs. The populations of *Lespedeza*, especially *L. procumbens* and *L. repens*, are more dense and widespread there than in any other area I have visited. . . . In 1959, 84 one-half acre plots were established by the Station in annually-burned pinelands which are rich in lespedezas. These plots are burned at varying intervals. In the plot burned annually in the summer are plants of *L.*

anqustifolia, L. hirta, L. procumbens, L. repens, L. stuevei, and *L. virginica;* it is difficult to find an area two meters square which is devoid of lespe-dezas. . . . The plot to be burned every seventh summer was a virtual jungle of mixed hardwoods 8–14 feet tall with no *Lespedeza* except for an occasional plant at the perimeter.

Kurz (1944) showed that with a few years and absence of fire the pine–bluestem savanna changed to hardwood and shrub stages and eventually into a beech–magnolia climax.

The Tall Timbers fire ecology plots established in 1959 to study the effects of fire at varying intervals from annual burning to total fire exclusion were studied by Vogl (1972). He compared the grassland found on these experimental areas and surrounding annual burned lands with that of western and midwestern prairies. He showed:

. . . that the understory vegetation of the open pinelands or savannas of the Southeast contain floristic elements comparable to that of tall grass prairies elsewhere. These southern understories are as rich in prairie species, and as diverse in total numbers, if not more diverse, than the prairies of the Great Plains. . . . Annual and biennial burning maintained the highest grass cover, which declined sharply starting with a fire every third year. Plots burned only once in the 12-year study period supported essentially the same low amounts of grass cover as the unburned plots. . . . Although the forb species' compositions were not recorded, *the burned plots were dominated by typical prairie herbs while the unburned plots largely contained shade-tolerant herbs characteristic of hardwood forest* [italics Komarek].

The quick succession of shrubs and deciduous trees into this "derived" grassland leads not only to fast changes in populations of quail but also in other grassland inhabitants ranging from insects to small mammals and birds. In the late 1930's and 1940's, experiments were conducted at Sherwood and Birdsong Plantations, in Grady County, Georgia (Komarek, 1939; E. V. Komarek and B. Komarek, unpublished). As the grassland plots changed because of fire exclusion, grassland rodents such as the cotton rat (*Sigmodon h. komareki*), harvest mouse (*Reithrontomys h. humulis*), and old field mouse (*Peromyscus p. polionotus*) disappeared and after several years of fire exclusion the golden mouse (*Ochrotomys n. aureolus*) inhabited the heavy deciduous growths that had developed instead (Komarek, 1939). Likewise the pine-woods sparrow (*Aimophila aestivalis*) disappeared and was replaced by the yellow-breasted chat (*Icteria virens*). They remained about 4 years and left the area when the deciduous brush had turned into a hardwood sapling grove. Baker (1972) confirmed these results for the cotton rat and the golden mouse.

The impact of change in plant succession due to fire is also evident in some earthworms. Thus, *Diplocardia mississippiensis* occurs primarily in slash pine savannas in extremely large numbers so that a million-dollar fishing bait industry is based on the 3-year rotation of burning on the Apalachicola National Forest. The people who gather the worms insist that if fire is not used regularly the earthworms disappear (E. V. Komarek, unpublished) or their numbers are reduced to economically unprofitable levels.

E. Coastal Marshes and Prairies

The fire relationships in coastal marshes are imperfectly known. In the early 1940's, O'Neal (1949) found that the old custom of burning marshes in Louisiana was beneficial and that fire played an important part in the welfare of the muskrat (*Ondatra z. rivalicus*). He recommended regular controlled burning and stated that "without annual burns neither normal nor peak rat populations can be reached."

He found that marsh three-cornered grass (*Scirpus americanus*), the principal food of the muskrat, disappeared rather quickly unless the succession was "recycled" on a very regular basis, in most instances annually by fire.

Lynch (1941) working with geese-wintering grounds in the Louisiana and Texas marshes found that most of the wintering food for "Blue Geese" (*Chen c. caerulescens*) was also three-cornered grass, and it too had to be burned regularly. Some 4 million acres of coastal marshes from Texas to Mississippi are the most important wintering grounds for geese in the eastern United States. Thus proper management of these marshes is extremely important. Today these coastal marshes managed by the U.S. Fish and Wildlife Service and the state game and fish agencies concerned are burned on a regular basis (Perkins, 1968).

Very few research studies have been conducted on the relationships of fire to other marshes on the Gulf and Atlantic coasts. However, some research has been done in the St. Marks Wildlife Refuge in northern Florida. In that area there is a problem of encroachment of hardwoods, beginning with willows that take over the freshwater marsh edges and shallow areas so necessary for water birds (Zontek, 1966). Robertson (1962) and Klukas (1972) have reported on studies of fire ecology of the everglades.

It is interesting to note that burning off the marsh for both muskrats and geese was an old established custom by trappers and hunters and was universally condemned in the past without any scientific evidence, simply as a matter of policy and education (O'Neal, 1949).

VIII. The Piedmont Plateau Forest and Grasslands

The Piedmont forest is characterized by stands of loblolly and shortleaf pines as well as several species of oaks and other hardwoods. It is ecologically, as well as silviculturally and economically, the main belt of the above pines, even though they are both found in some areas of the Coastal Plain. The fire relationships of loblolly pine have been well covered by Wahlenberg (1960). Unfortunately, we do not have as comprehensive an account for shortleaf pine. Excellent bibliographies are available (Hanley, 1955, 1962).

Oosting (1944) listed ten species of oaks (*Quercus phellos, Q. alba, Q. lyrata, Q. stellata, Q. velutina, Q. coccinea, Q. rubra, Q. borealis maxima,* and *Q. marilandica*) as the major oak species in this oak–pine type forest. Many other deciduous species are also present. There appears to be constant tension between the various hardwood components and the pines (Kozlowski, 1949). Fire apparently plays an important role in determining the successional stage, for one of the major silvicultural and wildlife management programs is the constant encroachment of these hardwoods. What the original conditions were like is in doubt, for much of the Piedmont Plateau went through an intensive farming phase in the past and relics of the original cover are lacking.

That the responses to burning are similar to those in the Coastal Plain forests is evident from the literature (Cushwa *et al.,* 1970; Cushwa and Redd, 1966; Cushwa and Cooper, 1966; Lotti *et al.,* 1960; Oosting, 1944; Wahlenberg, 1960). However, the needles of both loblolly and shortleaf pines lie flat to the ground, are not very flammable, and do not carry fire too readily. Cushwa and Redd (1966), reporting on a prescribed burn, found seven times more plants valuable to game on the cut-burned area than on the unburned area. Over 44 lb of desirable seed were gathered per acre on the burned area and only $5\frac{1}{2}$ lb on the unburned area. They pointed out that wild legumes increased substantially by burning. They also noted an increase in quail and doves on the burned area.

Cushwa and Cooper (1966) reported on burning for wildlife purposes on the Hitchiti Experimental Forest in the lower piedmont of Georgia. They found significantly more wild plants that were favored by wildlife in burned than in the unburned areas. Legumes also were more abundant on the burned area. Cushwa *et al.* (1970) reported that infrequent burns of low intensity could not be expected to alter species composition of southern pine stands.

All too often investigators in the past have expected miraculous results with one light burn after years of fire exclusion. When plants have been forced out by competition over a period of years, considerable time will be needed to restore them by regular burning. At times certain plants appear in striking abundance, particularly some annuals, seed of which lies buried in the ground for years until the opportune time— usually a hot and severe fire. Unfortunately, there is not the backlog of research in this region that has been accumulated in the Coastal Plain. However, the use of prescribed burning and research data on effects of fire in the Piedmont Plateau are increasing yearly for the management of forests and wildlife.

IX. Appalachian Mountains, Mesophytic Forests, and Grasslands

This region, characterized by deciduous hardwood forests, located on the Appalachian highlands and westward, is an area that due to meteorological and topographical conditions has a pattern of precipitation conducive to the development of these kinds of forests.

Although considerable lightning is present the opportunity for catastrophic fires under natural conditions is relatively small. The combination of a more even precipitation pattern, as contrasted to the erratic one of the Coastal Plain, along with a mesic broad-leaved forest that has developed as a consequence of more regular rainfall distribution is not conducive to large widespread fires. Man, however, has greatly altered the vegetative cover in this region. Overcutting of the forest put much more flammable fuel on the surface of the ground than could ever have occurred naturally. In any hardwood-cutting operation there is a great deal of slash, and in early forest-cutting periods only the high grade clear logs were utilized and consequently there was a great deal of flammable fuel during drought periods.

Following lumbering, settlers moved in and burned much of this residual fuel to clear the land and for livestock-grazing purposes. They considered trees as weeds that neither they nor their livestock could eat. After all, man is a grassland animal, not a forest animal.

Because of the amount of fuel on the surface and the disregard for potential timber values, many of the remaining forest stands were subjected to intense fires which rarely occurred under more natural conditions. Thus, it can be said that man introduced into this mesophytic forest a more devasting kind of fire. Because of this, the fact that fire was part of the region originally has been overlooked. Lightning fires do occur throughout the Appalachian Mountain region when conditions are conducive to burning and small area fires must have been a common occurrence. Komarek (1966, 1967b), Barden (1974), and Wilhelm (1972) have documented such lightning fires in the national forests and national parks in the Appalachian Mountains. Barden (1974) has shown that these fires, when allowed to burn, cover only relatively small areas. Thus, the fire mosaic in this region was, and is, where not unduly disturbed by man, a pattern of small burns interspersed over the landscape at irregular intervals, as contrasted with the more frequent and widespread ones in the Coastal Plain.

The sprouting capacity of many hardwoods following injury makes them well-adapted to a fire environment. The pines that occur in this region (*Pinus rigida, P. pungens, P. virginiana, P. strobus,* etc.) are considered to be pioneer species in forest succession in the absence of fire. They have many advantages and features that make them fire-adapted to a degree, such as mineral soil for regeneration, and two have serotinous cones. However, all have relatively short needles that when cast, lie close to the ground and thus possess less combustibility. Similarly hardwood leaves do not burn readily except shortly after leaf fall and before winter rains and snows. This does not mean, however, that fires cannot burn at other times, for hot fires that burn not only the litter but the accumulated duff and organic matter down to mineral soil can occur during droughts or extended dry periods.

That fire was a normal component of the broad-leaved ecosystem is shown by responses of certain grasses, legumes, forbes, etc., in many regions where burns occur. In the 1930's (1934–1941) a hardwood forest was converted to quail land by burning. The first burn, a moderately hot one, with much lumbering debris on the surface intermixed with grasses and weeds, induced an abundance of legumes of several genera (*Lespedeza, Meibomia, Strophostyles, Cassia,* etc.). This response occurred in the delta of the Mississippi River in northern Mississippi on a very black, deep soil except where large hardwoods shaded the ground heavily. After several years of annual burning, the area developed into good quail-hunting land with an excellent mixture of bluestem and other grasses as well as legumes, and resembled a midwestern prairie in many respects. The scattered trees gave the area a "parklike" effect similar in landscape to the hunting lands of the Thomasville, Georgia and Tallahassee, Florida region that have been burned on an annual basis for many years (E. V. Komarek, unpublished).

A similar response occurred on a hunting plantation in pin oak (*Quercus palustris*) woods near Stuttgrat, Arkansas (E. V. Komarek, unpublished). Again a landscape developed reminiscent of the open hunting lands of the Tallahassee Red Hill region and elsewhere in the Gulf Coast region.

Thor and Nichols (1974) and DeSelm *et al.* (1974) have observed comparable effects of burning in open hardwood forests in Tennessee.

Braun (1950), writing on original conditions in the "bluegrass" section of Kentucky, quoted from Daniel Boone (1784) as follows:

> . . . We found everywhere abundance of wild beasts of all sorts, through this vast forest. The buffaloes were more frequent than I have seen cattle in the settlements, browsing on the leaves of the cane, or cropping the herbage on those extensive plains, fearless, because ignorant of man.

And further:

> . . . Where no cane grows there is an abundance of wild rye, clover, and buffalograss, covering vast tracts of country, and affording excellent food for cattle . . . .

From the studies on cane and canebreaks in the Coastal Plain it would appear that fire would have had to be a factor in the longtime abundance of the canebreaks so often mentioned in the literature of western Kentucky and Tennessee and elsewhere. The response of typical prairie or savanna plant species, grasses, legumes, and other forbs and herbs, indicates that there was much more grassland in the deciduous forest regions than is commonly acknowledged. These species endemic to grasslands, not dense forests, certainly did not develop without a history of fire in the past. Man could not have brought all of these species into this environment, although a great many have been introduced.

Another factor pointing out that there must have been some grassland in this deciduous forest region is the fact that it is the *natural range* of several species of grassland animals, such as buffalo.

We have lost sight of the fact that because of man's overburning and burning under unnatural conditions, with much greater fuel accumulations, fire was a naturally recurring factor. Likewise, under a policy of complete fire exclusion on all lands we have also lost much of our original flora and fauna and replaced them with trees. The usefulness of fire in commercial hardwoods may be limited, and fire probably is not at all useful with some species. However, this should not obscure the fact that studies of fire ecology in these regions may be very rewarding. In establishing fire policies, allowances should be made for such investigations.

Because of past catastrophic fires there is the fear that the use of prescribed burning may induce more wildfires. Wahlenberg (1960) emphasized that the use of fire has not increased wildfires but has in fact decreased them in the South. Although his statements refer primarily to loblolly pine forests, the basic philosophy can be extended to the deciduous forests of the Southeast and elsewhere.

Another more recent fear is that the smoke from prescribed burning or even from fires set for research purposes might add objectionable substances to the atmosphere. However, fire is a natural force and the smoke from forest fires had been a part of the atmosphere long before man (Komarek, 1970; Vines, 1974; Schaeffer, 1974; Komarek, 1973). No material from forest or grass fires has been found in quantities that might be harmful, health-wise, to man (Cooper, 1971, 1973, 1974; Ward and Lamb, 1970; Darley *et al.*, 1966).

The fear of wildfires or impairment of air quality should not lead to regulations that would prohibit either the use of prescribed burning or research into fire effects. Fire has been and will continue to be part of the environment of the deciduous forest region and research on its possible beneficial effects for forest and wildlife management is urgently needed. The deleterious effects of fire in the Southeast have been too long overemphasized and have hindered much fire research, particularly in the deciduous forest region. There is less fire research in this area than in any other part of the country. This is in striking contrast with the lower South which perhaps has the largest amount and longest history of such ecological research. Present trends, however, in the broadleaved forest belt may change this shortly for much fire ecology research has now been initiated in much of the hardwood region and will continue if not prohibited by adverse regulations.

References

Ahlgren, I. F., and Ahlgren, C. E. (1960). Ecological effects of forest fires. *Bot. Rev.* **26**, 483–533.

Anonymous. (1962). "Tall Timbers Research Station Fire Ecology Plots," Bull. No. 2. Tall Timbers Research Station, Tallahassee, Florida.

Arnold, K. (1964). Project skyfire lightning research. *Proc. 3rd Annu. Tall Timbers Fire Ecol. Conf.* pp. 121–130.

Baker, W. W. (1972). Changes in small mammal populations in annually burned pineland after fire exclusion. *52nd Annu. Amer. Soc. Mammal.*, 1972.

Barden, L. (1974). Lightning fires in southern Appalachian forests. *Proc. 13th Annu. Tall Timbers Fire Ecol. Conf.* (in press).

Biswell, H. H., Southwell, B. L., Stevenson, J. W., and Shepherd, W. O. (1942). Forest grazing and beef cattle production in the coastal plain of Georgia. *Ga., Coastal Plain Exp. Sta., Circ.* **8**, 1–25.

Biswell, H. H., Shepherd, W. O., Southwell, B. L., and Boggess, T. S., Jr. (1943). Native forage plants of cutover forest lands in the coastal plain of Georgia. *Ga., Coastal Plain Exp. Sta., Bull.* **37**, 1–43.

Braun, E. L. (1950). "The Deciduous Forests of Eastern North America." McGraw-Hill (Blakiston), New York.

Bryson, R. A., Baerreis, D. A., and Wendland, W. M. (1970). The character of late-glacial and post-glacial climatic changes. *In* "Pleistocene and Recent Environments of the Central Great Plains" (D. Wakefield and J. Knox Jones, Jr., eds.), Misc. Publ. No. 3, Dept. of Geology, University of Kansas, Lawrence, Kansas, pp. 53–74.

Burton, G. W. (1944). Seed production of several grasses as influenced by burning and fertilization. *J. Amer. Soc. Agron.* **36**, 523–529.

Campbell, R. S. (1946). For better cattle on forest ranges. *Progr. Farmer* **61**, 19.

Campbell, R. S., and Cassady, J. T. (1951). Grazing values for cattle on pine forest ranges in Louisiana. *La., Agr. Exp. Sta., Bull.* **452**, 1–31.

Campbell, R. S., Epps, E. A., Moreland, C. C., *et al.* (1954). Nutritive values of native plants on forest range in Louisiana. *La., Agr. Exp. Sta., Bull.* **488**, 1–18.

Campbell, R. S., Halls, L. K., and Morgan, H. P. (1963). Selected bibliography on southern range management. *U.S., Forest Serv., S. Forest Exp. Sta., Res. Pap.* **SO-2**, 1–62.

Cassady, J. T. (1953). Herbage production on bluestem range in central Louisiana. *J. Range Manage.* **6**, 38–43.

Cassady, J. T., and Mann, W. F., Jr. (1954). The Alexandria Research Center. *U.S., Forest Serv., S. Forest Exp. Sta., Booklet.* pp. 1–49.

Clewell, A. F. (1966). Natural history, cytology and isolating mechanisms of the native American Lespedezas. *Tall Timbers Res. Sta., Bull.* **6**, 1–39.

Cooper, R. W. (1971). The pros and cons of prescribed burning in the South. *Forest Farmer.* **31**, 10–12 and 39–40.

Cooper, R. W. (1973). The impact of prescribed fire on the environment. *Southeast. Sect. Meet., Soc. Amer. Forest.*, *1973*, pp. 1–8.

Cooper, R. W. (1974). Status of prescribed burning and air quality in the South. *Proc. 13th Annu. Tall Timbers Fire Ecol. Conf.* (in press).

Croker, T. C. (1968). Longleaf Pine: An annotated bibliography 1946 through 1967. *U.S., Forest Serv., Res. Pap.* **SO-35**, 1–52.

Cushwa, C. T. (1968). Fire: A summary of literature in the United States from the mid-1920's to 1966. *U.S. Forest Serv., Southeast Forest Exp. Sta., Paper*, pp. 1–117.

Cushwa, C. T., and Cooper, R. W. (1966). The response of herbaceous vegetation to prescribed burning. *U.S., Forest Serv., Res. Note* **SE-53**, 1–2.

Cushwa, C. T., and Redd, J. B. (1966). One prescribed burn and its effect on habitat of the Powhatan game management area. *U.S., Forest Serv., Res. Note* **SE-61**, 1–2.

Cushwa, C. T., Hopkins, M., and McGinnes, B. S. (1970). Response of legumes to prescribed burns in loblolly pine stands of the South Carolina piedmont. *U.S., Forest Serv., Res. Note* **SE-140**, 1–6.

Darley, E. F., Burleson, F. R., Mateer, E. H., Middleton, J. T., and Osterli, V. P. (1966). Contribution of burning of agricultural wastes to photochemical air pollution. *J. Air Pollut. Contr. Ass.* **16**, 685–690.

Davis, K. P. (1959). "Forest Fire, Control and Use." McGraw-Hill, New York.

DeSelm, H. R., Clebsch, E. E., Thor, E., and Nichols, G. M. (1974). Behavior of the herb, shrub and seedling layer in controlled burn hardwoods in middle Tennessee. *Proc. 13th Annu. Tall Timbers Fire Ecol. Conf.* (in press).

Duvall, V. L., Johnson, A. W., and Yarlett, L. L. (1968). Selected bibliography on southern range management, 1962–67. U.S., Forest Serv., S. Forest Exp. Sta., Res. Pap. SO-38, 1–30.

Fraps, G. S., and Fudge, J. F. (1940). The chemical composition of forage grasses of the east Texas timber country. Tex., Agr. Exp. Sta., Bull. 582, 1–35.

Greene, S. W. (1931). The forest that fire made. Amer. Forests 37, 583–584.

Greene, S. W. (1935a). Relation between winter grass fires and cattle grazing in the longleaf pine belt. J. Forest. 33, 338–341.

Greene, S. W. (1935b). Effect of annual grass fires on organic matter and other constituents of virgin longleaf pine soils. J. Agr. Res. 50, 809–822.

Grelen, H. E., and Duvall, V. L. (1966). Common plants of longleaf pine–bluestem range. U.S., Forest Serv., S. Forest Exp. Sta., Res. Pap. SO-23, 1–96.

Halls, L. K., Southwell, B. L., and Knox, F. E. (1952). Burning and grazing in Coastal Plain forests. Ga., Coastal Plain, Exp. Sta., Bull. 51, 1–31.

Halls, L. K., Hale, O. M., and Southwell, B. L. (1956). Grazing capacities of wiregrass–pine ranges of Georgia. Ga., Agr. Exp. Sta., Tech. Bull. [N.S.] 2, 1–38.

Hanley, G. P. (1955). Shortleaf pine bibliography. U.S., Forest Serv., Southeast Forest Exp. Sta., Pap. 48, 1–60.

Hanley, G. P. (1962). A revised shortleaf pine bibliography. U.S., Forest Serv., Southeast. Forest Exp. Sta., Pap. 155.

Harper, R. M. (1914). Geography and vegetation of northern Florida. Fla., Geol. Surv., 6th Annu. Rep. pp. 266–279.

Hilmon, J. B., and Hughes, R. H. (1965). Forest Service research on the use of fire in livestock management in the South. Proc. 4th Annu. Tall Timbers Fire Ecol. Conf. pp. 261–275.

Hughes, R. H. (1957). Response of cane to burning in the North Carolina coastal plain. N.C., Agr. Exp. Sta., Bull. 402, 1–24.

Hughes, R. H. (1966). Fire ecology of canebrakes. Proc. 5th Annu. Tall Timbers Fire Ecol. Conf. pp. 149–158.

Hughes, R H., and Dillard, E. V. (1960). Vegetation and cattle response under two systems of grazing cane range in North Carolina. N.C., Agr. Exp. Sta., Bull. 412, 1–27.

Jackson, J. E., and Burton, G. W. (1962). Influence of sod treatment and the placement on utilization of urea nitrogen by coastal bermuda grass. J. Agron. 54, 47–49.

Klukas, R. W. (1972). Control burn activities in Everglades National Park. Proc. 12th Annu. Tall Timbers Fire Ecol. Conf. pp. 397–425.

Komarek, E. V. (1939). A progress report on southeastern mammal studies. J. Mammalogy. 20, 292–299.

Komarek, E. V. (1964). The natural history of lightning. Proc. 3rd Annu. Tall Timbers Fire Ecol. Conf. pp. 139–183.

Komarek, E. V. (1965). Fire ecology—grasslands and man. Proc. 4th Annu. Tall Timbers Fire Ecol. Conf. pp. 169–220.

Komarek, E. V. (1966). The meteorological basis for fire ecology Proc. 5th Annu. Tall Timbers Fire Ecol. Conf. pp. 85–125.

Komarek, E. V. (1967a). Fire—and the ecology of man. Proc. 6th Annu. Tall Timbers Fire Ecol. Conf. pp. 143–170.

Komarek, E. V. (1967b). The nature of lightning fires. Proc. 7th Annu. Tall Timbers Fire Ecol. Conf. pp. 5–41.

Komarek, E. V. (1968). Lightning and lightning fires as ecological forces. *Proc. 8th Annu. Tall Timbers Fire Ecol. Conf.* pp. 169–197.

Komarek, E. V. (1969). Fire and animal behavior. *Proc. 9th Annu. Tall Timbers Fire Ecol. Conf.* pp. 161–207.

Komarek, E. V. (1970). Controlled burning and air pollution: An ecological review. *Proc. 10th Annu. Tall Timbers Fire Ecol. Conf.* pp. 141–173.

Komarek, E. V. (1971a). Lightning and fire ecology in Africa. *Proc. 11th Annu. Tall Timbers Fire Ecol. Conf.* pp. 473–511.

Komarek, E. V. (1971b). Effects of fire in wildlife and range habitats. *U.S., Forest Serv., Prescribed Burning Symp. Proc.* pp. 1–160.

Komarek, E. V. (1972). Ancient fires. *Proc. 12th Annu. Tall Timbers Fire Ecol. Conf.* pp. 219–240.

Komarek, E. V. (1974). Further remarks on controlled burning and air pollution. *Proc. 12th Annu. Tall Timbers Fire Ecol. Conf.* pp. 279–282.

Komarek, E. V., Komarek, Betty B., and Carlysle, Thelma C. (1973). The ecology of smoke particulates and charcoal residues from forest and grassland fires: A preliminary atlas. *Misc. Publ. No. 3. Tall Timbers Research Station.* pp. 1–75.

Kozlowski, T. T. (1949). Light and water in relation to growth and competition of Piedmont forest tree species. *Ecol. Monogr.* **19**, 208–231.

Kurz, H. (1944). Secondary forest succession in the Tallahassee red hills. *Fla. Acad. Sci.* **7**, 1–42.

Langdon, O. G., Bomhard, M. L., and Cassady, J. T. (1952). Field book of forage plants on longleaf pine–bluestem ranges. *U.S., Forest Serv., S. Forest Exp. Sta., Occas. Pap.* **127**, 1–117.

Lay, D. W. (1956). Effects of prescribed burning on forage and mast production in Southern pine forests. *J. Forest.* **54**, 582–584.

Lay, D. W. (1957). Browse quality and the effects of prescribed burning in Southern pine forests. *J. Forest.* **55**, 342–347.

Lemon, P. C. (1949). Successional responses of herbs in the longleaf-slash pine forest after fire. *Ecology* **30**, 135–145.

Leukel, W. A., and Stokes, W. E. (1939). Growth behavior and relative composition of range grasses as affected by burning and the effect of burning on natural grass stands and upon the establishment of improved grasses. *Fla., Agr. Exp. Sta., Annu. Rep.* pp. 54–55.

Little, S., and Somes, H. A. (1956). Buds enable pitch and short-leaf pine to recover from injury. *U.S., Forest Serv., Northeast Forest Exp. Sta., Pap.* **81**, 1–14.

Lotti, T., Klawitter, R. A., and LeGrande, W. P. (1960). Prescribed burning for understory control in loblolly pine stands of the Coastal Plain. *U.S., Forest Serv., Southeast. Forest Exp. Sta., Pap.* **116**, 1–19.

Lynch, J. J. (1941). The place of burning in management of the Gulf coast refuges. *J. Wildl. Manage.* **5**, 454–458.

McClure, F. A. (1959). Bamboo as a source of forage. *Proc. Pac. Sci. Congr., 8th, 1953* Vol. 4B, pp. 609–664.

McClure, F. A. (1963). A new feature in bamboo rhizome anatomy. *Rhodora* **65**, 134–136.

McDaniel, S. (1971). The genus *Sarracenia* (Sarraceniaceae). *Tall Timbers Res. Stat., Bull.* No. 9, pp. 1–36.

Manson, W. G., Burton, G. W., and Williams, E. G. (1974). Effects of burning on soil temperature and yield of coastal bermuda grass. *J. Amer. Soc. Agron.* (in press).

O'Neal, T. (1949). "The Muskrat in the Louisiana Coastal Marshes." Pub. Fed. Aid Sect., La. Dept. Wildlife and Fish, New Orleans, Louisiana.

Oosting, H. J. (1944). Comparative effect of surface and crown fire on the composition of a loblolly pine community. *Ecology* **25**, 61–69.

Perkins, C. J. (1968). Controlled burning in the management of muskrats and waterfowl in Louisiana coastal marshes. *Proc. 8th Annu. Tall Timbers Fire Ecol. Conf.* pp. 269–280.

Requa, L. E. (1964). Lightning behavior in the Yukon. *Proc. 3rd Annu. Tall Timbers Fire Ecol. Conf.* pp. 111–119.

Robertson, W. B. (1962). Fire and vegetation in the Everglades. *Proc. 1st Annu. Tall Timbers Fire Ecol. Conf.* pp. 67–80.

Schaefer, Vincent J. (1974). Some physical relationships of fine particle smoke. *Proc. 13th Annu. Tall Timbers Fire Ecol. Conf.* pp. 303–307.

Shepard, W. O., Dillard, E. V., and Lucas, H. L. (1951). Grazing and fire influences in pond pine forests. *N.C., Agr. Exp. Sta., Tech. Bull.* **97**, 1–57.

Shepard, W. O., Dillard, E. V., and Lucas, H. L. (1953). Best grazing rates for beef production on cane range. *N.C., Agr. Exp. Sta., Bull.* **384**, 1–23.

Smart, W. W. G., Jr., Matrone, G., Shepard, W. O., Hughes, R. H., and Knox, F. E. (1960). Comparative consumption and digestibility of cane forage. *N.C., Agr. Exp. Sta., Tech. Bull.* **140**, 1–8.

Smith, L. F., Campbell, R. S., and Blount, C. F. (1955). Forage production and utilization in longleaf pine forests of south Mississippi. *J. Range Manage.* **8**, 58–60.

Smith, F. H., Beeson, K. C., and Price, W. E. (1956). Chemical composition of herbage browsed by deer in two wildlife management areas. *J. Wildl. Manage.* **20**, 359–367.

Smith, L. F., Campbell, R. S., and Blount, C. F. (1958). Cattle grazing in longleaf pine forests of south Mississippi. *U.S. Forest Serv., S. Forest Exp. Sta., Occas. Pap.* **162**, 1–25.

Spurr, S. H. (1964). "Forest Ecology." Ronald Press, New York.

Stoddard, H. L. (1931). "The Bobwhite Quail; Its Habits, Preservation and Increase." Scribner's, New York.

Stoddard, H. L. (1962). Use of fire in pine forests and game lands of the deep southeast. *Proc. 1st Annu. Tall Timbers Fire Ecol. Conf.* pp. 31–42.

Stoddard, H. L., and Komarek, E. V. (1941). The carrying capacity of southeastern quail lands. *Trans. N. Amer. Wildl. Conf.* **6**, 477–484.

Stoddard, H. L., Beadel, H. L., and Komarek, E. V. (1961). "The Cooperative Quail Study Association," Misc. Publ. Tall Timbers Research Station, Tallahassee, Florida.

Suman, R. F., and Halls, L. K. (1955). Burning and grazing affect physical properties of Coastal Plain forest soils. *U.S., Forest Serv., Southeast. Forest Exp. Sta., Res. Notes* **75**, 1–2.

Taylor, A. R. (1971). Lightning: Agent of change in forest ecosystems. *J. Forest.* **68**, 477–480.

Taylor, A. R. (1969). Lightning effects on the forest complex. *Proc. 9th Annu. Tall Timbers Fire Ecol. Conf.* pp. 127–150.

Thor, E., and Nichols, G. M. (1974). Some effects of fires on litter, soil and hardwood regeneration in middle Tennessee. *Proc. 13th Annu. Tall Timbers Fire Ecol. Conf.* (in press).

Vines, R. G. (1974). Bush-fire smoke and air quality. *Proc. 13th Annu. Tall Timbers Fire Ecol. Conf.* pp. 303–307.

Vogl, R. J. (1972). Fire in the Southeastern grasslands. *Proc. 12th Annu. Tall Timbers Fire Ecol. Conf.* pp. 175–198.

Wahlenberg, W. G. (1946). "Longleaf Pine, Its Use, Ecology, Regeneration, Protection, Growth and Management." Charles Lathrop Pack Forest. Found., Washington, D.C.

Wahlenberg, W. G. (1960). "Loblolly Pine, Its Use, Ecology, Regeneration, Protection, Growth and Management." Seeman Printery, Inc., Durham, North Carolina.

Ward, D. E., and Lamb, R. C. (1970). Prescribed burning and air quality—current research in the south. *Proc. 10th Annu. Tall Timbers Fire Ecol. Conf.* pp. 129–141.

Wells, B. W. (1932). "The Natural Gardens of North Carolina." Univ. of North Carolina Press, Chapel Hill.

Wells, B. W. (1942). Ecological problems of the southeastern United States coastal plain. *Bot. Rev.* 8, 533–561.

Wenger, K. F. (1958). Silvicultural characteristics of loblolly pine. *U.S., Forest Serv., Southeast. Forest Exp. Sta., Pap.* 98, 1–32.

Wheater, R. J. (1971). Problem of controlling fires in Uganda National Parks. *Proc. 11th Annu. Tall Timbers Fire Ecol. Conf.* pp. 259–275.

Wilhelm, G. (1972). Fire ecology of the Shenandoah National Park. *Proc. 12th Annu. Tall Timbers Fire Ecol. Conf.* pp. 445–488.

Williams, M. B. (1938). Annotated bibliography on the effects of fire on forests. *U.S., Forest Ser., Southeast Forest Exp. Sta.* p. 1.

Zontek, F. (1966). Prescribed burning on the St. Marks National Wildlife Refuge. *Proc. 5th Annu. Tall Timbers Fire Ecol. Conf.* pp. 195–201.

. 9 .

Effects of Fire on Temperate Forests: Western United States

Harold Weaver

I. Introduction

As we walked onto the beach at Bandon, Oregon that evening in late August 1933, we beheld to the north a tremendous wall of yellow smoke, thousands of feet high. It extended out over the ocean, seemingly to infinity, and slightly to the right of the setting sun. Obviously a big fire had blown up somewhere. Was the smoke from the fire near Forest Grove, in the big Douglas-fir (*Pseudotsuga menziesii*) forest in northwestern Oregon, extensively reported in the newspapers for the past several days, or was it from a new fire, somewhere in southwestern Oregon?

We were two Bureau of Indian Affairs foresters detailed from Klamath Indian Agency in southern Oregon to make a timber cruise of a "Public Domain" Indian allotment near Bandon. With automobiles and highways of the period the journey had taken most of the day, and it had been oppressively hot, particularly in the Rogue River valley. On occasional walks in the roadside forest the twigs and needles crackled sharply underfoot. It was obvious that fire danger was extreme.

We watched the awesome sight from the beach until after sunset, then returned to the hotel, where we were told radio news had reported that the fire in northwestern Oregon had blown up and was running wild. No other large fires had been reported. We had been watching drift smoke from this fire, over 150 miles away. This was the Tillamook fire, the most damaging and one of the most extensive of historic times.

After leaving Bandon the following morning, the road passed for miles through a young, even-aged forest in which Douglas-fir predominated, and we continued to encounter this age class during the cruise of the allotment property. Though the young trees had already grown to sawlog size, the market of the times was so depressed that they had little or no commercial value. Therefore a sampling designed to determine approximate board foot volume and number of trees per acre appeared adequate. We attempted, however, by a systematic survey, to locate, mark, and measure all the larger, more valuable trees.

There were comparatively few of these, scattered singly and in small groups; they consisted of Douglas-fir and Port Orford cedar (*Chamaecyparis lawsoniana*), then the most valuable commercial species. Some of these trees, particularly the cedars, bore sizable fire scars. Most of this older age class lay fire killed and windfallen on the forest floor, partially decayed and covered with mosses, ferns, and salal brush (*Gaultheria shallon*). These great windfalls and the steep, sharp ravines made progress slow.

It was obvious that the young, even-aged forest had originated after a great fire and that the scattering of large trees was what survived that fire. A local rancher, who had shown us our starting corner, said that the fire had burned 60 or 70 years before. Holbrook (1944) reported that the Coos fire burned in 1868. The area we had observed may have been burned by this fire or by another that burned at about the same time.

This history serves to introduce an aspect of the fire picture that has not received adequate attention. The great fires of historic times and countless other fires have not always been man-caused, and they have not always been unmitigated, irreversible catastrophes, as frequently reported by the press. Usually in the Douglas-fir region, as illustrated by the 1868 Coos burn, they have been followed by renewal and by the start of a new forest where exposed mineral soil and abundant overhead light have favored reestablishment of Douglas-fir. In long, continued absence of fire disturbance, after several centuries, the forest reverts to more shade-enduring but less valuable tree species.

There can be no doubt that the white man has greatly increased the incidence of fires and that some of his activities have had a pronounced effect on the intensity with which the fires have burned. Particularly destructive have been fires burning in young, restocking Douglas-fir stands on former burns and logged areas. Periodic, devastating fires burned countless centuries before the arrival of the white man, however, and the great Douglas-fir forest that he found and valued so much owed its very existence to these fires.

Throughout the western states, in other forest regions as well as in the Douglas-fir region, fire has been of great ecological significance in the development and perpetuation of our most valuable forest tree species. Also the future of these species depends to a considerable extent on skilfull application of fire as a management tool. There sometimes is extreme reluctance to accept this by people who have been taught that fire is an unmitigated evil in the forest.

How is it known that fire has been such an important ecological factor? It is known from the tree-ring and fire-scar records, some of which extend back many hundreds and even thousands of years before arrival of the white man; from historical records left by early explorers, naturalists, and pioneers; and from stand composition and age–class structure where periodic fires have continued to more recent date and where prescribed burning tests have been conducted. Finally it is known from changes that have occurred since white settlement, particularly since the advent of attempted total fire exclusion.

The western states have such a diversity of climate, topography, tree species, and fuels that it can be expected that there will be considerable diversity in incidence and intensity with which the fires burn. Because of the great areas that they cover it is appropriate to discuss first the dry, or xeric, ponderosa pine region. Then will follow a more detailed discussion of the moister, or more mesic, Douglas-fir region, and finally a discussion of the much smaller redwood region.

II. The Ponderosa Pine Region

Over vast, scattered forest areas of the western states, ponderosa pine (*Pinus ponderosa*) is best adapted to a regime of frequent periodic burning. Fires spread more readily over the forest floor through dry needles of this species than through the fallen needles and debris of any of its associated species, and, when weather and fuel moisture conditions are favorable, these fires spread into other forest types, particularly those occurring at higher elevations.

This pine is also one of the most important timber trees. As recently as 1962, for instance, it was second only to Douglas-fir (*Pseudotsuga menziesii*) in volume of commercial timber cut in the whole United States (Forest Service, 1965). In forestry and timber industry parlance the area where it grows over the eleven western states is termed the "Western Pine region."

Ponderosa pine occurs in several racial forms and is subject to greater variations in temperature and precipitation than nearly any other North American tree (Johnston, 1970). Its occurrence on dry sites with nut pines and certain of the junipers is indicative of its adaptability to arid conditions, and for extensive stretches it also borders on open grasslands and sagebrush. It attains its maximum development, however, on relatively moist but well-drained soils of the western slope of the Sierra Nevada range in California and the Siskiyou Mountains of southern Oregon (Harlow and Harrar, 1958).

A. THE RECORD OF TREE RINGS

That fires burned at frequent periodic intervals in the ponderosa pine forest is attested to by the tree-ring and fire-scar record. The tree rings show age of the tree and record vicissitudes of its life. When occasional fire scars are formed at the base of a tree, usually from exceedingly hot fires burning in heavy forest debris, new wood subsequently grows from the still-living edge of the cambium layer toward the center of the wound. The next fire frequently interrupts the healing process and sometimes enlarges the wound. After occurrence of a number of fires, one can frequently observe, on fire-scarred trees, a whole series of interrupted healing calluses. By dating the rings to points where calluses occur it is possible to determine with considerable accuracy the dates when the various fires burned (Fig. 1). A number of tree-ring and fire-scar studies have been conducted in the ponderosa pine region.

Keen (1937) found on one area in eastern Oregon that fires had burned during the years 1824, 1838, 1843, 1863, 1883, and 1888. On another area Keen (1940) found a large pine that had weathered 25 fires, which occurrred at approximately 18-year intervals.

Soeriaatmadja (1966) has also found evidence of frequent fires in eastern Oregon. In one of the dry fringe types at lower elevation, the average interval between fires was 14.2 years. At higher elevations in more mesophytic situations, fires occurred at less frequent intervals.

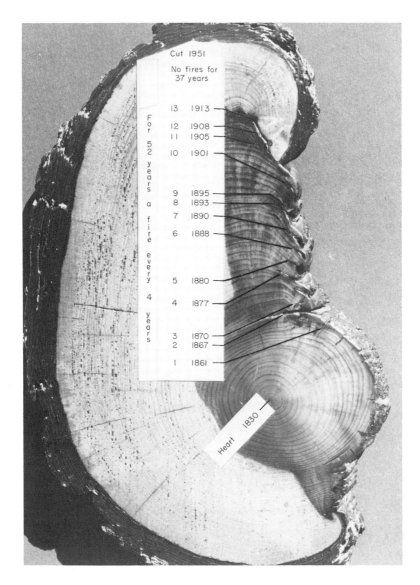

Fig. 1. Ground-line section from a ponderosa pine (*Pinus ponderosa*) cut near Turkey Creek in the Fort Apache Indian Reservation, Arizona, in 1951.

Wagner (1961) concluded that fires occurred at an average interval of once each 8–10 years in the Sierra Nevada of California. Reynolds (1959) reported on conditions resulting from burning conducted by the aboriginal Indians in the Yosemite portion of the Sierra Nevada. The Indians employed fire intensively to facilitate acorn gathering in their oak orchards, to prevent other trees from becoming established in these orchards, and to keep the forest free of thickets that might conceal enemies.

In eastern Washington, I found from one fire-scarred stump section that 23 fires had burned during the period 1740–1920, inclusive. This indicated an average frequency of one fire each 8 years (Weaver, 1955). In the Southwest I dated various fire-scarred stump sections, selected at random, with the assistance of the Tree Ring Laboratory of the University of Arizona. These indicated that fires burned as frequently as once each 4–5 years and that the average interval between fires was once each 6–7 years (Weaver, 1951).

Show and Kotok (1924) mentioned Huntington's investigation of the giant sequoias (*Sequoia gigantea*) which indicates that in the localities of the Sierra Nevada where these trees grow in association with the pines and other species of the mixed-conifer type there is evidence of fires occurring as far back as the year 245.

Studies recently conducted in the vicinity of the sequoia groves have indicated that fires probably occurred as frequently as once each 7–9 years (Kilgore, 1971b). Muir (1901), when exploring the sequoia groves of the Sierra Nevada in 1875, found evidence of past recurrent fires that ". . . throw a sure light back on the post-glacial history of the species, bearing on its distribution."

Everywhere throughout the ponderosa pine region there may be observed the fire-scar and tree-ring record. Throughout the major portion of the forests, fires occurred as frequently as fuel accumulated in sufficient quantity to support combustion over the forest floor, whenever weather conditions were favorable and whenever lightning strikes or Indians caused them to start.

B. OBSERVATIONS OF EARLY EXPLORERS

How did the ponderosa pine forest appear in earlier days? Accounts of explorers and scientists of the times are helpful. King (1871) described an ascent of the Sierra Nevada in 1864 as follows:

> At last, after climbing a long, weary ascent, we rode out of the dazzling light of the foot-hills into a region of dense woodland, the road winding through avenues of pines so tall that the late evening light only came down to us in scattered ray. . . . Passing from the glare of the open country into the dusky forest, one seems to enter a door, and ride into a vast covered hall. . . . You are never tired of gazing down long vistas,

where, in stately groups, stand tall shafts of pine. Columns they are, each with its own characteristic tinting and finish, yet all standing together with the air of relationship and harmony. . . .

Again he described how: "Wherever the ground opened level before us, we gave our horses the rein, and went at a free gallup through the forest. . . ."

Muir (1894) reported that "the inviting openness of the Sierra woods is one of their most distinguishing characteristics. The trees of all species stand more or less apart in groves, or in small, irregular groups, enabling one to find a way nearly everywhere, along sunny colonnades and through openings that have a smooth, parklike surface, strewn with brown needles and burs."

A number of early explorers, including Army officers and scientists, described early-day conditions in the ponderosa pine forests of Arizona. For instance, Lt. Edward Beale (1858) described a portion of northern Arizona as follows:

> . . . we came to a glorious forest of lofty pines. . . . The country was beautifully undulating . . . every foot being covered with the finest grass. . . . The forest was perfectly open and unencumbered with brush wood, so that the traveling was excellent.

C. E. Dutton (1881), in his "Physical Geology of the Grand Canyon District," described the Kaibab Plateau in the 1880's as follows: "The trees are large and noble in aspect and stand widely apart . . . . Instead of dense thickets where we are shut in by impenetrable foliage, we can look far beyond and see the tree trunks vanishing away like an infinite colonnade. The ground is unobstructed and inviting."

Plummer (Langille *et al.*, 1903) reported that "in this portion of the Cascade Mountains the yellow pine is an eastern slope tree. . . . Its forests are generally open, without much litter or under growth, and for these reasons are almost immune from fire."

How did early-day fires burn in the ponderosa pine and mixed-conifer forests? In the fall of 1875 John Muir (1901) observed a great forest fire burning in the Sierra Nevada. He described how

> It came racing up the steep chaparral-covered slopes of the East Fork canyon . . . in a broad cataract of flames, now bending down low to feed on the green bushes, devouring acres of them at a breath . . . the lurid flapping surges and the smoke and terrible rushing and roaring hiding all that is gentle and orderly in the work. But as soon as the deep forest was reached the ungovernable flood became calm like a torrent entering a lake, creeping and spreading beneath the trees where the ground was level or sloped gently, slowly nibbling the cake of compressed needles and scales with flames an inch high, rising here and there to a foot or two on dry twigs and clumps of small bushes and brome grass.

Muir felt safe in closely observing this fire because "there was no danger of being chased and hemmed in, for in the main forest belt of the Sierra, even when swift winds are blowing, fires seldom or never sweep over the trees in broad, all-embracing sheets as they do in the Rocky Mountain woods. . . . Here they creep from tree to tree with tranquil deliberation, allowing close observation."

Remnants of the primeval forest, where frequent periodic fires have continued to recent date, help us to visualize these conditions. In May 1943 a fire burned over approximately 15,000 acres of ponderosa pine forest in the remote Malay Gap portion of the San Carlos Indian Reservation in the White Mountains of east-central Arizona. Previous fires had occurred on an average of once each 6–7 years, and this latest fire acted like the fire observed by Muir, many years before in the Sierra Nevada. It crept from tree to tree, burning leisurely through the ground cover of dry needles, exfoliated bark scales, twigs, and dry muhly grass (*Muhlenbergia virescens*). Only occasionally, in dry, deadened snags and windfalls did the fire burn fiercely (Weaver, 1951; Cooper, 1960) (Fig. 2).

Another fire burned on the Malay Gap area in June 1946, under more severe conditions. This fire acted much like the one that had burned 3 years previously until it burned into an area of pine from which fire had been excluded for many years. There it crowned through the tree tops and caused considerable damage.

Fire control on the Malay Gap area became more effective about the year 1947. An area reported at about 3000 acres was prescribed burned in November 1966. Aside from this area, most of Malay Gap had been without surface fires, so characteristic of earlier days, for a period of 28 years. Then, in June 1971, a wildfire driven by swift winds covered approximately 10,000 acres. Because of the uninterrupted fuel accumulation of 28 years this latest fire was very destructive over portions of Malay Gap, except on the areas adequately treated by prescribed burning.

Burning intensities and actions of the earlier surface fires of the primeval ponderosa pine forest have also been simulated on areas where prescribed burning has been applied, particularly on areas that have been burned several times (Biswell, 1959, 1960; Cooper, 1960; Kallander, 1969; Weaver, 1957, 1961b, 1967a,b) (Figs. 3 and 6).

C. FIRE EFFECTS BY FOREST TYPES

Throughout the ponderosa pine region there is irrefutable evidence of the correctness of the conclusions of Jepson (1923), with particular reference to the Sierra Nevada forest, that "Indeed the main silvical

Fig. 2. Typical scene in the ponderosa pine (*Pinus ponderosa*) forest of Malay Gap, San Carlos Indian Reservation, Arizona, August 1949. This forest previously had experienced frequent periodic surface fires.

features, that is, density, reproductive power and dominance of types, are in great part expressions of the periodic fire status."

General effects of fire in the primeval ponderosa pine forest have already been discussed. It is appropriate now to discuss effects in more detail, particularly with reference to stand composition and stand or age–class structure. It is obvious, because of great variations in climate, topography, and elevation over the region, that fires will not always behave in the same manner. For instance, more moist or mesic sites may be too damp for fires to burn during normal years.

Kilgore (1971b) in discussing fire frequencies in the sequoia groves in the Sierra Nevada has stated that "The more mesic east and north

Fig. 3. This young ponderosa pine (*Pinus ponderosa*) stand on the Fort Apache Indian Reservation, several miles east of McNary, Arizona, has been treated by at least three prescribed burns since 1950. It also was thinned in 1964 and the slash from thinning was broadcast burned. Younger age–class trees originated in 1919. Groups of larger trees were reserved during heavy selective logging in mid-1920's. This photograph taken June 1972.

slopes do not burn as readily as the more xeric south and west slopes. Because of this, when they do burn, they may burn more intensely than those that burn more frequently." Soeriaatmadja (1966), in his study of past fire frequencies in ponderosa pine on the less rugged east slope of the Cascade range in Oregon, found that elevation was the variable that determined more mesic or xeric conditions and consequent fire frequencies.

On more mesic sites the more shade-enduring and less fire-resistant tree species have a better chance of survival. Where intervals between fires are of sufficient duration, types of less fire-resistant trees may develop. During exceptionally severe years, however, fires may completely destroy these types. These areas may then be seeded by more fire-resistant species, which in turn, in long continued absence of fire, will be invaded by understory trees of the shade-enduring and less fire-resistant species. Eventually these trees may completely recapture the site.

Following is a brief discussion of fire effects by forest types, beginning with types consisting of more fire-resistant trees.

1. *Ponderosa Pine–Mixed Conifer*

These two types are discussed together because under primeval conditions effects of surface fires were similar. These fires made it possible for the pines and other fire-resistant trees to maintain themselves as components of the mixed-conifer type, and it was in this type that they attained most rapid growth, greatest size, and highest quality. With exclusion of fire these most valuable species are gradually disappearing from the mixed-conifer type. This will be discussed in more detail later in this chapter.

a. FIRE HAZARD. Frequent, periodic fires maintain hazard at a low level. The Sierra Nevada forest observed by Muir and the open forest observed by Lt. Beale in northern Arizona were that way because of frequent periodic fires of the past. As has been indicated, this condition has continued to prevail on limited, isolated areas where periodic fires have continued to recent date (Weaver, 1951; Cooper, 1960). As has also been indicated, prescribed burning has reestablished this condition on areas where it has been applied repeatedly over a period of years (Kallander, 1969; Weaver, 1967a,b).

b. AGE CLASSES AND STAND STRUCTURE. Periodic fires cause development of uneven-aged stands, comprised of even-aged groups of trees of various age classes. The fires do prevent establishment of reproduction under the larger trees, where it is not needed and where it definitely proves deleterious if it is permitted to continue growing. Paradoxically as it may appear, the fires also cause clearings wherein seedlings become established and by means of which the ponderosa pine forest perpetuates itself. This occurs where single mature trees or groups of trees have been deadened by lightning strikes (Komarek, 1968; Pearson, 1949), insects, disease, or windthrow. The recurrent fires gradually reduce the snag and windfelled remains of these trees to ash beds, in the process

causing a clearing in the surrounding grass and shrub cover. Tree seedlings germinate in these favorable spots, and by the time that grass
and a mat of fallen needles can become reestablished, many of them
are of sufficient size to survive light surface fires (Figs. 4, 5, and 6).
Ponderosa pine seedlings are more fire resistant and are more apt to
survive than are most of the associated species. Thus there develop
new, even-aged groups of pines (Cooper, 1960; Weaver, 1951, 1967a,b).

How did more extensive even-aged stands of ponderosa pine originate?
A specific case history illustrates how this may come about.

In 1893 a severe epidemic attack of the defoliating pine butterfly
(*Neophasia menapia*) developed on over 150,000 acres of ponderosa
pine in Cedar Valley of the Klickitat River drainage in southern Washington. These insects were so numerous that they made the pines appear
white. Streams were choked by the dead butterflies and travelers reported that horses and men were covered from head to foot with webs

Fig. 4. June 1972 photograph showing new growth starting on terminal and lateral
shoots of ponderosa pine (*Pinus ponderosa*) seedlings that survived the Black River
fire of June 1971, on the Fort Apache Indian Reservation. Note partially burned
remnants of mature trees previously deadened by lightning or pine beetles. Additional
new seedlings will germinate in this opening.

Fig. 5. August 1949 photograph showing a later stage in the development of an even-aged group of ponderosa pines (*Pinus ponderosa*), in the Malay Gap area of the San Carlos Indian Reservation. These young trees had experienced several surface fires, the latest in May 1943.

of the larvae. Defoliation continued through 1895, when the epidemic ended, probably due to a wasplike parasite (*Theronia fulvescens*) (Keen, 1952).

Tree mortality was spotted and varied considerably; from 20 to 90% of the trees were killed. An entomologist who visited the area at the time reported that weakening of the trees through defoliation had encouraged a severe epidemic attack of the western pine beetle (*Dentroctonus brevicomis*), and he attributed most of the mortality to this insect (Keen, 1952).

A tree-ring and fire-scar study indicated an average fire frequency of one each 8 years during the 82-year period from 1807 to 1889, inclusive, and fires continued subsequent to the epidemic until 1914. In fact, after the epidemic, incidence of fires greatly increased, for stockmen and settlers were opposed to "brush" and windfalls that obscured the view and impeded travel. They fired the snags and windfalls of the deadened trees. Pine seedlings germinated in the ash beds resulting from this burning and many of them survived to become extensive, even-aged stands that now have grown to large pole and small sawlog size. Many of the trees in these stands have scars that show that there was extensive thinning by low intensity fires while they were still quite small in the late 1890's and early 1900's (Weaver, 1961a).

Under primeval conditions, lightning strikes, insects, and other agents would start deadening groups of these trees, and the stand eventually would become unevenly aged through the process described above. As Cooper (1960) has indicated: "The clumped pattern of ponderosa pine is the result of a cyclical pattern of community development, governed by fire and by the intolerant nature of the species."

c. STAND COMPOSITION. Periodic fires assure continued dominance of ponderosa pine and other fire-resistant species. Of ponderosa pine's associated species, western larch (*Larix occidentalis*), sequoia, sugar pine (*Pinus lambertiana*), western white pine (*Pinus monticola*), and lodgepole pine (*Pinus contorta*) are also fire-resistant or fire-climax trees. These trees are intolerant of shade and the seedlings demand mineral soil and abundant overhead light.

Various species of true fir (*Abies* spp.) and incense cedar (*Libocedrus decurrens*) are very shade tolerant. Sudworth (1908) indicated that white fir (*Abies concolor*) is particularly shade tolerant. Seedlings of these trees germinate and grow on shaded duff and debris of the forest floor. The trees are very susceptible to killing by fire, and they thrive in long, continued absence of fire. Seedlings are killed by creeping, light intensity fires.

Douglas-fir is considered an intolerant tree in comparison with its associated species of the Douglas-fir region, but in the ponderosa pine region it is intermediate in this category. It definitely is more tolerant of shade than is ponderosa pine (Sudworth, 1908), and in continued absence of fire much of the ponderosa pine forest in the Pacific Northwest is being taken over by this tree (Weaver, 1959, 1961a).

d. SEED BED PREPARATION AND SOIL NUTRIENTS. It should be emphasized that the frequent periodic fires of the primeval ponderosa pine forest and the mixed conifers, described herein, did not result in heavy, universal burning over the entire soil surface. Most of the surface was very lightly burned and only at intervals, where deadened snags or windfalls were consumed, was there prolonged, hot burning of the soil surface. These more severely burned spots provide ideal seed beds for ponderosa pine and other associated fire-climax species.

It has been found that hotly burned spots in the sequoia–mixed-conifer forest provides soft, friable soil on which the light-weight Sequoia seeds fall and in which they are buried. Also, highest survival of sequoia seedlings has been found on these spots (Hardesveldt and Harvey, 1967). It has also been found that germination and survival of seeds of other mixed-conifer species are best on these spots (Kilgore, 1971b). Seedling ponderosa pine and Jeffrey pine (*Pinus jeffreyi*) are favored by seedbed conditions after burning (Valmis *et al.*, 1956) Sugar pine is more shade tolerant than the other pines, but its seedlings also benefit from conditions following fire (Kilgore, 1971b).

Kilgore (1971b) quotes Hare as indicating that light burns frequently increase soil pH, stimulate nitrification, and improve soils chemically. There is increased availability of phosphorus, potassium, calcium, and magnesium in the ash deposit. He also quotes Behan as indicating that hot fires may volatilize soil nutrients and that dissolved nutrients may be lost in runoff from rains following fires. As has already been explained, however, only a very small percentage of the soil surface is subjected to such burning.

As has also been explained, ponderosa pine seedlings thrive best on more hotly burned spots. Seedlings planted in soil samples taken from the top 10 inches of burned and unburned plots showed greatest growth in more hotly burned areas. This may be due to greater availability of nutrients or sterilization of the soil or both (Valmis *et al.*, 1956).

Another example of the quick conversion of nutrients tied up in dead forest fuels, such as windfalls, snags, and heavy forest litter, to new living tissue may be found in growth release of surviving trees following fire. Near Nespelem, Washington, the average diameter of 40-year-old surviving ponderosa pines on a burned area was 7.4 inches, compared to 1.7 inches of pines of similar age on an immediately adjacent unburned area (Weaver, 1947). Also, in another 40-year-old stand near Nespelem, over a 22-year period during which three successive prescribed burns were applied, the average healthy crop tree inside of the burn grew to 7.5 inches d.b.h. and 40.7 ft in total height. On immediately adjacent unburned check plots, the average crop tree of the same age grew to 6.2 inches d.b.h. and 35.2 ft in total height (Weaver, 1967a).

Sequoias have shown marked growth increases following fires (Hardesveldt, 1964).

e. OTHER PLANTS. Periodic fires improve forage and range, by reducing heavy needle and debris mats that inhibit grass and other desirable plants, by releasing nitrogen that encourages growth of these plants, and by encouraging growth of nitrogen-fixing plants, such as lupines (*Lupinis* spp.). Burning seems to greatly stimulate the growth of pinegrass (*Calamagrostis rubescens*), and in the summer following burns seedheads of this species appear by the countless millions. Large, woody shrubs are deadened back to the ground line, and subsequent sprouts

Fig. 6. August 1966 photograph showing heavy pinegrass (*Calamagrostis rubescens*) ground cover in a ponderosa pine stand that had been prescribed burned in October 1965 on the Colville Indian Reservation in northern Washington. This fire also had deadened hardwood shrubs and had killed smaller pine and Douglas-fir (*Pseudotsuga menziesii*) seedlings under the larger trees, but it spared several saplings in the opening near the man. Surface fires on this area had occurred once each 7–8 years until the last wildfire in July 1920. Logging slash was broadcast burned in October 1944.

are more tender and more available to browsing animals (Weaver, 1967b) (Fig. 6).

Various desirable shrubs, such as deerbrush (*Ceanothus integerrimus*), are fire dependent, and, because of fire exclusion, these species have become increasingly scarce during the past 50 years in the sequoia–mixed-conifer stands (Kilgore, 1971b).

f. AESTHETICS. Periodic fires greatly improve the forest aesthetically. Anyone who has seen beautiful, parklike ponderosa pine stands that have experienced frequent periodic fires can appreciate this. This is of particular importance in our national parks, where "Vignettes of primitive America" should be preserved (Leopold *et al.*, 1963). What beauty is there in monotonous, dense, stagnating, debris-littered jungles, where there is nothing to attract animals or most bird life? (Fig. 7).

Fig. 7. Dense, debris-littered jungle of ponderosa pine (*Pinus ponderosa*) and incense cedar (*Libocedrus decurrens*) reproduction on a west slope below the sequoia groves on the summit of Redwood Mountain, Kings Canyon National Park, California, June 1967.

2. *Douglas-Fir*

Over the greatest, drier portion of the ponderosa pine region the Rocky Mountain form of Douglas-fir is a common associated species. The coast form of Douglas-fir does occur in mixture with ponderosa pine in southwestern Oregon, northern California, and the west slope of the Sierra Nevada as far south as the San Joaquin drainage (Sudworth, 1908).

In more mesic situations Douglas-fir sometimes occurs largely to the exclusion of other species, particularly in the northern Rocky Mountains, where, over extensive areas, ponderosa pine is not even a component. The even-aged characteristic of many of these stands appears to indicate that they have originated after hot crown fires. Previous devastating insect attacks may have provided dead, flammable material that helped feed these fires. Such insects could have been the Douglas-fir tussock moth (*Hemerocampa pseudotsugata*), the spruce budworm (*Choristoneura fumiferana*), or the Douglas-fir bark beetle (*Dendroctonus pseudotsugae*) (Keen, 1952).

Douglas-fir is more shade tolerant than ponderosa pine (Sudworth, 1908). Over much of the Pacific Northwest portion of the ponderosa pine region, Douglas-fir is the climatic climax species, while ponderosa pine is the principal fire-climax species (Daubenmire, 1952). With attempted total exclusion of fire over the past 60–70 years, many hundreds of thousands of acres of ponderosa pine are being converted to dense reproduction stands in which Douglas-fir predominates (Weaver, 1943, 1959, 1961a, 1967b; West, 1969).

3. *Western Larch*

This fire-climax tree is a common associate of Douglas-fir, ponderosa pine, and other species in northwestern Montana, northern Idaho, northern Washington, the Blue Mountains of Oregon and Washington, and the east slope of the Cascades through Washington and south into Oregon as far as the Metolius River drainage (Sudworth, 1908).

It is very intolerant of shade throughout life, and in middle age and maturity is highly resistant to fire because of thick basal trunk bark and the narrow, thinly branched crown. In silvical characteristics it appears very similar to lodgepole pine, its principal competitor, and following hot burns the composition of the reproduction that follows frequently depends on which species seeds first, the larch or lodgepole (Sudworth, 1908). Langille *et al.* (1903) in discussing restocking east of Mt. Hood, in Oregon, reported that "Tamarack has done more than any other species to restock the immense burns that have taken place in this part of the reserve."

The tree's light foliage does not prevent the growth of associated pines, or of Douglas-fir and other species. For this reason larch is normally a component of mixed stands that foresters refer to as the "Fir–Larch" type. Pure stands do sometimes occur, however, and I have observed extensive, nearly pure stands in the Kettle River range of northern Washington. These followed the hot Dollar Mountain fire of 1930.

4. *Lodgepole Pine*

In writing about the Yellowstone National Park, John Muir (1901) had this to say:

> . . . the lodgepole pine, which, though . . . thin-skinned . . . and . . . easily killed by fire, takes pains to store up its seeds in firmly closed cones, and holds them from three to nine years, so that, let the fire come when it may, it is ready to die and ready to live again in a new generation. For when the killing fires have devoured the leaves and thin resinous bark, many of the cones, only scorched, open as soon as the smoke clears away; the hoarded store of seeds is sown broadcast on the cleared ground, and a new growth immediately springs up triumphant out of the ashes. Therefore, this tree not only holds its ground, but extends its conquests after every fire.

This tree normally occurs at higher elevations than ponderosa pine, though in the pumice soils of the Klamath and Bend regions of Oregon it comprises extensive pure stands and frequently is a competitor of ponderosa pine. It also occurs in extensive pure stands at higher elevations, particularly in the Rocky Mountains.

At about 80 years it becomes subject to devastating epidemic attacks by the mountain pine beetle (*Dendroctonus monticolae*) (Keen, 1952). This creates deadened material that feeds the fierce fires that follow. Man has tried to prevent and to suppress these fires.

5. *Western White Pine*

This fire-climax tree reaches its best development on more mesic sites of northern Idaho and northwestern Montana. It also occurs in the high Cascades of Washington and Oregon, the Puget Sound and Olympic regions of Washington, and the high Sierra Nevada of California. As has been previously indicated, fires occur at less frequent intervals on more mesic sites, but, when they do occur, they burn very intensely. White pine seedlings germinate in the ashes and exposed mineral soil, and with abundant overhead light make rapid growth.

6. *True Firs*

Forests of the true firs occur at elevations above the mixed-conifer type. As has been indicated, however, white fir is an important component of the mixed-conifer type, and, being very tolerant, its reproduction together with that of incense cedar has developed in dense stands under the pines and sequoias.

Extensive stands of true fir occur at higher elevations of the Cascades of Oregon and Washington. At about the turn of the century great areas of this type were deadened by hot fires, particularly in the vicinity of Mt. Hood in Oregon (Langille *et al.*, 1903), north of the Columbia Gorge, and near Mt. Adams in Washington. High altitude Douglas-fir appears to have extended its type because of these burns, but, where seed sources were entirely destroyed, restocking has been very slow. Young growth of true firs is gradually restocking, however, and now the Indians are complaining that it is beginning to reduce the yield of huckleberries (*Vaccinium* spp.), particularly near Mt. Adams.

In the Sierra Nevada, at elevations higher than optimum growth for concolor fir, another true fir, red fir (*Abies magnifica*), forms extensive, nearly pure stands. The National Park Service, on the Kings Canyon National Park, has studied the effects of fire on this high mountain forest and has concluded that they are normally very mild. Fires seldom are able to make destructive runs because of the nature of the surface fuels and the prevalance of natural breaks, such as bare granite outcrops and cliffs, and mesic stream bottoms. Seedlings of red fir are easily deadened by fire, but the ashes provide admirable seed beds for lodgepole pine (Kilgore, 1971a).

The Park Service has concluded that suppressive action against fires in this type, at the higher elevations, is of questionable value. Unless the fires threaten other forest types or adjoining properties no action is now being taken to stop them.

7. Spruce

Engelmann spruce (*Picea engelmannii*) and Colorado blue spruce (*Picea pungens*) comprise dense forests at higher elevations of the Rocky Mountains and in the mountains of Arizona. These forests are normally so damp that they seldom burn.

During the exceptionally dry year of 1904, however, in the White Mountains of east central Arizona, fires originating in the ponderosa pine type burned up through the true firs and into the spruce forest, deadening approximately 20,000 acres on the Fort Apache Indian Reservation (Kallander, 1969). These severe burns restocked to quaking aspen (*Populus tremuloides*). Gradually over the years seedlings of blue spruce have invaded under the aspens, and eventually these will recapture the site.

In conclusion to this discussion of effects of fire on forest types of the ponderosa pine region, it should be indicated that whole volumes can be written on this complex and very interesting subject. This discussion of necessity is very abbreviated.

Space has not permitted discussion of such interesting fire-type trees as the knobcone pine (*Pinus attenuata*) or of the effects of fire on associated hardwood species, such as the black oak (*Quercus californica*), or of the various junipers (*Juniperus* spp.). Most of the discussion has concerned effects of fire on the ponderosa pine and mixed-conifer types because these are most extensive and include our most valuable and useful commercial tree species. Also changes caused by the white man are most profound and most evident in these types.

D. SUBSEQUENT EUROPEAN INFLUENCES RELATED TO FIRE

The white man brought many new influences to the ponderosa pine region to disturb the conditions previously described. One of the most important, and in many instances the first to be applied on a large scale, was extremely heavy grazing pressure by great herds of domestic livestock. This disturbed the density of the native forest floor grasses, in many instances entirely destroyed them, and, through trampling, exposed expansive areas of dust beds of mineral soil admirably suited for germination of coniferous seedlings. With periodic good seed years, combined with occasional growing seasons of favorable moisture, seedlings frequently did become established by the countless millions (Pearson, 1949; Weaver, 1947, 1967a).

Clear-cut logging caused great changes, particularly near the large mining camps and newly constructed railroads. This cutting, followed by broadcast slash burning, sometimes resulted in extensive even-aged stands of young ponderosa pine. On other extensive areas, however, it resulted in poor restocking and in brush fields, particularly where the logging had eliminated seed sources, and subsequent fires destroyed the young, restocking pines (Weaver, 1967c).

Next, and even more important, came attempts to exclude fire completely on the assumption that this was prerequisite to forest management. This started with establishment of the national parks and national forests.

Some of the most extensive overstocked thickets of ponderosa pine reproduction on poorer sites date back to the early 1900's, and it is obvious that fire has been excluded from their inception. On better growth sites considerable growth has occurred in dense reproduction stands, but even here it is obvious that useful growth is but a fraction of what it could be were it concentrated on selected crop trees through elimination of competition. It is obvious also in many of the more advanced of these stands that there actually was considerable thinning caused by low intensity fires in the early 1900's, when the trees were quite small (Weaver, 1961a).

Overstocking is continuing on many of the more recently cutover areas because prescribed burning is not employed and because other methods of control are impractical.

Also as a result of attempted total fire exclusion, reproduction of more shade-enduring associated species has captured the ground under the pine overstory over many hundreds of thousands of acres of the better growth sites, particularly in the California Sierra Nevada and in the Pacific Northwest. As has been indicated, reproduction of Douglas-fir, white fir, and incense cedar is particularly aggressive. Over extensive fringe areas in Arizona, reproduction of alligator juniper (*Juniperus pachypholea*) is invading under the pines (Kallander, 1969).

As early as 1902, Langille (Langille *et al.*, 1903), in discussing reforestation east of Mt. Hood, reported that "In the yellow-pine forests most of the young growth is red or white fir. . . ." He failed to recognize or to mention the role of periodic surface fires in reversing this trend.

The western pine beetle is continuing to kill the larger pines of the overstory. This is because of growth rate reduction of these trees, resulting from competition with the reproduction jungles for limited soil moisture. This renders them particularly susceptable to insect attacks. Inevitably, with harvesting continuing from beetle attacks, then light selective logging, then beetles again, the type will be changed from ponderosa pine to one of associated species. It has already happened over extensive areas (Weaver, 1967c).

Dense, even-aged stands of ponderosa pine, from about 8 to 20 inches in diameter and about 70 years of age, are sometimes decimated by epidemic attacks of the mountain pine beetle (Weaver, 1967c).

The great increase in fire hazard is the most ominous change since earlier days. The very success of foresters in suppressing fires has radically changed conditions described by Muir and other early observers. Great advances have been made in fire prevention and suppression, and fewer fires escape control. When they do, however, and they still do and will continue to, they usually are devastating. Uninterrupted fuel accumulations of the past 40–70 years (Fig. 8) together with development of reproduction and brush thickets have made it extremely difficult to control such fires, and costs of control may properly be described as fantastic.

In the 22-year period from 1940 to 1961 in California, for instance, it is reported that fires burned over 700,000 acres of commercial forest land in the ponderosa pine and mixed-conifer types. These fires were not in the category of the fire observed by Muir in 1875. Destructive fires have also recently burned in other portions of the region (Kallander, 1969; Weaver, 1961b). In 1971 severe fires burned in east central Arizona in ponderosa pine from which fire had been excluded for many years (Fig. 9).

Fig. 8. June 1972 photograph of a portion of the ponderosa pine (*Pinus ponderosa*) forest in the Maverick district of the Fort Apache Indian Reservation. This particular area had never been treated by prescribed burning.

The Leopold Committee (*Leopold et al.*, 1963) in discussing adverse effects of overstocked reproduction on other plant associations, on wildlife, on protection, and on recreation on national parks in the Sierra, reported that "Today much of the west slope is a dog-hair thicket of young pine, white fir, incense cedar, and mature brush . . . a direct function of over-protection from natural ground fires." The committee then recommended research toward methods designed to return portions of these parks to conditions comparable to primeval conditions encoun-

Fig. 9. June 1972 photograph of a portion of the timber killed by the Black River fire of the previous year. Larger fire-killed pines had been salvage-logged.

tered by the first white men. In the committee's opinion the parks should represent ". . . vignettes of primitive America."

The Park Service has made an excellent start in complying with the committee's recommendations (Kilgore, 1971a,b), and important supporting research has been conducted by the University of California on Whitaker's Forest on Redwood Mountain immediately adjacent to Kings Canyon National Park (Biswell and Weaver, 1968). The Bureau of Indian Affairs has conducted extensive prescribed burning projects on the Fort Apache and Huala-pai Indian Reservations in Arizona. These have been very effective but have been seriously underfinanced, and it has been impossible to conduct them on a fully adequate scale (Kallander, 1969; Knorr, 1963). Extensive and very expensive precommercial thinning projects are being conducted throughout the ponderosa pine region, but little or nothing is being done to reduce hazard. The problems of overstocking and increasing fire hazard must be corrected, and, in my opinion, this can only be accomplished by restoring periodic fire, under rigid control, to its rightful place as an ecological factor in ponderosa pine and the mixed conifers. Until this can be accomplished on an adequate scale, I can entertain little optimism concerning the future of forestry in the ponderosa pine region.

III. The Douglas–Fir Region

The role of fire in the Douglas-fir region has already been briefly discussed in the introduction to this chapter.

> It is the most magnificent forest in North America. . . . Its granduer in an artistic sense is beyond description, and can be fully appreciated only by one who abides for weeks or months in its perpetual twilight. Great fir trees, rising from 150 to 250, and even 300, feet above the ground, stand in closed ranks, their rugged trunks from 6 to 8 or 10 feet, and even more, in diameter, shaggy with mosses and lichens of many subdued tints of brown, green and yellow.

Thus wrote geologist Israel C. Russel in 1898.

This is one of the important timber-growing regions of the world. It includes most of the timbered portions of western Oregon and Washington from the forests of true firs (*Abies* spp.) and mountain hemlock (*Tsuga mertensiana*) along the crest of the Cascades to the narrow belt of forest of western hemlock (*Tsuga heterophylla*) and Sitka spruce (*Picea sitchensis*) along the Pacific coast. Southward along the coast in northern California, redwood (*Sequoia sempervirens*) becomes predominant. Further inland, in southwestern Oregon and northern California, sugar pine and ponderosa pine are important associates (Isaac, 1943).

A. THE ROLE OF FIRE

In order to understand the role that fire plays in the Douglas-fir forest it is appropriate, first, to briefly discuss some of the tree's silvical requirements.

1. *Silvical Requirements*

Sudworth (1908) noted that the tree is "moderately tolerant, becoming less so with age; endures more shade than western yellow pine, sugar pine, western white pine, and lodgepole pine, but less than western hemlock, western red cedar, white and alpine fir. . . ." He also noted that ". . . pure mineral soil, or a mixture of the latter and humus, best for germination and development of seedlings; reproduction rare on thick duff or vegetable matter, but abundant in humid regions after layer has been burned off or broken up by logging; unburned logged areas are commonly restocked by its northern associates, western hemlock and red cedar."

Munger (1911) stated that "Douglas fir seed germinates in almost any situation that the seed chances to fall, provided there is sufficient moisture, but it is only the seedlings whose roots quickly come in contact with the mineral soil and which receive direct light that survive."

Isaac (1943) stated, "It is a matter of common observation that no Douglas-fir saplings are ever found under the full canopy of a virgin stand, though young trees of the more shade-tolerant hemlock, cedar, and balsam firs are common. Furthermore, it is extremely significant that in the spring after every seed crop, germinating seedlings of Douglas-fir are abundant everywhere under the virgin forest cover and even under the densest of young stands of seed-bearing size. By fall, all these seedlings have disappeared because they cannot become established in the full shade of the mother forest. Douglas-fir seedlings must have enough overhead light not only to survive but also to compete successfully with the ground cover and the more tolerant and numerous seedlings of hemlock et al."

Before a new forest of Douglas-fir can become established something must clear or deaden the mother forest, provide a mineral soil seedbed, and temporarily eliminate or reduce the competition from other vegetation and advanced reproduction of other, more shade-tolerant associated species. Under primeval conditions fire is the only agent that could accomplish this.

2. Results of Primeval Fires

Muir (1918), from observations in the Puget Sound region in the late 1800's, wrote as follows

> Looking now at the forests in a comprehensive way, we find in passing through them . . . that some portions are much older than others, the trees much larger . . . while in the younger portions, where the elevation of the ground is the same as to sea-level and the species of trees are the same as well as the quality of the soil . . . the trees seem to be and are mostly of the same age, perhaps from one hundred to two or three hundred years, with no grey-bearded, venerable patriarchs—forming tall, majestic woods without any grandfathers. . . . Then, perchance, we come upon a section farther up the slopes towards the mountains that has no trees more than fifty years old, or even fifteen or twenty years old. These last show plainly enough that they have been devastated by fire, as the black, melancholy monuments rising here and there above the young growth bear witness. Then, with this fiery, suggestive testimony,

Fig. 10. Even-aged Douglas-fir (*Pseudotsuga menziesii*) stand in the Silver Falls State Park in the foothills of the Cascades east of Salem, Oregon. This stand started following logging and broadcast slash burning early in the present century. Reproduction under the Douglas-firs consists of lowland white fir and western hemlock.

on examining those sections whose trees are a hundred years old or two hundred, we find the same fire-records, though heavily veiled with mosses and lichens, showing that a century or two ago the forests that stood there had been swept away in some tremendous fire at a time when rare conditions of drought made their burning possible [Fig. 10].

Isaac (1943) noted, "The presence of extensive immature even-aged stands of Douglas-fir on the coming of the white man to the region is evidence that devastating fires had occurred periodically here and there, for in comparison with its associates Douglas-fir is very aggressive in reestablishing itself after burns. Areas which have not been visited by fire in the last 500 years are rare in the region and usually carry forests that have practically advanced to the climax type of tolerant species—hemlock, cedar, and balsam firs [Fig. 11]."

It is obvious that fire has played a very important role in development and perpetuation of Douglas-fir over countless centuries. Its role differed somewhat from the role it played in the more xeric ponderosa pine

Fig. 11. Decadent stand of large Douglas-firs (*Pseudotsuga menziesii*) south of Quinault Lake in the Olympic National Forest, Washington, September 1957.

region. There it is a rare year indeed when fires will not spread readily over the needle-carpeted forest floor under the ponderosa pines. In the more mesic Douglas-fir region, in contrast, years when fires will spread occur less frequently. When extreme drought conditions develop, however, fires are apt to burn with great intensity, frequently developing into the devastating crown fires noted by Muir and Isaac.

Douglas-fir fires sometimes spread as exceedingly hot surface fires, only occasionally causing openings in the overstory. This is one of the explanations for the patchwise occurrence of younger age classes in virgin stands noted by Isaac (1943). I have also noted such patchwise occurrence of younger age classes along some of the trails in the Columbia Gorge. Thick bark of the older trees is heavily charred by fire, sometimes to many feet above the ground. Most of this charring is reported to have been caused by fires that burned in 1902. The younger age classes apparently date from about that time.

3. *Effects on Associated Forest Types*

Fires that occurred in the high altitude true fir–mountain hemlock forests along the crest of the Cascades have already been discussed briefly in connection with the ponderosa pine region (Section II,C,6).

As would be expected, devastating fires occurred at less frequent intervals in the damp hemlock–spruce forest along the Pacific coast. That such fires did sometimes occur, however, is attested to by the presence of Douglas-fir stands within sight of the Pacific Ocean. One such area supports the even-aged Douglas-fir stand on Cascade Head in Oregon. It is reported to have originated following a fire about 1849 (Morris, 1934).

4. *Causes of Primeval Fires*

What caused fires to start in the Douglas-fir region under primeval conditions? Lightning has always been a major cause of fires, and even in the Douglas-fir region it probably started many. Indians burned extensive portions of the region on an annual and systematic basis. Morris (1934) noted, "The first printed record of extensive fires observed by white men in the Pacific Northwest is that of David Douglas, the botanist for whom Douglas-fir was named. . . . mention is made time and time again of recently burned-over land he encountered along the west side of the Willamette Valley from Fort Vancouver south to the Umpqua River in September and October, 1826." Only Indians could have started these fires, for there were very few white men, and they were at Fort Vancouver.

Applegate (1914), in discussing Indian burning in Willamette Valley as late as 1844, reported, "It was a custom of these Indians, late in the autumn, after the wild wheat, *Lamoro sappolil,* was fairly up, to burn off the whole country. The grass would burn away and leave the sappolil standing, with the pods well dried and bursting. Then the squaws . . . would go with their baskets and bats and gather in the grain." The sappolil was tar-weed, probably *Madia sativa.* Other reasons for burning were reported by Clarke (1905): "One was to keep down all undergrowth, so that hunters could see game from a great distance; another that no hostile war party could approach unseen." By 1852, Clarke reported, "Then the hills and prairies had already commenced to grow up with a young growth of firs and oaks, because the Calapooias no longer were there to burn off the face of the country and keep it clear to the vision."

B. CHANGES CAUSED BY WHITE SETTLEMENT

1. *Fire Incidence*

Morris (1934) noted that "Then the white men drove the Indians from the fertile valleys and prevented annual burning where crops or buildings would be endangered. At the same time the white men, traveling in the mountains at a distance from the settlements, habitually left camp fires burning and sometimes even set the woods afire intentionally. Land-clearing fires were kept burning near the settlements during the most dangerous fire weather, and often spread beyond control."

Forest cover surveys conducted by the Pacific Northwest Forest Experiment Station during the 1920's and 1930's determined the acreage of even-aged stands of timber of each 10-year class up to 100 years. Morris (1934) reported,

> The acreage found covered by timber 40, 50, 60, and 70 years of age indicates that forest fires have devastated more and more land since white settlement began. In a decade approximating 1845 to 1855, for example, about seven times as much land was deforested as in any of the three previous decades. Now the 10 years 1845–1855 are the period when white people first migrated to the Willamette Valley in large numbers, and also the period to which hearsay assigns the great fires of the coast country.

Plummer (1902), in describing forest conditions in the Washington Cascades, reported, "A wagon road follows the South Fork to Snoqualmie Pass . . . and it is along or near this route that the forests of the

Snoqualmie watershed have suffered by fire. The large burn on Denny Mountain was first fired in 1865, and in 1894 a second fire destroyed much of the restocking and included new areas of the forest." Langille *et al.* (1903), in describing forest conditions in the Oregon Cascades, reported, "The most destructive fires have taken place south of Mt. Hood along the old Barlow road and southward on the western slope to Salmon River. Some of these are said to have occurred as early as 1852, when the sections in the vicinity of Goverment Camp were burned." These men reported on many other burned areas in the Cascades.

Morris (1934) described the great fires that occurred west of the Cascades during the period subsequent to white settlement. During the average year it appeared that the air would be thick with smoke and the sun obscured during late summer and early fall. Sometimes the smoke would interfere with steamboat navigation on the rivers and would even delay seagoing ships from attempting the crossing of the dangerous bar at the mouth of the Columbia River.

The greatest and most destructive fires, however, occurred during 1849, 1868, 1902, and 1933. Approximately 500,000 acres, or more, burned in the Oregon coast range between the Siletz and Siuslaw rivers, probably during 1849. The 1868 fires were widespread and very destructive. Several of these burned in the vicinity of Coos Bay and south along the coast.

The 1902 fires were most destructive in terms of loss of human life and property. According to Holbrook (1944) the 1902 fires destroyed 700,000 acres of timber in the states of Oregon and Washington. These fires finally awakened the public to the need for forest conservation.

The Tillamook fire started on August 14, 1933. According to Morris, "The climax was reached in a 30-hour period on August 25 and 26, during which the fire covered 225,000 acres, or 85 percent of the total area burned." Appearance of the tremendous smoke cloud from this blow-up was mentioned in the introduction to this chapter.

2. Logging

Logging has also caused great changes in the Douglas-fir region. From a small start during pioneer settlement, considerable impetus was provided by the market resulting from the 1849 gold discoveries in California and later by expanding world markets. Methods evolved from horse- and bull-team skidding of logs to tidewater along Puget Sound and the lower Columbia River to powerful steam and later Diesel-skidding donkeys and to high-lead logging. Steam-logging railways have mostly been superseded by the modern Diesel truck. Many millions of acres of the lower elevations have been clear-cut and logging has moved back into the rugged slopes and canyons of the Cascades and coast ranges.

The former clear-cut areas and former burn areas now support millions of areas of excellent second growth, even-aged Douglas-fir stands, but other millions of acres are poorly stocked and support mostly red alder (*Alnus oregoni*). Isaac (1943) noted this condition and stated, "Under the impetus of the laws of both Oregon and Washington slashings were burned broadcast following logging and this practice has had a profound effect upon natural regeneration. . . ." Results were not always favorable.

Logging in old-growth Douglas-fir and in the coast type results in tremendous volumes of highly flammable slashings, and it was these that gave some of the great fires their initial impetus. Following the 1902 fires, state laws required that slashings be burned under such weather conditions and at such times that standing timber and other property would not be endangered.

Because of past fire history and silvical requirements of Douglas-fir, it would appear that clear-cutting followed by slash burning would favor reestablishment of this species. This does not necessarily follow. When crown fires of late summer deadened standing Douglas-firs, Isaac found that unless the fire was exceptionally hot, many of the green cones in the tops still contained viable seeds. Immediately following the fire, these scorched cones opened and shed their seeds into the fresh beds of ashes. Logging, in contrast, may have occurred before the cones were ripened, or the hot slash fire may have consumed all cones and seed. On natural burns standing snags of fire-deadened trees provided partial shade for the young seedlings during hot summer periods. On clear-cut areas such shade was lacking. Seed crops fail during certain years, and other causes may contribute to restocking failures (Isaac, 1943). Unless Douglas-fir seedlings become established early, brush species take over, and natural regeneration becomes almost impossible (Munger, 1911).

Isaac (1943) concluded that intensely hot slash burning was deleterious to establishment of Douglas-fir regeneration. Morris (1970), however, found that severe burns occurred on a very small percentage of the total areas slash burned. He also found that slash burning greatly reduced probable rate of spread and resistance to control of subsequent fires from 5 to 16 years.

Recent procedures on Forest Service, Bureau of Land Management, State and Forest Industry lands have involved clear-cutting on smaller, isolated blocks, or areas, slash burning to reduce hazard and improve seed beds, and prompt planting, usually with a superior strain of Doug-

Fig. 12. Clear-cut blocks on the southwestern slopes of the Olympic Mountains, Olympic National Forest, Washington, September 1964. Excellent planted Douglas-firs (*Pseudotsuga menziesii*) in foreground have been supplemented by wild seedlings of western hemlock (*Tsuga heterophylla*) and Sitka spruce (*Picea sitchensis*). All the clear-cut blocks have planted and wild reproduction, and restocking is successful. On distant ridge in left background is a recently slash burned and planted clear-cut.

las-fir seedlings adapted to the site. Once these seedlings become established and partial shade develops, natural seedlings of western hemlock, western red cedar (*Thuja plicata*), and white fir also become established, thus assuring a minor mixture of these species in the stand. In view of the timber-growing potentialities of good Douglas-fir sites and the silvical requirements of the species, this appears good ecological procedure (Fig. 12).

The more recent trend toward shelter-wood cutting, apparently motivated by environmental concerns, will probably assure conversion to reproduction stands in which Douglas-fir will be poorly represented. This trend together with air quality limitations on slash burning have already caused an alarming increase in the acreage of unburned slash. Also the trend toward use of skyline, balloon, and helicopter logging in steep terrain will cause problems in slash disposal, for even with presently available ground equipment it is difficult to burn slash on such areas. In view of the silvical characteristics of Douglas-fir and past fire history of the Douglas-fir region, it appears to me that any suggestion that fire can be eliminated as a management and protection tool is completely unrealistic.

IV. The Redwood Region

"California might have done without her gold mines but not without the resources of the Redwood belt . . . . Redwood is logged and milled into lumber on a large scale . . . . The greater portion of the product is consumed in California. By means of it have been built the farms, factories, railroads, cities and towns of California in great degree." This was written by Dr. Willis Linn Jepson, the famous California botanist (Jepson, 1923).

Redwood extends in a narrow, irregular belt, 5–35 miles wide, from the Chetco River, in Curry County, Oregon, southward to Salmon Creek Canyon in the Santa Lucia Mountains in southern Monterey County, California. There is a transverse break along the headwaters of the Mattole River in southern Humbolt County, California, and from Sonoma County southward the trees occur in detached and irregular areas (Roy, 1966). The original redwood timber area is estimated in excess of two million acres (Stone *et al.*, 1972).

Redwood almost always dominates in mature stands throughout its range but is usually mixed with other conifers and broad-leaved trees. Douglas-fir is a common associate. Others are western hemlock, grand fir (*Abies grandis*), western red cedar, tan oak (*Lithocarpus densiflorus*), and, from Humbolt Bay northward, Sitka spruce (Jepson, 1923; Roy, 1966).

On moist river flats and gentle slopes, below 1000 ft elevation and encompassing the best sites, redwood frequently occurs in pure stands, with mature trees 6–16 ft in diameter and heights up to 300 ft and taller. Merchantable volumes average from 125,000 to 150,000 ft, board measure, per acre. Limited areas have produced as high as 500,000 ft or more per acre. Age of mature redwoods is 500 to 1300 years (Jepson, 1923; Roy, 1966).

Although the redwood forests are always in close proximity to the ocean, they cannot tolerate the buffeting of ocean winds and the salt spray and for this reason prefer the more sheltered slopes and valley floors or the screening of other trees, such as Douglas-fir and Sitka spruce. The forests are coextensive with the heavy summer fog belt (Jepson, 1923; Roy, 1966; K. N. Boe, personal communication, 1972).

A. Silvical Characteristics and the Role of Fire

Jepson (1923) stated, ". . . *Sequoia sempervirens* reproduces by crown sprouts with remarkable persistence. It is the only strictly coniferous species which has this habit in any marked degree. The tree has no tap-root but a large number of heavy lateral roots which lie near the surface of the ground at their point of origin. A most advantageous position to generate by adventitious buds a circle of sprouts about the stump." According to K. N. Boe (personal communication, 1972), "Present terminology considers these buds around the root crown to be dormant buds rather than adventitious."

Sudworth (1908) wrote as follows concerning redwood: "Moderately tolerant of shade except in early youth . . . has marked characteristics of intolerant trees; a thin open crown, rapid loss of side branches, and the eager bending of crowns toward openings in crown cover; seedlings not able to come up in shaded places . . . . Stump sprouts often exist under densest shade for one hundred years, growing very slowly in diameter during this time, but recovering completely and growing rapidly when released from suppression." He also noted that the tree is a fairly prolific seeder, but only a small percentage of the seeds are perfect. Also in regard to seedlings, Jepson (1923) wrote: "While seeds are produced in enormous quantities seedlings are a rarity in the redwood belt, the densely shaded forest and the ground litter of foliage, often one foot thick, offering unfavorable conditions for germination." Roy (1966) quotes other authorities to effect that areas where slash burning was medium to heavy had 5 to 10 times as much redwood seedling reproduction as unburned or lightly burned surfaces. Fritz (1931) noted, "A fire will kill off all seedlings growing in the duff. Some of those on mineral soil may sprout from the burl at the root collar of each plant. . . ."

Though the redwood forest is normally very damp, there are periods of low humidity and high temperature when fires will burn readily. Fritz (1931) studied the fire history revealed by over one hundred fire-scarred stumps in Humbolt County in 1928. He found that during the past 1100 years there were at least 45 severe fires on that particular area or an average of at least four each century. These were fires set by aboriginal Indians, and there is possibility that some may have been started by lightning. Fritz felt that the redwood forest had persisted in spite of the fires. He particularly blamed past fires for the prevalence of heart-rot in butt logs of many of the large trees.

Some later researchers have concluded that fire was a very important ecological factor in the redwoods. Stone *et al.,* (1972) stated, ". . . the primeval redwood forest was a mosaic of ecosystems supporting redwood that existed prior to the arrival of the white man. Fire was an integral part of the environment. This resulted in a forested mosaic of successional sub-climaxes held or renewed in this mosaic by fire. . . . Redwood is favored over other species in the presence of fire by its thick, essentially fire-resistant bark, by its capacity to sprout along its stem and replace its branches when killed by fire, by its capacity to sprout from its root crown following destruction of the rest of the tree. . . ."

With reference to the prevelance of heart-rot, Stone *et al.* (1972) continued:

> Redwood can only sprout when its root crown is left in the ground. Thus, anything that increases the percentage of root crowns left in the ground when the redwood falls, increases the number of trees that can replace themselves by sprouting; and heart-rot does just this. Heavily infected trees generally break off above the ground when they fall and fail to pull up their root crowns in the process. Uninfected trees, on the other hand, almost invariably pull up their root crown when they fall.

Boe (1966), however, found that in redwood stands logged selectively or by the shelterwood system and along exposed margins of mature timber adjacent to small clear-cuts, more than three-fourths of the red-woods that subsequently fell were uprooted. Of the trees that were broken off, 60% had no heart-rot. Rot does not appear a particularly large factor affecting uprooting versus breaking off in cutover stands. In primeval redwood stands that have not been opened up by logging, however, it appears to me that it would be considerably more important.

Jepson (1923) noted, "Circles resulting from the death of aged trees are also found in virgin forests. In most cases the original stump has wholly disappeared as a result of repeated fires and we have the shallow tree-encircled hollows . . . [Fig. 13]."

Stone *et al.* (1972) further stated, "Other species are favored over redwood by the exclusion of fire. Redwood seedlings, unlike western hemlock and associated species, cannot readily establish themselves on the undisturbed forest floor."

B. CHANGES CAUSED BY WHITE SETTLEMENT

Aside from the white man's logging, Stone *et al.*, (1972) concluded, "The major impact of his presence has been the vigorous suppression of fire over the last 50 years. This has not yet, however, resulted in any significant successional changes."

Logging started in the early 1850's, following the gold discoveries in California, and it has continued on a large scale. Apparently most

Fig. 13. The beginning of a "tree-encircled hollow," such as described by Jepson. December 1957 photograph of second-growth redwoods (*Sequoia sempervirens*) from sprouts that have developed around a stump, following earlier logging and slash burning near Santa Cruz, California.

of the remaining primeval redwood areas are concentrated in the California state redwood parks, preserved in large part through the efforts of the Save-the-Redwoods League and the newly established Redwood National Park. Elsewhere, on industrial forest lands, conversion to second-growth stands is continuing rapidly. This involves clear-cutting in blocks from 10 to 100 or more acres in area, some progressive clear-cutting, seed-tree cutting, and selective logging. Under all systems of cutting, natural regeneration is obtained by windblown seed and red-wood sprouts. Reinforcement restocking is obtained by aerial seeding and planting. Fire is used to burn concentrations of logging slash for hazard reduction and preparation of seedbed (K. N. Boe, personal communication, 1972).

The primeval redwood forest maintained itself in an environment that included rain, windstorms, cooling summer fogs, and occasional fires. Redwood survives fire where other tree species are killed. It sprouts from dormant buds around the root crown if aerial portions are killed, or a new crown develops along the bole of the tree if only a portion of the crown is fire killed. Redwood produces abundant quantities of seed regularly, and seedlings become established readily on burned as well as mineral soil. Maintenance of primeval stands in state and national parks should involve management practices that include fire and probably cutting and scarification.

V. Alaska

This is but a brief discussion of the effects of fire on temperate forests of the western portion of "The Lower 48 States." Only brief mention can be made of its effects on the extensive forests of Alaska.

According to Lutz (1956), the forests of the vast interior of Alaska have experienced extensive and repeated fires during prehistoric and historic times. Man has caused most of these fires, but lightning has also started many. As in the other forest regions, the incidence of fires and size of areas burned increased very greatly with advent of the white man.

White spruce (*Picea glauca*) is the most commercially valuable forest tree species but is very susceptible to killing by fire. As a result extensive potential white spruce sites are occupied by paper birch (*Betula papyrifera*) or quaking aspen or, where fires have occurred too frequently, by various herbaceous or shrub species. On most of the burned areas there is exposed mineral soil, and, with adequate to abundant seed supplies, most of these burns are restocking to paper birch, aspen, or spruce, either pure or in mixture. Black spruce (*Picea mariana*), a physiographic climax species on poorly drained areas, regenerates more readily after fires, provided the fires do not occur too frequently.

Lutz concluded that there has been far too much forest destruction from uncontrolled wildfires since advent of the white man. He recognizes the possibility of prescribed burning, however, particularly in connection with possible improvement of moose ranges.

The dense coastal western hemlock—Sitka spruce forests of southeastern Alaska were viewed from a cruise ship as it sailed southward from Juneau toward Petersburg one sunny afternoon in early June 1970. They appeared almost continuous from tidewater up to the timberline, except for some smooth, glaciated granite headlands and some distant clear-cuts. One could imagine that there has been little change in general appearance since John Muir made his canoe voyage along this coast in October and November 1879 (Muir, 1915). Apparently fire very rarely burns in these forests.

References

Applegate, J. (1914). "Recollections of My Boyhood." Review Publ. Co.

Beale, E. F. (1858). Wagon Road from Ft. Defiance to Colorado River. 35th Congress, 1st Sess., House Exec. Doc. 124, p. 49.

Biswell, H. H. (1959). Man and five in ponderosa pine in the Sierra Nevada of California. *Sierra Club Bull.* **44**, 44–53.

Biswell, H. H. (1960). Danger of wild fires reduced by prescribed burning in ponderosa pine. *Calif. Agri.* **14**, 5–6.

Biswell, H. H., and Weaver, H. (1968). Redwood Mountain. *Amer. Forests* **74**, 20–23.

Boe, K. N. (1966). Windfall after experimental cuttings in old-growth redwood. *Proc. Soc. Amer. Foresters, 1965* pp. 59–62.

Clarke, S. A. (1905). "Pioneer Days of Oregon History." J. K. Gill Co.

Cooper, C. F. (1960). Changes in vegetation, structure and growth of southwestern pine forests since white settlement. *Ecol. Monogr.* **30**, 129–164.

Daubenmire, R. (1952). Forest vegetation of northern Idaho and adjacent Washington and its bearing on concepts of vegetation classification. *Ecol. Monogr.* **22**, 301–330.

Dutton, C. E. (1881). The physical geology of the Grand Canyon district. *U.S. Geol. Surv., 2nd Annu. Rep.* pp. 136–137.

Forest Service. (1965). Timber trends in the United States. *Forest Resour. Rep.* **17**, 95, 119, 123, 125, and 175–177.

Fritz, E. (1931). The role of fire in the redwood region. *J. Forest.* **29**, 939–950.

Hardesveldt, R. J. (1964). Fire ecology of the giant sequoias: Controlled fire may be one solution to survival of the species. *Natur. Hist., N.Y.* **673**, 12–19.

Hardesveldt, R. J., and Harvey, H. T. (1967). The fire ecology of sequoia regeneration. *Proc 7th Annu. Tall Timbers Fire Ecol. Conf.* pp. 65–78.

Harlow, W. M., and Harrar, E. S. (1958). "Textbook of Dendrology," 4th ed., pp. 96–101. McGraw-Hill, New York.

Holbrook, S. H. (1944). "Burning an Empire," pp. 108–111. Macmillan, New York.

Isaac, L. A. (1943). "Reproductive Habits of Douglas Fir." Chas. Lathrop Pack Forestry Found.

Jepson, W. L. (1923). "The Trees of California," pp. 13–23 and 156. Univ. of California Press, Berkeley.

Johnston, V. R. (1970). "Sierra Nevada," pp. 42–43. Houghton, Boston, Massachusetts.

Kallander, H. (1969). Controlled burning on the Fort Apache Indian Reservation, Ariz. *Proc. 9th Annu. Tall Timbers Fire Ecol. Conf.* pp. 241–250.

Keen, F. P. (1937). Climatic cycles in eastern Oregon as indicated by tree rings. *Mon. Weather Rev.* **65**, 175–178.

Keen, F. P. (1940). Longevity of ponderosa pine. *J. Forest.* **38**, 597–598.

Keen, F. P. (1952). Insect enemies of western forests. *U.S. Dep. Agr., Misc. Publ.* **273**.

Kilgore, B. M. (1971a). The role of fire in managing red fir forests. *Trans. N. Amer. Wildl. Natur. Resour. Conf.* **36**, 405–416.

Kilgore, B. M. (1971b). "The Role of Fire in a Giant Sequoia–Mixed Conifer Forest," Paper. AAAS, Philadelphia, Pennsylvania, 1971.

King, C. (1871). "Mountaineering in the Sierra Nevada" (reprinted by Norton, New York, pp. 48, 49, and 54, 1935).

Knoor, P. N. (1963). One effect of control burning on the Fort Apache Indian Reservation. *Proc. Watershed Symp., 7th, Ariz. Watershed Program* pp. 35–37.

Komarek, E. V. (1968). Lightning and lightning fires as ecological forces. *Proc. 8th Annu. Tall Timbers Fire Ecol. Conf.* pp. 169–198.

Langille, H. D., Plummer, F. G., Dodwell, A., Rixen, T. F., and Leiberg, J. B. (1903). Forest conditions in the Cascade Forest reserve. *U.S., Geol. Surv., Prof. Pap.* **9**, Ser. H, 36, 41 and 78.

Leopold, A. S., Cain, S. A., Cottam, C. M., Gabrielson, I. N., and Kimball, T. L. (1963). Wildlife management in the national parks. *Amer. Forests* **69**, 32–35 and 61–63.

Lutz, H. J. (1956). Ecological effects of forest fires in the interior of Alaska. *U.S., Dep. Agr., Tech. Bull.* **1133**, 90–95.

Morris, W. G. (1934). Forest fires in Oregon and Washington. *Oreg. Hist. Quart.* **35**, 313–339.

Morris, W. G. (1970). Effects of slash burning on overmature stands of the Douglas-fir region. *Forest Sci.* **16**, 258–270.

Muir, J. (1894). "The Mountains of California," p. 163. Houghton, Boston, Massachusetts.

Muir, J. (1901). "Our National Parks," pp. 68–69, 291–292, and 307–308. Houghton, Boston, Massachusetts.

Muir, J. (1915). "Travels in Alaska," pp. 114–196. Houghton, Boston, Massachusetts.

Muir, J. (1918). "Steep Trails," pp. 237–238. Houghton, Boston, Massachusetts.

Munger, T. T. (1911). The growth and management of Douglas fir in the Pacific northwest. *U.S., Forest Serv., Circ.* **175**.

Pearson, G. A. (1949). Management of ponderosa pine in the Southwest. *U.S., Forest Serv., Agr. Mon.* **6**, 37, 74, 78, 110, and 117.

Plummer, F. G. (1902). Forest conditions in the Cascade Range, Washington. *U.S., Geol. Surv., Prof. Pap.* **6**, Ser. H, 21.

Reynolds, R. (1959). Effect upon the forest of natural fires and aboriginal burning in the Sierra Nevada. Masters Thesis, University of California, Berkeley.

Roy, D. F. (1966). Silvical characteristics of redwood. *U.S., Forest Serv., Pac. Southwest Forest Range Exp. Sta., Res.* **PSW 28**.

Russel, I. C. (1898). A preliminary paper on the geology of the Cascade Mountains in northern Washington. *U.S., Geol. Surv.* **92**.

Show, S. B., and Kotok, E. I. (1924). The role of fire in the California pine forests. *U.S., Dep. Agr., Bull.* **1294**.

Soeriaatmadja, R. E. (1966). Fire history of the ponderosa pine forests of the Warm Springs Indian Reservation, Oregon. Thesis, Oregon State University, Corvallis.

Stone, E. C., Grah, R. F., and Zinke, P. J. (1972). Preservation of the primeval redwoods in the Redwood National Park. *Amer. Forests* **78**, Part I, 50–55; Part II, 48–59.

Sudworth, G. B. (1908). Forest trees of the Pacific slope. *U.S., Forest Serv.*

Valmis, J., Biswell, H. H., and Shultz, A. M. (1956). Seedling growth on burned soils. Calif. Agr. **10**, 13.

Wagner, W. W. (1961). Past fire incidence in the Sierra Nevada forests. *J. Forest.* **59**, 734–747.

Weaver, H. (1943). Fire as an ecological and silvicultural factor in the ponderosa pine region of the Pacific slope. *J. Forest.* **41**, 7–14.

Weaver, H. (1947). Fire—nature's thinning agent in ponderosa pine stands. *J. Forest.* **45**, 437–444.

Weaver, H. (1951). Fire as an ecological factor in the Southwestern ponderosa pine forests. *J. Forest.* **49**, 93–98.

Weaver, H. (1955). Fire as an enemy, friend and tool in forest management. *J. Forest.* **53**, 499–504.

Weaver, H. (1957). Effects of prescribed burning in ponderosa pine. *J. Forest.* **55**, 133–137.

Weaver, H. (1959). Ecological changes in the ponderosa pine forest of the Warm Springs Indian Reservation in Oregon. *J. Forest.* **57**, 15–20.

Weaver, H. (1961a). Ecological changes in the ponderosa pine forest of Cedar Valley in southern Washington. *Ecology* **42**, 416–420.

Weaver, H. (1961b). Implications of the Klamath fires of September, 1959. *J. Forest.* **59**, 569–572.

Weaver, H. (1967a). Some effects of prescribed burning on the Coyote Creek test area, Colville Indian Reservation. *J. Forest.* **65**, 552–558.

Weaver, H. (1967b). Fire and its relationship to ponderosa pine. *Proc. 7th Annu. Tall Timbers Fire Ecol. Conf.* pp. 127–149.

Weaver, H. (1967c). Fire as a continuing ecological factor in perpetuation of ponderosa pine forests in western United States. *Advan. Front. Plant Sci.* **18**, 137–154.

West, N. E. (1969). Successional changes in the Montane Forest of the central Oregon Cascades. *Amer. Midl. Natur.* **81**, 265–271.

. 10 .

Effects of Fire on Chaparral

Harold H. Biswell

I. Introduction

"Chaparral" is the term applied to the broad-sclerophyll scrub or brushland type of vegetation in California. This term originated from *chabarra*, a Basque word for a scrub oak of the Pyrenees. The early

Spanish explorers spelled it *chaparro* and applied the name to the temperate Mediterranean scrub of California (Shantz, 1947). This type of vegetation also occurs in Arizona and in Mexico, where it covers large areas on the west coast and smaller ones on the east coast.

Chaparral was defined by Cooper (1922) as a broad-sclerophyll scrub community, dominated by many species belonging to genera unrelated phylogenetically but of a single, constant ecological type. The most important features of chaparral species are the root system, extensive in proportion to plant size, the dense, rigid branching, and especially the leaf, which is small, thick, heavily cutinized, and evergreen. Exceptions to this definition are few, but oddly enough the leaves of chamise (*Adenostema fasciculatum*), the most important species, are short and needlelike in fascicles of 10–15 leaves, instead of the broad-scelerophyll type (McMinn, 1939).

Brushlands of the chaparral type are known under different names in various parts of the world, but they are all similar in many respects. In the Mediterranean area of France they are known as *garrigue* or *garigue* on the drier, calcareous soils and as *macchia* or *maquis* on siliceous soils. In Spain they are called *tomillares;* in the Balkans, *phrygana;* in Chile, *matarral;* and in south Africa, *macchi* and *fynbosch.* In South Australia they are called *mallee-scrub,* consisting mostly of Eucalyptus; *mulga scrub,* when composed mostly of *Acacia;* and *brigalow,* a scrub of *Acacia harpophylla. Heath* is used for ericoid vegetation, such as that of South Africa and southeast France. The discussions in this chapter are based largely on the chaparral in California.

The climate of broad-sclerophyll brushlands is mainly of the Mediterranean type, with mild, moist winters, when growth is active, followed by a prolonged drought of 5–6 months, when not more than 20% of the year-long precipitation occurs. In some of the chaparral areas of California, the rainfall may not be more than 2 cm during the hot summer months. These features make areas of chaparral very susceptible to fire and its spread.

The effects of fire on chaparral have been of interest to many. Geographers and plant ecologists are interested because chaparral is largely a fire-induced type with a remarkable capacity to persist with recurring fires. Conservationists are concerned with the relationships between chaparral and wildfires and severe erosion. Stockmen and sportsmen want to know the browse values of new growth following fires and the possibility of replacing chaparral, in suitable areas, with grass or other more palatable plants. Foresters are interested in areas in which chaparral interferes with timber production. Water users are seeking ways to manage chaparral and replace it with other vegetation that requires less water but will provide equally satisfactory control of erosion and floods. Some, but not all, fire-control officers see controlled fire as an effective means of reducing wildfire hazards. Last, but not least important, are people who live in or near chaparral and fear for their lives and property when wildfires are raging in strong winds. To further all these interests requires some understanding of the interrelationships between fire and chaparral.

II. Kinds of Chaparral

The chaparral brushlands may be subdivided into three broad categories: (1) climax, or true, chaparral; (2) woodland–grass chaparral; and (3) forest chaparral (Biswell, 1957). The latter two are considered successional in development.

A. CLIMAX, OR TRUE, CHAPARRAL

This chaparral may be characterized as follows: (1) Most areas in which it is found have been predominantly chaparral as far back as the records go (Figs. 1, 2, and 3). (2) It is largely a fire induced type. (3) The soils on which it grows are relatively shallow and lack surface fertility and water-holding capacity, but the underlying rocks may be

Fig. 1. Climax chamise chaparral in northern California.

Fig. 2. Sharp boundary between chamise chaparral (right) and annual grassland (left) on Mt. Diablo. Photograph by W. S. Cooper taken in 1914.

deeply fractured, thus permitting deep penetration of roots. (4) Landscapes are moderate to extremely steep and rugged and show severe erosion patterns. (5) Little herbaceous vegetation grows in the under-

Fig. 3. Photograph taken in August 1954 (same area as shown in Fig. 2), showing remarkable stability of some of the boundaries of chamise chaparral. Three fires burned in this area between 1914 and 1954.

story (Zinke, 1961). Exceptions occur: Some areas of chaparral have fairly deep soils and a topography that permits mechanical clearing of shrubs for agricultural production or conversion to grasslands in range-improvement programs (Love and Jones, 1952; Bentley, 1967).

A few of the more common shrubs in climax chaparral are the following:

SPROUTING SPECIES

Chamise
 (*Adenostema fasciculatum*)
Scrub oak
 (*Quercus dumosa*)
Interior liveoak
 (*Quercus wislizenii* var. *frutescens*)
Leather oak
 (*Quercus durata*)
Canyon oak
 (*Quercus chrysolepis*)
Eastwood manzanita
 (*Arctostaphylos glandulosa*)
Woolyleaf manzanita
 (*Arctostaphylos tomentosa*)
Chaparral pea
 (*Pickeringia montana*)
California laurel
 (*Umbellularia californica*)

Yerba santa
 (*Eriodictyon californicus*)
Chaparral whitethorn
 (*Ceanothus leucodermis*)
Bear bush
 (*Garrya fremontii*)
Coffee berry
 (*Rhamnus californica*)
Red berry
 (*Rhamnus crocea*)
Evergreen cherry
 (*Prunus ilicifolia*)
Black sage
 (*Salvia mellifera*)
Christmas berry or toyon
 (*Heteromeles arbutifolia*)

NONSPROUTING SPECIES

Wedgeleaf ceanothus
 (*Ceanothus cuneatus*)
Hoaryleaf ceanothus
 (*Ceanothus crassifolius*)
Wavyleaf ceanothus
 (*Ceanothus foliosus*)
Gregg ceanothus
 (*Ceanothus greggii*)
Whiteleaf manzanita
 (*Arctostaphylos viscida*)

Common manzanita
 (*Arctostaphylos manzanita*)
Stanford manzanita
 (*Arctostaphylos stanfordiana*)
Littleberry manzanita
 (*Arctostaphylos sensitiva*)
Bigberry manzanita
 (*Arctostaphylos glauca*)

A single shrub species may sometimes form an almost pure chaparral (Hedrick, 1951; Vogl and Schorr, 1972). More frequently, however, the various species grow in an almost infinite number of combinations, with sprouting and nonsprouting species intermixed. Variations in cover are influenced by such factors as soil type and exposure, altitude, age since the last burn, and frequency of burning. The single most abundant and widespread shrub is chamise. In any community where this species makes up as much as 20% of the cover, the brushland is known as chamise chaparral (Horton and Kraebel, 1955). This type covers about 2.95 million hectares of the total of 4.04 million hectares of climax chaparral in California. Other shrubs most frequently associated with chamise are species of *Arctostaphylos, Ceanothus,* and *Quercus.*

Cooper (1922) studied the distribution of climax chaparral species throughout California. He recorded the important and widespread species in 87 different localities, with the percentages of occurrence as follows: chamise, 86; manzanitas, 57; Christmas berry, 30; wedgeleaf ceanothus, 29; scrub oak and leather oak, 26; western mountain mahogany, 22; interior liveoak, 13; coffee berry, 12; and canyon or maul oak, 12.

Cooper made special mention of distribution of manzanitas. He reported that woolyleaf manzanita was by far the most important species, ranging over the whole region of climax chaparral. Bigberry manzanita is abundant in the southern half of the state, and common manzanita in the northern half; several others are prominent locally. Manzanitas often occupy the more moist, north-facing exposures, while chamise occupies the drier south-facing exposures. At higher altitudes, manzanitas may replace chamise (Wilson and Vogl, 1965). On south-facing slopes of coastal exposures, common shrubs are California sagebrush (*Artemisia californica*), black sage (*Salvia mellifera*), white sage (*Salvia apiana*), California buckwheat (*Eriogonum fasciculatum*), and deerweed (*Lotus scoparius*).

Climax chaparral is composed chiefly of evergreen plants, but a few deciduous species occur—for example, poison oak (*Rhus diversiloba*), which is fully deciduous, and western mountain mahogany (*Cercocarpus betuloides*) which may be more than half deciduous. Sampson (1944), in discussing plant successions on burned chaparral lands in northern California, referred to deciduous species with thin leaves as "soft brush," as opposed to "hard" chaparral for drought-tolerant, evergreen species.

B. WOODLAND–GRASS CHAPARRAL

Woodland–grass chaparral lies mainly between the forested areas at higher elevations and the lower grasslands. In general, this type was formerly maintained as grassland by frequent fires, which favored grasses at the expense of shrubs. As a result of fire suppression, however, shrubs gradually invaded and increased in density to form a chaparral vegetation cover, especially along the western slope of the Sierra Nevada (Biswell and Street, 1948) (Fig. 4).

The principal features of woodland–grass chaparral may be sum-

Fig. 4. Woodland–grass chaparral where wedgeleaf ceanothus is spreading into annual grassland.

marized as follows: the soils are relatively deep as compared with those under climax chaparral, the relief is more gentle than that of climax chaparral, and considerable herbaceous vegetation is found in open stands of shrubs. Because the woodland–grass areas are used primarily for grazing of domestic livestock, invasions of many unpalatable shrubs have been of concern to stockmen.

Among the more common shrubs in woodland–grass chaparral are the following:

SPROUTING SPECIES

Interior liveoak (scrub form)
  (*Quercus wislizenii*)
Western mountain mahogany
  (*Cerocarpus betuloides*)
Coyote bush
  (*Baccharis pilularis*)

Leather bush
  (*Fremontia californica*)
Poison oak
  (*Rhus diversiloba*)

NONSPROUTING SPECIES

Wedgeleaf ceanothus
  (*Ceanothus cuneatus*)
Stanford manzanita
  (*Arctostaphylos stanfordiana*)

Whiteleaf manzanita
  (*Arctostaphylos viscida*)
Mariposa manzanita
  (*Arctostaphylos mariposa*)

C. Forest Chaparral

At higher altitudes, chaparral may occur on large acreages as a stage of secondary succession in coniferous forests. Following disturbances that open up the forest stand, such as logging and wildfires, seedlings of both sprouting and nonsprouting shrubs may appear in great abundance from germination of viable seeds stored in the duff and soil (Biswell and Schultz, 1958a). In most places, shrubs grow in admixture with trees (Fig. 5). Opening up a forest stand enables the shrubs to increase in density and occupy the space vacated by trees. The topography varies from gentle to steep. Although in such areas the general objective in forest management is to grow crops of timber, these areas are also important for wildlife, water, scenery, and outdoor recreation.

Some of the more common chaparral shrubs in forests include the following:

SPROUTING SPECIES

Deerbrush
  (*Ceanothus integerrimus*)
Littleleaf ceanothus
  (*Ceanothus parvifolius*)
Mountain whitethorn
  (*Ceanothus cordulatus*)
Canyon oak
  (*Quercus chrysolepis*)

Greenleaf manzanita
  (*Arctostaphylos patula*)
Snowbrush
  (*Ceanothus velutinus*)
Chinquapin
  (*Castanopsis sempervirens*)
Huckleberry oak
  (*Quercus vaccinifolia*)

NONSPROUTING SPECIES

Wedgeleaf ceanothus
  (*Ceanothus cuneatus*)
Whiteleaf manzanita
  (*Arctostaphylos viscida*)

Pinemat manzanita
  (*Arctostaphylos nevadensis*)

Many deciduous shrub species occur in forested areas, including snowberry (*Symphoricarpos mollis*), service berry (*Amelanchier alnifolia*), Sierra currant (*Ribes nevadensis*), Sierra gooseberry (*Ribes roezlii*), California rose (*Rosa californica*), Sierra plum (*Prunus subcordata*), western choke-cherry (*Prunus demissa*), and bitter cherry (*Prunus emarginata*). Mountain misery (*Chamaebatia foliolosa*) forms dense stands in many areas of the ponderosa pine and mixed-conifer forests.

Of the forest chaparral species listed above, only three are common to climax chaparral: wedgeleaf ceanothus, whiteleaf manzanita, and canyon oak. Deerbrush may appear in the higher and moister portions of climax chaparral as well as in the ponderosa pine and lower mixed-conifer forests.

Fig. 5. Forest chaparral where ponderosa pine and whiteleaf manzanita grew together after a logging operation.

Distribution of the various shrubs in forested areas is related to altitude. Among the ceanothus species, wedgeleaf occurs at lower altitudes in both woodland–grass and climax chaparral and in the lower forest zones. At increasing elevations we find deerbrush, followed by mountain whitethorn and littleleaf ceanothus.

III. Occurrence of Fires in Chaparral

Fires have burned through chaparral and other vegetation of California for at least 100,000 years (Jepson, 1925). Features of the Mediterranean-like climate make the areas within its range very susceptible to fire occurrence and spread. Nearly rainless summers dry out the vegetation and soils; daytime temperatures are high and humidity low during the summer, and high winds blow from the interior deserts and valleys to the Pacific Ocean during the dry months.

Early fires undoubtedly were caused mainly by lightning, which is still one of the primary causes of forest fires. Thus fire has always been a natural feature of the environment. In climax chaparral, for example, frequent fires kept the plant cover at a young stage of development, and because of this periodic fuel consumption, fires did not burn with the intensity that is evident today. The frequency of lightning fires is indicated by the fact that in 1972, 1750 lightning fires occurred in the national forests of California.

Below this forest chaparral zone, in the climax and woodland–grass chaparral, the number of lightning-set fires is surpassed by those started either accidentally or intentionally by man.

A. BURNING BY INDIANS

Reynolds (1951) assumed that lightning alone could not have accounted for all the fires which occurred in the central Sierra Nevada. He therefore concluded that frequent, widespread, and knowledgeable burning was performed by the Indians and that this cultural practice probably extended the range of those plants on which the Indians' economy depended. The Indians set burns to prepare feeding grounds for game and to make hunting easier; to facilitate collection of seeds, bulbs, berries, and fiber plants; and to increase yield of useful plants, for example, manzanitas, the berries of which were used to make cider (Jepson, 1921; Sampson, 1944; Sauer, 1950; Reynolds, 1951; Stewart, 1956). The Indians also burned in forested areas to keep them open for easier travel and to reduce the danger of lightning fires in late summer (Gianella, 1972).

It seems reasonable that on deer winter ranges in the upper portions of the woodland–grass areas, Indians burned the grass to keep wedgeleaf ceanothus from being destroyed by intense summer fires, because they recognized that this chaparral shrub was a valuable winter browse. The fires could have been set in early summer at a time when they would creep through and burn the dry grasses without killing any shrubs except small seedlings.

That Indians frequently burned forested areas was borne out by various explorers and naturalists who observed this practice (Miller, 1887). For example, Galen Clark, for many years the guardian of Yosemite, and Dr. L. H. Bunnell, a member of the 1851 Yosemite discovery party, saw and described Indian burning, and Joaquin Miller wrote in 1887:

> In the spring . . . the old squaws began to look for the little dry spots of headland or sunny valley, and as fast as dry spots appeared, they would be burned. In this way the fire was always the servant, never the master. . . . By this means, the Indians always kept their forests open, pure and fruitful, and conflagrations were unknown.

The subject of burning by Indians is largely one of academic interest, of course, and there are some who believe that this source of fire was an insignificant factor in forest development. Burcham (1960), for instance, maintained that Indians burned only in grasslands, and that the extent and frequency of burning were limited. He suggested that the failure of burning to become a widespread practice was due to the small populations of Indians and their lack of communication. These may not be valid conclusions, however, because a single fire could carry for miles through the dry vegetation in late summer, perhaps from river to river and into the mountains above, when no one was there to put it out.

B. Burning by Early Settlers

When Europeans arrived in California and took up mining, lumbering, and grazing, fires in forested areas became intense and damaging (Biswell, 1967).

Miners used fire to remove slash after cutting timber for mining props and fuel, and to clear the landscape to facilitate mining activities. These slash fires must have been intense, but fortunately they did not kill all trees, many of which remained to reseed the landscape. Today some of the best pine stands in California are found in areas where miners did heavy burning.

Early-day lumbermen also used fire destructively. They had little or no concept of sustained-yield forestry; their idea was to cut and get out. Very often, therefore, they cut all the salable trees on an area and then burned to get rid of the slash. This practice resulted in high-intensity fires that killed most of the trees that remained after logging. Many of the heavily cut and burned areas turned to chaparral (Show and Kotok, 1924).

Sheepmen were a third group that burned annually and killed many trees in their efforts to open up forest stands and to improve understory grasses and other conditions for grazing.

The destructive fires of the early settlers caused much concern to thoughtful observers and no doubt helped stimulate the conservation movement which followed around the beginning of the twentieth century. In 1872, legislation was passed to prevent setting of fires, but it was largely ineffective. In about 1905 the U.S. Forest Service adopted a firm policy of virtual fire exclusion on its lands, and a similar policy, finally adopted in 1924 by the California Division of Forestry, covered private lands (Clar, 1959). As a result of these policies, fuels have increased to a point at which any unwanted fire may become very intense and difficult to control. Costs of controlling wildfires have increased steadily over the years and are now extremely high, as are costs of repairing damage after such fires.

C. PRESCRIBED OR CONTROLLED BURNING

In 1945, the California legislature authorized the state Division of Forestry to issue control-burning permits for purposes of brush–range improvement. By 1971, ranchers and sportsmen had burned a net total of 775,402 ha, of which 305,480 ha have been reburned (California Division of Forestry, 1971). In addition, 77,133 ha were burned by escaped fires. Other burning has been practiced outside of the fire season, when permits were not necessary. The U.S. Forest Service and Bureau of Land Management have also practiced prescribed burning on lands under their jurisdiction. More recently the National Park Service initiated a program of burning in California to restore fire as a natural process and an important ecological factor in maintenance of natural landscapes (Kilgore, 1972; Kilgore and Briggs, 1972).

In selected places and under proper management and control, prescribed burning has been a useful tool in chaparral to improve conditions for livestock grazing and wildlife habitat (Figs. 6, 7, 8, and 9). There have also been associated benefits, such as more effective and less costly

Fig. 6. An area of chamise chaparral that was burned by light fires to create edge and new browse for deer—an ideal habitat.

Fig. 7. A dense stand of woodland–grass chaparral where the shrubs gradually increased in abundance as a result of fire exclusion.

Fig. 8. Same areas as that shown in Fig. 7 after it was control burned.

Fig. 9. Same area as in Fig. 7 after 3 control burns.

Fig. 10. Prescribed burning upslope in chamise chaparral. Fires are set in the most flammable fuels in such a way that they will burn uphill and then go out, because the fuel on the north slope is too moist to burn. This is one form of strip burning.

wildfire control, increased water yields, and improved access for hunting and recreation (Biswell and Schultz, 1958b).

In prescribed burning, an important consideration is the fuel that carries the fire. In climax chaparral, the entire plant cover burns, although the fuels in some spots are more flammable than those in others, and may burn under moister conditions (Fig. 10). In woodland–grass chaparral, dry, herbaceous vegetation is the chief fuel that carries fire from one area of brush to another (Fig. 11). In forest chaparral, the fuel is varied. If the mixture is manzanita in ponderosa pine, the pine needles serve as fuel to carry the fire (Fig. 12). Enough needles fall every year to carry a surface fire.

IV. Adaptations of Chaparral to Fire

Broad-sclerophyll brushlands throughout the world have burned so often yet survive and recover so quickly that plant geographers regard them as fire-induced or fire-adapted types (Jepson, 1930; Shantz, 1947; Naveh, 1967). Chaparral is remarkably well adapted to recurring fires

Fig. 11. Prescribed burning in woodland–grass chaparral. The dry grass carries the fire from one clump of chaparral to the next.

for two reasons: many of the shrubs stump-sprout after fire, some reproducing further by layering or from underground stems; and seeds are produced at an early age. These seeds may lie dormant in the duff and soil for extremely long periods of time, and they have high resistance to fire. In addition to these adaptations, the brushlands have evolved toward characteristics that make them highly flammable and dependent on recurring fires for restoration and for optimum growth and health. This is a reciprocal relationship, because the frequent fires depend on the fuels that feed the flames.

A. STUMP SPROUTS, LAYERING, AND UNDERGROUND STEMS

Most of the important species of chaparral stump-sprout vigorously with rapid recovery and growth. Such species probably have an advantage over those that reproduce only from seed, because the sprouts rapidly outgrow the seedlings. Sprouting may be considered an adaptation to recurring fires.

Fig. 12. Prescribed burning in ponderosa pine to reduce fire hazards and chaparral. Pine needles carry the surface fire.

Chamise does not sprout as vigorously as some other shrubs. Often about a fourth of the plants fail to sprout after fire. The older plants are more susceptible to fire kill than are younger, middle-aged plants. For every plant that fails to sprout after fire, however, many seedlings germinate to take its place. This might be a more beneficial adaptation for fire survival than complete sprouting, since it serves to weed out the old plants and restore the stand to young, vigorous growth.

In addition to sprouting as a means of regeneration, some shrubs increase and spread in cover density by layering—i.e., from rooting of branches—and others by means of underground stems. Layering is particularly common among shrubs at higher elevations where snow may force the stems close to the ground to bring them into contact with soil. A few scattered seedlings after fire can grow in a few years to become broad thickets of dense chaparral. This happens often in forested areas where the shrubs have long been shaded out and new plants arise from seed after disturbance of a forest. Among the important shrubs that layer readily are mountain whitethorn, greenleaf manzanita, snowbrush, huckleberry oak, and pinemat manzanita. All manzanitas probably reproduce by layering if the stems somehow touch the soil. However, most manzanitas at lower elevations seldom layer because of their upright growth and lack of soil contact.

Yerba santa commonly reproduces from underground stems. It may appear in abundance from seed after fire. Where the soil is fertile and abundant moisture is present beneath ash beds, the underground stems may grow out as much as 2.5 m in one summer, giving rise to new plants every 20–25 cm. Normally, however, the plants do not grow this rapidly. In dense chaparral, plants of this species survive for only about 25 years, but it is not known how long the seed will remain viable.

B. Seed Production, Dormancy, and Resistance to Fire

Most chaparral species begin to produce seed within 3–5 years after a fire, and usually the brushlands will not burn more often than this unless they are reseeded to flammable grasses, such as domestic ryegrass. Some people regard heavy seed production by nonsprouting shrubs as indicative of true fire-type species.

Another adaptation of chaparral species to fire is the ability of seeds to remain dormant and viable for long periods in the duff and soil (Quick, 135, 1956, 1959, 1961, 1962; Gratkowski, 1962). Zavitkovski and Newton (1968) inferred that seeds of snowbrush might remain viable in forest litter up to 575 years. Thus, in forested areas where chaparral shrubs have long since disappeared as a result of shading or old age, they may reappear after a fire, which activates the dormant seed.

Another adaptation to fire is the production of heat-resistant seeds (Quick, 1959; Sampson, 1944; Stone and Juhren, 1951). Chaparral seedlings normally spring up all over a burned area, sometimes as many as 7 million per hectare. In chamise chaparral, there have been as many as 3000 seedlings on one square meter of soil. Tolerance to heat has been studied by subjecting seeds to boiling water and to different oven temperatures for varying periods of time. In one study, seeds of deerbrush were boiled from 1 to 20 min, and those of mountain whitehorn from 5 to 30 min. After these treatments, the seeds were immersed in cold water, planted in autoclaved sand, stratified to obviate any embryo dormancy, and germinated in a greenhouse (Quick, 1959). Twelve percent of the deerbrush seeds germinated after being boiled for 20 min. Twenty-five percent of the mountain whitehorn germinated after being boiled for 25 min, but none germinated after being boiled for 30 min.

Tests of resistance of chaparral seeds to dry oven heat for periods of 5 min show that many species will germinate after being exposed to temperatures of 127°–138°C, and chamise and whitethorn chaparral will withstand temperatures of 138°–149°C (Sampson, 1944). The higher temperatures apparently favor germination of some species, and heat from fires is the main factor in inducing that germination.

Seeds in heavy duff may also be destroyed by fire. Where debris was piled and burned in a forested area, no new seedlings came in the center of the burned pile, but numerous seedlings of deerbrush grew on the edges where the fire was light and not all the duff had been removed down to mineral soil. On the basis of these observations, many chaparral species growing where the duff is relatively shallow, as under chamise, and particularly where the seeds have become covered with soil, would apparently have few seeds destroyed by fire since the soil temperature at the 2 cm level might not exceed 65°–95°C.

The seeds of some species are very dependent on fire for germination. The heat and action of fire scarify the seed coat, enabling the seed to adsorb moisture. Seeds of deerbrush and mountain whitethorn, for example, depend on fire for germination, whereas those of chamise and western mountain mahogany do not. Germination of chamise after fire is greatly increased, but this is thought to result from removal of the old vegetation and destruction of phytotoxic substances. However, the increased germination of chamise seeds after fire may be the direct effect of fire in combination with other improved conditions.

C. FLAMMABILITY OF FUELS

It has been suggested that, through the evolutionary process, plant communities subjected to frequent fires over thousands of years have developed features that make them highly flammable. Such fire-dependent plant communities burn more readily than those less dependent on recurring fires, because natural selection has favored development of characteristics that make them more flammable and at the same time more fire-tolerant (Mutch, 1970). An excellent example of this is chamise, the most abundant and widespread shrub in climax chaparral. It grows rapidly after fire and in about 15 years begins to develop dead branches in a peculiar fashion. A lateral twig, 15–30 cm back from the twig tip, continues to grow, and the old twig branch dies, producing fine dry fuels. Furthermore, half or more of the chamise branches and stems are less than 1 cm in diameter. The fuels have a high surface area-to-volume ratio (Countryman and Philpot, 1970), fuel-bed porosity is high, and the leaves are resinous. All these features contribute to high flammability, and an interesting interaction results: the chamise chaparral depends on recurring fires of 15–20 years' frequency for optimum health and stability, and such fires depend on the chamise chaparral that produces the fuels to carry the fires.

V. Plant Successions after Fire in Climax Chaparral

Plant succession after fire is affected by such factors as growth habits and normal life spans of plants, phytotoxic substances in soils and plants, competition, browsing by animals, and time and frequency of fires.

A. RECOVERY OF CLIMAX CHAPARRAL

When fire burns through climax chaparral it consumes much of the plant cover and initiates new successions (Fig. 13). These have been

Fig. 13. A dense cover of new shrubs near the end of the first growing season after a control burn in woodland–grass chaparral. Counts showed nearly 600,000 seedlings of wedgeleaf ceanothus and 100,000 seedlings of yerba santa per hr. Competition from a light seeding of annual ryegrass would have thinned the chaparral seedlings.

Fig. 14. A dense stand of wedgeleaf ceanothus after the area was burned by a wildfire.

studied extensively by Cooper (1922), Sampson (1944), Miller (1947), Hedrick (1951), Biswell *et al.* (1952), Stone and Juhren (1951), Horton and Kraebel (1955), Sweeney (1956), Gibbens and Schultz (1963), Wilson and Vogl (1965), Hanes and Jones (1967), Hanes (1971), Vogl and Schorr (1972), and others. The various species of shrubs and trees quickly regenerate and usually rapidly regain dominance on a burned area (Fig. 14).

The particular cover that will develop after a fire can be predicted reasonably well from knowledge of the plant growth at the time of burning. A few "new" species may appear, however, as a result of germination of seeds long stored in the litter and soil, the plants themselves having died out long ago. Examples include wavyleaf ceanothus, wedgeleaf ceanothus and, on southern exposures, deerweed which disappear in about 20–25, 50, and 3–5 years, respectively, but are quickly reestablished from seeds in the litter when exposed areas are reburned.

When fire burns through vigorous sprouting species, such as eastwood manzanita and scrub oak, in well-stocked stands, those species may be expected to reappear in about the same numbers as before, and recover finally to their former densities. If the cover should be one of dense, nonsprouting species, such as wedgeleaf ceanothus, seedlings will no doubt appear in dense stands from seeds stored in the duff and soil.

In areas where sprouting and nonsprouting species are thoroughly intermixed, somewhat similar mixtures may be found after fire. Generally, however, successions are toward strongly sprouting species with recurring fires, because the sprouts grow rapidly and compete more effectively with the seedlings of nonsprouting species.

B. GROWTH AND SURVIVAL OF SPROUTS AND SEEDLINGS

Shrubs that stump-sprout after fire have an advantage over those that reproduce only from seed, because the sprouting plants have well-developed roots and grow much faster than do seedlings. In the north coast range, stump sprouts attained heights of 50–100 cm in the first full growing season after a late summer fire (Biswell *et al.*, 1952). Maximum heights of seedlings in the same general area at the end of the first full growing season were only 7–15 cm. Another study was made in the foothills of the central Sierra Nevada on July 7, after a fire in late September of the preceding year. Sprouting species averaged 60 cm in height while seedlings of wedgeleaf ceanothus, which does not sprout, averaged only 6.35 cm (Biswell and Gilman 1961). In southern California, measurements by Horton and Kraebel (1955) were similar; chamise sprouts grew 48 cm the first growing season after fire, while seedlings averaged only 7 cm. Sprouts usually grow rapidly for a few years after fire and then slow to a gradual increase for perhaps 20–25 years after which the shrubs reach their maximum growth. The seedlings continue to grow slowly for many years in competition with the taller growing sprouts.

The start of sprout growth after fire has been observed to vary with the season of burn. After chamise was burned in April and May, sprouting began $3\frac{1}{2}$ and $2\frac{1}{2}$ weeks later, respectively. By the end of the summer the sprouts on the April burn were as tall as those that had grown all spring and summer from a burn of the previous fall, but those on the May-burned areas were only one-half as tall. On another area burned in mid-July, measurements a month later showed the sprouts of interior liveoak to be 20 cm tall, those of red berry, 15 cm, and poison oak, 18 cm. Two months later these same sprouts had grown to 40, 25, and 33 cm, respectively. On a nearby area burned in mid-August there were no sprouts one month later, because sprouts usually do not appear on late summer and fall burns until the next spring.

Some shrubs stump-sprout almost 100%; others, less vigorously. Nearly every plant of eastwood manzanita and scrub oak sprouts after fire, but only 70–75% of chamise plants produce sprouts.

Chamise normally grows in mixed stands, very often with nonsprouting species of ceanothus. The space made available by chamise plants failing to sprout after fire is filled in with seedlings of both chamise and the associated species—possibly wedgeleaf ceanothus in northern California and hoaryleaf ceanothus in southern California.

Sampson (1944) found that sprouts of chamise weighed 1753 kg per ha at the end of the first season of growth, and increased a little over 1100 kg per ha per year to a weight of 12,786 kg at the end of the eighth year. The old stand had a cover of 31,539 kg of dry-weight materials per ha. Growth of the mixed chaparral was similar, but always a little less. At the end of 8 years it had produced 11,885 kg of dry-weight material per ha and the old stand had 31,078 kg. Thus, if the shrubs continue to produce about 1100 kg per ha per year, about 20–25 years will be required for a burned-over area to recover its preburn weight completely.

C. NORMAL LIFE SPANS

Very little is known about the normal life spans of most chaparral species. Hedrick (1951) provided some information in a study of chamise brushland that had burned 93 years earlier. He saw very little evidence that the chamise was either dying out or being replaced by other vegetation. Many old plants had become defoliated and the stems had died, but these were replaced by new crown sprouts (Fig. 15). First observations indicated that the chamise cover was made up of a number of uneven-aged plants. However, critical examination of the root crowns revealed fire scars, indicating that the plants had survived the fire of 93 years ago. There was also an additional set of plants that had appeared as seedlings after the fire. Hedrick sampled the amount of debris on the ground beneath the shrubs and concluded that it reached its maximum amount in about 25 years. This agrees with Sampson's (1944) measurements about the time required for chamise chaparral to reach its maximum weight. The year of the fire was determined by counting growth rings on healthy stems of scrub oak scattered through the chamise stand. This species and some of the other sprouting trees and shrubs may live for centuries if burns are only occasional.

Fig. 15. A 93-year-old stand of chamise in which nearly all the wedgeleaf ceanothus had died of old age. The soil is nearly bare and erosion is active.

Many dead plants of wedgeleaf ceanothus in various stages of decay were found in the 93-year-old brushland. Observations there and elsewhere have shown that plants of this species begin to die in about 50 years. This brushland was mainly one of chamise with wedgeleaf ceanothus, and over the years the latter declined, leaving the remaining stand almost entirely in chamise. When such brushlands are again burned, however, wedgeleaf ceanothus seedlings will appear in abundance and the brushland will be restored to about the same floristic composition as that following the fire of 93 years ago. In the north coast ranges, yerba santa and wavyleaf ceanothus may appear in abundance after fire. In dense chaparral these disappear from the stand in 20–25 years, when a chamise brushland reaches full maturity. The seeds remain viable so that seedlings appear in abundance after the next fire. Thus, for some components of the chaparral type, recurrent fires are responsible for maintenance of the species in the vegetative cover. Without this environmental input, some species die out in a relatively short time, but are renewed in the cover by the next fire.

D. INCREASE AND DECLINE OF HERBACEOUS VEGETATION

Herbaceous plants usually appear in some abundance in areas of climax chaparral following fires that consume the shrub cover. They reach their peak in 1–5 years, then decline and give way to the growth and reestablishment of shrubs. Finally, in mature chaparral, herbaceous plants are sparse and include only those growing around rock outcroppings or in other small openings where the chaparral fails to reach maximum density. A few perennial herbs persist as bulbs. The presence, abundance, absence, and successions of herbaceous plants on burned and unburned areas of chaparral have been studied by Sampson (1944), Hedrick (1951), Sweeney (1956), Muller et al. (1968), and Vogl and Schorr (1972).

Some of the species that occur on burned areas may not be found in nearby, unburned chaparral. They probably came from seeds lying dormant in the litter and soil since the last fire, which may have been some 40 or 50 years earlier. Plants that appear on burned areas but are rare elsewhere have been called "burn species" and "fire followers." A fire in chaparral does four things favorable to the herbaceous species: (1) it consumes the shrub cover; (2) it destroys phytotoxic materials produced by the shrubs; (3) it prepares a good seedbed high in nutrients and moisture; and (4) it reduces for a while the competition from shrubs.

Some plants appear for only one year after a fire and then disappear. Occasional fires are therefore essential for their survival (Sweeney, 1956). An excellent example is whispering bells (Emmenanthe peduliflora), a broad-leaved annual widely distributed over burned areas of climax chaparral. Others include horned snapdragon (Antirrhinum cornutum), a phacelia (Phacelia suaveolens), and a monkey flower (Mimulus rattanii) (Sweeney, 1956). Other herbaceous species, such as tuberous skullcap (Scutellaria tuberosa) and a navelseed (Navarretia mellita), appear the first year after fire and increase in abundance for 2–5 years before declining.

Few annual grasses appear on burned areas the first year after fire, except around rocky outcrops or other open spots where they existed previously. The seeds of annual grasses do not lie dormant for long periods as do the seeds of some of the broad-leaved annuals; furthermore, the grass seeds are more easily destroyed by fire. Although annual grasses may be sparse the first year, they often increase rapidly for the next 2–3 years. In northern California the most common grasses on burned chaparral lands are common foxtail (Festuca megalura), red brome (Bromus rubens), silver hairgrass (Aira carophyllea), and nitgrass (Gastridium ventricosum), all characteristic of areas low in soil fertility and moisture.

Along with the annuals, the first season's growth also includes several bulb-forming perennials. These species persist in the mature chaparral, but on a very low level. With the occurrence of fire, they undergo vigorous growth and flowering. Some examples of these are purplehead brodiaea (*Brodiaea pulchella*), soap plant (*Chlorgalum pomeridianum*), death camas (*Zygadenus fremontii*), and several species of mariposa lily (*Calochortus* spp.) (Muller *et al.*, 1968).

The herbaceous cover reaches peak density within 1–4 or 5 years after fire. Working in dense eastwood manzanita chaparral in southern California, Vogl and Schorr (1972) found the peak of herbaceous vegetation to occur the first year after fire. They found a total of 26 annual and perennial species on a burned-over area. From a cover of 15.5% on the 1-year-old burn the species decreased to 7.1% on the 2-year-old burn to only 0.2% in an unburned chaparral where the dominant species was wild cucumber (*Mara macrocarpa*). In a somewhat similar study in northern California, maximum density of herbaceous plants was reached in 2 years after the fire and was equal to 17 times that of the prefire density (Sampson, 1944). Conspicuous downward trends in the amount and density of herbaceous vegetation occurred in the third, fourth, and fifth years after burning, and approached in sparseness those of the preburn area.

E. Growth Inhibitors

Considerable interest has been shown in the general absence of herbaceous vegetation in mature chaparral and the bare zone that often exists between chaparral and adjacent grasslands. Possible explanations for this situation include such factors as lack of top-growth space; competition for light, moisture, and soil nutrients; rodent, rabbit, and bird predation and activity; and presence of phytotoxins or allelopathic substances (Sampson, 1944; Went *et al.*, 1952; Landers, 1962; Muller *et al.*, 1964, 1968; Muller and Muller, 1964; Muller, 1965, 1966; Muller and Del Moral, 1971; Naveh, 1960; McPherson and Muller, 1967, 1969; Bartholomew, 1970, 1971). Several factors and interactions among them probably are involved.

Volatile or water-soluble phytotoxic substances have been found in a number of chaparral species, such as sages, manzanitas, and chamise. In studies of chamise, water-soluble toxic substances have been found in the leaves, flowers, litter, and roots (McPherson and Muller, 1969). These substances are produced as a normal product of metabolism, and accumulate in the shrubs during the dry months. With the autumn rains, the toxins are carried to the soil, where they concentrate and subse-

quently inhibit seed germination and growth of herbaceous species. Little
new growth takes place until a fire destroys the toxins. Late summer
fires destroy the toxins while they are still in or on the shrubs. Following
a fire, there is abundant growth of herbaceous vegetation and regeneration of shrubs until the shrub toxins again reach levels capable of
preventing seed germination and growth of understory herbaceous vegetation. Hence, certain facets of the chaparral fire cycle—the sudden abundance of herbaceous vegetation following fire and the disappearance
of this vegetation in later years—may be partly linked to these toxic
substances.

F. EFFECTS OF SEEDED GRASSES ON BRUSH SEEDLING SURVIVAL

Burned chaparral lands often are seeded to grasses. The grass estab-
lishes a protective cover which conserves soil and water and prevents
destructive floods. Annual ryegrass (*Lolium multiflorum*) is used most
often because seeding is easy, seeds are relatively cheap, germination
percentage is high, early growth is rapid, and the grass tillers freely,
soon providing dense cover.

In the summer of 1970, 241,086 ha of chaparral were burned by wildfires, mainly in southern California. Of this, 169,520 ha were seeded
to annual ryegrass. Stands are not always uniform because the seeds
may be blown from the ridges and drift, or they may be washed away
in the first autumn rains. Furthermore, this species usually disappears
from the cover in 3–5 years.

Following controlled fires in woodland–grass chaparral, several annual
and perennial species may be seeded, but for a different purpose. This
seeding is done mainly for forage, for competition to reduce the number
of brush seedlings, and to provide fuel for a reburn within 2–3 years.

In competition, the advantage lies with those plants that are first
present. Grasses and other herbaceous species on burned-over brushlands
emerge in the autumn with the first effective rains, but the brush seedlings do not emerge until the following spring, mostly in February and
March. By that time, the grasses have already developed considerable
cover density and root depth so that they will undergo rapid tillering
and growth during the next few weeks of favorable growing conditions.
When the annual ryegrass becomes dry in about mid-May, its fine, fibrous
roots extend downward approximately 1 m. They completely occupy
the soil and dry it to such an extent that no moisture remains for the
brush seedlings that have roots no deeper than 0.3 m in the grass cover.
Thus, the brush seedlings die from lack of water. Where the brush seedlings grow without competition from grasses, their roots may attain a

depth of 1 m within a period of about 3 months, thus keeping in contact with soil moisture (Schultz and Biswell, 1952; Schultz *et al.*, 1955).

Hardly any brush seedlings survive the summer drought period where the annual ryegrass reaches a density of 39% cover at maturity (Schultz *et al.*, 1955). The same situation prevails where resident annuals reach 65% cover density. This shows that annual ryegrass is more competitive than the resident annuals and causes greater mortality of brush seedlings. Resident annuals, unless seeded, seldom grow in stands with as much as 65% cover density. It appears that a dense and uniform cover of annual ryegrass can virtually eliminate the seedlings on a burned-over brushland. This may be favorable or unfavorable depending on whether the objective was conversion from brushland to grassland or maintenance of a dense stand of chaparral on steep slopes as watershed cover.

G. Effects of Reburns

A chaparral cover can be greatly altered by reburns one to several years after the original fire. The general overall effect is a large kill of sprouting plants of some species and almost complete removal of seedlings. Most chaparral seedlings appear in the early spring following a summer or autumn burn; a few appear the second year, and almost none the third year. Those that appear the second year are usually weak, and survival is low.

In northern California, a reburn 1 year after the original fire killed 43% of the sprouted chamise plants that had been browsed by deer. In this case the first fire had killed 44%. The reburn was made in a seeded area where grass carried the fire. Thus, of 391 chamise plants on the plots under study, only 13% survived the two fires (Hedrick, 1951). On the same plots, the reburn killed 99% of 707 chamise seedlings. After the reburn, however, 156 new seedlings appeared the next spring, but most of these died later from grass competition.

In another set of plots, a fire had killed 23% of an original 296 chamise plants. A reburn in seeded grass 2 years later killed another 19%. On the same plots, 97.5% of 860 2-year-old chamise seedlings were killed by the reburn. On a third set of plots, a reburn 3 years later killed 24% of the chamise plants, 100% of the seedlings of wedgeleaf ceanothus, and 49% of the sprouts of yerba santa. From these data it may be concluded that reburns 1–3 years after a fire result in high mortality of sprouted chamise shrubs and a very high to complete kill of seedlings (Hedrick, 1951).

When shrubs are heavily browsed and weakened by deer or livestock, mortality from reburns is increased. One method of opening up stands

Fig. 16. The area shown in Figs. 2 and 3 as it appeared in August 1972. It was burned by wildfire in 1962, reseeded and heavily browsed by deer, and burned by another wildfire in 1968. The combination of reseeding, heavy browsing, and reburning has tended to destroy the sharp boundary shown in Figs. 2 and 3.

of chamise to maintain this open state is to burn, reseed to annual ryegrass, browse heavily, and reburn (Fig. 16).

VI. Plant Successions after Fire in Woodland–Grass Chaparral

Woodland–grass chaparral results from the spread and increased density of broadleaf sclerophyll shrubs in grasslands. Some of the nonsprouting shrubs will become established in stands of resident grasses. For these shrubs, such as wedgeleaf ceanothus, chaparral whitethorn, mariposa manzanita, and whiteleaf manzanita, fire is not absolutely essential for seed germination. Many seedlings appear each spring in the vicinity of mature plants. Most die during the summer from competition and lack of moisture, but a few live. If rains should come in late spring and if the grasses are closely grazed, an abnormally large proportion of the chaparral seedlings will survive the summer drought. Since survival is closely related to late rains, increase in chaparral shrubs is periodic.

When a stand of scattered shrubs with sparse grass understory burns, a good seedbed free of vegetative competition is achieved and abundant shrub seedlings appear and rapidly increase in density.

Seedlings of some of the sprouting species, such as chamise and yerba santa, withstand very little competition. Yerba santa never seems to start from seed in competition with grass; after becoming established on heavily burned spots free of competition, however, it can spread into grasslands by sprouts from underground stems. Both this species and chamise can survive on ash beds where nonsprouting shrubs grew. Thus, aggressive nonsprouting species pave the way for sprouting species.

Sometimes nonsprouting shrubs are so heavily browsed that no seeds are produced for many years (Fig. 17). After a fire in wedgeleaf ceanothus, no seeds were produced in the next 15 years, and a reburn competely destroyed the shrub cover. Since this was on a deer winter

Fig. 17. Wedgeleaf ceanothus in an area that had been burned 15 years earlier. The shrubs were so heavily browsed by deer that no seeds were produced. A fire through this area would destroy the shrub cover and greatly reduce its carrying capacity for deer. The effects of fire and deer and livestock browsing cannot be totally separated.

range, the reburn greatly reduced the browse value of the area. Thus, if the objective is to produce and maintain wedgeleaf ceanothus on deer winter ranges, reburns must not occur before seeds are produced and stored in the litter and soil; otherwise, the shrub cover can be destroyed.

Occasional fires can result in the spread and thickening of shrubs in grasslands, but frequent, recurring fires can destroy the chaparral and bring back a former grassland cover.

VII. Plant Successions after Fire in Forest Chaparral

Species of chaparral occur in limited amount throughout most of the conifer forests, particularly those of the Sierra Nevada. Richards (1959) reported that the ground cover on the west side of the Sierra Nevada above 1500 m elevation was comprised of 17.2% brush. The chaparral may be sparse at any one time because of shading from trees. Wildfires increase chaparral in forested areas by conditioning dormant seeds for germination, destroying trees and opening up forest stands, making room for lower growing shrubs. They also aid nutrient release and mineral cycling, and prepare seedbeds relatively free of competition from other plants.

As mentioned earlier, several species of forest chaparral depend heavily on fire for germination, for example, deerbrush, mountain whitethorn, and greenleaf manzanita. Seedlings of these species are seldom seen except where there has been fire. At lower elevations in ponderosa pine, however, whiteleaf manzanita and wedgeleaf ceanothus may become dense after logging operations even without fire.

If wildfires are only moderately intense and if not all trees are killed, tree seedlings become established from seeds produced by the survivors. If a second wildfire occurs after the shrubs have produced more seeds, and if the fire kills the remaining trees, natural return to trees may be very slow. Many manzanita brushfields in northern California originated after early-day logging operations more than 100 years ago, and the trees show little tendency to invade and take over. There is no nearby source of seeds, and if such areas are seeded, competition is so great that few, if any, seedlings survive.

Where wildfire burns through deep litter and duff, many chaparral seeds may be destroyed. On Whitaker's Forest, in Tulare County, California where forest debris was piled and burned, brush seedlings did not appear under the piles, but on the edges, where the fire was light and the duff only partly consumed, there were large numbers of seedlings.

Recurring fires in forested areas are very effective in reducing chaparral, especially in ponderosa pine and other types where the fuels are highly flammable and subject to frequent surface fires. A prescribed fire in mature manzanita under ponderosa pine in the north coast area killed 93% of the native brush plants (Biswell and Schultz, 1958a). A second burning, under slightly drier conditions, killed the remaining shrubs. Where the pine was thinned and the understory debris burned in February, manzanita seedlings soon appeared in great numbers—thousands per hectare. Some died, but more emerged in the next 3 years. Most of the seedlings had been heavily browsed by deer and had made only little growth, but some protected by fencing reached a maximum height of about 1 m in 3 years. After 3 years the plots were reburned to kill brush seedlings. Treatments of this kind, if repeated, can gradually deplete the seed in the soil. The fire killed all but one of the thousands of manzanita seedlings, and that one survived because there were no pine needles nearby to carry fire to it. A third and fourth burn completely eliminated manzanita seeds from the soil.

In some forested areas, such as the national parks, where fire suppression has been very successful and shade-tolerant trees have increased, the shrubs are vanishing. For example in a giant sequoia forest in the King's Canyon National Park, no chaparral seedlings of deerbrush or greenleaf manzanita were found on a 16 ha tract. Later, one-half of the area was prescribe burned. The next spring a sampling showed 16,000 deerbrush seedlings per ha on a portion lightly burned, and 540 per ha on a heavily burned portion. No seedlings were found on the entire 8-ha control plot (Kilgore and Biswell, 1971).

In Whitaker's Forest, a 0.3-ha area was cleared of trees, and the debris was burned in piles. The next year there were 955 deerbrush plants, 134 whiteleaf manzanita, 81 gooseberry shrubs, 130 wild rose, and 8 littleleaf ceanothus seedlings on the plot. Before clearing, no plants of these species were found, nor was there any indication that they had ever grown in the area. It was known, however, that shrubs of these species had been dense in the area following heavy logging between 1875 and 1879 and firing in the summer of 1880. In this case it was clear that the shrubs had completely disappeared and the seeds stored in the duff and soil had produced the large numbers of new plants (Lawrence and Biswell, 1972).

VIII. Soil Erosion after Fire

Soil erosion usually accelerates to some degree following fire in chaparral, depending on such factors as erodibility of soil, steepness of slope, time, amount, and intensity of rainfall, percentage of plant cover, severity of fire, and length of time since the last burn. A few examples will be discussed.

Effects of wildfires on erosion in climax chaparral have been studied extensively on the 6900-ha San Dimas Experimental Forest on the south slope of the San Gabriel Mountains, near Los Angeles (Anderson and Trobitz, 1949; Anderson et al., 1959; Sinclair, 1954; Krammes, 1960, 1965; Krammes and Rice, 1963; Rice and Foggin, 1971). The chaparral is highly susceptible to burning, especially when it reaches 30 years of age, because of the dense fuel accumulations and seasonal dry windy periods. The soils are generally granitic and very erodible. Fifty-nine percent of the slopes are steeper than 70%, the average being 68%, near the angle of repose for the unconsolidated materials. Normal erosion, even in the absence of fire, is very high.

Anderson et al. (1959) and Krammes (1965) studied movement of gravitational debris on steep slopes covered by old chaparral. Some of their plots were on rejuvenated slopes where recent geological uplift caused renewed stream-channel downcutting. On such slopes, they found that an average of 8000 kg of debris per ha per year moved down the slopes to the stream channels, where it could later be flushed downstream during intense storms. Movement in dry seasons exceeded that in wet seasons on *all* slopes. When dry, the noncohesive soils on the steep slopes started creeping or sliding with every breeze or passage of a deer. Wetting made the soils more cohesive and less erosive. Where slopes are not so steep, wet-season movement equals or exceeds the dry-season movement. In October, 1959, a wildfire swept through seven of the nine study sites. Debris movement accelerated immediately, and on the south-facing slopes reached 10 times that of the already high prefire rate (Fig. 18). In this case nearly 90% of the debris movement came during the dry period.

In another case, lightning started a fire at noon on July 20, 1960, which swept over most of the San Dimas experimental watershed. As with most large wildfires in mature and old chaparral, the fire burned with great intensity. Debris movement started immediately at rates similar to those following the October, 1959, fire.

After the 1960 fire, several cultural treatments were tested for their value in control of erosion after fire (Rice and Foggin, 1971), including converting chaparral to grass. November and December of 1965–1966 were very wet months, and the winter of 1968–1969 was one of the most severe that southern California had ever experienced, with several large rainstorms from January to March. In each year, soil slippage was heavy. In the winter of 1965–1966, chaparral areas produced 21 m^3 per

Fig. 18. Dry-creep erosion may be severe immediately after intense wildfire in climax chaparral on slopes above the angle of repose. Photography courtesy of Pacific Southwest Forest and Range Experiment Station, U.S. Forest Service.

ha of debris slippage, and the areas converted to grass, 156 m³ per ha—7.4 times that from chaparral. In 1969, erosion, mainly from slippage, under the burned chaparral amounted to 298 m³ per ha of debris and that from grass, 843 m³ per ha. These results indicate that on very steep slopes, especially where the soils are unconsolidated, it is not wise to replace deep-rooted shrubs by more shallow-rooted grasses. Furthermore, the greater the percentage of cover of chaparral, the more deeply penetrating roots there are to hold the soil in place.

Studies in other areas show that soil erosion after chaparral fires occurs mostly in the early winter of the first year after fire, when the soil is still loose and not yet covered with new vegetation. After conversions to grass on slopes of less than about 35%, there may be little or no accelerated erosion (Pace and Fogel, 1967), and in the long run it may be less than that beneath mature chaparral, especially chamise, with its relatively poor soil-protecting qualities. After prescribed burning in ponderosa pine–manzanita chaparral, there was no increase in erosion related specifically to the fire (Biswell and Schultz, 1957). Most accelerated runoff and erosion in such places come from roads and trails and from activities that bare and compact the soil.

WATER REPELLENCY

Soil scientists have found that some soils under certain conditions are nonwettable. In fact, they repel water much the same as a duck's feathers (Debano, 1969).

Although water-repellent soils are found throughout the world, much of the literature about them has originated in California. In Florida such soils have been associated with orange orchards (Jamison, 1943), and in Australia with thick fungal layers in the soil (Bond, 1960). In chaparral, these soils are usually the result of compounds leached or volatilized from the litter of resinous vegetation (Debano, 1969; Debano and Rice, 1971; Krammes and Debano, 1965). In all cases, the origin is organic, but mineral properties of the soil and environmental influences, such as intense wildfires, can alter the water repellency.

Water repellency is caused by a group of hydrophobic, organic molecules that coat the soil particles. As the coat becomes more complete, the water repellency increases. Chaparral shrubs, such as chamise, western mountain mahogany, and scrub oak, produce litter that can make the soil water repellent (Debano et al., 1967). Because the climate in most areas of chaparral is cool in winter and dry and hot in summer, decomposition of the leaf litter is slow, and ground fuel loads naturally build up. Decomposition of the simpler chemical compounds in the litter, plus leaching by rainfall, help to carry water-repellent compounds into the soil surface, where they create a weak water-repellent layer.

It is thought that when intense wildfires occur, hydrophobic compounds liberated by pyrolysis of the litter are polymerized, diffuse downward into the soil, and condense on the cooler soil particles at some depth. In a very hot fire, this vaporization–condensation sequence occurs in a continuous fashion, and initially condensed particles may be revaporized and move deeper into the soil (Fig. 19). Sometimes the entire water-repellent layer is transferred beneath the surface, while the surface water repellency is destroyed.

After summer and fall wildfires, the soil is usually dry, and water repellency is at its maximum. A situation is thus created in which runoff and erosion may be severe with the first winter rains, especially if the rains are heavy. Raindrops ball up on the soil surface, infiltration is reduced, and runoff and erosion are increased. Where a subsurface repellent zone has formed, mud flows are likely. The rain infiltrates and saturates the surface soil, but the subsurface, repellent layer prevents

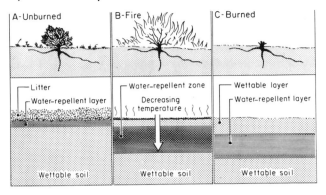

Fig. 19. Diagram illustrating location of water-repellent layer under chaparral; unburned (left), intensely burned (right), moderately intense fire (center) Photograph courtesy Pacific Southwest Forest and Range Experiment Station, U.S. Forest Service.

further water penetration, and acts like a concrete pavement (Debano *et al.,* 1967). The saturated surface soil has a lowered resistance to shear stress and tends to flow down the slope, clogging streams and blocking roads in its path (Fig. 20).

Soil characteristics give clues to the potential severity of water repellency. Coarse-textured soils, such as sands or sandy loams, common to many areas of chaparral, are affected more than are fine-textured clay soils, because they have a lower surface area per unit mass and are thus more easily coated by a given amount of hydrophobic material.

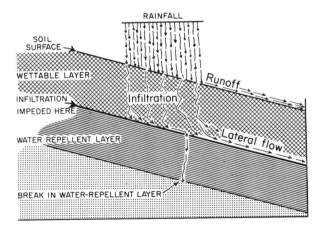

Fig. 20. Effect of rain on erosion. With intense rains the upper, wettable soil layer becomes saturated and is subject to slippage erosion. Photograph courtesy of Pacific Southwest Forest and Range Experiment Station, U.S. Forest Service.

Soil moisture content at the time of the burn is also very important. Wet soil restricts penetration of heat into the soil, and without large temperature gradients water repellency is not highly intensified. Water repellency is but one of the factors involved in the complex processes of infiltration and erosion. The proportion of large pores in the soil, water-holding capacity of litter and its protective influence against raindrop splash, moisture content of the soil, and intensity and duration of rainfall all contribute to the infiltration process. In combination, they may either aggravate or alleviate the water-repellent nature of soil in the chaparral zone.

IX. Discussion and Fire Management Implications

Fire in chaparral is both natural and inevitable. It has always occurred and probably always will, because vegetation becomes extremely dry near the end of a long, hot, nearly rainless summer. At that time, also, humidity may be extremely low and winds high. These conditions make fire control extremely difficult.

Climax chaparral is adapted to fire (Jepson, 1930), and periodic fires every 15 years or so appear necessary to maintain its good health. Unfortunately, fire has generally been considered by most land management agencies to be a wholly destructive agent and not an integral part of chaparral ecosystems. Most of the effort in management has therefore been toward fire exclusion, i.e., going to battle with wildfires (often with bulldozers), building unsightly, largely ineffective fuel breaks on ridgetops, and cutting access roads and trails, all of which favor increased runoff and erosion (Fig. 21). Many of these activities only compound and add to the severity of the wildfire problem; they do not solve it.

In spite of man's capability to fight wildfires with the most modern aerial and ground equipment and the best trained firemen in the world, it seems certain that wildfires in chaparral cannot be completely prevented (Pillsbury, 1963). Good intentions and large and expensive efforts in fire control do result in fires at less frequent intervals. But with longer time between fires, the fuels continue to build up (Dodge, 1972) and become more widespread, and when fires do get out of control, the toll in human life, natural resources, and costs is enormous. For example, in 1955, from August 27 to September 13, 436 wildfires laid waste 67,135 ha of chaparral and 57,151 ha of timberland (U.S. Forest Service, 1955). Again, in 1970, from September 15 to November 15, 1260 fires burned nearly 243,000 ha, mostly of chaparral, plus 885 homes. Fourteen lives were also lost. The total estimated costs were: fire fighting, $12

Fig. 21. Bulldozers, roads, and trails contribute to severe erosion on steep slopes in climax chaparral. Photograph courtesy of Pacific Southwest Forest and Range Experiment Station, U.S. Forest Service.

million; watershed loss, $100 million; property loss, $106 million; rehabilitation, $15 million—a total of $233 million (Wilson, 1971). The 70,800-ha Laguna fire in San Diego County was the second largest single fire ever recorded in California history (Wilson and Dell, 1971). In 1961 the Bel Air holocaust in Los Angeles blackened more than 2400 ha of chaparral and burned 505 buildings and residences valued at 30 million. In the Loup and Canyon fires in Los Angeles County in 1966 and 1968, 20 firefighters were killed.

Wildfires in chaparral present a dilemma as well as a severe land management problem. Nobody wants the disastrous wildfires during the windy periods in late summer, but they do occur. What, then, needs to be done? Perhaps land managers concerned with chaparral should rethink their fire-control problems and policies and ask whether virtually complete fire protection is practicable and economical. This was done in Australia, and a program of prescribed burning was started, which seems to be very effective (Vines, 1968). Since fire is a part of nature and probably will always burn over the chaparral-covered watersheds, would it not be wise to work in harmony with nature? Why not use fire under our terms, in a program of prescribed burning, and make fire the servant and not the master? Prescribed burning could be done in selected places and at selected times, probably with long-term results better than those obtained from attempted total fire exclusion and periodic, intense wildfires.

Prescribed burning and fire control should be integrated into a fire management program where benefits could be thoroughly tested and evaluated. More research is still required, and it could be carried out best in conjunction with such a program.

In the management of chamise chaparral with slopes of more than 30 or 35%, the aim should be dense stands of shrubs with both soil-building and soil-protecting qualities. Some shrubs are better in these respects than others. For example, all the ceanothus species, as well as western mountain mahogany have root nodules with nitrogen-fixing capability (Vlamis et al., 1964; Delwiche et al., 1965). As a result, these plants are high in nitrogen and protein and thus help increase soil fertility. The importance of soil fertility in maintaining better growth and watershed protection in chaparral has been discussed by Hellmers et al. (1955) and Hellmers and Kelleher (1959). Furthermore, the leaf mulch beneath ceanothus protects the soil against the force of raindrops and surface runoff. Scrub oak, another species valuable for buildup of nitrogen, produces a superb mulch. At the other extreme is chamise, which is extremely poor in soil-protecting and soil-building qualities. Studies by Zinke (1969) show that soil nitrogen under chamise decreased with time. Furthermore, the phytotoxins produced by chamise inhibit growth of other species in the understory. The canopy is usually thin and open, so raindrops can beat down on bare soil. Unfortunately, the most abundant species in climax chaparral thus contributes to soil deterioration.

When other species such as ceanothus, scrub oak, and western mountain mahogany grow with chamise, every effort should be made to maintain and increase their abundance. Nonsprouting species, such as wedge-leaf ceanothus, will be removed from chaparral by fires so frequently that the plants cannot produce and store seeds in the duff and soil in the intervals between fires. On the other hand, these species are relatively short-lived, and when fire is withheld for about 50 years they tend to disappear from the cover. Thus, both frequent fires and long periods without fire are detrimental to these valuable shrubs. Not only should the frequency of fires be regulated to favor the valuable species, but every effort should be made to experiment with planting and seeding of these species in suitable places.

Shrubs grow rapidly after fire for about 15 years, after which time growth slows and debris begins to build up. Young growth is less flammable than old growth, so a prescribed burn every 15–20 years would prevent large fuel accumulations and allow time for nonsprouting species to produce seed in the interval between fires. Rotational burning about every 15 years might be a highly valuable procedure in chaparral fire management. Under such a system, 5–7% of the chaparral would be prescribe burned each year, under conditions when the fires would not burn uniformly but would produce a patchy network of burned and unburned slopes.

Intense wildfires burning under the driest conditions completely remove the plant cover, including all debris. Sometimes the stems of shrubs are burned into the soil. During intense fires, rocks explode from the heat. All these reactions increase creeping or sliding of soils. Under the lighter fires possible under prescribed burning, the soil is not so completely bared and creep erosion should be less severe. Erosion accelerates on steep slopes after prescribed burning, but the extent of erosion at any one time is less than the massive soil movements that occur after wildfires.

Probably the best time to prescribe burn in chaparral in southern California would be March, April, and early May, when the vegetation is not exceedingly dry, and strong desert winds are not so frequent. In the forest chaparral, prescribed burning can be done from the last half of February through March. This amount of burning in such a short time could probably be accomplished by using aerial ignition, as is being done in Australia (Packham and Peet, 1967). With this method, 4000 to 8000 ha could be burned in one afternoon, with roads and natural barriers serving as the boundary fire breaks. We need further research on environmental conditions under which such burning can be done.

The practice of seeding ryegrass on steep slopes of burned areas is questionable. This grass is a heavy user of moisture and nitrogen and is very competitive with chaparral seedlings. High density of grass at maturity will kill most of the chaparral seedlings, whether chamise or some of the excellent quality shrubs, such as wedgeleaf ceanothus. Annual ryegrass tends to disappear in 3–5 years, leaving behind thinned-out chaparral and greatly reduced nonsprouting species, such as wedge-leaf ceanothus. Thus, although the reseeding of burned areas may appear desirable to some, it may do more harm than good in the long run. This needs more investigation in those areas where the objective is to grow dense stands of chaparral on steep slopes.

The practice of using fire in both woodland–grass chaparral and forest chaparral should be continued and expanded in order to encourage grass and timber, improve wildlife habitat, increase water yield, and reduce wildfire hazards. Finally the training given those who will carry out prescribed burning should equal that given those who fight wildfires. We cannot always let fire take its course, but we can regard fire as an integral part of chaparral ecosystems and use it to advantage.

Acknowledgment

Thanks are due to Winifred Kessler for assistance in preparing this chapter.

References

Anderson, H. W., and Trobitz, H. K. (1949). Influence of some watershed variables on a major flood. *J. Forest.* **47**, 347–356.

Anderson, H. W., Coleman, C. B., and Zinke, P. J. (1959). Summer slides and winter scours—dry–wet erosion in southern California mountains. *U.S., Forest Serv., Pac. Southwest Forest Range Exp. Sta., Tech. Pap.* **36**, 1–12.

Bartholomew, B. (1970). Bare zone between California shrub and grassland communities: The toll of animals. *Science* **170**, 1210–121.

Bartholomew, B. (1971). Role of animals in suppression of herbs by shrubs. *Science* **173**, 462–463.

Bentley, J. R. (1967). Conversion of chaparral areas to grassland: Techniques used in California. *U.S., Forest Serv., Pac. Southwest Forest Range Exp. Sta., Agr. Hand.* **328**.

Biswell, H. H. (1957). The use of fire in California chaparral for game habitat improvement. *Proc. Soc. Amer. Foresters, 1957* pp. 151–155.

Biswell, H. H. (1967). Forest fire in perspective *Proc. 7th Annu. Tall Timbers Fire Ecol. Conf.* pp. 43–63.

Biswell, H. H., and Gilman, J. H. (1961). Brush management in relation to fire and other environmental factors on the Tehama deer winter range. *Calif. Fish Game* **477**, 357–389.

Biswell, H. H., and Schultz, A. M. (1957). Surface runoff and erosion as related to prescribed burning. *J. Forest.* **55**, 372–374.

Biswell, H. H., and Schultz, A. M. (1958a). Manzanita control in ponderosa. *Calif. Agr.* **12**, 12.

Biswell, H. H., and Schultz, A. M. (1958b). Effects of vegetation removal on spring flow. *Calif. Fish Game* **44**, 335–348.

Biswell, H. H., and Street, J. E. (1948). Wedgeleaf ceanothus, range brush: Increase studied and control method recommended. *Calif. Agr.* **2**, 3 and 12.

Biswell, H. H., Taber, R. D., Hedrick, D. W., and Schultz, A. M. (1952). Management of chamise brushlands for game in the north coast region of California. *Calif. Fish Game* **38**, 453–484.

Bond, R. D. (1960). The occurrence of microbial filaments in soils and their effect on some soil properties. *CSIRO Aust. Div. Soils, Div. Rep.* No. 10/60.

Burcham, L. T. (1960). "The Influence of fire on California's Pristine Vegetation." Agr. Ext. Ser., University of California, Berkeley.

California Division of Forestry. (1971). "Brushland Range Improvement. Annual Report," pp. 1–22. CDF, Sacramento, California.

Clar, C. R. (1959). California government and forestry. Calif. Div. Forest.

Cooper, W. S. (1922). "The Broad-Sclerophyll Vegetation of California," Carnegie Institution, Washington, D.C.

Countryman, C. M., and Philpot, C. W. (1970). "Physical Characteristics of Chamise as a Wildland Fuel." Pac. Southwest Forest and Range Exp. Sta., Berkeley, California.

Debano, L. F. (1969). Water repellent soils; a worldwide concern in management of soil and vegetation. *Agr. Sci. Rev.* **7**, 4–18.

Debano, L. F., and Rice, R. M. (1971). "Fire in Vegetation Management: Its Effect on Soil," pp. 327–347. Amer. Soc. Div. Eng., Bozeman, Montana.

Debano, L. F., Osborn, J. F., Krammes, J. S., and Letey, J. (1967). Soil wettability and wetting agents . . . on current knowledge of the problem. *U.S., Forest Serv., Pac. Southwest Forest Range Exp. Sta., Res. Pap.* **PSW-43**, 1–13.

Delwiche, E. E., Zinke, P. J., and Johnson, E. M. (1965). Nitrogen fixation by ceanothus. *Plant Phys.* **40**, 1045–1047.

Dodge, M. (1972). Forest fuel accumulation—a growing problem. *Science* **177**, 139–142.

Gianella, B. (1972). Natural and primitive areas. *Indian Historian* **5**, 11–14.

Gibbens, R. P., and Schultz, A. M. (1963). Brush manipulation on a deer winter range. *Calif. Fish Game* **49**, 95–118.

Gratkowski, H. J. (1962). Heat as a factor in germination of seeds of *Ceanothus velutinus* var. *laevigatus* t. & g. Ph.D. Thesis, Oregon State University, Corvallis.

Hanes, T. L. (1971). Succession after fire in the chaparral of southern California. *Ecol. Monogr.* **41**, 27–52.

Hanes, T. L., and Jones, H. (1967). Postfire chaparral succession in southern California. *Ecology* **48**, 259–264.

Hedrick, D. W. (1951). Studies on the succession and manipulation of chamise brushlands in California. Ph.D. Thesis, pp. 1–113. Agr. and Mech. College of Texas, College Station.

Hellmers, H., and Kelleher, J. M. (1959). *Ceanothus leucodermis* and soil nitrogen in southern California mountains. *Forest Sci.* **5**, 275–277.

Hellmers, H., Bonner, J. F., and Kelleher, J. M. (1955). Soil fertility: A watershed management problem in the San Gabriel Mountains of southern California. *Soil Sci.* **85**, 189–197.

Horton, J. S., and Kraebel, C. J. (1955). Development of vegetation after fire in the chamise chaparral of southern California. *Ecology* **36**, 244–262.

Jamison, U. C. (1943). The slow reversible drying of sandy surface soils beneath citrus trees in central Florida. *Soil Sci. Soc. Amer., Proc.* **7**, 36–41.

Jepson, W. L. (1921). The fire-type forest of the Sierra Nevada. *Intercollegiate Forest Club Annu.* **1**, 7–10.

Jepson, W. L. (1925). "Flowering Plants of California." Ass. Students Store, University of California, Berkeley.

Jepson, W. L. (1930). The role of fire in relation to the differentiation of species in the chaparral. *Proc. Int. Congr. Bot., 5th, 1929* pp. 114–116.

Kilgore, B. M. (1972). Fire's role in a sequoia forest. *Naturalist* **23**, 26–37.

Kilgore, B. M., and Biswell, H. H. (1971). Seedling germination following fire in a giant sequoia forest. *Calif. Agr.* **25**, 8–10.

Kilgore, B. M., and Briggs, G. S. (1972). Restoring fire to high elevation forests. *J. Forest.* **70**, 266–271.

Krammes, J. S. (1960). Erosion from mountainside slopes after fire in southern California. *U.S., Forest Serv., Pac. Southwest Forest Range Exp. Sta.*

Krammes, J. S. (1965). Seasonal debris movement from steep mountainside slopes in southern California. *U.S., Dep. Agr., Misc. Publ.* **970**, 85–89.

Krammes, J. S., and Debano, L. F. (1965). Soil wettability: A neglected factor in watershed management. *Water Resour. Res.* **1**, 283–286.

Krammes, J. S., and Rice, R. M. (1963). Effect of fire on the San Dimas Experimental Forest. *Proc. Annu. Watershed Symp.* pp. 31–34.

Landers, R. W. (1962). The influence of chamise, *Adenostoma fasciculatum*, on vegetation and soil along chamise–grassland boundaries. Ph.D. Dissertation, pp. 1–194. University of California, Berkeley.

Lawrence, G., and Biswell, H. H. (1972). Effect of forest manipulation on deer habitat in giant sequoia. *J. Wildl. Manage.* **36**, 595–605.

Lewis, H. (1961). Chaparral lands of southern California. *In* "Man, Fire and Chaparral," pp. 13–17. Agr. Exp. Sta., University California, Berkeley.

Love, R. M., and Jones, B. J. (1952). Improving California brush ranges. *Calif., Agr. Exp. Sta., Circ.* **371**, 1–38.

McMinn, H. E. (1939). "An Illustrated Manual of California Shrubs." Stacey, San Francisco, California.

McPherson, J. K., and Muller, C. H. (1967). Light competition between ceanothus and salvia shrubs, *Bull. Torrey Bot. Club* **94**, 41–55.

McPherson, J. K., and Muller, C. H. (1969). Allelopathic effects of *Adenostoma fasciculatum. Ecol. Monogr.* **39**, 177–198.

Miller, E. H. (1947). Growth and environmental conditions in southern California chaparral. *Amer. Midl. Natur.* **37**, 379–420.

Miller, J. (1887). Paper read before American Forestry Congress. *Rep. Amer. Forests Congr.* **10**, 25–26.

Muller, C. H. (1965). Inhibitory terpenes volatilized from *Salvia* shrubs. *Bull. Torrey Bot. Club* **92**, 38–45.

Muller, C. H. (1966). The role of chemical inhibition (allelopathy) in vegetational composition. *Bull. Torrey Bot. Club* **93**, 332–351.

Muller, C. H., and Del Moral, R. (1971). Role of animals in suppression of herbs and shrubs. *Science* **173**, 462–463.

Muller, C. H., Muller, W. H., and Haines, B. L. (1964). Volatile growth inhibitors produced by aromatic shrubs. *Science* **143**, 471–473.

Muller, C. H., Hanawalt, R. B., and McPherson, J. K. (1968). Allelopathic control of herb growth in the fire cycle of California chaparral. *Bull. Torrey Bot. Club* **95**, 225–231.

Muller, W. H., and Muller, C. H. (1964). Volatile growth inhibitors produced by *Salvia* species. *Bull. Torrey Bot. Club* 91, 327–330.

Mutch, R. W. (1970). Wildland fires and ecosystems—A hypothesis. *Ecology* 51, 1046–1051.

Naveh, Z. (1960). The ecology of chamise (*Adenostoma fasciculatum*) as affected by its toxic leachates. *Bul. Ecol. Soc. Amer.* 41, 56–57.

Naveh, Z. (1967). Mediterranean ecosystems and vegetation types in California and Israel. *Ecology* 48, 445–459.

Pace, C. P., and Fogel, M. M. (1967). Increasing water yield from forest, chaparral, and desert shrub in Arizona. *Proc. Int. Conf. Water Peace* 5, 753–764.

Packham, D. R., and Peet, G. B. (1967). "Developments in Controlled Burning from Aircraft." CSIRO, Australia.

Pillsbury, A. F. (1963). "Principles for the Management of Steep Chaparral Lands Undergoing Urbanization in the City of Los Angeles," pp. 1–14. City of Los Angeles Office of Civil Defense.

Quick, C. R. (1935). Notes on the germination of ceanothus seeds. *Madrono* 3, 135–140.

Quick, C. R. (1956). Viable seeds from the duff and soil of sugar pine forests. *Forest Sci.* 2, 36–42.

Quick, C. R. (1959). Ceanothus seeds and seedlings on burns. *Madrono* 15, 79–81.

Quick, C. R. (1961). How long can a seed remain alive? *Yearb. Agr., U.S. Dep. Agr.* pp. 94–99.

Quick, C. R. (1962). Resurgence of a gooseberry population after fire in mature timber. *J. Forest.* 60, 100–103.

Reynolds, R. D. (1951). Effect of natural fires and aboriginal burning upon the forests of the central Sierra Nevada. M.S. Thesis, University of California, Berkeley.

Rice, R. M., and Foggin, G. T. (1971). Effect of high intensity storms on soil slippage on mountainous watersheds in southern California. *Water Resour. Res.* 7, 1485–1496.

Richards, L. G. (1959). Forest densities, ground cover slopes in the snow zone of the Sierra Nevada west-side. *U.S., Forest Serv., Pac. Southwest Forest. Range Exp. Sta., Tech. Pap.* 40, 1–21.

Sampson, A. W. (1944). Plant succession on burned chaparral lands in northern California. *Calif., Agr. Exp. Sta., Bull.* 685, 1–144.

Sauer, C. O. (1950). Grassland climax, fire, and man. *J. Range Manage.* 3, 16–21.

Schultz, A. M., and Biswell, H. H. (1952). Competition between grasses reseeded on burned brushlands in California. *J. Range Manage.* 5, 338–345.

Schultz, A. M., Launchbaugh, J. L., and Biswell, H. H. (1955). Relationship between grass density and brush seedling survival. *Ecology* 36, 226–238.

Shantz, H. L. (1947). "The Use of fire as a Tool in the Management of the Brush Ranges of California. California State Board of Forestry, Sacramento, California.

Show, S. B., and Kotok, E. I. (1924). The role of fire in the California pine forests. *U.S., Dep. Agr., Bull.* 1294, 1–80.

Sinclair, J. D. (1954). Erosion in the San Gabriel Mountains of California. *Trans., Amer. Geophys Union* 5, 264–268.

Stewart, O. C. (1956). Fire as the first great force employed by man. *In* "Man's Role in Changing the Face of the Earth" (W. L. Thomas, ed.), pp. 115–133. Univ. of Chicago Press, Chicago, Illinois.

Stone, E. C., and Juhren, G. (1951). The effect of fire on the germination of the seed of *Rhus ovata*. *Amer. J. Bot.* **38**, 368–372.

Sweeney, J. R. (1956). Responses of vegetation to fire. *Univ. Calif., Berkeley, Publ. Bot.* **28**, 143–250.

U.S. Forest Service. (1955). "California Aflame," pp. 1–3. U.S. Forest Serv., San Francisco, California.

Vines, R. G. (1968). The forest fire problem in Australia—a survey of past attitudes and modern practice. *CSIRO, Aust. Sci. Teacher's J.* pp. 1–11.

Vlamis, J., Schultz, A. M., and Biswell, H. H. (1964). Nitrogen fixation by root nodules of western mountain mahogany. *J. Range Manage.* **17**, 73–74.

Vogl, R. J., and Schorr, P. K. (1972). Fire and manzanita chaparral in the San Jacinto Mountains, Calif. *Ecology* **53**, 1179–1188.

Went, F. W., Juhren, G., and Juhren, M. C. (1952). Fire and biotic factors affecting germination. *Ecology* **33**, 351–364.

Wilson, C. C. (1971). Co-mingling of urban and forest fires. *Fire Res., Abstr. Rev.* **13**, 35–43.

Wilson, C. C., and Dell, J. D. (1971). The fuels build-up in American forests: A plan of action and research. *J. Forest.* 471–475.

Wilson, R. C., and Vogl, R. J. (1965). Manzanita chaparral in the Santa Ana Mountains, California. *Madrona* **18**, 47–62.

Zavitkovski, J., and Newton, M. (1968). Ecological importance of snowbrush, *Ceanothus velutinus*, in the Oregon Cascades agricultural ecology. *Ecology* **49**, 1134–1145.

Zinke, P. J. (1961). Physical and biological aspects of chaparral management. *In* "Man, Fire and Chaparral," pp. 21–25. Agr. Exp. Sta., University of California, Berkeley.

Zinke, P. J. (1969). "Biology and Ecology of Nitrogen," pp. 40–53. Nat. Acad. Sci., Washington, D.C.

. 11 .

Fire in the Deserts and Desert Grassland of North America

Robert R. Humphrey

I. Introduction

A. PREVALENCE OF FIRES IN DESERT AREAS

Because of the inescapably close correlation between prevalence of fires and amount of fuel, deserts are characteristically less affected by fire than are most ecosystems. And, as a corollary, the more arid the desert, the less fuel is produced and the less frequent and severe are any fires that may occur. Conversely, however, even though fire frequency and severity may be relatively low in any rating scale, their effect on the ecosystem may be extreme. Thus, as in portions of the desert grassland of the Southwest, even occasional fires may prevent attainment of the potential overstory climax of woody species, maintaining a grass subclimax that has often been mistaken for a true climatic climax.

B. FIRE AS IT AFFECTS GRASSES AND WOODY PLANTS

With a few notable exceptions such as *Eucalyptus, Sequoia,* and some pines and oaks, fires are more destructive of woody species than of most grasses. This is particularly true when the woody taxa, or ligniphytes, are shrubs or small stature trees. This differential susceptibility to damage is a direct result of morphological differences in the two life forms.

Most perennial grasses die back each year to about ground level, the following year's growth developing either from peripheral or central buds in the crown of the plant or from stolons or rhizomes. Thus, unless the plant has been completely killed, and this is not usually the case, only a single year's growth, most of which has already died, is merely removed by the fire. In addition, grass fires, because of the rapidity with which they burn, are generally classed as "cool" and do not typically burn deeply into the grass crowns.

By contrast, the killed portion of a ligniphyte usually represents many years' growth which, even if the plant were a sprouter, i.e., had the ability to sprout from the base after fire, would usually require several years to regrow to a point where it would again flower and produce seed. Thus, repeated fires, even when they do not kill woody taxa outright, keep them in a juvenile, nonfruiting stage. And, as will be seen later, many of them are nonsprouters and are completely killed. For a fuller discussion of these differences the reader is referred to the chapter "Fire" in Humphrey (1962).

C. SPROUTERS AND NONSPROUTERS

Shrubs are often classified by "pyrecologists" as either *sprouters* or *nonsprouters:* those that sprout from the base after a fire or those that fail to sprout and thus are killed. This sounds like an easy, simple classification. Unfortunately, however, like most such alternatives this one, too, has its problems. For example, big sagebrush (*Artemisia tridentata*) is very nearly a nonsprouter while horsebrush (*Tetradymia canescens*) and most species of rabbitbrush (*Chrysothamnus*) are strong sprouters that are harmed only temporarily by burning. Bitterbrush (*Purshia tridentata*), on the other hand, and mesquite (*Prosopis velutina*) sometimes sprout and sometimes do not. It has also been observed in the desert grassland of Arizona that burroweed (*Aplopappus tenuisectus*), although typically a nonsprouter, may sprout prolifically if burned while the plants are in the flush of new growth. As a conclusion, it would seem highly probable that most shrub species are predominantly either sprouters or nonsprouters but that many fall somewhere in between and that many even of the nonsprouters possess a latent sprouting ability under certain conditions of growth.

II. Four Deserts and the Desert Grassland

Four broad areas are usually recognized as being both geographically and floristically distinct North American deserts. These are the Great Basin, or Sagebrush, Desert, the Mojave Desert, the Sonoran Desert, and the Chihuahuan Desert. A fifth dry and distinct area, the desert grassland, is often classified either as desert or as semidesert.

The general geographical location of each of the four deserts is shown on the accompanying map. The desert grassland, because of its intricacy and interspersion in part with adjacent vegetation types, has not been shown.

All four deserts and the desert grassland have aridity as a common characteristic. All five also have a variety of low-growing woody plants as a common growth form. Although often floristically distinct, all except the Great Basin Desert have several of the same species in common.

III. Great Basin Desert

A. LOCATION

The Great Basin Desert (also sometimes known as the northern desert shrub or the Sagebrush Desert) is essentially coterminous with the Great Basin Physiographic Province. As indicated in Fig. 1, it includes most of Nevada, southern Oregon, about the southern half of Idaho, extends into southwestern Wyoming, covers roughly the western half and the southeastern quarter of Utah, and dips irregularly into northern Arizona.

B. CLIMATE

Precipitation in the area falls largely in the winter and spring months but has less extreme seasonal highs and lows than the other deserts. Much of it occurs as snow. In part because of the more equable distribution, in part because the bulk falls during the cooler months when evaporation is least, and in part because the relatively high elevation and northerly latitude are conducive to a cooler climate and consequent lower evaporation rate, the limited precipitation is more effective than an equal amount would be in the hotter, more southerly deserts.

C. VEGETATION

The Great Basin Desert is a high- or cold-desert area characterized by a variety of shrubs, largely gray in color, with a generally sparse understory of perennial or annual grasses intermingled with a variable stand of highly seasonal forbs. Plant communities dominated by one of two species, big sagebrush (*Artemisia tridentata*) or shadscale (*Atriplex confertifolia*), are encountered more commonly than any others in the area. The sagebrush is a reliable indicator of deep, well-drained soils and is the principal fuel source for free-running fires in this ecosystem. Other sagebrush species, principally those classified as *Artemisia rigida* and *A. nova*, occur in fine-textured, poorly drained sites. The shadscale also occurs largely on low-lying, poorly drained soils but most characteristically on those with a high alkaline content.

In addition to sagebrush and shadscale there are several other locally dominant shrubs or halfshrubs whose distribution is determined by local soil and soil-moisture conditions. Each of these is usually the principal species in its own ecosystem. Most of them provide too little fuel to carry fire readily and, as a consequence, will not be discussed here. They are covered in considerable detail by Shreve (1942) and by Shantz (1924).

D. PREVALENCE OF FIRES

It may be generalized that the Great Basin Desert is more subject to burning, either as a result of lightning or man, than the other three desert areas. This is a consequence of the prevalence of big sagebrush, often with an understory of grasses that provides additional fuel, rather than because of any excess of lightning storms or people.

The Great Basin Desert lies adjacent to or is intermingled with such vegetation types as juniper or juniper-pinyon pine, the Palouse grassland, the mixed-grass or short-grass plains, or, in the Southwest, the desert grassland. These often supply considerable fuel in the form of either perennial or annual grasses that serve to carry any fires into the adjacent desert areas. And, as is true of most contiguous vegetation types, only occasionally is there a sharp species differentiation between the taxa that characterize these types and those of the Great Basin Desert types. As a consequence, many of the fuel-providing grasses may extend for long distances into the so-called desert. In addition, a number of grasses such as cheatgrass (*Bromus tectorum*), bluebunch wheatgrass (*Agropyron spicatum*), and squirrel-tail (*Sitanion hytrix*), among others, are characteristic components of much of the Great Basin Desert. Thus, either because of invasion from contiguous types or because of the suitability of much of this desert to grass growth, extensive portions of it do produce a ground cover that is adequate to carry fire from shrub to shrub.

Shrubs are dominant in most of the Great Basin Desert. Most of these are highly flammable, particularly during the hot, arid summer months. Moreover, they often grow together, not infrequently to heights of 5 ft or more. As a consequence, they provide a fuel base that is often well suited, even without grasses, to carry a spreading fire over extensive areas.

E. GENERAL RESPONSES

In an Idaho study made near the northern limits of the Great Basin Desert, Blaisdell (1953) evaluated the effect of various intensities of burning on forage production and the response of various grasses, shrubs, and forbs to fire. Because the observations made during this study are representative of fire effects over extensive portions of the desert they are quoted at some length as follows.

During the early part of the first growing season after burning . . . damage to vegetation far outweighs the benefits. Perennial grasses and forbs are clearly lowered in vigor; old clumps are badly broken up and remaining plants are small and scattered. Although rhizomatous species are apparently damaged less than others, even these have poor vigor. Shrubs are represented by only a few sprouts. Much bare ground is exposed, but an abundant growth of annuals may fill many of the openings. As the season progresses, new shoots of rhizomatous grasses and forbs appear and tuft-forming species begin to stool out. Although greater vigor is apparent in most plants, scarcely any flower stalks are produced. Perennial grasses and forbs on burns remain green about 2 weeks longer than on unburned areas.

Despite the injurious effects of burning, rhizomatous species are often able to produce an increased amount of herbage by the end of the first year, but production of most other species is still below the original level. . . .

During the second year perennial grasses and forbs continue to increase and vigor is high. Sprouting shrubs are larger, but are still an inconspicuous part of the vegetation. The most noticeable feature of the burn during the second year is the abundant flower stalk production of almost all grasses and forbs. . . . The reason for this phenomenon is not known, but it may be related to a temporary increase in mineral nutrients and increased soil moisture. At any rate, this profuse flowering of grasses and forbs is typical of 2-year-old burns on sagebrush-grass ranges and supplies a source of seed for revegetation of areas that may not be supporting a full plant cover.

Total herbage production of grasses and forbs generally reaches its maximum about the third year after burning, largely as a result of increases in the fire-resistant rhizomatous species. Although this increased production may persist indefinitely, more often it declines in subsequent years. This general decline in grass and forb production is accompanied by an increase in shrubs and many nonrhizomatous herbaceous perennials.

Of the individual grass species occurring in the upper Snake River Plains, thickspike wheatgrass [Agropyron dasystachyum], plains reedgrass [Calamagrostis montanensis], and bluebunch wheatgrass [A. spicatum] are apparently least damaged by burning, for within 3 years these had recovered and were producing considerably more herbage on burned than on unburned areas. Other grasses are slower to recover. . . .

The rapid increase of . . . [thickspike wheatgrass and plains reedgrass] after they overcome the initial setback from burning and the temporary decrease of most of the other grasses suggests that repeated burning of such range might produce a fire subclimax dominated by coarse rhizomatous grasses. The fact that certain decreases in bluegrass [Poa spp.] and fescue

[*Festuca idahoensis*] were still evident after 12 and 15 years . . . suggests that burning sagebrush-grass ranges that have an herbaceous understory dominated by such finer bunchgrasses might result in a permanent reduction in total grass yield or a shift to a higher proportion of coarse rhizomatous grasses.

Many forbs, especially the rhizomatous ones, make rapid recovery from burning and produce an increased amount of herbage within 3 years. Others, particularly the suffrutescent species, are slower to recover, but none of the perennial forbs are permanently damaged, and many apparently benefit from burning. . . .

Shrubs are . . . more damaged by burning than either grasses or forbs. Not only is all of the current herbage destroyed by fire, but the above-ground, woody parts are either killed or completely consumed, resulting in destruction of stored reserves. . . . However, rabbitbrush and horsebrush [*Tetradymia canescens var. inermis*] sprout profusely following burning and are only temporarily injured. . . . Bitterbrush plants that sprout grow rapidly, and some exceed their original size. However, part of the recovery of this species comes from new plants established from seed, especially on burns of heavy intensity where most of the old plants are killed.

Big sagebrush, which is completely killed and must re-establish itself entirely from seed, is much slower to recover. . . . On the upper Snake River Plains . . . the absence of sagebrush is often an indicator of past burns.

F. Subecosystem Responses

Perhaps because extensive portions of the Great Basin Desert rarely if ever burn, there has been little research on the effects of fire in the component or subsystems of this very extensive ecosystem. Some of the more clearly delineated of these, which do not indicate merely successional stages, include the following.

1. Big Sagebrush

Two ecosystems within the Great Basin Desert are most susceptible to being swept by wildfires. These are primarily the one dominated by big sagebrush and secondarily the one where antelope bitterbrush (*Purshia tridentata*), often intermixed with big sagebrush, is dominant. The sites where one or the other of these occurs as the dominant species are typically well drained with a deep soil and a climate that permits growth of fire-supporting grasses and forbs. Competition from the sagebrush may largely preclude the establishment of any other vegetation except a sparse understory of annual cheatgrass (*Bromus tectorum*), but in these situations the sagebrush supplemented by the grass usually provides ample fuel to carry a fire.

Because big sagebrush is a nonsprouter, has almost no value as forage for cattle, and competes effectively with valuable forage species for the limited moisture, cattlemen usually look on a sagebrush fire as a blessing. The species does have grazing value for sheep, particularly in winters when deep snows cover the shorter vegetation. In general, however, sagebrush fires benefit both cattlemen and sheepmen, either by releasing the suppressed grasses and forbs or by providing a clean, ash-covered seedbed for reseeding to more desirable species. Such reseeding is necessary in some areas, particularly in the potentially highly productive sagebrush alluvial flats where competition from the brush and/or long years of overgrazing have all but eliminated other species.

Rather typically on the upland sites, however, there are enough grasses and forbs to make a rapid recovery under the impact of removal of the competing shrubs combined with the fertilization afforded by the ashes. The grasses, with their inherent ability to withstand fires and to recover quickly will, unless overgrazed, soon produce more forage than the combination of sagebrush and grass yielded prior to burning. There tends to be a reduction in grass forage production for the first 2 years after a fire but under a program of no grazing the first year and light grazing the second, forage production is usually greater than ever by the third year. Ultimately the sagebrush will return in the absence of fire because it represents the largest growth form and best adjusted species capable of establishment as the climax under the prevailing climate.

Despite the relatively greater incidence of fires here, it should not be inferred that fires occur commonly or throughout the big sagebrush ecosystem. Extensive portions of this desert either support other species of *Artemisia* or other taxa that provide too little fuel to carry a fire. *Artemisia nova* and *A. rigida*, for example, are small in stature and occur typically in poorly drained sites with few grasses or other potentially flammable growth. As compared with the extensive stands of big sagebrush, these are often restricted in area and do not provide an opportunity for extensive burns.

2. Antelope Bitterbrush

From almost any point of view the antelope bitterbrush ecosystem is harmed rather than benefited by burning. *Purshia tridentata*, as distinct from other species or varieties of *Purshia*, might be classified as a variable or uncertain sprouter that is often completely killed by fire. It is, in addition, a valuable forage species used by both deer and all kinds of domestic livestock. Because it usually occurs as a dominant on light, sandy soils, exposure of these soils to wind erosion following burning sometimes results in rather severe soil blowing.

Burned areas will usually soon be revegetated, first to a cover of Russian thistle (*Salsola kali*) and or to Jim Hill mustard (*Norta altissima*), followed within a year or two by cheatgrass. After many years during which grazing has been rigidly controlled, these annuals will gradually be replaced by perennial grasses and forbs beneath a new stand of bitterbrush. These sites, however, are comparatively fragile and usually suffer much more than big sagebrush sites from the combination of grazing and burning.

3. Rabbitbrush

In much of the Great Basin Desert various species of rabbitbrush (*Chrysothamnus*) occur as a common component of the vegetation. The response of these to fire is highly variable. In general, however, they tend to be sprouters and, as a consequence, are harmed much less by burning than are big sagebrush or antelope bitterbrush. One of the commonest species, and one that usually indicates range deterioration, is big rabbitbrush (*C. nauseosus*). Although usually a prolific sprouter, Robertson and Cords (1957) cite two instances in which burning resulted in an almost complete kill of this species. In contrast, they also mention another instance where it made a 95% recovery.

Pechanec *et al.* (1954), in their classic study of sagebrush burning, mention only a single *Chrysothamnus* species, downy rabbitbrush (*C. puberulus*). They indicate, however, that it was undamaged by fire. Shantz (1924) recorded that little rabbitbrush (*C. stenophyllus*) and related species might occupy the ground almost to the exclusion of other plants where drought or fire have killed the sagebrush. Although not mentioning fire, he stated that where sagebrush had been destroyed, big rabbitbrush was often dominant.

Additional research is needed on the response to fire of the various *Chrysothamnus* species under a variety of phenologic and climatic conditions. For the present, however, the evidence points to fire as an uncertain and usually not a practical means of control except in those situations where the rabbitbrush occurs intermixed with a strongly dominant and unwanted stand of big sagebrush.

4. Shadscale

Shadscale (*Atriplex confertifolia*) occurs extensively and widely as essentially pure stands throughout the Great Basin Desert. Because the plants are short and usually occur as open, almost pure stands that provide little fuel, shadscale communities rarely burn.

5. Budsage and Saltsage

Two other subsystems, budsage (*Artemisia spinescens*) and saltsage (*Atriplex corrugata* and *A. nuttallii*), although limited essentially to the northern portion of the Great Basin ecosystem, are similar to shadscale in the general growth-form characteristics of their dominant taxa. And, like the shadscale communities, these also, rarely burn.

6. Winterfat

Winterfat (*Eurotia lanata*), although not usually occurring to the exclusion of other shrubs and grasses, is, nonetheless, typically a strong dominant and usually may be accepted as indicating a distinct ecosystem where it does grow. Its distribution is not limited to the Great Basin Desert and it is a characteristic species of many arid sites in the West from Canada into Mexico.

Winterfat is highly relished by herbivores ranging from rabbits, deer, and antelope to sheep, cattle, and horses. As a consequence, it is often so closely grazed that little is left that might serve as fuel. Although I can find no literature on fires in this ecosystem it is probably not immune from burning. And, because it sprouts and branches freely in response to grazing, I am hazarding the guess that it would probably also sprout after fire.

G. ADEQUACY OF THE ANALYSIS

Space does not permit more complete analysis of fire in the Great Basin Desert. The treatment given here is more in the nature of a partial synopsis than a complete discussion, but if it serves as a stimulus to a later, relatively thorough analysis its incompleteness will have served a useful end.

IV. Mojave Desert

A. LOCATION

The smallest of the four North American desert areas, the Mojave Desert, lies largely in extreme southern Nevada and adjacent California (see Fig. 1). A small fringe area extends into northwestern Arizona where it is hardly distinguishable from the northwestern portion of the Sonoran Desert with which it intermingles. The northern boundary is also indistinct, adjoining as it does the floristically similar southern edge of the Great Basin Desert.

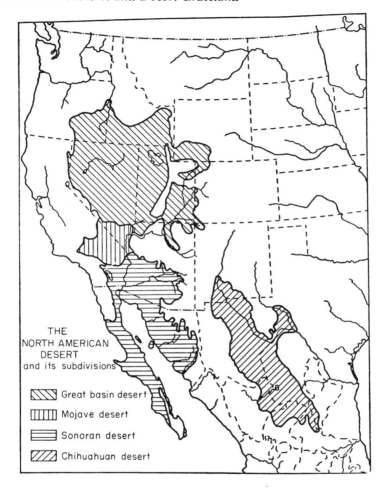

Fig. 1. The North American Desert and its subdivisions.

B. CLIMATE

The Mojave Desert probably has an average aridity greater than that of any other of the North American deserts. Limited portions of the Sonoran Desert around the head of the Gulf of California may receive even less rainfall and be as hot, but the combination of low rainfall and high temperatures over most of the Mojave Desert area would seem to give it the dubious distinction of being the most arid. As Shreve (1942) points out, the aridity is accentuated by the almost total lack of any rain in the summer months, the little that does fall being largely restricted to the winter and spring. The extreme aridity is suggested by the fact that this desert embraces Death Valley.

C. VEGETATION

Two extreme xerophytes, creosotebush (*Larrea tridentata*) and white bursage (*Franseria dumosa*), characterize the greater part of the Mojave Desert. Much of the area has a high soil-salt content that restricts both the variety and volume of plant growth. Because of the similarity of these factors with portions of the Great Basin Desert there is a considerable overlapping of species in the two ecosystems.

A characteristic plant of the Mojave Desert that is unique to this desert is the large Joshua tree (*Yucca brevifolia*) (Fig. 2). A smaller but distinctive related species, the mojave yucca (*Yucca schidigera*) is also widespread.

Most of the area has only internal drainage with the consequence that the occasional runoff drains into an almost countless number of normally dry lake beds. These are devoid of perennial vegetation and even contain no water most of the time. The high salt content of their soils and the occasional inundation preclude the possibility of the establishment of vegetation except seasonal stands of winter–spring annuals around the edges.

Fig. 2. Joshua tree (*Yucca brevifolia*) forest in Mojave County, Arizona.

Probably because of the low rainfall that is largely restricted to a single season, there are few grasses in the area. A perennial grass that is sometimes abundant, big galleta (*Hilaria rigida*), is found primarily in areas of blowing sand, along the edges of the occasional washes, or in swales with fine-textured alluvium. Except in the swales, big galleta occurs too sparsely or infrequently to be of any importance in carrying fire.

D. Prevalence of Fires

General

In the Mojave Desert, even more than in the other North American deserts, there is little in the published record concerning fires. This is understandable since the area is sparsely settled, fires are a rarity, and the few fires that do occur cause little apparent damage to the various aspects of the ecosystem that affect man and his economics.

E. Subecosystem Responses

1. *Creosotebush–Bursage*

An open zone of creosotebush and white bursage typically lies between the bare lake beds and the more rugged mountainous terrain above. The creosotebush and bursage are too sparse and the creosotebush plants are too open to carry a fire of themselves. Only in those occasional years when exceptionally heavy winter rains have produced an abnormally heavy stand of annuals is there any possibility of fire in this community. Shreve (1942) has estimated that 70% of the Mojave Desert falls within this *Larrea–Franseria* community, the remainder being either dry lake beds, areas of dune sand, or mountains. As the lakes and the sand dunes never burn, and the *Larrea–Franseria* ecosystem is essentially nonflammable, fires are most likely to occur in the mountains or in restricted alluvial flats where moisture accumulation permits a rather dense growth of big galleta and annuals.

2. *Mountainous Areas*

The mountains of the Mojave Desert are rugged, desert ranges supporting for the most part a sparse mixture of shrubs, forbs, and a few grasses interspersed between extensive areas of bare rock. In their higher portions they usually have a scattered growth of low-growing oaks and junipers. The fires that may occasionally occur usually cover limited areas but may be highly destructive of the more common suffrutescent species such as bladdersage (*Salazaria mexicana*) and shrubby buckwheat (*Eriogonum fasciculatum*). Because of the sparsity of perennial grasses, these mountain burns are slow to revegetate, and water erosion may result in considerable soil losses, particularly during the first year after a fire.

3. *Big Galleta Swales*

Small area fires that are usually man-caused not infrequently burn the big galleta flats adjacent to roads or settlements. Damage to the grass may vary from slight to severe, depending on grass dormancy. If the plants are extremely dry, with little or no new growth, damage may be severe because of burning out of the live center. Conversely, when the plants contain considerable green material, burning is more superficial and the living centers are little harmed. In this latter situation, any damage to the grass is usually temporary. Regardless of severity of damage, fires here may be of benefit by helping to control such competing shrubs as mesquite (*Prosopis juliflora*) and catclaw (*Acacia greggii*). Mojave yucca, which is often a characteristic component of the vegetation, is usually little harmed by the fire.

4. *Blackbrush*

Blackbrush (*Coleogyne ramosissima*), although not typical of most of the Mojave Desert, does occur as almost pure stands locally as well as in the southern Great Basin Desert. Perhaps because of aggressive competition for moisture by this species there are usually few forbs or grasses in blackbrush stands that might aid in carrying fire (Fig. 3). Despite this and the fairly wide spacing of the plants, blackbrush communities have been known to burn under conditions of high temperature, high wind velocity, and low relative humidity. My not entirely conclusive observations have been that blackbrush is not a sprouter and, in the event of a clean burn, is almost entirely destroyed by fire. No specific studies relative either to damage or method and rate of recovery after burning have come to my attention.

V. Sonoran Desert

A. Location

The most southwestern of the North American deserts, the Sonoran Desert, occupies roughly the southwestern third of Arizona; the south-

Fig. 3. Typical blackbrush (*Coleogyne ramosissima*) community with scattered Joshua trees in the background. The absence of vegetation between the shrubs inhibits the spread of fire.

eastern corner of California; most of the peninsula of Baja California, Mexico; and the greater part of the state of Sonora, Mexico (see Fig. 1). Except for a narrow strip in Baja California that lies in the Pacific Ocean drainage, the entire area drains into the Colorado River or directly into the Gulf of California.

B. CLIMATE

The mean annual rainfall ranges from a low of about 2–3 inches in the area around the head of the Gulf of California to about 12 inches in parts of Arizona. Annual distribution is largely bimodal, with a pattern of winter and summer rains alternating with spring–fall periods of drought. Almost none of the precipitation falls as snow although all except the southern portions of the desert do receive some frost in most years.

Fig. 4. Representative Sonoran Desert community in southern Arizona showing a wide variety of plant species but a sparse cover little conducive to burning.

C. Vegetation

The Sonoran Desert is characterized by a large variety of woody species. To a much greater extent than in the other deserts, many of these are low growing trees. A second outstanding difference between this and the other North American deserts is an abundance of cacti in the Sonoran Desert (Fig. 4).

Because of the extensive area included within the Sonoran Desert (nearly 120,000 square miles) and the consequent climatic diversity, its floristic composition is highly variable. As indicated above, the characteristic species are predominantly woody and include many cacti. These vary widely, however, over the approximate 850-mile north–south extent of the desert with regard to species composition and density, as well as to frost and drought tolerance. Comparatively few plant taxa are distributed throughout the desert, but there is a small minority that tend to tie it all together. These include ironwood (*Olneya tesota*), creosotebush, jojoba (*Simmondsia chinensis*), triangle bursage (*Franseria deltoidea*), and the genus *Fouquieria,* represented over most of the Sonoran Desert by the ocotillo (*F. splendens*).

Although the area contains several typical perennial grasses, the plants usually occur too sparsely to provide a reliable fuel base for carrying fire. In years of exceptional rainfall extensive areas may support a good cover of annual grasses or forbs, and, when fires do occur, these (the grasses in particular) usually provide the necessary fuel.

D. Vegetation Characteristics as Related to Fire Frequency

General

Despite the extensive geographical area included within the limits of the Sonoran Desert and the severe lightning storms that characterize the summer rainy season, fires occur here only rarely. This is largely because of the dominance of widely spaced open-branched trees and shrubs that are poorly suited to burning, either separately or collectively. The few perennial grasses that characterize most subecosystems within the Sonoran Desert usually occur locally as a few plants beneath some of the shrubs or trees and do not provide the continuity of cover necessary to carry fire. Perennial forbs usually occupy the same type of ecological niche and, in any event, occur too sparsely to be of any importance as fuel.

In most years annual forbs and grasses do not become sufficiently dense to carry fire. In addition, they are small in size and because of their small biomass produce little fuel.

Most fires in the Sonoran Desert occur in the ecotone between the desert and desert grassland, a variable but often wide area with sufficient precipitation to grow a fairly dense stand of annuals during exceptionally wet years. Some areas of this sort are of doubtful ancestry and, although they may be classified as Sonoran Desert today, probably more correctly represent the deteriorated, lower fringes of the desert grassland. As grazing has killed the grasses or fire control has permitted invasion of certain of the Sonoran Desert ligniphytes there has been an apparent shift in the desert boundary. This feature will be discussed in some detail in Section VII.

E. Tobosa Flats

Tobosa flats, e.g., poorly drained, alluvial swales that occur primarily in the northern Sonoran Desert and also in the Chihuahuan Desert, constitute an exception to the general nonflammability of the Sonoran Desert perennial vegetation. Although tobosa grass (*Hilaria mutica*) does grow on both upland and lowland sites, only in the swales is the

Fig. 5. Pristine tobosa-grass (*Hilaria mutica*) flat on the Papago Indian Reservation in southern Arizona. Occasional fires historically maintained such ecosystems essentially free of woody plants (cf. with Fig. 6).

stand usually dense enough to carry a fire. During the last 50–100 years the vegetal composition of these "flats" has been gradually changing from essentially pure grass to a grass–shrub mixture (Figs. 5 and 6). As this change does not seem to be correlated with a reduction in grass density, some other factor, probably absence of fire, must be causative.

Because of the rather limited area of tobosa grass flats they must have burned less frequently than the more extensive desert grassland in which a fire, once started, could run for many miles before encountering a natural nonflammable barrier. In addition, surrounded, as tobosa flats usually are, by relatively sparse vegetation, they were, and are, less subject to burning than areas such as the desert grassland that adjoined other flammable types.

Tobosa flats are almost never burned today and most of them are being invaded by ligniphytes, usually mesquite and whitethorn (*Acacia constricta*). The relatively high moisture content of the soils of these sites is conducive to ready establishment and rapid growth of the invad-

Fig. 6. Tobosa-grass (*Hilaria mutica*) flat near Tucson protected from burning for many years. Note the invasion by woody species, largely velvet mesquite.

ing shrubs, and there is no reason to suppose that they will not continue to be so invaded in the absence of fire.

F. Response of Specific Taxa to Fire

The knowledge of the response of individual Sonoran Desert species to fire has been obtained largely from studies involving those sometime ecotonal taxa that have been exposed to fires from, and adjacent to, the desert grassland. The following lists the principal among these and their responses to fire.

1. (Velvet mesquite (*Prosopis juliflora* var. *velutina*): a facultative sprouter; thin-barked, with low resistance to the heat of fires (see additional discussion in Humphrey, 1962, Chapter 9).

2. Creosotebush (*Larrea tridentata*): a facultative sprouter; thin-barked, small-stemmed, and highly susceptible to fire damage or killing.

3. Burroweed (*Aplopappus tenuisectus*): an obligate nonsprouter with a dense crown of resinous leaves and stems; burns readily. Despite its

Fig. 7. Prickly pear (*Opuntia*) killed by a ground fire. Bacterial rot and foraging animals usually finish the job the fire begins.

susceptibility to fire the genus is difficult to control because of the abundant production of wind-disseminated seeds.

4. Snakeweed (*Gutierrezia lucida*): fire reactions essentially identical with those of burroweed. The synonym "matchweed" suggests its almost explosive flammability.

5. Cholla cactus (*Opuntia fulgida*): most of the plants, particularly in young stands, may be largely controlled by fire. After burning of the spines the plants are unprotected from grazing animals. Fire is sometimes used in severe drought years to supplement the limited forage with cactus, a practice that effects some degree of control of the cholla.

6. Prickly pear (*Opuntia* spp.): the effect of burning is variable, depending on fire intensity. In a hot fire the plants are completely killed (Fig. 7). The stems are incombustible of themselves and without adequate fuel will be little harmed.

7. Bisnaga, barrel cactus (*Ferocactus* spp.): plants 1 or more feet tall are rarely killed since only the spines are combustible. Burning off of the spines, however, makes the plants susceptible to destruction

Fig. 8. Barrel cactus (*Ferocactus*) from which most of the spines have been removed by fire, rendering the plant largely defenseless against foraging animals. This individual has been eaten by jackrabbits and cattle.

by cattle, horses, and rabbits (Fig. 8). Those plants less than about 1 ft tall may suffer 75% or higher mortality either as a direct result of burning or from subsequent grazing damage.

VI. Chihuahuan Desert

A. LOCATION

The Chihuahuan Desert lies largely in Mexico, principally in the states of Chihuahua and Coahuila, but including parts of Durango, Zacatecas, Nuevo Leon, and San Luis Potosi. The relatively small portion in the United States lies largely in south central New Mexico and southwestern Texas (see Fig. 1). In addition to the area delineated on the map there are smaller but almost as typical segments in southeastern Arizona and southwestern New Mexico (Shreve, 1942).

B. Climate

This is an area with great extremes in mean annual precipitation. These range from as little as 3 inches in parts of Coahuila to 12–16 inches in the higher elevations near the western and southern edges of the desert. About 65 to 80% of the rain falls during the summer months from June through September. This results in a long, 8-month dry season in which the period from January to May is exceptionally dry. The lower areas typically receive light frosts particularly during December, January, and February. The higher elevations may be subjected to severe freezing during this same period (Shreve, 1942).

C. Vegetation

The Chihuahuan Desert is essentially an area of medium-size shrubs, few cacti, and a moderate understory of grasses at the higher elevations. Although creosotebush, ocotillo (*Fouquieria splendens*), and mesquite are common both here and in the Sonoran and Mojave deserts, other shrubs that are essentially restricted to the Chihuahuan Desert provide it with its primary floristic individuality. These are principally tarbush (*Flourensia cernua*), sandpaperbush (*Mortonia scabrella*), mariola (*Parthenium incanum*), and whitethorn (*Acacia vernicosa*).

The upper boundaries of the Chihuahuan Desert become indistinct because of its floristic intergradation with the desert grassland and the instability of the ecotone. Both fire and grazing appear to have played a part in the successional flexibility of this boundary.

D. Vegetation Characteristics as Related to Fire Frequency

General

Despite their geographical, climatic, and floristic differences, the Sonoran and Chihuahuan desert ecosystems are somewhat similar in respect to the prevalence of fire. Of the two the Chihuahuan Desert is slightly more susceptible to burning. A greater proportion of the component shrubs are low growing and have a relatively dense crown. In addition, the woody plants are more often interspersed with perennial grasses than in the Sonoran Desert. As a consequence the occasional fires that do occur have a greater opportunity to run.

Despite these characteristics, however, there are few fires in the Chihuahuan Desert. This is in part a result of a natural insufficiency of fuel and in part a side-effect of the close grazing and fire control that have gradually changed the composition of the vegetation from a grass–shrub mixture to one consisting primarily of shrubs.

Like the Sonoran Desert, the Chihuahuan Desert is most subject to burning in the ecotone where it borders on the desert grassland. And, similarly, the original boundaries of the Chihuahuan Desert have probably shifted to include former grassland areas during the past 100 years as grazing and fire control have favored the natural woody plant climax (Humphrey, 1953, 1958).

E. Tobosa Flats

As in portions of the Sonoran Desert, so also in the Chihuahuan Desert, tobosa flats are a characteristic of poorly drained, alluvial areas that provide ample fuel for burning. I know of no published records of the effect of such fires, which probably go largely unnoticed because of their limited areas. Fires in these grasslands are probably less prevalent than formerly, and, as in the Sonoran Desert, the tobosa swales are being invaded by woody species, largely mesquite and whitethorn.

F. Response of Specific Taxa to Fire

Because fires in the Chihuahuan Desert have not been a topic of research either in the United States or Mexico, it is difficult to obtain factual data on any aspects of burning in this ecosystem. Included in the knowledge gap is information on the reaction of individual taxa to fire. In the absence of more precise information the following limited observations are submitted.

Allthorn (*Koeberlinia spinosa*): probably a sprouter, although rarely growing in situations where it might be exposed to fire.

Creosotebush: as indicated previously, this is a facultative sprouter that has extended its original range into adjacent desert grassland areas during the last 75–100 years.

Honey mesquite (*Prosopis juliflora* var. *torreyana*): a sprouter that, because of its characteristic multistemmed habit and frequent occurrence in areas of blowing sand, is either highly resistant to fire damage or is rarely exposed to burning.

Mariola, sandpaper bush, and tar bush: sprouting reactions unknown.

Tobosa grass (*Hilaria mutica*): a species that, like most grasses, is highly resistant to fire damage. However, a slow burning fire in a protracted period of severe drought can be highly damaging. Intentional burning of tobosa flats should be restricted to periods when rains have initiated regrowth in crowns of the plants, thus reducing damage to the live center.

Whitethorn (*Acacia vernicosa*): a sprouter that is set back only temporarily by fire. The species may even be stimulated to produce supplemental plants from roots and additional stems at the original crown by this kind of natural pruning.

VII. Desert Grassland*

A. LOCATION

The desert grassland lies primarily in southeastern Arizona, south central and southwestern New Mexico, and southwestern Texas. The area does not include the entire portions of the states mentioned but occurs as local grasslands rather widely interspersed with other types. Although to a considerable extent occupying extensive plains areas, it also typically lies as broad belts around the bases of the many southwestern mountain ranges, principally at elevations of 3000–3500 ft. South of the border it extends far into Mexico and, as Clements (1920) suggests, the center of the desert grassland probably lies in Mexico.

B. CLIMATE

Although the desert grassland is the most arid of all North American grassland regions, it is more correctly classified as semidesert than as a true desert. Mean annual precipitation ranges from about 12–18 inches in the west to 20–30 inches in the east. Over most of the area this falls largely during two seasons, summer and winter. Wind velocities and summer temperatures are high and consequently evaporation rates are also high.

C. VEGETATION

The desert grassland concept used here corresponds essentially with that of Shantz' desert grassland and desert savanna (Shantz 1924). Shreve's desert grassland transition (Shreve, 1917) is identical in part but also includes rather extensive areas, largely in eastern New Mexico and the Staked Plains area of western Texas that are classified

---

* Excerpted in part from "The Desert Grassland" (Humphrey, 1958). The reader is referred to this publication for a more complete discussion.

Fig. 9. Excellent-condition desert grassland with *Yucca elata* in southern New Mexico. The good ground cover of perennial grasses provides a good fuel base for the occasional fires that still occur.

as short grass or tall grass by Shantz. Clements' desert plains (Clements, 1920) includes much the same area as Shantz' desert grassland.

Three genera, *Bouteloua, Hilaria,* and *Aristida,* provide most of the grass species in the type. Although *Bouteloua* is represented by many perennial species, *B. eriopoda, B. gracilis,* and *B. curtipendula* are probably the most abundant. In the genus *Hilaria,* three species, *H. mutica H. belangeri,* and *H. jamesii,* are most common. Although several species of *Aristida* are prevalent, four of the most common are *A. divaricata, A. hamulosa, A. glabrata,* and *A. longiseta.* Other grass genera characteristic of at least part of the area are *Andropogon, Eragrostis, Heteropogon, Leptochloa,* and *Trichachne* (Fig. 9).

The woody plant growth is extremely diversified, being derived in part from the southern desert shrub below, in part from the chaparral, pinyon-juniper, and oak woodland above. Although shrubs, low-growing trees, and cacti were always present to some extent in the desert grassland, they were originally largely restricted to drainages that supported little grass or to rocky or shallow soil areas.

In spite of the large number of woody taxa, only a few comprise the bulk of the trees and shrubs; three varieties of *Prosopis juliflora* are outstanding. As classified by Benson (1941), these are the varieties: *velutina, torreyana,* and *glandulosa.* Other shrub taxa that are locally or generally abundant include *Larrea tridentata, Acacia, Opuntia, Yucca, Flourensia cernua, Aplopappus,* and *Gutierrezia.*

D. VEGETATION CHANGES IN RELATION TO FIRE

Prior to the introduction of large numbers of domestic livestock into the Southwest, fires at rather frequent intervals appear to have been a characteristic feature of much of the desert grassland (Bracht, 1848; Bray, 1901, 1904a,b; Nunez, 1905; Cook, 1908; Thornber, 1907, 1910; Griffiths, 1910; Wooton, 1916; Humphrey, 1953, 1958). Although the buildup of large livestock numbers (primarily cattle) in the Southwest during the latter part of the nineteenth century somewhat antedated the development of an acute fire-control consciousness, there was, nonetheless, a reduction in fires dating back to this large increase in livestock. Many of the ranges were overstocked; some of them heavily so, by about 1880 (Wagoner, 1952). The resultant removal of forage or potential fuel must have reduced both the area and intensity of range grassland fires. Even so, the historical record shows that they still occurred frequently enough to have been an important factor in determining the plant composition of the desert grassland.

Within about the last 25–50 years the change in dominant life form has been so marked in some portions of the desert grassland that few or no vestiges of the original vegetation remain. And, were it not for early photographs, there would be little record today of many of the changes that have taken place. These changes have resulted in no little confusion in attempts to classify certain of these areas. For example, velvet mesquite is often popularly considered to be a characteristic dominant of the Sonoran Desert. Although a component of the Sonoran Desert vegetation, its presence on upland sites as the dominant woody species, very often with an understory of burroweed is a reliable indication of former grassland. Figures 10 and 11 show one instance of a complete change from the grasses that were at one time dominant to the mesquite and burroweed that are there today.

The historical and recent records of fires in the desert grassland and a knowledge of the physiological responses of grasses and woody plants to fire seem to me to leave little doubt that the invasion of the area by ligniphytes can be attributed largely to too few fires and too many cattle and horses.

Fig. 10. Area along the United States–Mexico international boundary west of Tucson as it appeared in 1893. The line of trees in the background is growing on a major drainage channel. Compare this with Fig. 11.

Fig. 11. Same view shown in Fig. 10, as it appeared in 1956, 63 years later. The grasses that dominated in 1893 are all gone, replaced by burroweed (*Aplopappus tenuisectus*) and velvet mesquite (*Prosopis juliflora* var. *velutina*).

All who have studied the situation, however, do not agree with this. Hastings and Turner (1965) conclude that "there is no reason to suppose that fires used to sweep the desert grassland frequently or on a large scale." As a consequence they "reject the hypothesis that fire suppression has been a primary cause of the changes" in plant composition. In place of fire they propose the hypothesis that the shrub increase and grass decrease are primarily a result of a gradual trend toward higher temperatures and lower precipitation. Space does not here permit a detailed discussion of Hastings and Turner's conclusions, but their publication will be of no little interest to those who wish to explore their analysis more deeply.

E. Vegetation Characteristics as Related to Fire Frequency

The line of demarcation between the desert grassland and the Sonoran and Chihuahuan deserts seems to have been determined in large part by the ability of the vegetation to carry fire. Where precipitation or soil moisture permitted the growth of enough grass to carry fire, periodic burning kept the shrubs in check. Where fuel was inadequate the shrubs grew unchecked. Thus, it might be said that shrubs in this area occurred where there was not enough precipitation to grow enough grass to carry enough fire to kill the shrubs.

Fires in timbered areas leave signs that may persist for decades or even centuries; the evidence of grassland fires is usually soon gone. As a consequence, while historical records are only one of several sources of information on early fires in forests, they are almost the only means of determining the occurrence of former grassland fires. Fortunately, the historical record is fairly adequate and generally appears to be reliable.

F. The Historical Record

1. *The Early Record*

The earliest account of grassland fires in the Southwest that has come to my attention is an account by Cabeza de Vaca relating to his travels in southeastern Texas in 1528. In the words of de Vaca (Nunez, 1905):

> The Indians go about with a fire brand, setting fire to the plains and timber so as to drive off the mosquitos, and also to get lizards and similar things which they eat, to come out of the soil. In the same manner they kill deer, encircling them with fires, and they do it to deprive the animals of pasture, compelling them to go for food where the Indians want.

The U.S. Department of Agriculture had a number of workers in the Southwest in the late nineteenth and early twentieth centuries studying range problems. The observations of these men are particularly valuable, coming as they did, at a time when the pressures of settlement were beginning to be expressed in vegetational changes. Three of the early workers in southern Arizona, David Griffiths, botanist with the Arizona Agricultural Experiment Station and the U.S. Department of Agriculture; E. O. Wooton, with the U.S. Department of Agriculture; and J. J. Thornber, botanist at the Arizona Station, were all active in southern Arizona about the beginning of the twentieth century, and all studied or commented on the effect of fires in the desert grassland. W. L. Bray and O. F. Cook were equally active with the U.S. Department of Agriculture in Texas at about the same time.

The conclusions reached by these early workers as set down in their own words provide us with a credible record. Note, as we examine these, the degree to which they agreed on the general effect of grassland fires, even though they did not all work in the same areas.

First, quoting Bray (1901) in his observations and conclusions on Texas ranges:

> Regarding the establishment of woody vegetation it is the unanimous testimony of men of long observation that most of the chaparral- and mesquite-covered country was formerly open grass prairie.... Apparently under the open prairie regime the equilibrium was maintained by more or less regular recurrence of prairie fires. This, of course, is by no means a new idea, but the strength of it lies in the fact that the grass vegetation was tolerant of fires and the woody vegetation was not. It was only after weakening the grass floor by heavy pasturing and ceasing to ward off the encroaching species by fire that the latter invaded the grasslands.

Bray was familiar with the Edwards Plateau of Texas at a time when parts of it were still primarily grassland in spite of the rapid encroachment of shrubs. Note, for example, his comments on this area at about the turn of the century (Bray, 1904a):

> If the Edwards Plateau were an uneroded highland, its vegetation would, under natural condition, be open grass prairie. As a matter of fact . . . it is in process of transformation from a grass prairie to timberland. This transformation is being hastened by the interference of man. Both agriculture and grazing have operated to prevent the recurrence of prairie fires, which, so long as they were periodic, kept the field swept clean of woody vegetation. The grass throve under this burning; seedlings of trees were killed.

And Bray (1904b):

> This struggle of the timberlands to capture the grasslands is an old warfare. For years the grass, unweakened by overgrazing of stock, and with the

fire for an ally, held victorious possession. Now the timber has the advantage. It spreads like infection. From the edge of the brush each year new sprouts or seedlings are pushed out a few feet farther, or, under the protection of some isolated live oak or briar or shrub, a seedling gets its start, and presently offers shelter for others. This has been going on all along, but in former days these members of the vanguard and the scattered skirmishers were killed by the prairie fires, and the timber front was held in check or driven back into the hills.

Cook (1908) also writing primarily of Texas, and at about the same time, concluded:

Before the prairies were grazed by cattle the luxuriant growths of grass could accumulate for several years until conditions were favorable for accidental fires to spread. With these large supplies of fuel, the fires that swept over these prairies were very besoms of destruction not only for man and animals, but for all shrubs and trees which might have ventured out among the grass, and even for any trees or forests against which the burning wind might blow. That such fires were evidently the cause of the former treeless condition of the southwestern prairies is also shown by the fact that trees are also found in all situations which afford protection against fires. . . . Nor is there any reason in the nature of the climate or the soil why trees should not thrive over the vast areas of open prairie land.

In the same publication he comments as follows on the use of fire by the early white settlers:

Settlers in south Texas early adopted the practice of burning over the prairies every year; partly to protect their homes against fires, partly to give their cattle readier access to the fresh growth of grass. The fires were often set near the coast, the strong breeze which blows in from the Gulf spreading the flames over many square miles. While the grass was still abundant these annual burnings were able to keep the woody vegetation well in check. . . .

The conclusions reached by Bray and Cook in Texas had their essential counterpart in southern Arizona studies at about the same time (Griffiths, 1910):

The probability is that neither protection nor heavy grazing has much to do with the increase of shrubs here, but that it is primarily the direct result of the prevention of fires. . . . Previously, before the country was stocked, it probably produced more grass than it does now, and was frequently burned over, the fire extending down as far as vegetation would permit. Such burning did comparatively little injury to the grasses, but was very destructive to all small shrubs; consequently, these were able to exist only along the sandy washes where the grasses were least productive, and upon the lower areas, where fires did not molest them. . . . The main factor, through, in the opinion of the writer has been that of fire.

It is firmly believed that were it not for the influence of this factor the grassy mesas would to-day be covered with brush and trees. . . . In short, the same laws apply here that govern in our great prairie states . . . where the treeless plains were kept so by frequent fires.

Thornber (1907), writing of the same area, noted that even at that time fires were not infrequent, two of them occurring within a space of about 5 years. One of these he recorded as burning unchecked for 2 days. A variety of shrubs was killed by these fires including creosote-bush, burroweed, mormon tea (*Ephedra trifurca*), desert hackberry (*Celtis pallida*), velvet mesquite, and palo verde (*Cercidium*). In conclusion he stated: "That such fires burning over the mesas and foothills have not been uncommon in times past may be judged by the fact that in many places abundant remains of charred stumps of at least 10 years duration are frequently met with."

Roughly 30 years later I had the idea that some vestiges of charring might still be found if one looked in the right places. With this in mind, I examined a number of the older mesquite trees along some of the major washes in the same area where Thornber's observations had been made. Thirty-two trees with a basal diameter of 12 inches or more were examined, 17 with a diameter of 12–14 inches, and 15 with a diameter greater than 14 inches. Out of the 17 in the smaller size class only one bore visible charcoal scars as contrasted with two-thirds of those in the 14-inch-plus class (Humphrey, 1953). These trees were growing on the rock banks or bottoms of washes where grass growth had apparently been too sparse to burn with enough heat to kill the trees. The fires had, however, damaged the trees enough to leave evidence of their occurrence when I examined them in 1935.

Although fires in this area (The Santa Rita Experimental Range, which lies about 40 miles south of Tucson) are almost unknown today, the written record indicates that they occurred rather frequently as late as 1916. Wooton (1916) has provided us with the last known statement of the prevalence of fires on this experimental range:

The complete protection of the reserve for a number of years has resulted in a rather heavy crop of dry grass, which burns readily, especially in the dry hot weather of May or June, just before the summer rains begin. Several such fires have occurred, due to lightning, carelessness of passers, or incendiarism. . . . In June, 1914, occurred one of the largest and hottest fires, which burned over about four sections [2560 acres] of the heaviest grass.

Although there are no known photographs showing fires in this area, there are several that show the vegetational changes that have taken place (Figs. 12 and 13).

Fig. 12. Cutting hay on the Santa Rita Experimental Range in 1903. Trees in the background occur largely along drainages. Compare this photograph with Fig. 13 (U.S. Forest Service photo.)

Fig. 13. Same area shown in Fig. 12 but rephotographed in 1964, 61 years later. An intermediate photo taken in 1941 only 38 years after the first, showed an equally dense stand of woody plants and little grass, much as in 1964 (U.S. Forest Service photo.)

2. More Recent Studies

A study was made in this area (Reynolds and Bohning, 1956) to determine the effect of fire on specific species of shrubs and grasses

The ground vegetation, although rather sparse, was sufficiently heavy to permit a fire to run and give a fairly complete burn. The effects of the fire, as expressed in percent mortality of individual species, showed the following: burroweed, 88; barrel cactus (*Ferocactus wislizeni*), 67; jumping cholla (*Opuntia fulgida*, 44; cane cholla (*O. versicolor*), 42; prickly pear (*O. engelmannii*), 28; and velvet mesquite, 9. *Calliandra eriophylla*, the most valuable browse plant in the area, was burned to the ground, but none of the plants was killed. Four years later the *Calliandra* plants had a greater crown density than on the adjacent unburned check area.

It might seem, offhand, that the kills reported in this study might not be sufficient to maintain a range essentially free of woody plants. The authors point out, however, that difficulty was experienced in obtaining a uniform burn due to inadequate fuel in the vicinity of mesquite trees. Previously, when fires occurred frequently, there were apparently no mature mesquites on this and similar areas. As a consequence, the grasses grew unhampered by competition from trees, and the fires would have been hot enough to be much more destructive than they now are. An additional feature of importance was the relative frequency of occurrence of fires, a frequency that prevented most of the woody plants growing to larger than seedling or near seedling size. These would have been more readily killed than larger, more mature plants. In addition, only occasionally would they have matured sufficiently to produce seed. This would have further reduced the potential number of new plants.

Cable (1967) studied the effects of burning in an area similar to and adjacent to the one investigated by Reynolds and Bohning. His observations were in part as follows: "Mesquite seedling establishment was significantly higher on the unburned area than on the burned area. Burroweed was easily killed by burning," the kill ranging between 92 and 98%. From 32 to 64% of the cactus plants were killed by one burn, few by a second.

Photographs of this area taken in 1903 show an open grassland with a dense stand of grasses capable of carrying a hot fire (Fig. 12). By the time of Cable's 1967 study the area had for many years been dominated by a variety of woody species. The remaining grasses were often too sparse to carry a fire. Thus fire effectiveness as a medium for controlling most woody plants must have been far different than it had been some 50–60 years earlier.

A nearby study by Blydenstein (1957) indicated the relationship between velvet mesquite mortality and size. He concluded that although a single fire does not result in a high mesquite mortality it does reduce the size of many of the larger trees by killing them back to ground level and that there is a high correlation between tree size and fire damage, i.e., fire is most efficient in controlling newly established stands.

G. CONCLUSIONS

There still remain in the desert grassland extensive areas that are essentially free of woody plants. The question is sometimes raised whether these uninvaded grasslands may not refute the universality of the fire-control theory for the desert grassland. Whether they do or not cannot be stated with certainty. It is unfortunate that the answer to a question of this sort must be based on opinion, no matter how well-considered such an opinion may be. However, with the assumption that an opinion based on many years of study may be largely objective and consequently of value, my opinion and supporting facts are given for what they may be worth.

In contrast to much of the Great Plains area, most of the open, grassy portions of the desert grassland appear to have a soil and a climate suited to growth of woody plants. Insofar as data are available no caliche or other hardpan layer seems consistently to typify these grasslands. The precipitation is similar to that on adjacent areas that do support shrubs. As further evidence that these grasslands are capable of growing ligniphytes, many of them are being encroached upon by trees and shrubs from nearby drainages, hillsides, or other areas.

Occasional fires still occur in the desert grassland. Although infrequent, these fires do act as a deterrent to shrub invasion. An additional point that may explain in part today's still rather extensive ligniphyte-free desert grassland ranges, and one that is rarely considered is that very little time has elapsed in the brief period since the days of the Indian and uncontrolled fires. Fires have been controlled in this area largely only during the last 60–70 years, a short period in terms of invasion and establishment of the trees and shrubs involved.

Initial invasion of the open grassland may be seen in many areas even today. This invasion is usually linked with cattle movements from mesquite-infested areas. There seems to be little doubt that lack of invasion is often due to failure of seeds to be transported to the open grasslands. Recent invasions often are an indication of a change in the grazing pattern that results in seeds being brought in where none were before.

The fire history of the desert grassland raises the question: Is fire a practical tool to use today in controlling woody plant invasion in this type? The answer cannot be given as a clear-cut yes or no. Many aspects of the total picture have changed during the last 75–100 years. When the desert grassland ranges were grazed only by deer, antelope.

and other game animals, much of the annual production of forage was left ungrazed and served as a source of fuel that carried fires readily and produced a flame hot enough to kill most woody plants that might lie in its path. By comparison, today the grasses usually do not have either the density or volume to carry a hot fire, or often, any fire at all. Grazing removes much of the growth that is produced, further reducing the possibility of obtaining an effective burn.

The size of the plants that must be controlled today adds to the difficulty of obtaining effective control by fire. Previously, when fires were frequent, it is probable that few trees or shrubs in the grassland survived long enough to develop much beyond the seedling stage. As small plants they were highly susceptible to killing by fire and rarely lived to be old enough to set seed. This restriction of a potential seed source must, in itself, have been of no little importance in maintaining the grassland in an essentially shrub-free state.

It takes little imagination to visualize the change that has taken place in the potential for control of the woody plants by fire. Instead of the occasional small shrublike plants lying in the path of the hot fires of former years, today we have mature trees, often with stem diameters up to 10 and 12 inches and possessing relatively thick, fire-resistant bark. The former hot fires that might have been effective in killing even these large trees are today too weak to be lethal. On the other hand, they will kill new reproduction, thus not only reducing the rate of shrub invasion but actually reversing the trend. In many situations recurrent burning at 5- to 10-year intervals would probably be an effective and cheap method of control. A long-term study is needed to determine not only the broader aspects of this problem but also specific items such as the most desirable fire frequency, effectiveness with various kinds of grasses and woody plants, and the relationship between fire and size of woody plants.

References

Benson, L. (1941). The mesquites and screwbeans of the United States. *Amer. J. Bot.* **28**, 748–754.

Blaisdell, J. P. (1953). Ecological effects of planned burning of sagebrush–grass range on the upper Snake River plains. *U.S., Dep. Agr., Tech. Bull.* **1075**.

Blydenstein, J. (1957). The survival of velvet mesquite (*Prosopis juliflora var. velutina*) after fire. M.S. Thesis, University of Arizona, Tucson.

Bray, W. L. (1901). The ecological relations of the vegetation of western Texas. *Bot. Gaz., (Chicago)* **32**, 99–123, 195–217, and 262–291.

Bray, W. L. (1904a). Forest resources of Texas. *U.S., Dep, Agr., Forest. Bull.* 47, 1–71.

Bray, W. L. (1904b). The timber of the Edwards Plateau of Texas; its relation to climate, water supply, and soil. *U.S. Bur. Forest. Bull.* 49.

Cable, D. R. (1967). Fire effects on semidesert grasses and shrubs. *J. Range Manage.* 20, 170–176.

Clements, F. E. (1920). "Plant Indicators. The Relation of Plant Communities to Process and Practice." Carnegie Institution, Washington, D.C.

Cook, O. F. (1908). Change of vegetation on the south Texas prairies. *U.S. Dep. Agr., Bur. Plant. Ind., Circ.* 14.

Griffiths, D. (1910). A protected stock range in Arizona. *U.S. Dep. Agr., Bur. Plant. Ind., Bull.* 177.

Hastings, J. R., Turner, R. M. (1965). "The Changing Mile: An Ecological Study of Vegetation Change with Time in the Lower Mile of an Arid and Semiarid Region." Univ. of Arizona Press, Tucson.

Humphrey, R. R. (1953). The desert grassland, past and present. *J. Range Manage.* 2, 173–182.

Humphery, R. R. (1958). The desert grassland—A history of vegetational change and an analysis of causes. *Bot. Rev.* 28, 193–253 (republished by the Univ. of Arizona Press, Tucson, 1968).

Humphrey, R. R. (1962). "Range Ecology." Ronald Press, New York.

Humphrey, R. R. (1963). Role of fire in the desert and desert grassland areas of Arizona. *Proc. 2nd Annu. Tall Timbers Fire Ecol. Conf.* pp. 45–62.

Nunez, Cabeza de Vaca, (1905). "The Journey of Alvar Nunez Cabeza de Vaca and his Companions from Florida to the Pacific, 1528–1536" (edited with an introduction by A. F. Bandelier).

Pechanec, J. F., Stewart, G., and Blaisdell, J. P. (1954). Sagebrush burning, good and bad. *U. S., Dep. Agr., Farmers' Bull.* 1948.

Reynolds, H. G., and Bohning, J. W. (1956). Effects of burning on a desert grass-shrub range in southern Arizona. *Ecology* 37, 769–777.

Robertson, J. H., and Cords, H. P. (1957). Survival of rabbitbrush (*Chrysothamnus* spp.) following chemical, burning, and mechanical treatments. *J. Range Manage.* 10, 83–89.

Shantz, H. L. (1924). The natural vegetation of the United States: Grassland and desert shrub. *In* "Atlas of American Agriculture," Part I.

Shreve, F. (1917). A map of the vegetation of the United States. *Geogr. Rev.* 3, 119–125

Shreve, F. (1942). The desert vegetation of North America. *Bot. Rev.* 8, 195–245.

Thornber, J. J. (1907). *Ariz., Agr. Exp. Sta., Annu. Rep.* 18.

Thornber, J. J. (1910). The grazing ranges of Arizona. *Ariz., Agr. Exp. Sta., Bull.* 65.

Wagoner, J. J. (1952). History of the cattle industry in southern Arizona. *Ariz., Soc. Sci. Bull.* 20.

Wedel, W. R. (1957). The central North American grassland: Man-made or natural? *Soc. Sci. Monogr.* 3, 36–39.

Wooton, E. O. (1916). Carrying capacity of grazing ranges in southern Arizona. *U.S., Dep. Agr., Bull.* 367.

. 12 .

Effects of Fire in the Mediterranean Region

Z. Naveh

I. Introduction

The important role of fire in the Mediterranean landscapes has often been overlooked or misunderstood because in recent times fire has operated in close combination with uncontrolled grazing and other human interference that led to land deterioration and desiccation. The object of this chapter is to present a balanced and unprejudiced view of the past and present impact of fire on Mediterranean ecosystems on the basis of current knowledge and to emphasize problems that need further study.

II. Natural Fire Conditions in the Mediterranean Region

A. MEDITERRANEAN "FIRE BIOCLIMATES"

The Mediterranean region, located in the basin of the Mediterranean Sea, is characterized by a more or less dry and hot summer and wet and mild winter climate. Similar Mediterranean-type climates, classified by Koeppen (1923) as Cs dry summer subtropical "olive" climates, are also found in other regions of the world between latitude 32° and 40° north and south, respectively, of the equator on the western side of the continents, namely in California, Cape Colony, central Chile, and southwestern Australia.

The UNESCO (1963) Mediterranean bioclimatic classification, which uses a xerothermic index of hot weather drought, can serve as a good indication of the high degree of fire susceptibility in these regions. The xerothermic values are derived from the sum of monthly indices of the degree of drought of a dry month—defined as a month in which total precipitation P (in mm) is equal to or less than twice the monthly temperature ($T = C°$; $P \leq 2T$)—by multiplying number of days without rain, mist, or dew (or half a day with mist or dew) by a coefficient of relative atmospheric humidity H, ranging from 1 ($H = 40\%$) to 0.5 ($H = 100\%$). Thus, the monthly xerothermic index, x_m, for July with 4 days of rain, 8 days with mist and dew, and a mean atmospheric humidity of 65%, is $x_m = [31 - (4 + \frac{8}{2})] \times 0.8 = 18$.

In the driest Mediterranean desert and subdesert climates with xerothermic indices of $300 < x > 200$, the natural vegetation canopy in general is not dense and continuous enough to carry through fires over large areas. However, in xero- and thermo-Mediterranean climates ($150 < x > 200$) and accentuated thermo-Mediterranean climates ($125 < x > 150$), and to a lesser degree also in attentuated thermo-Mediterranean climates ($100 < x > 125$), the native woody and herbaceous vegetation as well as rain-fed cereal fields are very susceptible to fire throughout the dry period, lasting from 4 to 8 months. Therefore these can be called true Mediterranean fire bioclimates. Ombothermic diagrams of locations typical for these climates are presented in Fig. 1. Their chief centers are North Africa, the Levant, and the southern and

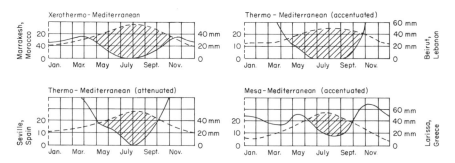

Fig. 1. Annual temperature and rainfall relations of four Mediterranean locations.

eastern Mediterranean European coasts, namely, Greece, Spain, southeastern France, southern Italy, Sicily, west and southern Turkey, and their adjacent islands.

A climatological factor that greatly enhances the fire hazard at the beginning and end of the dry season (in May–June and September–October is especially common in the eastern and southern Mediterranean Basin) is a heat wave accompanied by very low humidity, called a *sirocco, khamsin,* and, more recently, *sharav* (a Hebrew word meaning "heat of the land") (Winstanley, 1972). That most wildfires of maqui in Israel occur on such *sharav* days has been shown by Naveh (1973).

B. MEDITERRANEAN FIRE VEGETATION TYPES

The close connection among these Mediterranean bioclimates and fire-induced and maintained sclerophyll vegetation types of maqui (called *macchia* in Italy, *xerovuni* in Greece, *choresh* in Israel, *fynbos* in South Africa, *mallee scrub* in Australia, and *chaparral* in California and Chile) and garigue, or garrigue, has been recognized by early plant geographers (Griesebach, 1872). But most attention has been given to the physiological adaptive value to drought in the epharmony in physiognomy and morphology of these: mostly evergreen nanophanerophytes, with small leptophyll or nanophyll sclerophyllous leaves and extensive root systems, and medophanerophytes and low chamaephytes, classified by Ruebel (1930) as *durilignosa* communities.

Walter (1968) summarized current knowledge of the ecology of this duriligneous vegetation of the Mediterranean region and concluded that it is composed of innumerable variants of different degradation and regeneration stages. These are very remote from the original, pristine forests that apparently once covered the fertile lowlands, cultivated now for several thousand years, and their botanical remnants are retained only on untillable, rugged hill and mountain lands. Such wildlands constitute 40–60% of the total land surface of the above described Mediterranean fire bioclimatic regions and are used chiefly as rough grazing lands or are converted into planted pine forests (Tisdale, 1967; Naveh, 1968). For detailed information on the extent of the Mediterranean forest and shrublands the reader is referred to LeHouerou (1973). On the basis of bioclimatologic criteria, resembling those mentioned above and used by the Emberger School of Phytogeography, LeHouerou (1973) distinguished among 13 major vegetation types. These range from high mountain coniferous forests, dominated by species of *Abies, Pinus, Juniperus,* or *Cedrus,* to sclerophyll woodlands and maqui, in more typical Mediterranean conditions, dominated by deciduous or evergreen oaks and other trees and shrubs, to spiny or aromatic, xerophytic dwarf shrublands and semidesert grasslands in the driest conditions.

In Israel Zohary (1962) followed Eig (1927) in the distinction between "maquis" (dominated by sclerophyll vegetation, chiefly evergreen, dense, low tree and shrub cover up to a height of 4 m, i.e., *Quercus calliprinos, Pistacia lentiscus, Ceratonia siliqua,* and *Phillyrea media*) and "garigue" (dominated by lower chamaephytes and nanophanerophytes, up to a height of 1 m, i.e., *Calycotome villosa, Satureja thymbra, Salvia,* and *Cistus*). This is contrary to the French definition in which garigue is composed of sclerophyll shrubs such as *Quercus coccifera* on calcareous soils, up to a height of 2 m, whereas maqui is both higher and denser and is not restricted to calcareous soils (Trabaud, 1973). Eig (1927) also introduced a third term, *batha,* derived from the Bible, for dwarf shrubland, up to 50 cm in height, dominated by *Poterium spinosum* and *Thymus capitatus* and sometimes composed of mixtures of dwarf shrubs, perennial grasses, hemicryptophytes, geophytes, and therophytes. The Spanish *tomilares* (dominated by *Thymus*) and the Greek *phrygana* come closest to this type. This is the lowest maqui degradation stage but, according to Zohary (1962), also forms stable "border climax" formations on the xeric ecotones between the Mediterranean and Irano-Turanian regions in Israel.

Among recent, detailed descriptions of these Mediterranean degradation types, those of Knapp (1965) for Kephalina in Greece and Freitag (1971) for southeastern Spain are also typical for vegetation of Mediterranean fire bioclimates.

Using Israel as a model for man's impact on the Mediterranean landscape, Naveh and Dan (1973) showed that his active intervention lasted for more than 50,000 years—longer than the presently prevailing climatic pattern. It proceeded throughout various phases of changing land use patterns and anthropogenic biofunctions of aggradation and degradation cycles. The latter dominated and were characterized not only by destruction of the original vegetation canopy and upper soil mantle but also by progressive deterioration of such controlling factors as topography, water regime, soil, and surface climate, leading to overall, man-induced desiccation of the landscape.

One of the main conclusions from study of recent biofunctions in Israel was that during the long phase of agricultural decay and population decline in the last centuries of Ottoman rule a new equilibrium was established on those noncultivated xero- and thermo-Mediterranean upland ecosystems, which were neither overgrazed and heavily coppiced nor completely protected but were moderately grazed and occasionally burned and coppiced. This man-maintained equilibrium among tree,

Fig. 2. Maqui tree (*Quercus calliprinos*) regeneration on ancient terrace wall with rich shrub, herb, and grass strata in western Galilee, Israel. These terraces were abandoned after the Crusader period and are moderately grazed and occasionally burned.

shrub, herb, grass, and geophytic strata contributed much to the biological diversity and attractiveness of the Mediterranean landscape and is one of its main assets for recreation and tourism (Fig. 2). It is now endangered by population explosion and increasing pressure of traditional, pastoral land use (Fig. 3) and also by the accelerating speed of urban sprawl, neotechnological erosion, and despoliation and pollution. A similar situation has been described in Greece by Agnostopolous (1967) and prevails also in other Mediterranean countries. It is reflected

Fig. 3. Degraded maqui with stone remnants of eroded terrace walls. These terraces, near an Arabic village, have been cultivated but were neglected and are now heavily overgrazed by cattle and goats.

in the disrupted patterns of shrubland and forest fires, which will be described in Section III.

III. Fire History of Mediterranean Ecosystems

In his review on the role of fire in management of California brush-lands, Shantz (1947) stated that the Mediterranean brushland is a "fire type" and "that this type was ever free from fire seems unlikely." He showed that in the Mediterranean region this type was subjected to repeated fires during historical times and by natural causes for hundreds and thousands of years; fires "have fixed the physiological habits of these plants and plant communities to such a degree that without fire they could not survive."

In Shantz's opinion, fire carried this type far beyond its natural area into grassland on the drier and lower sites and into forest on the higher and wetter sites: "the destruction of much of the forest by fire has resulted in the great extension of macchia over most of the area."

A. FIRE IN GEOLOGICAL AND PREHISTORICAL TIMES

At present, lightning seems to play only a minor role in naturally caused wildfires in the Mediterranean region, in contrast to California (Komarek, 1967b). According to Susmel (1973) it accounts for only 0.6 to 2.4% of the area burned annually. But there is no reason to suppose that natural fires, caused by lightning as well as by volcanic eruptions, have not raged since the late Pliocene and early Pleistocene and especially since the desiccation of the last interglacial period, when the present climatic fluctuations between wet and dry seasons and the Mediterranean flora and fauna finally became established (Butzer, 1964). Fire may have acted, therefore, as a dominant environmental agent together with drought in the evolution of the Mediterranean flora in a way similar to that recognized by Axelrod (1958) for evolution of the Madro-Tertiary geoflora of California.

As emphasized by Sauer (1956), Stewart (1956), Oakley (1961), and Komarek (1967a), fire was the first forceful tool for energy input and ecosystem manipulation used deliberately by primitive man. The findings of wooden ash and hearths in meso-Mediterranean and sub-Mediterranean bioclimatic locations in Spain (Howell, 1962) and Greece (Higgs *et al.*, 1967) and in drier subhumid Mediterranean locations in Israel in final Acheulian and Levalloiso-Mousterian levels at the el Tabun, Mount Carmel caves (Garrod and Bate, 1937) provide archeological evidence that this is true also for the Mediterranean region. The rich faunal collections in the Carmel caves point to the existence of advanced and diversified hunter–gatherer economies of the Upper Pleistocene "Palestinian Neanderthal" man. He may have used fire to open the dense forest and brush thicket and thereby created ecotones and secondary successions, rich in edible shrubs, grasses, herbs, and tuber plants for man and game. His use of fire also facilitated hunting and gathering.

Tchernov's (1968) paleontological analyses of Quaternary fauna of these caves showed a steady increase in Mediterranean steppe- and garigue-dwelling rodents from the Acheulian levels onward and the appearance of rock-dwelling rodents from the Upper Paleolithic-Natufian period onward. This replacement of more mesic Mediterranean wood and maqui dwellers was interpreted by Tchernov (1972) as an indication of increasing aridity. However, in view of the climatic fluctuations occurring during these periods it seems more plausible to regard the progressive desiccation of habitats as a result of the increasing impact of man on his surroundings through fire. As emphasized by Naveh and Dan (1973), man-induced fire became the first controlling factor of the microsite and ecosystem modification in this anthropogenic biofunction by opening the dense forest and brush cover and creating more and more exposed and rocky habitats.

Thus, after natural fire operated for many thousands of years as a major force in the biological evolution of this region, it also became the first vehicle of the cultural evolution which in turn effected the further evolution of biota and shaped their landscapes for at least 50,000 years.

B. FIRE IN HISTORICAL TIMES

The Bible, Talmud, and the classical Greco-Roman sources provided ample proof for the abundance of both natural and man-caused fires in Mediterranean lands in early historical times.

Fire was mentioned several times in the Bible in connection with lightning as "the fire of God" and with the "heat of the summer drought." Thus in the book of Job 1:16, "the fire of God is fallen from heaven and hath burned up the sheep, and the servants and consumed them," and at the same time, the "great wind from the wilderness" caused fire which destroyed the house and children of Job (1:19) . . . and the prophet, Joel (1:19) says: "Lord, to thee I will cry for the fire hath devoured the pastures of the wilderness, and the flame hath burned all trees in the field."

The wide acquaintance of the ancient Jews with fire, its ecology and effects on plants, is demonstrated by many citations in the Bible and Talmud. Among the burned trees mentioned are the cedars of Lebanon which were used around 2600 B.C. as timber for export to Pharaoh Snefru in Egypt. Some other plants mentioned in the Bible are listed in Table I.

Probably the first man-made fire was set by Samson in revenge (Judges 15:5). He used firebrands between the tails of 300 foxes to burn the shocks, corn, vineyards, and olives of the Philistines. Another cause of fire, mentioned in Exodus (12:16) and still very common today, is carelessness during burning of stubble fields and thistles, which was then already a common method for field clearing and preparation and for obtaining ashes for use as fertilizer (Felix, 1963).

Liacos (1973) cited several sources in "Iliad" from Homer as well as Thoukidides which showed that naturally started fires were very common in ancient Greece. He also cited Virgil, Kassianos Vassus, and Xenophon on beneficial effects of stubble burning and wood ashes as fertilizers and for control of weeds and pests.

The use of fire by pastoralists in forest and woodlands was mentioned by Virgil ("Aneid" X, 405–411) as "the scattered fires set by the shepherds in the woods, when the wind is right." According to LeHouerou (1973), there is evidence of ancient periodic burning of forests in southern France for improvement of pastures, and Semple (1931) described this as a system practiced by shepherds and goatsmen in the dry season in all Mediterranean countries since the downfall of the Byzantine Empire and the invasion of Arab Bedouin pastoralists from the desert. Such pastoral burning thus became part of the above-mentioned multivariate anthropogenic biofunction in which burning was closely interwoven with grazing and cutting.

TABLE I

PLANTS MENTIONED IN THE BIBLE IN CONNECTION WITH FIRE[a]

Biblical name	Source	Scientific name[b]
Thorns	Nahum 1:10	*Poterium spinosum*
Thistles	Exodus 22:6	
Thorns	Isaiah 33:12	
Thorns	Psalms 118:12	
Stubble	Isaiah 47:14	
Stubble and chaff	Isaiah 5:24	
Stubble and chaff	Joel 2:5	
Thorns and briers	Isaiah 9:18	*Daucus maximus* (?)
Thorns and briers	Isaiah 10:17	
Bush	Exodus 3:2	*Rubus sanctus, Acacia tortilis, Zysyphus Spina christi*
Pasture of the wilderness	Joel 1:19,20	
Trees of the field	Joel 1:19	
Green and dry trees	Ezechiel 20:47	
Forest	Jeremiah 21:14	
Forest	Ezechiel 20:47	
Forest	Isaiah 10:17	
Forest thicket	Isaiah 9:18	
Wood	Psalms 83:14	
Bramble	Judges 9:15	*Lycium europaeum* or *Rhamnus palaestina*
Cedars of Lebanon	Judges 9:15	*Cedrus libani*
Cedars of Lebanon	Zechariah 11:1	*Cedrus libani*
Oak	Isaiah 6:13	*Quercus calliprinos* or *ithaburensis*
Teil tree	Isaiah 6:13	*Pistacia*
Olive tree	Jeremiah 11:16	*Olea europea* or *O. europea* var. *Oleaster europea* var.
Olive tree	Judges 15:5	
Vine	Judges 15:5	*Vitis vinifera*
Corn	Exodus 22:6	
Flax	Judges 15:14	*Linum usitatissimum* var. *vulgaris*

[a] Based on "Historical Aspects of Fire in Israel" (A. Derman and Z. Naveh, unpublished manuscript).

[b] According to Felix (1957).

In the last century such "Brandkulturen" also were described by explorers of Palestine and Syria (Anderlind, 1886) and more recently by Rickli (1942) in Corsica and other western Mediterranean countries as well as by several botanists cited by Shantz (1947).

C. FIRE IN PRESENT TIMES

The extent of wildfires at present in the Mediterranean region was described by LeHouerou (1973), who estimated that each year close to 200,000 ha of Mediterranean forest and shrublands are burned, causing direct damage of at least $50,000,000. In the "red belt" of southern France and Corsica, each piece of forest is burned every 25 years, and, in Greece, from 1956 to 1971, the average number of wildfires per year amounted to 612 on 10,500 ha, with damage estimated at $7,360,000 (Liacos, 1973). The main forest species affected are *Pinus halepensis* and *P. brutia* of the lower Mediterranean zones.

In Israel there were more than 724 wildfires in 1973, of which 285 caused considerable damage, chiefly to planted pine forests, natural grass pastures, and cereal fields. In 1973 the Jewish National Fund Forestry Department spent $300,000 (10% of its budget) on fire prevention, but, because of the rapid increase in forest recreation (from 50,000 visitors in 1958 to 1,100,000 in 1972), the number of fires in the highly flammable dense pine afforestations is rising steadily.

A similar situation prevails in other Mediterranean countries with increased forest use by tourists. At the same time, increasing pressure on upland pastures in countries and regions with traditional Mediterranean land use, especially in the Levant and North Africa, has led to such heavy overgrazing and defoliation of woody plants that very little fuel has been left for fires, even in the dry season. Such a situation is described by Naveh and Dan (1973) in the Hebron Mountains and southern Judean hills where renewal of the erosive degradation cycle led to bare soil, rocks, and "Asphodelus deserts."

It should be kept in mind, however, that in many countries and locations where traditional pastoral use has been abandoned and in nature reserves and parks in maqui and woodland, which are protected from fire, the accumulation of dry grass fuel, litter, and debris as well as brush encroachment will lead finally to hotter and more devastating wildland fires. Hence, fire exclusion policies may lead to similar undesirable situations, as already encountered in the California chaparral areas (Hanes, 1971).

D. CONTROLLED BURNING

In contrast to California, where prescribed burning has become an important tool in brush, range, and forest management (Biswell, 1967), rational use of fire in Mediterranean wildlands is still in its infancy.

Large-scale trials and farm operations of maqui brush conversion in

Israel have shown that controlled burning followed by reseeding of perennial grasses, rotational deferred grazing, and selective arboricidal control of undesirable woody resprouters can lead to a manyfold increase in pasture output of 1000–1500 Scandinavian Feed Units/ha/year (Naveh, 1960b, 1968). At the same time such treatments can also lead to increased water yields from the converted watersheds and catchment areas (Soil Conservation Division, Israel, 1964). However, heavy initial investments and lack of manpower are preventing further application of these improvement methods in Israel.

In recent years, extensive studies on effects of fire on Mediterranean shrub ecosystems were conducted in southern France (Trabaud, 1973), and use of controlled burning in management of coniferous forests, maqui, and high mountain grasslands was studied in Greece (Liacos, 1973) (Figs. 4 and 5). It is hoped that the results of these studies will lead to reevaluation of the entrenched pattern of wholesale condemnation of fire, which has hampered both study and use of fire in management of Mediterranean wildlands.

IV. Effects of Fire on Mediterranean Ecosystems

A. Effect on Soil and Erosion

One of the chief objections to burning in Mediterranean lands is its alleged detrimental effect on soil fertility and stability (Shantz, 1947). However, studies in western Galilee in Israel (Naveh, 1960b) support the warning voiced by Sampson (1944b) in California against rash and broad generalizations and for the need for considering the effects of each burned area in terms of its specific ecological conditions and fire and grazing history.

These studies have shown that periodic burning does not impair the granular structure and high infiltration capacity of the brown rendzina and terra rosa soils, developing under dense maqui canopies, and, even after loss of about a fifth of the organic matter, a tenth or more is still retained in the upper 20 cm of these shallow soils and 13–16% in the upper 5 cm of the A_1 profile, which is affected by the fire. This can explain the almost complete absence of traces of runoff, soil splashing, movement, and erosion, even on fire-denuded slopes of 30–40%, after heavy rains in the first winter months after burning. The compacted ash layer of the incinerated litter debris and semidecomposed Aoo and Ao profiles, provides conditions similar to those described by Bently and Fenner (1958) in California after hot chaparral fires as "white ash

Figs. 4 and 5. Experimental prescribed burning of *Pinus halepensis* forest in Greece. Fig. 4. (top) Before pruning and burning. Fig. 5. (bottom) After pruning and burning. (Photographs courtesy of L. Liacos.)

seedbeds." They are ideal for rapid development of natural spreading and reseeded perennial grasses and other invading herbaceous plants.

On the other hand, these studies also showed that in less favorable conditions of low shrub cover, less fertile and more erodible soils, such as the highly calcareous pale rendzinas, the hazards of postfire soil erosion were much greater. This was especially so if these soils had been disturbed and compacted by uncontrolled grazing prior to burning. However, according to our experience in western Galilee in Israel, even here, under well-distributed rainfall of 600 mm and more, if livestock grazing is postponed until the second spring after a fire, the rapidly regenerating woody and herbaceous vegetation can ensure sufficient soil protection to prevent further degradation.

Unfortunately, the greatest damage to the soil–vegetation system is caused not by the fire itself but by the uncontrolled grazing and exploitive management following these wildfires in Mediterranean brush- and grasslands. This was also emphasized by Liacos in Greece (1973). In an attempt to evaluate the true role of fire on these ecosystems a clear distinction should be made between these different situations.

B. EFFECT OF FIRE ON VEGETATION

As stated by LeHouerou (1973), the effects of fire in the Mediterranean region are very diverse not only because of the great complexity of plant communities and interference of grazing and cutting with burning but also because of the different responses to type and intensity of fire, its season, and frequency. The latter determine the degree of "pyrophytism," namely, regeneration and/or reproduction after fire (Kunholz-Lordat, 1939, 1958). For this reason, the above "fire sequences," as well as those mentioned by Walter (1968) from Braun-Blanquet (1925) and Bharucha (1932), which are based on circumstantial field observations without control of these factors, should be treated with reservation. On the other hand, current research in France, Greece, and Israel is still in an early stage. We shall have to be content, therefore, with a broad summary of facts that seem to be most relevant for the understanding of the Mediterranean postfire auto- and synecological behavior and point to gaps in our knowledge in this respect.

In a comparison of Mediterranean ecosystems in California and Israel, Naveh (1967) emphasized striking similarities in response of individual plants as well as plant communities to fire, based on their great resilience and recuperative powers. This is manifested by a process of postfire "autosuccession" (Hanes, 1971) of the sclerophyll woody vegetation, lasting, according to regional climatic and microsite conditions, from 3 to 6 years. It is obscured by a short interlude of herbaceous plant dominance on the fire-denuded, temporarily opened niches. This autosuccession is based on the vegetative regeneration from basal stems and adventitious roots of the same, burned individuals and/or on propagation by seeds and thereby shifts in generation, and it is completed with full reencroachment of the woody canopy. The results of LeHouerou (1973) and Trabaud (1970, 1973) verify this process also for western Mediterranean shrub ecosystems. Naveh (1967) further postulated that fire played a similar role in the shrub ecosystems of California and Israel by converting the woody plants to a more rejuvenated and vigorous state and by mobilizing nutrients tied up in the highly lignified wood and dead and very slowly decomposing litter. The temporary opening of favorable niches, rich in nutrients and moisture, is combined in some instances with removal of phytotoxic and antibiotic agents in the litter and duff. This apparently improves biological and chemical soil conditions and leads to a short-lived increase in perennial grasses, geophytes, and therophytes. The presence of such fire-nonstable, germination-inhibiting allelopathic agents has been shown in the chamise chaparral in California by Naveh (1960a) and McPherson and Muller (1969). The action of allelopathic agents was also shown in reseeding trials in maqui in Israel (Naveh, 1960b) and by scarce and scattered distribution of herbaceous plants under certain shrubs and trees, even where light and space were not limiting. But further experimental study is needed on allelopathic agents and especially on the role of fire in their destruction and inactivation, as well as in fire-induced changes in soil microbiota.

The great resilience of Mediterranean ecosystems to burning can, therefore, be comprehended best as complex cybernetic feedback responses to fire: positive feedback responses of increased, postfire vegetative and reproductive activity overcome the fire stress and negative feedback responses, enabling avoidance of fire damage and stress in space and time. Plants that are equipped chiefly with positive feedback mechanisms have been called "active pyrophytes" by Kuhnholtz-Lordat (1939, 1958) and Trabaud (1970) and "passive pyrophytes" when equipped with negative feedback mechanisms. However, in many cases both types of responses may act simultaneously and balance each other. This is especially true for herbaceous plants, as will be described below.

1. Responses of Woody Plants to Fire

There is probably no woody species, abundant naturally in the Mediterranean fire bioclimates described above that cannot survive fire with the help of such feedback mechanisms. In most cases these are positive, *vegetative*, namely, epicormic resprouting from dormant buds of the root crown and sometimes also from adventitious buds of lateral roots and stems and from old shoots that were not destroyed by the fire, and *reproductive,* namely, postfire volunteers from fire-stimulated or -initiated seed germination. Naveh (1973) subdivided Mediterranean woody plants of Israel into (1) *obligatory resprouters,* including most sclerophyll trees and shrubs that rely solely on vegetative regeneration, and (2) *facultative resprouters*—all chamaephytes, which regenerate both by resprouting and also by seeds and therefore do not depend solely on resprouting. The list of plants from Israel and the eastern Mediterranean region presented in Table II can be enlarged by many more species, mentioned by LeHouerou (1973), to embrace the whole Mediterranean range.

Examples of some of the most important obligatory resprouters are the evergreen Kermes oak, *Quercus coccifera,* and its east Mediterranean vicariant, *Quercus calliprinos,* and the evergreen shrub, *Pistacia lentiscus.* All these species are characterized by vigorous resprouting from root crowns and suckers from extensive, deep, and laterally branched root systems of the *"Olea"* type (Zohary, 1962), commencing shortly after the fire, even in summer and continuing throughout the whole year. They also resprout from adventitious roots, but *Pistacia lentiscus* also branches off laterally from prostrate, leafy twigs that take root. Thus *Pistacia* provides even more effective soil protection and competition to other plants than these oaks, as early as the first year after burning (Fig. 6). According to LeHouerou (1973), Kermes oaks have been in equilibrium for centuries with periodic burning of the pyro-stable vegetation, but, when fires are too frequent, they may be killed, leaving the ground to short open swards of *Brachypodium ramosum.* Its spiny, hard leaves and rigid, short branches affort protection from overgrazing to mature and nonburned oaks. But the lush, resprouting shoots and leaves are highly palatable and are browsed eagerly by goats and cattle in the first year after the fire. This also is true for other sclerophyll resprouting trees and shrubs (Naveh, 1960b). It is possibly the synergetically acting defoliation pressure of frequent burning, followed by heavy browsing, that causes depletion of carbohydrate reserves in roots and thereby enhances the elimination of these shrubs. As mentioned above, controlled experimental studies in which both factors can be separated and kept constant are necessary to answer this question.

In a comparison of productivity of sclerophyll vegetation after fire in southern France, California, and South Australia, Specht (1969) showed that in the first 10 years of postfire autosuccession, the annual growth of 1500 kg/ha found by Long *et al.,* (1967) for *Quercus coccifera* garigue is similar to that of the chaparral in California and the Australian mallee. However, Specht's own data for a vigorous stand of *Quercus*

TABLE II

REGENERATION BEHAVIOR AFTER FIRE OF SOME COMMON
MEDITERRANEAN WOODY PLANTS IN ISRAEL

Name of plant	Resprouting	Spreading by seeds[a]
Trees		
Pinus halepensis	−	+
Quercus calliprinos	+	−
Quercus ithaburensis	+	−
Quercus boisseri	+	−
Ceratonia siliqua	+	−
Styrax officinalis	+	−
Laurus nobilis	+	−
Arbutus Andrachne	+	−
Rhamnus alaternus	+	−
Pistacia palaestina	+	−
Phillyrea media	+	−
Cercis siliquastrum	+	−
Shrubs		
Pistacia lentiscus	+	−
Rhamnus palaestina	+	−
Calycotome villosa	+	+
Dwarf shrubs		
Sacropoterium spinosa	+	+
Cistus salvifolius	+	+
Cistus villosus	+	+
Salvia triloba	+	+
Teucrium creticum	+	+
Majorana syriaca	+	+
Satureja thymbra	+	+
Thymus capitatus	+	+
Climbers		
Rubia tenuifolia	+	−
Smilax aspera	+	−
Tamus communis	+	−
Asparagus aphyllus	+	−
Clematis cirrhosa	+	−
Lonicera etrusca	+	−
Prasium majus	+	−

[a] Only plants with pronounced postfire germination.

coccifera with 100% coverage (Table III) showed an annual growth increment of 4000 kg/ha.

In Israel, 100% of the obligatory resprouting trees, shrubs, and climbers regenerated after a burn, whereas the degree of root regeneration of

Fig. 6. *Pistacia lentiscus* shrub resprouting from roots after fire on rocky terra rossa soil.

facultative resprouters varied from fire to fire and site to site, according to age, vitality, and successional status of these plants, and, in the case of *Calycotome villosa,* sometimes only 50% regenerated. Also the extent of volunteering by seeds is variable and ranges from a few scattered seedlings around burned mother plants to more than a hundred seedlings per square meter, as in the case of *Cistus cretica* and *C. salvifolius.* But here, heavy intraspecific competition and dying back are conspicuous (Naveh, 1960b, 1973).

LeHouerou (1973) mentioned 12 different *Cistus* species as typical active pyrophytes, spreading by seeds and creating pure stands after fire, and Shantz (1947) cited many sources to show that *Cistus* is one of the main unpalatable fire followers of "the degenerate fire climax" as consequences of fire and heavy grazing.

Most facultative resprouters depend on the beginning of the rainy season for initiation of postfire root regeneration, but they produce seeds again in the first summer after the fire, and their volunteers produce seeds as early as the second year after germination and thereby early postfire seed production becomes an additional positive feedback mechanism to postfire resprouting and germination. The differential timing and rate of regrowth of obligatory resprouters (*Quercus calliprinos, Pistacia palaestina,* and *P. lentiscus*) as opposed to facultative resprouters

TABLE III

Garrigue Vegetation (*Quercus coccifera–Brachypodium ramosum* Association)
near Montpellier, southern France[a]

	Age (years from burning)				
	1	2	6	7	13
Dry weight of tops (kg/ha):					
Quercus coccifera	3789	8398	23,155	30,841[b]	49,530
Brachypodium ramosum	262	313	26	Trace	Trace
Carex halleriana	89	99	Trace	Trace	Trace
Dorycnium suffruticosum	408	13	19	6	Trace
Miscellaneous spp.[c]	182	51	140	92	163
Total dry weight of tops (kg/ha)					
95% confidence limits	±969	±1973	±2717	±3624	±6747
Height (cm)	20–40	45–60	60–75	100–120	200
Mean basal diameter of stems (cm)	0.4	0.7	1.25	1.45	2.1
Percentage dry weight to					
fresh weight of *Quercus coccifera*	58	58	62	60	63

[a] Statistics derived from a study of garrigue (Specht, 1969) on eroded Rendzinoid soil 13 km northwest of Montpellier, France.

[b] 27 stems per m².

[c] Miscellaneous species: *Aphyllanthes monspeliensis, Asparagus acutifolius, Avena bromoides, Brachypodium phoeniciodes, Daphne gnidium, Echium pustulatum, Euphorbia segetalis, Festuca ovina, Fumana ericoides, Genista scorpius, Helianthemum hirtum, Hieracium wiesbaurianum, Iris chamaeiris, Phillyrea angustifolia, Phlomis lychnites, Pistacia lentiscus, Rhamnus alaternus, Rubia peregrina, Rubusulmifolius, Rumex intermedius, Sanguisorba minor, Sedum* sp., *Teucrium chamaedrys.*

(*Calycotome villosa* and *Poterium spinosum*) after burn in July is shown in Fig. 7.

Another important group of plants with both vegetative and reproductive feedback responses after fire are acidophil *Ericaceae: Erica arborea* shrubs, typical for more humid Mediterranean conditions and the evergreen, broad-leaved *Arbutus* trees, *A. andrachne* and *A. unedo*, its west Mediterranean vicariant. Both of these are vigorous resprouters from root crowns, but not from suckers, and at the same time they also regenerate by seeds and can therefore dominate in burned forest and maqui (LeHouerou, 1973).

A special status with respect to response to fire is occupied by the two most abundant and highly flammable coniferous trees in typical Mediterranean fire bioclimates: *Pinus halepensis* and its close relative *P. brutia*, which are also widely used as the chief trees for afforestation. Their positive feedback response is restricted to propagation from seed

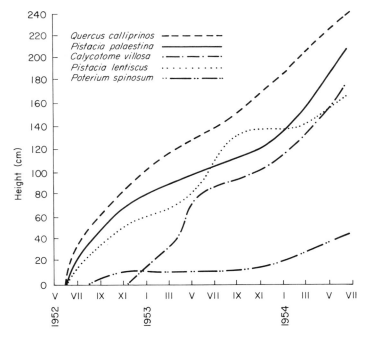

Fig. 7. Regrowth of obligatory resprouters (*Quercus calliprinos, Pistacia palestina,* and *Pistacia lentiscus*) and facultative resprouters (*Calycotome villosa* and *Poterium spinosum*) following burning in July.

cones, which burst from the heat and are spread around the burning mother trees. Like the "pioneer" volunteering dwarf shrubs, their heliophilous seedlings are capable of establishing themselves on poor, exposed, and rocky sites, in contrast to the obligatory resprouters requiring more favorable, sheltered, and humus-rich seedbeds not provided by the fire. In these Aleppo pines the lack of resprouting ability is compensated fully by postfire germination of hundreds of seeds benefiting from the temporary removal of the competition of the dense maqui understory and the lower ecological requirement for ecesis of the seedling. Similar to the case of the dwarf shrub seedling, germination is followed by a continuous process of natural thinning out which leaves a scattered, rejuvenated stand of pine trees under the regenerating shrub canopy or, in the case of burned planted pine forests, even more or less dense young, even-aged pine stands (Fig. 8). However, studies in Israel showed that regeneration can be prevented under more severe climatic conditions and heavy grazing pressure. Much more systematic field and laboratory research is needed for full understanding of the dynamics of postfire behavior of Mediterranean woody plants.

Fig. 8. Natural regeneration of *Pinus halepensis* from seeds 7 years after a wildfire. An unburned pine plantation is shown in the background.

The main negative feedback response to fire by woody plants is the development of defense mechanisms, such as thick bark, resisting the high temperatures. This is the case in *Quercus suber,* the cork oak, which according to LeHouerou (1973) is typical for fire stages of the *Q. faginea* forests in North Africa, Portugal, and Spain, where this oak has been killed out by burning. The cork oak forests and maqui also give way to maqui shrubs and trees with active pyrophytic behavior under too frequent burning regimes.

2. *Responses of Perennial Herbaceous Plants*

Under dense maqui and garigue stands, only an oppressed and relict understory of shade-tolerant and chiefly perennial plants—grasses, herbs, and geophytes—can be found. These occur chiefly along the tree and shrub edges and rock outcrops and openings, but they are much more abundant in open woodlands and comprise, together with annual grasses and herbs, the dense and highly fluctuating cover of open, Mediterranean grasslands. Their response to fire is very similar to that of the woody facultative resprouters and is regulated by a dual positive feedback mechanism of vegetative regeneration from fire-resistant, underground bulbs, corms, or stem bases, by reactivation of intercalary meristems and axillary buds, and through increased propagation by seeds.

From the few experimental data available it is apparent that typical fire followers, perennial grasses, such as *Oryzopsis miliacea,* not only resist high temperatures but even increase germination from seeds, up to temperatures of 90°C (Meiri, 1959). This grass belongs to an interesting group of shade-tolerant, calciphilic, and xeromorphic, erect-growing bunch grasses that regenerate, after burning in fall before the onset of rains, from the edges of the basal culms and from intercalary meristems and from lateral buds in the axils of charred leaves. In the second summer after maqui burned in the western Galilee in Israel, they did not enter dormancy but stayed in a stage of "semi-greening" throughout the whole summer. Naveh (1973) suggested that their striking phenoecological plasticity is of ecological advantage for utilizing flexible moisture and light regimes and may be a result of the evolution of these grasses as fire-induced edge plants of maqui and woodlands.

Other interesting plants are thermophilic, semiprostrate bunch grasses of tropical origin, like *Hyparrhennia hirta* and *Andropogon distachium.* These are abundant on sunny and rocky slopes in open shrub and batha stands and in frequently burned semiarid steppe—grasslands in Israel. In these grasses a similar mode of regeneration after fire is initiated even in the middle of the summer in semiarid regions. Similar to the many species of *Andropogon* of the United States (Daubenmire, 1968), they show a marked fire-stimulated increase in inflorescenses (Naveh, 1960b, 1973; J. Friedman, personal communication, 1973).

The number of perennial grasses establishing themselves after fire from burned mother plants and seeds is highly variable and in our studies in Israel reached several plants per square meter (Naveh, 1960b). Ample moisture and nutrient supply and lack of competition in the ash seedbeds of burned, dense maqui provide apparently ideal conditions for rapid development similar to that described first by Sampson (1944a), in California. This may explain the success of reseeding, not only of these typical maqui fire followers but also of other perennial grasses, such as *Phalaris tuberosa* and *Festuca arundinacea,* requiring, in general, deeper soils for establishment. In the second year after the burn, fresh weight production of forage of rocky and shallow slopes of reseeded *Phalaris tuberosa* averaged 6 kg/m^2 and of *Oryzopsis miliacea* 3.8 kg/m^2. These plants covered 30–40% of the slopes and, together with the annual pasture vegetation, provided in the following 3 years an average of 317 grazing days and 1580 Scandinavian Feed Units per hectare per year for beef cattle (Naveh, 1960b).

As can be seen in Figs. 9–13, the increase of perennial *Oryzopsis* and *Stipa bromoides* grasses in the first 3 years after the burn from volunteer (not reseeded) seedlings was striking. These shrub plots are typical for the shallow and rocky but fertile and well-structured brown rendzinas of the western Galilee, described above, and were completely protected from grazing. It is obvious that under these conditions these "active grass-pyrophytes" could spread freely and compete successfully with chamaephyte seedlings but not with the resprouting shrubs *Pistacia lentiscus* and *Rhamnus palaestina*, which rapidly re-covered the ground.

It is important to mention that, under usually prevailing conditions in the Mediterranean region, these plots would have been grazed heavily by goats and cattle soon after the burn and throughout most of the year. According to our observations on such grazed brush burns, all young seedling and regenerating perennial grasses are nibbled off and trampled down, but most of the aromatic and thorny facultative resprouting woody plants are rejected. The young lush shoots of resprouting shrubs like *Pistacia lentiscus* and, to a much greater extent, *Pistacia palestina, Ceratonia siliqua,* and *Quercus calliprinos* are browsed, and their regeneration is slowed down considerably. The "autosuccession" would, therefore, have been deflected in favor of *Salvia trilobia, Calycotome villosa,* and other unpalatable species present.

The chief comparable perennial grass, abundant in the *Quercus coccifera* garigue near Montpelier, in which Trabaud's (1973) elaborate burning studies were conducted, is *Brachyposium ramosum*. In his studies this species showed remarkable indifference to timing and frequency of burning and its "regeneration index" of postfire presence was 100% for all treatments. This grass is very similar in its shrublike regeneration behavior to the above-described *Oryzopsis miliacea* group and can be considered its ecological vicariant. Its "indifference" to fire may explain why it becomes dominant under frequent burning (but moderate grazing) regimes, as shown by Braun Blanquet (1925) and LeHouerou (1973).

The increase of perennial grasses after fire may be of economic importance for pasture improvement. But the increase of flowering *geophytes* on burned maqui, garigue, and batha, observed in Israel (Loeb, 1960; Naveh, 1960b, 1971), is of great importance for conservation of biological diversity and attractiveness of these Mediterranean upland ecosystems. Light measurements under different shrub and tree-cover types and densities on Mt. Carmel, reflecting previous fire and biotic disturbance history, revealed that under dense shrub and tree canopies where light intensity on clear spring days was less than 0.11 gm/cal/cm²/min, even sciophytic orchids such as *Cephalanthera longifolia* were smothered out. But on open, previously burned and disturbed low shrub cover of *Pistacia lentiscus,* with light intensities up to 1.3 gm/cal/cm²/min, many *heliophytic orchids,* such as *Ophris fuciflora,* colonized.

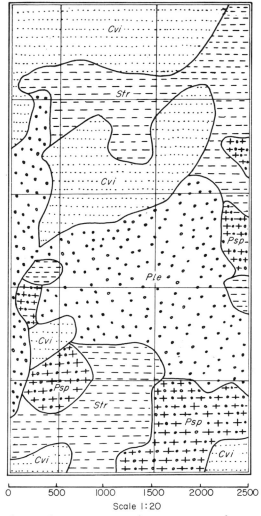

Scale 1:20

Figs. 9–13. Quadrants showing course of regeneration of maqui shrubs, dwarf shrubs, and perennial grasses during 30 months following burning. Western Galilee, Israel. Scales in millimeters.

Fig. 9. (above) January 2, 1951, before burning. See following pages for Figs. 10–13.

Key

Woody plants: *Ple, Pistacia lentiscus* ▦; *Rpa, Rhamnus palaestina* ▨; *Cvi, Calycotome villosa* ⬚; *Psp, Poterium spinosum* ⊞; *Str, Salvia triloba* ⊞; *STR, Satureja thymbra* ▩; *Eca, Ephedra campylopoda* ▤.

Perennial grasses: *Adi, Andropogon distachyus* ▥; *Oca, Oryzopsis caerviescens* ▨; *Omi, Oryzopsis miliacea* ▨.

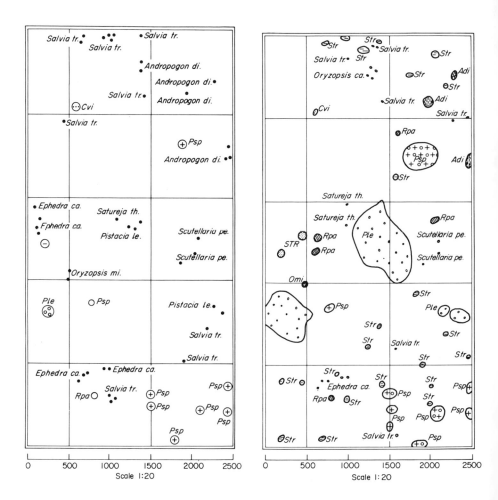

Fig. 10. March 30, 1951, after burning.

Fig. 11. July 28, 1951.

After burning of dense maqui in the western Galilee, many flowering geophytes appeared, among these, orchids, such as *Serapias vomeracea*, *Orchis papilionacens*, and *Ophris sintenissi*, which were not found under the adjacent closed brush canopy. The same is also true for several ornamental hemicryptophytes such as *Pterocephalus plumosus* and *Scutelaria peregrina* (Naveh, 1960b).

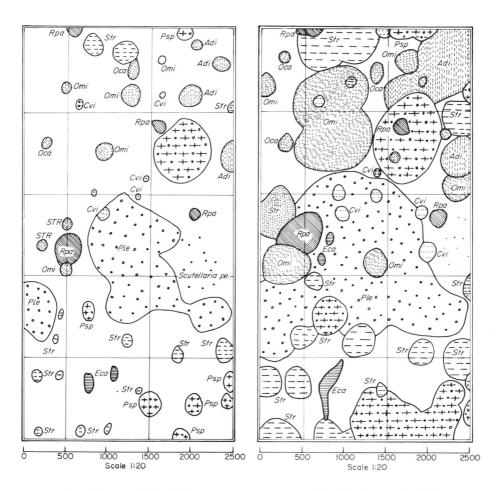

Fig. 12. August 28, 1952.   Fig. 13. March 10, 1953.

If the effect of fire on these plants is restricted to opening of the shrub canopy and the resulting increase in radiation intensity, this effect could also be achieved by mechanical clearing and thinning. But if fire also has a direct, stimulative physiological effect, as appears to be the case with germination of the perennial grass and reproduction and woody active pyrophytes, then it should become an indispensable tool in management of these ecosystems in protected areas for nature conservation and recreation (Naveh, 1971).

A good example of the detrimental effects of prolonged protection from fire on ecological diversity is the Um Rechan Maquis and Forest Reserve in the northern Samaria Mountains in Israel, which has been established for more than 50 years by the British Mandatory Forest Service and has turned since then into a monotonous, inaccessible, and stagnating brush thicket, comprised in its undisturbed parts almost exclusively of *Quercus calliprinos, Phillyrea media,* and *Pistacia lentiscus* (Naveh, 1971, 1973).

Only with the help of integrated ecosystem studies, concerned with postfire dynamics of nutrient cycling and energy flow in all trophic levels and not only with the botanical components, shall we be able to give a more conclusive answer on the role of fire in the maintenance of highest biological productivity, diversity, and neg-entropy in these Mediterranean shrub ecosystems.

3. *Response of Annual Plants*

In dense maqui and garigue stands, the distribution of annual plants is restricted to shallow soil patches and rock outcrops and to the edges of paths and openings. Also in batha with dense dwarf shrub cover their occurrence is scarce. But after fire, therophytes, like perennial grasses and facultative woody resprouters, can take advantage of the open, fertile niches and invade the burn in the first and second winter. Due to their rapid growth they provide heavy competition to the slower growing seedlings of perennial plants. As Litav *et al.* (1963) have shown, they are able under suitable conditions to inhibit or even prevent establishment of dwarf shrub seedlings. This also was the main reason for failure of reseeding with perennial grasses after controlled burning of open garigue and batha stands, which were rapidly invaded by these annuals (Naveh, 1960b).

But, on the other hand, they are smothered out by the resprouting perennials and especially by the obligatory resprouting trees and shrubs, and, in general, they decline rapidly in the third year after the fire. In Table IV, an example is given from the great number of highly variable therophytic components after a maqui burn in Israel. LeHouerou (1973) pointed out that these annuals are typical "anthracophytes" and "nitratophytes," colonizing the bare ground left after the fire, and therefore their response is merely opportunistic. However, as described in Section III, the Mediterranean region and its drier fringes in which many of these annuals have evolved have a long fire history. One can assume, therefore, that they should have also developed genotypic adaptations to such a fire-selected environment, enabling them to survive and overcome the recurring fire stress. This is manifested

TABLE IV

ABUNDANCE OF ANNUAL PLANTS IN THE FIRST AND SECOND SPRING
AFTER MAQUI BURN (MAZUBA, WESTERN GALILEE, ISRAEL)[a]

Plants	Relative abundance		Plants	Relative abundance	
	1953	1954		1953	1954
Grasses			Compositeae		
Aegilops peregrina	1	1	Anthemis spp.	1	2
Alopecurus myosuroides	1	1	Chrysanthemum		
Avena barbata		1	coronarium	1	1
Avena sterilis	2	4	Carthamus tenuis	1	2
Bromus madritensis	1	1	Cichorium pumilum	1	2
Bromus scoparius	1	1	Crepis palaestina	1	3
Bryza maxima	—	2	Centaurea cyanoides	1	1
Cynosurus coloratus	—	2	Rhagadiolus stellatus	1	2
Lamarckia aurea	1	1	Geropogon glabrum	—	1
Lolium rigidum	—	1	Silybum marianum	—	1
Phalaris minor	—	1			
Scleropoa rigida	—	1	Diverse Families		
Stipa tortilis	1	1	Anagallis coerulea*	1	1
			Caucalis tenella	—	1
Legumes			Cephalaria joppica*	2	3
Hippocrepis unisiliquosa	—	1	Chaetosciadium		
Lathyrus aphaca	—	1	trichospermum	1	1
Lotus peregrinus*	1	2	Chenopodium ambrosioides	1	1
Medicago polymorpha	—	1	Crucianella		
Medicago orbicularis	—	1	macrostachys	1	1
Medicago tuberculata	1	1	Convulvulus spp.	1	1
Onobrychis crista-galli	1	2	Erodium malacoides	1	1
Ononis Natrix	—	1	Eryngium creticum*	1	1
Scorpiurus muricata	—	1	Geranium purpureum	1	1
Trifolium campestre	1	3	Galium articulatum	1	2
Trifolium clypeatum	—	1	Lagoecia cuminoides	1	3
Trifolium cherleri	—	1	Lavatera punctata	1	1
Trifolium eriosphaerum	1	1	Plantago lanceolata	1	1
Trifolium purpureum	—	2	Plantago psyllium	1	1
Trifolium resupinatum	—	2	Scabiosa prolifera	—	1
Trifolium stellatum	—	1	Scandicium stellatum	1	1
Trifolium argutum	—	2	Scandix pecten-		
Vicia angustifolia	1	1	veneris	1	1
			Sinapis alba	—	1
			Specularia falcata	1	1

[a] Key: 1 = rare; 2 = occasional; 3 = abundant; 4 = very abundant; and * = sometimes very abundant after burn.

by the rapid recovery of Mediterranean winter and spring growing grasslands in the first spring after the fire and by their great resilience to recurrent fires during the summer, when all therophytes are dead and the perennial grasses and herbs are in a summer dormancy stage. In fact, among the most productive natural pastures in Israel are such open grasslands dominated by *Avena sterilis* and *Hordeum bulbosum* on rocky, basaltic soils of volcanic origin on the eastern slopes of the Galilee Mountains above the Lake of Tiberias (Seligman *et al.*, 1959). These, as well as the more xeric steppe grasslands descending farther south to the Jordan Valley, are very prone to frequent wildfires for 8 months or more because of the hot summer temperatures and the dry and strong, adiabatically heated western winds and the hot and very dry eastern *sharav* winds, mentioned in Section II.

As could be expected, in such summer-dry plants, all postfire feedback responses should be centered around reproductive behavior. This also is probably true with respect to other environmental stress factors, acting simultaneously as powerful selective agents, namely, drought and grazing pressure.

As a model for such "optimization" of survival mechanisms, *Avena sterilis*, one of the most common Mediterranean annual grasses can be used. This highly productive and palatable pasture plant is equipped with heavy spikelets which are shed at the onset of the dry season in March to May, according to regional climate and local site conditions, around the dried-off mother plants. With the help of an efficient boring mechanism of the hygroscopic awns and the callous basal tips, the dissemules are soon buried in the grass litter and penetrate the soil cracks during the summer. Thus in well-structured terra rossa and brown rendzina or basaltic soils, they may reach depths of several centimeters below the surface. In this way the seeds minimize the danger of being burned, eaten, or trampled, but maximize their chances for germination in a well-protected and favorable soil environment during the critical periods of the first winter months. The capacity to bury their dispersal units in one way or another is shared by all annual grasses and by many other annuals. They also include the perennial grass, *Hordeum bulbosum*, mentioned above. Most efficient in this respect are small disseminules, equipped with hygroscopic awns or other torsion mechanisms that can rapidly bury themselves in the soil. To this group belong *Stipa tortilis*, dominating above-mentioned xeric steppe grasslands, which may be burned almost every second or third year without showing any decline in *Stipa* abundance and dominance. In this case the seeds are dispersed in early spring before the soil becomes dry and impervious, and thus great reserves of seeds are built up in the upper soil layer.

However, the most significant pyrophytic adaptation and positive feed-back response is the postfire stimulation of germination. This has been found by Meiri (1959) not only for *Oryzopsis miliacea,* as mentioned already, but also for *Avena sterilis,* while studying the effect of fire on germination in a typical open Tabor oak woodland in the lower Galilee. In this case heating the seeds up to 105°C for half an hour caused increase in germination and the seeds retained viability up to tempera-tures of 125°C. Actual temperatures reached after fall burning of a dense grass stand, measured by Meiri (1959) were 350°C for several seconds above the soil surface, but at a depth of 1 cm they did not exceed 51°C in this dark, well-structured rendzina soil. In the following winter he found seeds of *Avena* and other grasses, germinating both above and close to the soil surface, also from seeds with charred awn tips, which had their caryposes buried vertically in the soil.

In our recent studies (A. Naveh and Z. Naveh, unpublished) 20% of *Avena* seeds, collected in the fall after a hot grass fire and more or less charred, retained viability in germination tests. But in the follow-ing winter 1972–1973, which had a severe drought spell, only seeds buried in the soil down to 5 cm germinated in the burned grassland. In the unburned plots, however, seedlings also protected by a dense layer of grass mulch, emerged from the soil surface.

On the other hand, during this dry winter in the semiarid steppe grasslands above the Jordan Valley, both on burned and unburned plots, only *Stipa tortilis* seeds covered with 1–2 cm of soil emerged (Z. Naveh, S. Shalmon, and Z. Zaith, unpublished). The great heat tolerance of annual grasses of Mediterranean origin, among these *Bromus* species, has also been shown in California by Wright (1931), Sampson (1944a), and McKell *et al.* (1962). Sweeney (1967) described very similar fire adaptations of annual plants in burned areas in California under similar temperature regimes. He mentioned also the breaking of seed coat dor-mancy by fire and polymorphism as fire adaptations, typical for legumes. Meiri (1959) found that heating seeds of *Medicago polymorpha,* one of the most abundant Mediterranean annual legumes, both in Israel and in California (Naveh, 1967) up to 90°C, increased water perme-ability of hard-coated seeds and thereby also their germination. In our present studies, seeds of charred annual clovers such as *Trifolium pilosum* lost their viability, but this was compensated by the large num-bers of seeds produced and their small size, enabling them to "hide" in soil cracks and beneath stones, and their capacity to retain viability for several years. In several *Medicago* species and in *Onobrychis,* a similar polymorphic postfire germination behavior is indicated and fur-ther studies will enable more conclusive answers on the great diversity of options chosen by these plants to overcome and/or avoid the combined hazards of a harsh environment and a rigorous regime of fire and grazing pressure. Sauer (1961, p. 262) claimed that: "mountainous areas in arid and semiarid lands do set up dry-season turbulence with lightning discharges into flammable plant matter. Such fires occur under conditions of low humidity and are accompanied by little or no rain." He also maintained that: "the capture of fire may well have come about as a discovery on food-collecting following after a natural fire. As the source of such a fire I should prefer vulcanism; Italy and the eastern Mediterranean, the East African Rift . . ." (of which the Jordan Valley is its northern-most tip).

The semiarid grasslands of the eastern upper Galilee and the Jordan Valley, with which we are concerned here, corresponded, without a doubt, to both conditions. Recent archeological findings at Ubaidiya (Stekelis *et al.*, 1969) in the upper Jordan Valley show great resemblance in the Lower Paleolithic Chellean hunting cultures of bed 2 of Olduvai Gorge in the East Africa Rift Valley. However, at these butchering sites, Leakey has not detected any traces of fire (Oakley, 1961). The chief differences in both semiarid environments are the discrepancies in the lower temperature–winter rain regime in Israel as compared with the summer–rain dry tropical savanna. The latter favored a dense cover of perennial grasses with small seeds as opposed to the chiefly winter annual grasses and legumes, with larger grains in Israel. In fact, these semiarid annual grasslands have served as centers of distribution of wild barley (*Hordeum spontaneum*) and the Emmer wheat (*Triticum dicoccoides*), the progenitors of our domesticated winter cereals (Zohary, 1969), and they are the natural surroundings of the late Mesolithic Yarmukian cultures which are among the cradles of cereal and livestock breeding economies in the Middle East (Whyte, 1961). One can only speculate insofar as both the colder climate and the presence of edible roasted grain seeds that could be collected after a wildfire have triggered here a more advanced cultural evolution. But in any case, the findings lend much support to Komarek's (1967a) contention of the origin of cereal grains in "fire environments."

V. Conclusions

In spite of great gaps in our knowledge of the effect of fire on Mediterranean ecosystems, it is obvious that fire has acted, not as a wholly destructive force, but as a powerful selective and regulatory agent throughout geological, biological, and cultural evolution of the Mediterranean landscape. Its action was closely interwoven with other environmental stress factors. Among these are not only climatic and edaphic stresses but also the traumatic defoliation disturbances caused for many millenia by grazing and browsing herbivores and, in more recent times, by domestic livestock and by cutting and coppicing, partly replacing fire and partly acting synergetically with it.

An attempt was made in this chapter to present the response of plants to fire in a cybernetic context of the plant–environment system, as adaptive feedback to the information being partly transformed or recoded as useful for survival of the gene pattern and blocking that part which threatens it. According to the Law of Requisite Variety (Ashby, 1954), those plant species in which this useful transmuted information as positive or negative feedbacks was large enough could survive and possibly also be predisposed for receiving further information from other defoliation disturbances.

The long and widespread recurrence of fire in xero- and thermo-Mediterranean bioclimates around the Mediterranean Basin leads to the conclusion that fire is of no less importance for the understanding of the complex structure, function, and dynamics of Mediterranean communities and ecosystems than climate and soils, to which most attention has been devoted. As a consequence, in the ordination of Mediterranean woody and herbaceous plants along environmental complex gradients or coenoclines (Whittaker, 1967), a special pyric, as well as biotic disturbance axis will be required to take account of this fact. In view of the vital need of these ecosystems as last resources of open landscapes for natural productivity, diversity, and environmental quality (Naveh, 1971), such understanding is not only of basic scientific interest, but of great practical importance.

Our findings bring additional evidence from the Mediterranean region which support the contentions of many distinguished anthropologists, geographers, and ecologists, aptly summarized by Komarek (1967a), on the great importance of fire in the evolution of human cultures. Fire has apparently served as a most beneficial tool in the skilled hands of our Mediterranean ancestors, but it has been abused greatly by later generations, and it is being now neglected and rejected. It is up to the ecologist and the enlightened land user to show how fire can be turned from a curse in Mediterranean lands to a blessing.

References

Agnostopoulos, G. D. (1967). Landscape planning in different areas: Assessments and applications. *Proc. Tech. Meet. IUCN, 10th 1966* pp. 52–73.

Anderlind, A. (1886). Ackerbau und Tierzucht in Syrien insbesondere Palaestina. Z.D.P.F. 9 pp. 62–63.

Ashby, W. R. (1954). "An Introduction to Cybernetics." Chapman & Hall, London.

Axelrod, D. I. (1958). Evolution of the Madro-Tertiary geoflora. Bot. Rev. 24, 433–509.

Bentley, J. R., and Fenner, R. L. (1958). Soil temperatures during burning related to postfire seedbeds in woodland ranges. J. Forest. 56, 737–774.

Bharucha, F. R. (1932). Etude écologique et phytosociologique de l'association a Brachypodium ramosum et Phlomis lynchitis des Garigues Languedociennes. Sigma Xi Quart. 18.

Biswell, H. H. (1967). The use of fire in wildland management in California. In "Natural Resources: Quality and Quantity" (S. V. Ciriacy-Wantrup and J. Parsons, eds.), pp. 71–86. Univ. of California Press, Berkeley.

Braun-Blanquet, J. (1925). Die Brachypodium ramosum—Phlomis lynchitis—Assozation der Roterdeboeden Suedfrankreichs, Festschr. C. Schroeder, Veroeff. Geobot. Inst. Ruebel H. 3.

Butzer, K. W. (1964). "Environment and Archeology: An Introduction to Pleistocene Geography." Methuen, London.

Daubenmire, R. (1968). Ecology of fire in grasslands. Advan. Ecol. Res. 5, 209–266.

Eig, A. (1927). On the vegetation of Palestine. Inst. Agr. Natur. Hist., Agr. Exp. Sta., Bull. 7.

Felix, J. (1957). "The Plant World of the Bible." Massada (in Hebrew), Tel Aviv.

Felix, J. (1963). "Agriculture in Eretz Israel in the Time of Mishna and Talmud." Hebrew University, Jerusalem (in Hebrew).

Freitag, H. (1971). Die natuerliche Vegetation des suedost-spanischen Trockengebiets. Bot. Jahrb. 91, 147–308.

Garrod, D. A. C., and Bate, D. M. A. (1937). "The Stone Age of Mt. Carmel," Vol. I. Oxford Univ. Press (Clarendon), London and New York.

Griesebach, A. (1872). "Die Vegetation der Erde nach ihrer klimatischen Anordnung." Engelmann, Leipzig.

Hanes, T. L. (1971). Succession after fire in the chaparral of southern California. Ecol. Monogr. 41, 27–52.

Higgs, E. S., Vita-Finci, C., Harris, D. R.. and Fagg, A. E. (1967). The climate, environment and industries of Stone Age Greece. Part III. Proc. Prehist. Soc. 33, 1–29.

Howell, F. C. (1962). Ambrona/Torralba: Acheulian open-air occupation sites in northern Spain. Annu. Meet., Amer. Anthropol. Soc. No. 15–18, pp. 1–13.

Knapp, H. (1965). "Die Vegetation von Kephalinia, Griechenland." Koenigstein.

Koeppen, W. (1923). "Die Klimate der Erde." de Grundè, Berlin.

Komarek, E. V. (1967a). Fire and the ecology of man. Proc. 6th Annu. Tall Timbers Fire Ecol. Conf. pp. 143–170.

Komarek, E. V. (1967b). The nature of lightning fires. Proc. 7th Annu. Tall Timbers Fire Ecol. Conf. pp. 5–41.

Kuhnholtz-Lordat, G. (1939). "La terre incendiée." Editions Mais Carrée, Nimes.

Kuhnholtz-Lordat, G. (1958). "L'ecran vert." Editions du Muséum, Paris.

LeHouerou, H. N. (1973). Fire and vegetation in the Mediterranean Basin. Proc. 13th Annu. Tall Timber Fire Ecol. Conf. (In press).

Liacos, L. G. (1973). Present studies and history of burning in Greece. Proc. 13th Annu. Tall Timber Fire Ecol. Conf. (in press).

Litav, M., Kupernik, G., and Orshan, G. (1963). Ecological studies on some sub-lithophytic communities in Israel. *Isr. J. Bot.* **12**, 41–54.

Loeb, H. (1960). Regeneration of maquis shrubland after burning. *In* "The Western Galilee." pp. 271–277. Sulam Zur-Gaaton (in Hebrew).

Long, G., Fay, F., Thiault, M., and Trabaud, L. (1967). "Essais de détermination expérimentale de la productivité d'une garrigue à *Quercus coccifera* C.N.R.S.," Doc. 39 (mimeo.). Centre d'Etudes Phytosociologiques et Ecologiques, Montpellier.

McKell, C. M., Wilson, A. M., and Kay, B. L. (1962). Effective burning of rangelands infested with medusahead. *Weeds* **10**, 125–131.

McPherson, J. K., and Muller, C. H. (1969). Allelopathic effects of *Adenostoma fasciculatum*. *Ecol. Monogr.* **39**, 177–198.

Meiri, A. (1959). The effect of burning on the germination of pasture plants. M.Sc. Thesis, Hebrew University, Rehovot (in Hebrew).

Naveh, Z. (1960a). The ecology of chamise (*Adenostoma fasciculatum*) as effected by its toxic leachates. *Proc. A.A.A.S. Annu. Meet. Ecolog. Sect., 1960* pp. 56–57.

Naveh, Z. (1960b). Agro-ecological aspects of brush range improvement in the maqui belt of Israel. Ph.D. Thesis, Hebrew University, Jerusalem (in Hebrew).

Naveh, Z. (1967). Mediterranean ecosystems and vegetation types in California and Israel. *Ecology* **48**, 445–459.

Naveh, Z. (1968). Multiple use of Mediterranean range land—New approaches to old problems. *Ann. Arid Zone* **7**, 163–176.

Naveh, Z. (1971). The conservation of ecological diversity of Mediterranean eco-systems through ecological management. *In* "The Scientific Management of Animal and Plant Communities for Conservation" (E. Duffey and A. S. Watt, eds.), pp. 605–622. Blackwell, Oxford.

Naveh, Z. (1973). The ecology of fire in Israel. *Proc. 13th Annu. Tall Timber Fire Ecol. Conf.* (in press).

Naveh, Z., and Dan, J. (1973). The human degradation of Mediterranean landscapes in Israel. *In* "Mediterranean Type Ecosystems. Origin and Structure." (F. di-Castri and A. Mooney, eds.), pp. 372–390. Springer, Heidelberg.

Oakley, K. P. (1961). On man's use of fire, with comments on toolmaking and hunting. *Viking Fund. Publ. Anthropol.* **31**, 176–193.

Rickli, M. (1942). "Das Pflanzenkleid der Mittelmeerlaender," p. 216. Surkamp. Bern.

Ruebel, E. A. (1930). "Pflanzengesellschaften der Erde." Murray, Bern.

Sampson, A. W. (1944a). Plant succession on burned chaparral land in northern California. *Calif., Agr. Exp. Sta., Bull.*

Sampson, A. W. (1944b). Effect of chaparral burning on soil erosion and on soil moisture relations. *Ecology* **25**, 191–195.

Sauer, C. O. (1956). The agency of man on earth. *In* "Man's Role in Changing the Face of the Earth" (W. L. Thomas, ed.), pp. 49–69. Univ. of Chicago Press, Chicago, Illinois.

Sauer, C. O. (1961). Sedentary and mobile bents in early societies. *Viking Fund Publ. Anthropol.* **31**, 256–266.

Seligman, N., Rosensaft, Z., Tadmor, N., Katznelson, J., and Naveh, Z. (1959). "Natural Pastures in Israel, Vegetation, Carrying Capacity and Improvement." Sifriat Hapoalim, Tel Aviv (in Hebrew with English summary).

Semple, E. C. (1931). "The Geography of the Mediterranean Region." Holt, New York.

Shantz, H. L. (1947). "The Use of Fire as a Tool in the Management of the Brush Ranges in California." Calif. State Board of Forestry.

Soil Conservation Division, Israel. (1964). "Conversion of Maqui into Pasture Parks as a Means of Enrichment of Ground Water" (mimeo.). Ministry of Agriculture, Tel Aviv (in Hebrew).

Specht, R. L. (1969). A comparison of the sclerophyllous vegetation characteristics of Mediterranean type climates in France, California and southern Australia. II. Dry matter, energy, and nutrient accumulation. *Aust. J. Bot.* **17**, 293–308.

Stekelis, M., Bar Yosef, O., and Schick, T. (1969). "Archeological Excavations at Ubaidiya 1964–66." Israel Academy of Sciences and Humanities. Jerusalem.

Stewart, O. C. (1956). Fire as the first great force employed by man. *In* "Man's Role in Changing the Face of the Earth" (W. L. Thomas, ed.), p. 115–133. Univ. of Chicago Press, Chicago, Illinois.

Susmel, L. (1973). "Sviluppi e problemi attuali del controllo degli incendi nella foresta mediterranean." FAO, Rome (cited by LeHouerou, 1973).

Sweeney, J. R. (1967). Ecology of some fire types in northern California. *Proc. 7th Annu. Tall Timbers Fire Ecol. Conf.* pp. 111–126.

Tchernov, E. (1968). "Succession of Rodent Faunas during the Upper Pleistocene of Israel." Parey, Berlin.

Tchernov, E. (1972). Faunal and climatical changes in the Quaternary in Israel. *In* "Climatic Variations and Botanical History of Israel" (J. Weisel and J. Cohen, eds.) (mimeo.).

Tisdale, E. W. (1967). "A Study of Dry-land Conditions and Problems in Portions of Southwest Asia, North Africa and the Eastern Mediterranean," Report 3 (mimeo.). Dry-land Res. Inst., University of California, Riverside.

Trabaud, L. (1970). Quelques valeurs et observations sur la phytodynamique des surfaces incendiées dans le bas Languedoc. *Nat. Montpel., Ser. Bot.* **21**, 231–242.

Trabaud, L. (1973). Etude expérimentale de la dynamique d'une garrigue de *Quercus coccifera* L. soumise à des feux contrôles. (Premiers results tats.). *Proc. 13th Annu. Tall Timber Fire Ecol. Conf.* (in press).

UNESCO (1963). Bioclimatical map of the Mediterranean zone. *Arid Zone Res.* Vol. 21.

Walter, H. (1968). "Die Vegetation der Erde," Vol. 2, pp. 64–78. Fischer, Stuttgart.

Whittaker, R. H. (1967). Gradient analysis of vegetation. *Biol. Rev. Cambridge Phil. Soc.* **42**, 207–264.

Whyte, R. O. (1961). Evolution of land use in southwestern Asia. *In* "A History of Land Use in Arid Regions" (D. Stamp, ed.), pp. 57–118. UNESCO, Paris.

Winstanley, D. (1972). Sharav. *Weather* **27**, 146–165.

Wright, E. (1931). The effect of high temperatures on seed germination. *J. Forest.* **29**, 679–687.

Zohary, D. (1969). The progenitors of wheat and barley in relation to domestication and agricultural dispersal in the Old World. *In* "The Domestication and Exploitation of Plants and Animals" (P. J. Ucko and G. W. Dimbleby, eds.), pp. 47–66. London.

Zohary, M. (1962). "Plant Life in Palestine." Ronald Press, New York.

. 13 .

Effects of Fire in Forest and Savanna Ecosystems of Sub-Saharan Africa

John Phillips

I. Introduction

This chapter deals primarily with effects of fire on forest and related ecosystems. Evidence is presented to support the view that fire is a bad master, but, when employed correctly, it is a good servant (Phillips, 1930b, 1936, 1965, 1972).

Below I define terms and concepts modified to express my hypothesis and philosophy:

Tropics and subtropics. For purposes of this study I include in the tropics and subtropics the equatorial regions and those extending to the Tropics of Cancer and Capricorn. For the sake of comparison I also discuss certain temperate-subtropical forests and related vegetation.

Forest is evergreen to mixed evergreen and partially deciduous; vegetation is 18–60 m tall, with crowns normally touching, overlapping, and stratified so as to create a closed to almost closed canopy. Tree, shrub, fern, forb, and other layers may number 3–5. Tall emergent trees occur in some types but do not form a closed canopy except very locally. Epiphytes and lianas vary greatly in kind and number. This physiognomic community is climax but is frequently associated with successional stages such as thicket, woodland, open woodland, and wooded grassland (Phillips, 1971a), today almost entirely secondary.

The significant successional types include the following.

Thicket. The principal physiognomic stage prior to the climax is dense woody vegetation 10–20 m high. The short trees and tall and other shrubs often are closely spaced, with stems and crown intertwined. The canopy is usually closed, the layers not marked in the shorter phases. Epiphytes and lianas may occur, but forbs and grass are rare to absent. While usually secondary today, climax types do occur in subhumid terrain.

A subtropical type, climax or seral, in the Cape Province, South Africa and on some of the higher tropical mountains is *Macchia* (South Africa: *Fynbos*), reminiscent of some types of chaparral.

Woodland, relatively advanced in the forest succession and also present as a proclimax in subhumid terrain, is a several-layered community of trees, tall and other shrubs or palms, 5–20 m high. The crowns either touch lightly or are separated 1–10 m. The subcanopy and other layers are usually lightly foliaged, permitting both direct sun and diffuse light to penetrate to the ground. Epiphytes and lianas are frequent to rare. Grass occurs but is less vigorous than in lighter canopied vegetation.

This type is common within the derived and subhumid wooded savanna, varying in composition, dominance, structure, and related biota: *evergreen* (rarely) to evergreen/deciduous to deciduous according to climate, soil, and floristics.

Open woodland. Variable in density and structure in the forest succession, this is a more widely spaced form of woodland. The trees and shrubs grow either singly or in single to several-layered tree and shrub clumps of varying area. A highly variable distribution and proportion of undershrubs, forbs, and particularly, grass occurs, the grass being relatively tall, dense, and luxuriant, contrasted with that in woodland. Deciduous to partially evergreen variants exist: denser, intermediate, and scattered as to the distribution of the crown or clump spacing, roughly 20–30, 30–60, and 60–100 m where this merges into wooded grassland the spacing exceeds 100 m.

Wooded grassland is comprised mainly of high, tall, medium, or short grass communities and associated forbs, with single trees and tall shrubs or small clumps scattered more than 100 m apart. In the forest succession the soils often have been considerably altered through exposure to insolation, grazing, and cultivation.

Derived wooded savanna. Keay (1953), Phillips (1959, 1965), Clayton (1961), and others considered thicket, woodland, open woodland, wooded grassland, and other physiognomic types in the forest succession to be derived from forest by cutting, fire, livestock, and shifting and other cultivation. This type varies floristically according to the nature of the climax forest or primary thicket stages and according to the severity and duration of disturbance. It ranges from open grass to woodland thicket and even to small relict patches of forest.

Although I have replaced the term "savanna," owing to its wide definition by woodland and the related terms mentioned, "savanna" is useful in general description when referring to woody vegetation other than thicket and forest (Phillips, 1971a).

This type may be evergreen, evergreen/deciduous, deciduous, tall to short, with trees and shrubs closely to very widely spaced and with a corresponding luxuriance, vigor, and volume of grass communities.

Subhumid wooded savanna (Phillips, 1959, 1961, 1965, 1966) ranges from thicket through woodland to wooded grassland and even local nonclimax open grassland within a succession not culminating in forest. The climax supporting this vast type in its most luxuriant facies, but not in the forest succession, is subhumid (Phillips, 1959, 1961, 1965, 1966). The soils are distinct from those in the forest succession (Phillips, 1959), because they have developed under markedly less humid conditions than those favoring forest. Although the effects of fire are comparable with those in certain phases of the derived wooded savanna, some marked differences exist.

Biotic community is constituted by the plants and animals present within any particular habitat; this is a basic entity in both natural and induced forms. Man has become a major associate, indeed the most powerful dominant in a high proportion of such communities (Phillips, 1931a). This entity and its habitat constitute the ecosystem (Tansley, 1935). Even if the original definition did not contain the term "habitat," its inclusion was inferred.

Ecosystem. Energy-generated and -generating, this biotic community, integrated with its reciprocally controlling and controlled environment, is widely accepted as a unit. Man has increasingly become associated in it as the master dominant (Tansley, 1935; Phillips, 1934/1935, 1965).

A *bioclimatic unit* (Phillips, 1959, 1961, 1963, 1964, 1965, 1966, 1968a,b, 1971a,b, 1972) may be a great region, a site, or even smaller entity differentiated by an interplay of climatic factors and biotic phenomena, so integrated as to permit development of natural vegetation and associated animal life to a mosaic of stages in dynamic equilibrium with the climate. These units are locally effectively modified by edaphic factors. I have applied this concept in Africa south of the Sahara, and in detail in Natal and elsewhere in southern Africa. The concept provides a useful first classification for study of ecological phenomena and therefore a basis for conservation or development of a terrain for crop, livestock, forestry, or other production (Phillips, 1959, 1961, 1965, 1966, 1968a, 1972).

Its significance has been improved by an associated concept:

Pedo-ecological unit. Study of the climate, vegetation geology, physiography (especially elevation, aspect, degree of slope, and ruggedness of topography), soils, and water relations reveals that this mosaic constitutes a natural unit for classification of ecological sectors of a terrain and for the later planning of its agricultural and other development. The unit is mappable (see Phillips, 1971b) and provides a practical basis for research and action. Even if tautologous, the term is more meaningful than biogeoclimatic unit or some variant thereof.

II. Some Major Types of Forest and Subhumid Wooded Savanna

Over the vast area of Trans-Saharan Africa much diversity exists in detail of floristics, physiognomy, nature and structure of communities, and ecological features in the types of forest and subhumid wooded savanna. For our purposes the vegetation is divided into the following more significant types.

FOREST
  I. Closed canopy: equatorial and tropical
     A. *Lowland to upland*
        (1) Moist evergreen
        (2) Moist evergreen and semideciduous
        (3) Evergreen (rarely evergreen and semideciduous), freshwater swamp, and riverine
     B. *Montane*
        (1) Evergreen broad-leaved (*Podocarpus* in mixture locally)
        (2) *Juniperus procera*
 II. Closed canopy: subtropical and temperate-subtropical
     A. *Lowland to upland*
     B. *Montane*
        (1) Evergreen broad-leaved (*Podocarpus* in association in some instances)
        (2) *Widdringtonia*

SUBHUMID WOODED SAVANNA
  I. Equatorial and tropical: lowland, upland to montane
 II. Subtropical: lowland to upland

For descriptions of features of vegetation types of Africa the reader is referred to the following: widely applicable papers—Aubréville (1949a,b, 1956), Richards (1952), Phillips (1959), Keay (1959), Schnell (1970, 1971), White (1975); Guinean, Nigerian, and Equatorial Africa—de Wildeman (1920, 1924), Chipp (1927), Lebrun (1935, 1936), Aubréville (1937, 1938, 1948, 1949a,b), Robyns (1938, 1950), Evans (1939), Richards (1939), Gossweiler and Mendonça (1939), Keay (1953), Lebrun and Gilbert (1954), Taylor (1952, 1960), Hopkins (1961, 1965), Grandvaux Barbosa (1970); and eastern and southern Africa—Sim (1907), Marloth (1908), Engler (1910–1925), Bews (1912 *et seq.*), Troup (1922), Phillips (1928a,b, 1930a, 1931b), Rea (1935), Adamson (1938, 1956), and Chapman and White (1970).

III. History of Fire in Forest and Derived and Subhumid Wooded Savanna

Fire undoubtedly occurred in vegetation in Africa millennia before being kindled by man. The prime cause was lightning. It is possible that falling igneous rock from time to time caused local burns (Henniker-Gotley, 1936; Phillips, 1965). In certain special habitats fire probably also was caused occasionally by fermentation in damp, adpressed vegetation, as in ill-packed hay ricks. Although these agents would have operated in derived and subhumid wooded savanna they are unlikely to have been so frequent and significant in the forest, particularly in highly humid and humid types.

Reference to the role of lightning is made by Johnstone (1906) for tropical Africa; Staples (1926) for Natal, South Africa; Phillips (1930b, 1936, 1965) and West (1965) for Africa; and Komarek (1964, 1966, 1967, 1968) for North America. It is noteworthy that a communication regarding lightning observations by satellite OSO-5 by Sparrow and Ney (1971) revealed that lightning is very much more frequent over the humid equatorial and tropical sectors of Africa than over the less humid subtropics. Despite this, burning induced by lightning in the humid equatorial and tropical forest is rare. There appear to be no published records of this. Conversely there are records of lightning causing fires in both the derived and the subhumid wooded savanna.

Volcanoes still occasionally cause fires locally in East and less often in West Africa, but, because eruption has for long been very local compared with that in strongly volcanic sectors of the world, the extent of destruction is slight.

Clarke (1959), in studying ancient man in parts of southern and central Africa, noted that the earliest evidence of primitive man south of the Sahara using fire was linked with the Chelles-Acheul and possibly the Fauresmith cultures, at almost the end of the earlier Stone Age. If this be so, it might imply that man began to use fire about 50,000–55,000 years ago (according to carbon-14 dating of wood). Whether this was for more than domestic purposes, implying nuturing the coals as he moved from camp to camp, or for honey hunting and driving of game, is not known. Certainly man used fire increasingly for these purposes in the course of time and, later, in local clearing of vegetation for cultivation.

We should relate this possible dating of man's use of fire to the bioclimatic units in which he would have been living. The Kalambo site on the border of Zambia and Tanzania studied by Clarke was not in forest but in wooded savanna, in which the grass, forb, and dead to moribund woody material could have been fired in the dry season with comparative ease. Burning would have been much more difficult, if not impossible, in the highly humid to humid and humid–subhumid forest unless man was prepared to kindle large fires and painstakingly spread the flames from point to point. Lacking an objective, such as protection from predations of lion, leopard, elephant, buffalo, or from the "spirits," he would not have exerted himself to do this. Modern Africans do not burn in the forest without an objective. It is plausible, therefore, that their forebears would have been even less interested in attempting to burn damp to moist material, except locally for warmth, cooking, and possible protection against dangerous beasts. The earlier users of fire were almost certainly hunters of game and honey, followed much later by simple pastoralists. When man first began to cultivate in Trans-Saharan African forests is not known, but probably this would have been many thousand years after his first use of fire.

Although it would have been a long time before the earlier users of fire would have burned forest locally to prepare land for cultivation, it is feasible that they would have burned its margins and in glades in the course of hunting and, late, for providing fresh grazing and browsing for livestock.

Firing in the highly humid forest is to this day confined to marginal burning and clearing, marginally and internally, for initiation and maintenance of local patches in a system of shifting cultivation. In the humid forest there has been much greater frequency of burning for this purpose and also, during the past century or so, in exploitation of timber. In humid–subhumid low, medium elevation and montane tropical forests and in the subtropical equivalents in southern Africa, burning has been much more common, not only because of more ecologically dry months (less than 25 mm rain per month) but also because man's activities have been more concentrated in much smaller areas.

Fascinating as the history of burning is, it must suffice to refer to observations recorded for several parts of Trans-Saharan Africa: Aubréville (1937, 1947a, 1949a,b, 1950) described the destructive role of fire in various types of forest. He believed that owing to a delicate equilibrium between the mesophytic forest communities and the habitat in terms of humidity, rain, and soil moisture, fire widely disturbed that equilibrium, thus encouraging development and extension of derived savanna. He also applied this argument to East Africa and southern Africa. While not wholly agreeing that circumstances militated against rehabilitation of forest because of the induced severity of the habitat, I support his conclusions that there has been intensive and extensive damage.

While Busse (1901, 1902, 1907, 1908), a critical observer in parts of the former German territories in East Africa (Tanganyika) and West Africa (Togo, Dahomey, Cameroon), was more concerned with fire in the wooded savanna, he also referred to its harmful effect on forest. Interesting comments were also made by Sienbenlist (1914) for Tanganyika and by Hutchins (1909) and Troup (1922) for Kenya.

The Portuguese literature contains references to destruction of forest in Angola and Moçambique. Classic contributions for Moçambique are those by Sim (1909) and Swynnerton (1918, 1921).

South Africa's restricted forest terrain has been greatly reduced in extent and quality by fire. References to the more important literature are given elsewhere (Phillips, 1930b, 1931b, 1959, 1963, 1964, 1965), but it is worth mentioning the observations by Pappe (1862), Brown (1887), Harison (1856–1874), Sim (1907, 1909), Fourcade (1889), and Bews (1917, 1918).

Derived savanna has a long history of burning, from the time of its origin through fire and other kinds of disturbance.

The history of fire in subhumid wooded savanna has been covered by French, German, British, Portuguese, Rhodesian, and South African observers. This is for much longer than that for the other prime types, because of the higher degree of flammability and the type probably having been the home of primitive and later man from much earlier times. A number of references is recorded by Busse (1901, 1902, 1907, 1908), Aubréville (1937, 1947a,b, 1949a,b, 1950), West (1965), Phillips (1930b, 1936, 1959, 1964, 1965, 1968b), and others.

IV. Kinds of Fire

The important kinds of fire in forest, thicket, and well-grown woodland are ground fire, crown fire, and the intensive burning applied during the several stages of preparation and maintenance of land for shifting cultivation (Phillips, 1931b, 1964). Firing on the ground with or against the wind has little bearing on burns in almost wind-free areas in forests but is most important in crown fires because of exposure to prevailing or gusty wind. Moreover, at some distance above, the ground fire creates local air currents and eddies. In the woodland and other more widely spaced types within both the forest and wooded savanna successions, ground fire is common, but in the closer facies of woodland, especially where crowns are festooned with *Usnea* and other lichens, fierce crown fires may occur during strong and dry winds. Burns with or against the wind have significantly different effects in these more open communities, the severity varying with local conditions.

Ground fire in forest other than the perhumid type burns deep into the litter and decomposed organic matter. When these fuels are sufficiently dry and readily flammable, the heat of the burn itself dries contiguous material, so that a slow fire may be kept alive for many days to weeks during dry spells. Where wind occurs or is locally generated by the heat, a burn with the wind may cause less severe damage over a wider area compared with one against it, when the effects may be reversed: a slower, more concentrated and deeper burn being more destructive, but over a smaller area. Ground fires in the derived and subhumid woodlands and related communities affect roots of woody and other species less than in the forest, due to the higher natural resistance of many species in these types, and to the stimulus which fire gives to growth of perennial grasses and woody plants. Moreoever, the effect of burning may often be comparatively superficial, owing to the relative lack of organic matter and the already somewhat indurated soil surface.

Crown fire in forest may create great havoc through killing or severely damaging crowns, boles, and whole trees. Falling branches and boles in turn ignite flammable material at lower levels and particularly at the ground or litter layer. In denser woodland, derived and other crown fires may be mildly harmful where the crowns are mature, but crown fire rarely if ever occurs in the absence of surface burning.

Fire fueled by man may burn for many days to weeks in sites in the forest which are being prepared, extended, or reconditioned for cultivation. Killing of roots, rhizomes, seeds, fruits, seedlings, and older plants often is widespread. Other effects are sterilization, baking, and covering the soil with deep ash. As physicochemical and biotic changes are wrought, the soil may be seriously altered, but should the fire be less severe and maintained for a shorter time, the effects may be even slightly beneficial, through contribution of minerals from the ash and removal of excessive accumulation of raw litter and other undecomposed organic matter. Although burning is used in preparing sites for cultivation in the other vegetation types, the severity of the burns usually is much less, because fuel is less abundant and the soil less vulnerable physically and biologically.

V. Influence of Fire in Forest Types, Related Derived Savanna, and Subhumid Wooded Savanna

The aerial and edaphic factors of the bioclimatic units which may be affected by fire are summarized in Tables I and II. The effects of fire on these factors and on the vegetation are shown in Tables IV to VII.

A. FLAMMABILITY AND POTENTIAL COMBUSTIBILITY

Flammability is defined for our purpose as the capacity for vegetation to be ignited at a flash point (e.g., immediately), whereas potential combustibility denotes the range of conditions contributing to its capacity to sustain fire in the absence of fuel provided by man, for instance, in the course of shifting cultivation, timber exploitation, and collection of fuel (Batchelder and Hirt, 1966).

Although it might be argued that in practice there is little difference between these capacities, Table V shows that even in the highly humid and humid primary ecosystems, flammability is slightly higher than potential combustibility, whereas in the secondary ones, derived wooded savanna and *macchia,* and also in the subhumid wooded savanna, potential combustibility is as high as and even higher than the capacity for flammability. Should man provide dead or dying material, potential combustibility is enormously increased even in the highly humid ecosystems.

The primary highly humid and humid forests are relatively slightly flammable and potentially combustible when not encumbered by dead and dying material or altered by establishment of secondary grass, forb, and other communities. Conversely, the derived stages such as woodland to open grassland and locally, in parts of South Africa, *macchia,* and the subhumid wooded savanna communities are much more flammable

TABLE I

A GROSS COMPARISON OF AERIAL FACTORS CONDITIONING THE MAJOR BIOCLIMATIC UNITS SUPPORTING FOREST AND DERIVED SAVANNA RELATED THERETO AND SUBHUMID WOODED SAVANNA IN TRANS-SAHARAN AFRICA[a]

Bioclimatic unit	Radiation[b]	Humidity[c]	Rain[d]	Drought[e]	Saturation[f] deficit in dry seasons	Evaporation
Highly humid or *perhumid* forest, low and medium elevations north and south of Equator, mainly north	Megatherm, range very slight; monotonously uniform throughout year	Uniformly high, range very slight	Very high to high, reliable, locally excessive	Nil to negligible to very short; EDM: 0(1-2)	Negligible	Negligible to slight under canopy, moderate (to high) on extensive exposed areas; less than rain
Humid forest, low and medium elevations north and south of Equator, mainly north	Megatherm, but above 762-914 m mega-mesotherm; range very slight throughout year	Uniformly high, with slight decrease in driest months and on drier days only; range moderate	Very high to high, normally reliable	Short and mild, but locally may be longer and more severe; EDM: 1-2-3 (4), rarely more	Slight to very slight to negligible	Slight under canopy, moderate (to high) on extensive exposed areas; less than rain
Humid *montane* forest (with a less humid variant: Humid-subhumid *montane* forest), 1219-3048 m, north and south of Equator, mainly north	Mesotherm, range moderate (rarely high)	High, but in driest coolest months moderate in some localities; range moderate	Moderate to high, locally moderate; reliable but occasionally irregular	Short to moderate: EDM: 3-4(5) but locally less: 1-2	Slight to moderate	Slight under canopy, moderate under full exposure (temporarily high when wind strong); almost equal to rain in some localities, usually less

Derived wooded savanna, derived from highly humid forest but more often from humid forest and humid montane forest, north and south of Equator, mainly north, locally converted to open grassland	Megatherm where derived from highly humid and humid forest, mesotherm where derived from humid montane forest; generally appreciably higher because of more open canopy	As for forest bioclimate in which it has been derived, with reduction in humidity when radiation is intense in order and duration; range moderate	Normally as for the forest bioclimate in which derived; impact of rain on soil severe as canopy much less dense and even "open" in many parts	Much as for the forest bioclimate in which derived, but more severe, because of greater exposure to sun and wind and thus subject to higher evaporation	Much as for the forest bioclimate in which derived, but may be more severe because of greater exposure to sun and wind and thus subject to higher evaporation	More severe than for the forest bioclimate in which the community has been derived because of greater exposure to sun and wind; greater than rain where vegetation is woodland, open woodland, and wooded grassland
Subhumid wooded savanna, low, medium, and upland elevations north and south of Equator; thicket woodland, open woodland to secondary grassland	Megatherm, but at higher elevations (914 m upward) megamesotherm; range moderate to great according to rainy and dry seasons; locally mesotherm in subtropics	High in rains, moderate in dry seasons; range moderate to wide	Moderate to high; reliable to fairly so to unreliable	Moderate to fairly long and fairly severe to mild; EDM: (3)4–5	Moderate to high	(Moderate) to high in dry season, moderate in rains; greater than rain

TABLE I (Continued)

Bioclimatic unit	Radiation[b]	Humidity[c]	Rain[a]	Drought[e]	Saturation[f] deficit in dry seasons	Evaporation
Macchia (*Maqui*; South Africa *F'ynbos*), subtropical, successional to forest in South Africa. Knysna; successional to forest on high mountains in East Africa: petty areas only	Mild mesotherm: more severe during warm, dry season (October–March); frost rare	High to moderate during rainy season (usually in "cool" season, April–September); moderate to slight during dry season	Fairly well distributed where successional to forest: moderate to high	Where *proclimax* to forest, slight to moderate	Where *proclimax* to forest, slight to moderate	Slight to moderate during "cool," rainy season; usually less than rain during rainy season

[a] Parentheses indicates occasional, not usual.

[b] Heat–light complex. Megatherm: above mean 23°C; mild mesotherm: mean of 15–20°C.

[c] Very high: 20 mm; high: 15–20 mm; moderate: 10–15 mm; slight: below 10 mm.

[d] Excessive: above 2489 mm (may attain over 5080 mm locally); very high: 1803–2489 mm; high: 1397–1803 mm; moderate: 991–1397 mm; slight: 610–991 mm; very slight: 406–610 mm; very, very slight: 20–406 mm.

[e] Ecologically dry months (EDM) (25 mm and less): negligible: below one month; very short: 1–2 months; short: 3 months; moderate: 4–5 months; fairly long: 6 months.

[f] Excessive: above 15 mm; very high: 10–15 mm; high: 7–10; moderate: 5–7; slight: 3–5; very slight: below 3; negligible: 1.

TABLE II

Bioclimatic units	Great soil units	Certain significant edaphic features
Highly humid, humid, humid/subhumid forest, at low and medium elevation, in the tropics and in the subtropics except the highly humid; also the relatively petty, widely scattered *montane* types in East, Northeast, Central, and West Africa: extent in West, Equatorial, East Africa (including Ethiopia) currently about 2.5–3 million km²—large portions now derived wooded savanna, due to shifting and other cultivation and other human activities	A. Ferrallitic soils B. Ferrisols C. Alluvium D. Hydromorphic soils	A. Strongly leached, low in plant nutrients, erodibility high in some forms/series on exposure to insolation, rain and wind B. Less leached, erodibility as above C. Moderate to moderately high in nutrients; erodibility by water high D. Nutrients variable, texture and drainage mediocre to poor
Fire occurs in the forest proper only during shifting cultivation, exploitation of forest and honey hunting		In general, the removal of forest canopy induces physical, and biotic deterioration
Derived wooded savanna within the successions in which the foregoing are *climax: Thicket* to *Woodland* to *Wooded Grassland* and local *Grassland*	Soil units as above	Soils are frequently deteriorated by several years of cultivation alternating with a number of years of wooded or other fallow (commonly termed "bush fallow"). Rehabilitation of thicket and forest communities induces soil amelioration
Fire occurs annually to frequently to occasionally, according to the local climate and volume of flammable material such as grass and luxuriant annual and perennial forbs and shrubs		

TABLE II (*Continued*)

Bioclimatic units	Great soil units	Certain significant edaphic features
Subhumid wooded savanna: *Thicket* to *woodland* to *wooded grassland* and local *grassland*. A most extensive type covering about 6 million km², north and south of the Equator, increasingly altered by man; fired annually or frequently where volume and condition of grass and other readily flammable material permit	A. Ferrallitic soils B. Ferrisols C. Fersiallitic soils D. Vertisols locally	A. Strongly leached: nutrients low B. Less leached: nutrients slightly more C. Slightly to moderately leached: nutrients much as in B D. Nutrients fair: more than in A, B, and C
		Soils, like those in the derived wooded savanna, are frequently deteriorated by cultivation. During rest periods, grass and wooded communities rehabilitate the structure and add to the nutrients in the course of time
Macchia (*Fynbos*, S. Africa), where successional to temperate-subtropical at low, medium and montane elevations occur, as in the Knysna subregion in the Cape Province	Bd: Lithosols Jd: Ferruginous (fersiallitic) soils and locally Fa: Pseudo-podsolic soils	Soils are leached and low in nutrients on cover of forest or *macchia* being removed

and combustible because of grass, forbs, and woody growth which desiccate to varying degree during ecologically dry periods.

B. REDUCTION IN EXTENT OF FOREST, RELATED ECOSYSTEMS, AND SUBHUMID WOODED SAVANNA

Many investigators (e.g., Aubréville, 1936, 1938, 1947b, 1949a,b, 1950, 1956; Robyns, 1950; Phillips, 1931b, 1936, 1959, 1965; Schnell, 1971) emphasized the long continued and severe influences of fire in Trans-Saharan Africa. Fire effects are commonly associated with shifting cultivation and also with forest exploitation and hunting of elephant, buffalo, bushpig, antelope, and leopard.

It is difficult to estimate the present and potential (largely past) areas of forest in this vast terrain. Aubréville (1949a, 1956) provided some useful data but tended to include within the potential forest zone some types which Phillips placed within the subhumid wooded savanna.

Aubréville's estimates, edited by Phillips, are summarized in Table III.

Hence, only about 60% of the potential area is under forest today. Moreover, probably over half of this is altered in varying measure, mainly through shifting cultivation and locally through timber exploitation.

Table V indicates the relative reduction of each of the main groups of ecosystems, according to (a) historic and (b) continuing deterioration. The rough proportion already converted to derived wooded savanna is estimated for the relevant groups. The estimated conversion of derived wooded savanna and subhumid wooded savanna to open grassland and of *macchia* to grassland or Karoo transition also is given.

The highly humid forest ecosystems have been moderately reduced, the humid forest rather more, and the montane forest the most. The conversion of derived wooded savanna to open grassland has been moderately severe while that of subhumid wooded savanna to open grass has been moderate only. *Macchia*, itself a stage leading to humid temperature–subtropical forest, has been severely affected.

Fire has played a significant part in reduction of forest and related ecosystems, but frequently shifting (and in parts increasing proportions of sedentary) cultivation have greatly accelerated deterioration of primary ecosystems. The quality and details of structure and composition of the ecosystems have also been significantly altered.

Because the effects vary according to the faciations of the dense climax or later proclimax types, more so in the secondary communities associated with these in the derived wooded savanna and just as much in the subhumid wooded savanna, these will be examined in the climax and secondary stages of the prime groups.

C. Effects on the Habitat

Climax, Subclimax, and Proclimax Closed Forest Canopy Units

The bioclimatic units are the highly humid lowland and upland forests of the tropics, the humid lowland, upland, and montane forests of the tropics and subtropics, and the temperate-subtropical lowland and upland humid forests of the Knysna, South Africa.

TABLE III

Grand blocks	Potential forest area (millions of ha)	Existing forest area (millions of ha)
Guinean		
(Guinea, Sierra Leone, Liberia	33	17
Ivory Coast, Ghana)		
Nigerian		
(Togo, Dahomey, Nigeria)	16	4
Equatorial		
(Cameroon, Gabon	63	43
Congo: Brazzaville,		
Central African Republic,		
Maiombe–Cabinda–Angola)		
Zaire	105	74
Eastern montane		
(Zaire, Rwanda, Burundi)	1	1
	218	139

Lesser forest areas[a]	Potential forest area (millions of ha)	Existing forest area (millions of ha)
Ethiopia and southern Sudan	9	4
Eastern Africa	8	3
(Uganda, Kenya, Tanzania,		
Moçambique, Malawi)		
Southern Africa	0.6	0.2
(Rhodesia, South Africa)		
	235.6	146.2

[a] Phillips, 1959, edited 1972.

a. AERIAL FACTORS. Damage by fire greatly increases light intensity at ground and other levels when the several layers of the closed forest are removed from an area of more than half of a hectare. This, together with greatly increased heat, not only stimulates growth of marginal vegetation but also affects seed germination and stimulates or retards growth of microflora and microfauna (see I. F. Ahlgren, Chapter 3). An increased frequency and speed of wind, a moderate increase in saturation deficit, and more severe insolation raise the evaporating power of the air considerably, according to season, height above ground, and, except on or near the Equator, aspect. The overall change in microclimate may be moderate to severe. In turn this change bears significantly on the local mosaic of bioclimatic and edaphic features.

b. SOIL. Owing to intrusion of radiation and wind and to the much greater impact of rain on the soil, organic matter is rapidly decomposed and lost to the air and washed from the site, should the slope be greater than 10°. Nitrogen is lost but there often is some addition of other prime nutrients (phosphorus, potassium, magnesium, and calcium) from the ashes. While changes in pH, for a short time increased, probably are not important for higher plants, they might temporarily stimulate growth of the microflora.

Deflocculation of soil colloids, induced by excessive amounts of ash might also occur, depending on drainage. Soil moisture usually increases somewhat until there is rapid and vigorous growth of annual and perennial vegetation. This is noteworthy particularly in humid–subhumid units and where the aspect is warm and dry, for instance on northern to western aspects in the Southern Hemisphere, some distance from the Equator.

Runoff is not accelerated unless the slope exceeds about 5–10° according to the soil series, the cover of secondary growth is insufficient to protect the surface, and the rate of fall is intense (50–75 mm per hr). Instances have been noted of moderately severe local erosion resulting from falls at the rate of 150–200 mm per hr in exploited humid forests in West Africa, whereas in exploited humid–subhumid faciations in the subtropics of South Africa, moderately severe to severe erosion has been observed during falls of the order of 50–75 mm per hr. Erosion within undisturbed forests is mild to negligible, except along extraction roads, cattle paths, and game trails. Hardening of the surface, which accelerates runoff and induces erosion, may be moderate to moderately severe in the exploited highly humid forests but is frequently more severe in humid forests, tropical and subtropical.

Deterioration of physical, chemical, and biotic features of the exposed soil may range from moderate to moderately severe and occasionally to severe in the highly humid and the humid forests of the tropics. Deterioration may be more severe in the lowland, upland, and montane forests of the subtropics, probably due to the lower humidity and the longer ecologically dry periods, when monthly rainfall is below 25 mm.

Evidently the highly humid forest and tropical lowland and upland humid forest habitats are altered less by fire than those of the tropical humid montane forest and the lowland, upland, and montane faciations in the subtropics. This is a matter of degree, according to intensity and frequency of burning in the various units. The highly humid forests experience far less frequent and severe burning than the humid tropical units, which again are subjected less to fire than those in the humid subtropics. Lightning-caused and man-induced fires are far more frequent where the potential combustibility is greater, because of the more rigorous microclimatic factors.

D. SECONDARY STAGES: WOODLAND AND OTHER COMMUNITIES

a. AERIAL FACTORS. *Derived wooded savanna* (see Table I), constituted by *secondary woodland, wooded grassland,* and *open grassland,* obviously would experience an overall order of intensity of radiation, humidity, rain, saturation deficit, evaporation, and drought similar to that experienced by the climax vegetation in the corresponding bioclimatic units. But the local microclimatic conditions differ in varying degree, because the original forest community, with its closed canopy, has been replaced by a much lighter (woodland) to a very much lighter (open woodland) one and by virtually no canopy, except for that of scattered trees in the wooded grassland. The soil of the grassland may have a luxuriant grass and forb canopy.

Radiation at all levels is more intense than in the forest, while microclimatic humidity is less and therefore saturation deficit and the evaporating power of the air are greater. Consequently drought is more serious because of the greater exposure to sun and wind and consequent modification of other site factors. The local climatic setting is, nonetheless, much less severe than that in even the more favored faciations of the subhumid wooded savanna, frequently erroneously mentally associated with the derived wooded savanna.

Fire destroys or impairs the cover in woodland, so as to convert it to open woodland and, in time, even to wooded grassland and open grassland. This results in marked increase in radiation, saturation deficit, and evaporation. The impact of rain is more severe and in turn bears on the change in several edaphic features noted later. Edaphic changes, together with increased radiation and saturation deficit, cause increased loss of available water. Significant combinations and permutations in aerial conditions, resulting from the gradual regrowth of vegetative cover, also are involved.

Table V summarizes broadly the comparative changes wrought by fire in derived savanna. During a restricted period these would be slight to moderate in severity for light, heat, wind, and humidity, integrated as a moderate increase in the evaporating power of the air.

Secondary *macchia*

In a totally different kind of secondary community, *macchia* successional to forest in the Knysna, the aerial factors other than amount and distribution of rain are more severe where macchia, ranging from below a meter to over 5 m in height, has replaced tall, closed forest. Accordingly, destruction or reduction of the dense but low macchia canopy further increases the severity of radiation, wind, and evaporating power of the air.

b. Soil. Derived wooded savanna: the contrast between the soil features under successional woodland protected and burned is very much less marked than when fire aids in conversion of forest to secondary communities. This is also evident when woodland is converted to open woodland and to wooded and open grassland.

As the amount of organic matter is rarely more than low to moderate, except in the most congenial localities, reduction by frequent fires is slight to moderate, although an exceptionally fierce fire is devastating. Nitrogen is only slightly reduced, but phosphorus, potassium, magnesium, and calcium are commonly slightly increased in the ash. Runoff is accelerated slightly to moderately but, during particularly torrential storms, may be severe, with surface, rill and gully erosion increasing at corresponding rates. Soil moisture may be increased for a time after the burn, because of death or reduction in growth of woody plants and grass, but as vigorous regrowth occurs, there is a related decrease. The surface may be indurated slightly to moderately, but the general physical characteristics are usually only slightly impaired. Soil organisms at very shallow depths are temporarily reduced slightly to moderately, but may be reduced for a longer period by slow but concentrated fire where organic matter is more abundant. Where accumulations of woody and other flammable materials are burned either gradually by slow fire or rapidly by a fierce one, the surface is baked.

Secondary *macchia*

The quality of the soil is generally poor (Table IV). Fire does slight damage except in long-protected communities, characterized by abundant raw and decomposing organic matter, which burns slowly or rapidly. Roots, seeds, microorganisms, and most other forms of life are killed or seriously injured. Trollope (1971) eradicated this community by biennial and quadriennial burning.

E. Subhumid Wooded Savanna

a. Aerial Factors. Although these are naturally comparatively rigorous in insolation and saturation deficit, reduction of the canopy and

TABLE IV

BROAD INTERRELATIONS OF BIOCLIMATIC UNITS AND SELECTED EDAPHIC FEATURES

Bioclimatic unit	Depth	Organic matter	Reaction	Erodibility when exposed, cultivated, or trampled by livestock	Drainage	Aeration	Major nutrients	Biota	General
HHF: Highly humid forest	(Shallow) to mediocre to moderate to (marked)	Moderate to high but raw and acid	pH 5.5 to 4.2	In some laterites moderate, in other types fairly high	Mediocre to (moderate)	Mediocre to (moderate)	Often strongly leached: Mediocre to moderate; N, P, K usually adequate, until soil exposed to insolation and rain	Mediocre to moderate	Variable in physical, chemical, and biotic qualities—very rapidly despoiled on exposure
HF: Humid forest	Shallow to mediocre to moderate to (marked)	(Mediocre) to moderate, raw, and acid but less so than in HHF	pH 6.5 to 4.5	Same as for HHF	Mediocre to moderate to (rapid)	Mediocre to moderate to (good)	Often strongly leached: Mediocre to moderate to (high); on the whole a higher status than in HHF; N, P, K adequate until soil exposed	(Mediocre) to moderate to (high)	Usually better quality than under HHF, but very rapidly despoiled on exposure
HMF: Humid montane forest	Shallow to mediocre to moderate (rarely marked)	Mediocre to moderate less raw and acid than in HF	pH 6.5 to 4.1	High, especially on steeper slopes	Moderate to (rapid)	Moderate to (good)	Moderately leached: Moderate to (high); N, P, K adequate until soil exposed	(Mediocre) to moderate to high	In best sites as high quality as in HF, but rapidly despoiled on exposure

DWS: Derived wooded savanna	Same as for original forest, but less where surface eroded	Mediocre to moderate according to the site and later incidence of disturbance by fire, livestock and cultivation; slight to negligible to nil locally	pH wide ranging according to site and history of disturbance: 6.0–4.5, where calcareous materials: higher	Moderate to high according to site and type	Mediocre to moderate to rapid	Mediocre to moderate	Often moderately strongly leached: N, P, K mediocre to moderate, but Locally adequate	(Mediocre) to moderate to locally high	Where forest long destroyed, much as for subhumid wooded savanna but locally of moderate to high quality; deteriorates rapidly on full exposure and burning
SHWS: Sub humid wooded savanna	Shallow to mediocre to moderate to marked	(Slight) to mediocre to (moderate)	pH 6.0 to 4.5	Moderate to high until return of cover which is more rapid than in the less humid regions	Mediocre to moderate to rapid	Mediocre to moderate	Often strongly leached, N, P, K mediocre to moderate	(Mediocre) to moderate to high	Highly variable in quality but in the absence of fertilizers are less productive than in more arid regions
Macchia (successional to forest)	Shallow to mediocre to moderate	Moderate to high, raw, acid	pH 5.0 to 4.1	Moderate but high on steep slopes	Mediocre locally moderate	Mediocre to moderate	Strongly leached, N, P, K, mediocre	(Mediocre) to moderate	Variable according to time since under forest; mediocre to (moderate) quality

TABLE V

EFFECT OF FREQUENT BURNING ON AERIAL FACTORS: CLIMAX AND RELATED CLOSED CANOPY STAGES[a]

Bioclimatic unit	Flammability	Potential combustibility	Reduction in area[b]	Effect on insolation		Wind flow (increase)	Evaporating power of the air (increase)	Humidity on burned sites (decrease)	Overall effect on micro or local climate
				Light (increase)	Heat (increase)				
Highly humid forest	Nil-Sl-(M)	VSl in severe ecologically dry months	a. M b. M c. M to derived wooded savanna	VH	VH	M-MH-H	MH-H	M	S
Humid forest (tropics)	Sl-M	Sl in EDM	a. MS b. M c. MS to derived wooded savanna	VH	VH	M-MH-H	MH-H	M	S
Humid forest	M-MH	Sl-M in EDM	a. MS b. MS c. S to derived wooded savanna	VH	H	M-MH-H	H	M	S
Humid montane forest (tropics)	M-MH	Sl-M in EDM	a. S b. MS c. MS to derived wooded savanna	VH	H	MH-H	H	M	S

Humid montane forest (subtropics)	Sl-M in EDM	M-MH	a. S b. S c. S to derived wooded savanna	VH	MH-H	H	M	S
Derived wooded savanna: woodland/open woodland	MH-H in EDM	MH	a. M b. M c. MS to open grassland	Sl-M	Sl-M	M	Sl-M	M
Subhumid wooded savanna: woodland/open woodland	V-VH in dry season	MH-(H)	a. M b. M c. M to open grassland	Sl-M	Sl-M	M	Sl-M	M
Macchia in South Africa	H-VH in dry spells	MH-H-VH	a. M b. M c. H: to open *macchia* with grass, or Karoo pioneers	VH	M	MH-(H)	M	S

[a] The following apply throughout wherever they occur: VH, very high; H, high; MH, moderately high; M, moderate; Sl, slight; Nil, negligible (to nil); MS, moderately severe; S, severe; VS, very severe; ( ) around a symbol implies *occasional* not usual; and NA not applicable.

[b] a. Historic; b. continuing; c. rough proportion already converted.

destruction of the grass and forb cover by fire render the conditions more severe, until a flushing of foliage and regrowth of grass shortly before the onset of the rains. This severity is accentuated during the blowing of hot, dry winds in West, East, Central, and Southern Africa.

b. SOIL. Fire has much the same effects on the soil as in the derived wooded savanna, but its influence is more marked when relatively newly created examples of this type are contrasted with the subhumid wooded savanna proper (Table VI). The severity of the burn is greater where the vegetation is conserved for several years.

F. EFFECTS ON VEGETATION COMPONENTS OF THE ECOSYSTEM

This section reviews the influence of fire on structure, composition, and broad ecology within bioclimatic units.

Structure and composition are so interlinked that we may accept the term "structure" as covering the floristic composition, abundance, spacing, canopy, stratification, life forms, and other morphological and social characteristics of a plant community.

Plant forms and biological stages in even a relatively simple vegetation component of a temperate biotic community are many, but they are even more numerous in tropical and subtropical forest and related ecosystems. Omitting lianas, epiphytes, lichens, mosses, liverworts, and fungi, it is still necessary to consider briefly at least nine forms and stages: woody vegetation both adult and adolescent, flowers, fruit or seed, seedlings, root suckers, and coppice as well as grasses and forbs (Table VII).

For brevity, climax infers climax, proclimax, subclimax, preclimax, postclimax, and all other dense types of communities with closed canopy. Secondary infers woodland, open woodland, wooded grassland, grassland, and mosaics of these; the canopy, although almost closed in some instances (as in woodland with secondary thicket), is not closely knit and in some, is petty, local, and even absent.

1. Adult and Adolescent Woody Vegetation

Harmful effects of fire are assessed in terms of the degree of functional blemish to stems, crowns, and occasionally root systems.

Highly humid forest. Fire is virtually harmless in the climax communities but, toward the end of exceptionally severe ecologically dry months (EDM), may be slightly harmful to these classes of trees and large woody shrubs along or near margins and glades. In secondary woodland and related communities fire may do moderate harm only, depending on the volume and the seasonal stages of grasses and forbs.

Humid forest—tropics. Almost no harm is wrought in the climax by early fire on the margin, but, at the end of the EDM, late burns may cause slight damage. Effects may then be moderate to moderately severe in secondary communities, where grasses and forbs are abundant and flammable.

Humid forest—subtropics. In this more restrictive climax, fire at the end of the EDM may cause slight damage along and near the margins, but its effect may be moderate to moderately severe in the secondary woodland and other stages, because of the abundant grass and forbs, occasionally fairly flammable. Hazards are greater than in the tropical faciation.

Humid montane forest—tropics. Despite frequent occurrence of gusty winds at higher elevation in rugged physiography, burns are only slightly harmful along the margins of the climax forest toward the end of the EDM. In the secondary communities, often rich in tall grass and forbs, the effects of early fires may, however, be moderately harmful to adolescent trees. Late fires during the EDM, may cause moderate to moderately severe damage to this class but less harm to adults.

Humid montane forest—subtropics. The effects on the climax are much as in the tropical faciation, marginal trees alone being liable to damage. Early fire causes moderate harm to adolescent trees in secondary communities, but a late burn does moderately severe damage to this class, with adults harmed only moderately. At the end of exceptionally severe EDM, the effects of fire on marginal climax communities and secondary stages are definitely more severe in the tropics. Because the comparatively much smaller areas of climax forest are usually widely scattered in extensive and flammable secondary vegetation, they are more vulnerable.

Humid forest—temperate-subtropics. Forests of the Knysna faciation, South Africa, are the largest example of this beautiful and valuable type. Although only slight to moderate damage is wrought to marginal trees in the climax during the wetter months, moderately severe and occasionally even severe harm results from fire during the somewhat drier months, particularly during hot, dry "Berg" winds. Occasionally these blow during the wetter months, February and March, when out-of-season fire may damage marginal growth.

Macchia (Fynbos). The commonest of the secondary communities associated with this type of forest, owing to accumulated dry matter and presence of oily secretions in some genera, is much more flammable and combustible than forest, and particularly so during the dry months. While slight to moderate damage may be wrought in adult and adolescent small trees and species of larger shrubs during the moist months,

TABLE VI

EFFECT OF FREQUENT BURNING ON EDAPHIC FEATURES: CLIMAX AND RELATED CLOSED CANOPY STAGES[a]

Bioclimatic unit	Loss in organic matter	Change in pH	Effect on nutrients	Surface temperature	Runoff	Erosion according to soil and slope	Soil moisture	Surface hardening in time	Physical features (general)	Biota (micro and other)	Overall deterioration of physical, chemical, and biotic features
Highly humid forest	MS-S	Often raised temporarily but falling later	N: severe loss; P, K, Ca: slight gain	Raised by insolation temporarily until vegetation develops	Sl-M-(MS) according to degree of slope	Sl-M	Temporary increase until vegetation reasserts itself vigorously	M	M-MS	M-MS	M-MS-(S)
Humid forest (tropics)	MS-S	Often raised temporarily but falling later	N: severe loss; P, K, Ca: slight gain	Raised by insolation temporarily until vegetation develops	Sl-M-(MS); according to degree of slope	Sl-M	Temporary increase until vegetation reasserts itself vigorously	M-MS	M-MS	M-MS-(S)	M-MS-(S)
Humid forest (subtropics)	MS-S	Often raised temporarily but falling later	N: severe loss; P, K, Ca: slight gain	Raised more during hot months	Sl-M-(MS); according to degree of slope	Sl-M-MS	Temporary increase until vegetation reasserts itself vigorously	M-MS	M-MS	M-MS-(S)	MS-S
Humid montane forest (tropics)	S-VS	Often raised temporarily but falling later	N: severe loss; P, K, Ca: slight gain	Raised more during hot months but fair rise during cooler months	M-MS-(S) according to degree of slope	MS-(S)	Temporary increase until vegetation reasserts itself vigorously	M-MS	M-MS	M-MS-(S)	MS-S
Humid montane forest (subtropics)	S-VS	Often raised temporarily but falling later	N: severe loss; P, K, Ca: slight gain	Raised more during hot months but only slight	M-MS-S	MS-S	Temporary increase until vegetation	M-MS	M-MS	M-MS-(S)	MS-S

Derived wooded savanna (tropics)	SI-M	Slightly raised, according to amount of ash; temporary	N: slight loss; P, K, Ca: slight gain	Raised temporarily until vegetation develops	SI-M	SI-M	A marked increase until vegetation reasserts itself vigorously	SI-M	SI	SI-M	SI-M
Derived wooded savanna (subtropics)	SI	Slightly raised, according to amount of ash; temporary	N: negligible to slight loss; P, K, Ca: slight gain	Raised until vegetation develops, but slight only in cooler months	SI-M	SI-M-(S)	A marked increase until vegetation reasserts itself vigorously	SI-M	SI	SI-M	SI-M
Subhumid wooded savanna (tropics)	SI	Slightly raised, according to amount of ash; temporary	N: slight loss; P, K, Ca: slight gain	Raised until vegetation develops, but slight only in cooler months	SI-M-MS	(SI)M	A marked increase until vegetation reasserts itself vigorously	SI-(M)	SI	SI	SI-(M)
Subhumid wooded savanna (subtropics)	SI-M-(MS) according to time since last fire	Raised very slightly; temporary	N: slight loss; P, K, Ca: slight gain	Raised temporarily until vegetation reasserts; but slight in cooler months	SI-M-MS	(SI)-M	A marked increase until vegetation reasserts itself vigorously	SI-(M)	SI	SI	SI-(M)
Macchia (Maqui) S. Africa: *Fynbos*	SI-M-(MS) according to time since last fire	Raised slightly according to amount of ash; temporary	N: slight loss; P, K, Ca, slight gain	Raised temporarily until vegetation reasserts; but slight in cooler months	M-MS	M-MS-(S)	A marked increase until vegetation reasserts itself vigorously	SI-(M)	SI-(M)	SI-M-(MS)	SI-M-(MS)

a See footnote a in Table V for explanation of codes.

TABLE VII

EFFECT OF FIRE ON FOREST VEGETATION, CLIMAX AND SECONDARY, AND ON SUBHUMID WOODED SAVANNA[a]

Bioclimatic unit	Timing of fires	Woody vegetation							Grass: basal cover	Forbs	Overall ultimate effect of unaided fire on woody growth	Sites prepared for shifting cultivation or fired after timber extraction
		Adult stages	Adolescent stages	Flowers	Fruit or seed	Seedling stages	Root suckers	Coppice				
Highly humid forest Climax–proclimax[c]	Early[b]	Nil	Nil	Nil	Nil	Nil	NA	Nil	NA	NA	Nil-(Sl) local on margins	M harm, rarely permanent; vigorous regeneration by coppice, suckers, and seedlings if site left alone for 2–3 years
	Late[b]	Nil-(Sl)	Nil-(Sl)	Nil-(Sl)	Nil-(Sl)	Nil-(Sl)	NA	Nil	NA	NA	Nil-(Sl) local on margins	M-(MS) harm, vegetative and seedling regeneration slower to reassert than on early burns
Secondary woodland to wooded grassland	Early[b]	Nil-(Sl)	Nil-(Sl)	Nil-(Sl)	Nil-Sl	(Nil)-Sl	(Nil)-Sl	Nil-Sl	Nil-Sl-(M: if strong fire)	Nil-Sl-(M)	Nil-Sl-(M: local)	M-MS harm, but rarely permanent; vegetative and seedling regeneration vigorous
	Late[b]	Sl-(M, local)	Sl-(M, local)	Sl-(M, local)	Sl-(M, local)	Sl-(M, local)	Sl-(M, local)	Sl-(M, local)	Sl stimulated	Sl	Sl-M	MS-(S) harm, regeneration vegetative and seedling less vigorous than on early burns

Humid forest: tropics Climax[d]	Early[b]	Nil	Nil	Nil	Nil	Nil	NA	Nil	NA	NA	Nil-SI-(M, on margins, local)	M-MS harm, rarely permanent; vigorous vegetative and seedling regeneration in 2–3 years
	Late[b]	Nil-(SI)	Nil-(SI)	Nil	Nil-SI	Nil	NA	Nil	NA	NA	Nil-SI-M (local on margins)	MS-(S) harm, regeneration vegetative and seedling less vigorous than on early burns
Secondary woodland to wooded grassland	Early[b]	Nil-SI	SI	Nil-SI	Nil-SI	SI	SI	SI	M-MS if strong fire	SI-(M) if strong fire	SI-M to woody species, M-MS to grass	M-MS harm, regeneration vegetative and seedling vigorous
	Late[b]	SI-M	M-(MS)	M-(MS)	M-(MS)	M-(MS)	M-(MS)	M-(MS)	Stimulated	M-(MS)	M-(MS) if strong fire	MS-S harm, regeneration vegetative and seedling less vigorous than on early burns
Humid forest: subtropics Climax-proclimax	Early	Nil-(SI)	Nil-(SI)	Nil-(SI)	Nil-(SI)	NA	Nil	NA	NA	NA	Nil-SI-(M) on margins	Now no shifting cultivation, in past much as for tropics. Now usually no timber extraction, in past much as for tropics
	Late	Nil-SI	Nil-SI	Nil-SI	Nil-SI	NA	SI	NA	NA	NA	Nil-SI-M on margins	As above
Secondary woodland-wooded grassland	Early	SI	M	M	M	M	M	SI	SI-M harm if strong fire	SI-M	(SI)-M	As above
	Late	M	M-MS	M-MS	M-MS	M-MS	M-MS	M-MS	Stimulated	M-(MS)	M-(MS)	As above

TABLE VII (Continued)

Bioclimatic unit	Timing of fires	Woody vegetation							Grass: basal cover	Forbs	Overall ultimate effect of unaided fire on woody growth	Sites prepared for shifting cultivation or fired after timber extraction
		Adult stages	Adolescent stages	Flowers	Fruit or seed	Seedling stages	Root suckers	Coppice				
Humid montane forest; tropics climax–proclimax[d]	Early[b]	Nil	Nil-Sl	Nil-(Sl)	Nil-(Sl)	Nil-(Sl)	NA	Nil	NA	NA	Nil-(Sl)[d]	M-MS harm, vigorous vegetative and seedling regeneration in 2–3 years
	Late[b]	Nil-Sl	Nil-Sl	Nil-Sl	Nil-Sl	Sl	NA	Nil	NA	NA	Nil Sl[d]	M-MS harm, vegetative and seedling regeneration less vigorous
Secondary woodland–wooded grassland	Early[b]	Sl	Sl-(M)	M	M	M	M	M	M harm	M	(Sl)-M	M-MS harm, vigorous regeneration, as above. Restricted exploitation in woodland only: harm done by fire Sl-(M)
	Late[b]	M	M-(MS)	M-(MS)	M-(MS)	M-(MS)	M-(MS)	M-(MS)	Stimulated	M-(MS)	M-(MS)	MS-S harm, less vigorous regeneration. Restricted exploitation in woodland only: harm by fire M-(MS)

Vegetation type	Phenology											Human impact
Humid montane forest: subtropics Climax–proclimax	Early[b]	Nil-(Sl)	Nil-(Sl)	Nil-(Sl)	Nil-(Sl)	Nil-(Sl)	NA	Nil-(Sl)	NA	NA	Nil-(Sl)	Now no shifting cultivation; in past much as for montane tropics. Now no extraction of timber, in past much as for montane tropics
	Late[b]	Nil-Sl / SI	Nil-Sl	Nil-Sl	Nil-Sl	Nil-Sl	NA	Nil-Sl	NA	NA	Nil-Sl (Sl)-M	As above / NA
Secondary woodland–wooded grassland	Early	M	M-(MS)	M-(MS)	M-(MS)	M-(MS)	M-(MS)	M-(MS)	M harm if strong fire / Stimulated	M-(MS)	M-(MS)	NA
	Late	M	M-(MS)	M-(MS)	M-(MS)	M-(MS)	M-(MS)	M-(MS)	Stimulated	M-(MS)	M-(MS)	NA
Humid lowland and upland forest: temperate–subtropical Knysna, South Africa	In moister months	Nil-(Sl)	Nil-(Sl)	Nil-(Sl)	Nil-(Sl)	Nil-(Sl)	NA	Nil-(Sl)	NA	NA	Nil-(Sl)-(M) on margins	No shifting cultivation at any time. Now restricted exploitation by State, no burning
Climax–proclimax[e]	In less moist months	Nil-Sl-(M) on margins	Nil-Sl-(M) on margins	Nil-Sl-(M)-(MS) on margins	Sl-(M)-(MS) on margins	Sl-(M)-(MS) on margins	NA	Sl-(M) on margins	NA	NA	Nil-Sl-M-(MS)-(S) on margins	As above
Secondary macchia successional to forest, Knysna	In moister months	Sl-M	Sl-M[f]	Nil-Sl-(M)-(MS) near margins	M	M	M	NA	SI-M harm if strong fire	SI-M stimulation	Sl-M stimulation	No shifting cultivation at any time, but fired by resident farmers: M harm
	In less moist months	M-MS-(S)	MS-S	VS to flowers[f] / in situ	MS-S	MS-S	MS-S	NA	SI-M harm, but marked increase in volume in following season	NA	Reduction and death ultimately in some species; increase in grass volume	No shifting cultivation at any time, but firing by resident farmers: M-MS-(S), harm

TABLE VII (*Continued*)

Bioclimatic unit	Timing of fires	Woody vegetation							Grass: basal cover	Forbs	Overall ultimate effect of unaided fire on woody growth	Sites prepared for shifting cultivation or fired after timber extraction
		Adult stages	Adolescent stages	Flowers	Fruit or seed	Seedling stages	Root suckers	Coppice				
Subhumid wooded savanna: tropics	Early (mild fire)	Sl-M	Sl-M	M	M	M	M	M	Sl-(M) if strong fire	Sl-M if strong fire	Sl-M	Although the grass is rarely fired early for shifting cultivation or depasturing, it may be burned to protect economic trees and their regeneration
	Late (hot fire)	M-(MS)	MS	MS	MS	MS	MS	MS	Stimulated	MS-S	M-MS, reduces woody growth	Lopping of crowns, high felling of stems and hot fire kill some trees and shrubs but merely retard others; grass. if not hoed, is often stimulated
Subhumid wooded savanna: subtropics	Early (mild fire)	Sl-M	Sl-M	M	M	M-(MS)	M-(MS)	M-(MS)	Nil-Sl-(M) if strong fire	Sl-M if strong fire	M	Rarely fired early for depasturing; this would intensify woody growth

| Late (very hot fire) | MS-S | MS-S | MS-S | MS-S | MS-S | MS-S | Stimulated | (MS)-S | (M)-MS-(S); reduces woody growth | Shifting cultivation is much less frequent than in tropics, because permanent to semipermanent cultivation, which almost completely removes woody growth and grass. Ultimately long continued burning completes the destruction of the original vegetation |

[a] See Table V, footnote *a* for explanation of codes.

[b] Early and late firing is frequently not possible, because of the variability in the number and degree of dryness of the EDM (Ecologically Dry Months: each less than 25 mm). Early implies fire in the early stages of the EDM; late implies fire toward the end of the EDM: the early fire is usually much milder than the late fire.

[c] Fire kindled and spread by man in order to clear forest for shifting cultivation; unaided spreading usually not possible, except during relatively rare severe EDM.

[d] Unaided spread usually local only, except during longer and more severe EDM.

[e] Fire rarely originates in closed forest, except in older glades (*Fynbos* "islands") and timber exploitation sites during hot, dry Foenlike "Berg" winds: Crown fires may travel rapidly owing to *Usnea* on older branchlets, while ground fires ignite dry litter and deep organic matter common in the "wet" and "moist" types of forest.

[f] While current flowers are destroyed, fire stimulates many dicotyledons and some monocotyledons to flower more vigorously the following season.

burns during the dry months, aided by the "Berg" winds, usually inflict moderately severe to severe damage.

Subhumid wooded savanna—tropics. Early burns cause slight to moderate damage to both adult and adolescent plants, whereas those during the height of the dry season may be moderately severe to both classes. Accumulations of grasses, forbs, and dead and dying woody material are highly flammable and combustible by late fires.

Subhumid wooded savanna—subtropics. The effects of early and late fires resemble those in the tropical facies, but locally and also occasionally over extensive terrain may be more severe, because of the greater flammability and combustibility of grass, forbs, and woody material. Late fires are considerably more severe than early ones.

2. *Regeneration: Flowers, Fruit, and Seed*

Table VII summarizes the effects of both early and late fires on both climax and secondary communities.

Highly humid forest. Marginal trees and shrubs in the climax communities may occasionally suffer slightly during the rare EDM but, in general, no harm is done to flowers or fruits. Early fires occasionally are slightly harmful in secondary communities but late ones may be moderately damaging.

Humid forest—tropics. Marginal trees and shrubs suffer almost as little in the climax communities as in the unit just discussed, but at the end of a more severe ecologically dry period, may suffer rather more locally. While early burns in secondary stages cause slight harm, unless there is local accumulation of material, burns toward the end of the dry season may be moderate to moderately severe for both flowers and fruits. The volume and degree of combustibility of the grass, forb, and woody growth are highly significant.

Humid forest—subtropics. Early and late fires along margins in the climax stages do almost as little harm as those in tropical faciations but, during exceptionally severe EDM, burning of a wealth of mildly combustible material may cause slight to moderate damage locally. In secondary communities with fair growth of grass and forbs, damage to flowers and fruits may be moderate from early fires but moderately severe from those toward the end of the dry period.

Humid montane—tropics. The effect of either early or late fires on marginal trees and shrubs in flower or fruit in the climax communities is much as in forest types already considered but, owing to more turbulent air conditions, there may occasionally be slight to moderate loss. In the secondary communities the effects of early burns along the margins are moderate but the harm caused by late burns may be moderately severe. More severe drought and presence of grass and other combustible growth may occasionally provide appropriate conditions for fire.

Humid montane forest—subtropics. Due to the small areas scattered in flammable secondary vegetation, the climax communities may sometimes undergo somewhat more damage than those in related faciations in the tropics, but often no or only slight loss results from marginal burns. In the secondary stages moderate loss is caused by early burns, that from late fires usually being moderately severe but occasionally severe, according to season.

Humid forest—temperate-subtropics. During the wetter months the climax forests at Knysna rarely undergo more than slight damage marginally, but during the drier "Berg" winds the harm done to flowers and fruit is moderate to moderately severe.

As the secondary community of the greatest extent is *macchia* of variable floristics, height, and vigor, fires have varying influence on flowers and fruits: that in the moister months being slight to moderate only, but that in the drier period and during "Berg" winds moderately severe, to severe, and even very severe. While witholding fire for 3 to 5 years may have a depressing influence on the luxuriance and beauty of the flowers in certain families (e.g., Ericaceae, Proteaceae, Bruniaceae, Compositae, Iridaceae, Liliaceae, Amaryllidaceae, and Restionaceae), appreciable damage may be caused to flowers of the season of the burn and flower buds of the subsequent season.

Where *Virgilia oroboides* Salt., a rapidly growing leguminous tree, forms a successional community with abundant hard seed on the surface or is lightly covered by soil, fire may destroy the existing community but stimulate seed germination, thus providing a successive stand (Phillips, 1926).

Subhumid wooded savanna—tropics. Early burns are responsible for moderate damage whereas later ones may be moderately severe to severe for trees and woody shrubs. Where exceptionally dry conditions and a wealth of high combustible grass and other material exist, damage may be very severe.

Subhumid wooded savanna—subtropics. While the effects of both early and late fires are much as in the tropical facies, the locally lower humidity and greater growth of grass, forbs, and other highly combustible material frequently encourage large fires which are much more severe than those in the tropics.

3. Regeneration: Seedlings, Rooted Suckers, and Coppice

With the exception of regeneration established on or relatively near the transition between forest and the secondary communities, early and late fires either have no effect or only very slight effect on seedlings and coppice. As rooted suckers are either absent or rare in the climax stages, this form of regeneration is not affected.

The data of Table VII may be summarized as follows:

Highly humid forest. Fire has no effect in the climax forest but in the secondary communities early burns do slight harm, whereas late ones may be moderately damaging locally to all types of regeneration.

Humid forest—tropics. The effects in both climax and the secondary stages are often as insignificant as in the foregoing ecosystems, but occasionally, toward the end of the more severe EDM, late fires may be slightly more effective in the climax and even moderate to moderately severe in secondary communities. Such effects occur during strong to gusty wind during long, dry periods.

Humid forest—subtropics. While the harm caused during the EDM may often be no more serious in the climax stages than in the tropical faciations, smaller areas set in extensive secondary ones tend to be injured because of hot fire in ecotones: Damage may be moderate to moderately severe. Secondary communities may frequently undergo no more than moderately severe injury but again, during long and severe EDM, may be severely damaged.

Humid montane forest—tropics. Although marginal regeneration of climax stages may be harmed only slightly during especially severe EDM, the secondary vegetation may be damaged moderately from early fire and moderately to moderately severely from burns late in dry spells. Late burns may occasionally have severe local effects.

Humid montane forest—subtropics. Because of greater isolation of this faciation in great expanses of secondary flammable vegetation, the effect of fire during severe EDM may be more serious than in tropical faciations. Aided by high desiccating wind during very severe ecological drought, fire may severely damage the secondary communities.

Humid forest—temperate-subtropics. Climax stages may escape fire but during desiccating "Berg" winds smaller areas may suffer appreciably. As the secondary stages are either *macchia* or communities in which *Virgilia oroboides* is dominant, fire during the moister months causes only moderate harm to seedlings and suckers but injury may be extensive during the EDM. As noted, *Virgilia* regenerates freely after fire has heated seeds on and near the soil surface (Phillips, 1926).

Subhumid wooded savanna—tropics. Early fires may do slight to moderate harm only, but where dry grasses and other materials have accumulated during several seasons, such fires may have locally severe influences on woody regeneration. Generally, however, it is fire in the late dry season which acts moderately severely to severely on regeneration. Foresters, therefore, favor early rather than late burning.

Subhumid wooded savanna—subtropics. Fire may be more harmful to regeneration stages than in the tropics, because of the frequently lower humidity during the late dry season and luxuriant growth of grass and forbs. In very dry years severe damage is wrought in communities which escaped fire for several years.

4. Communities of Grass and Associated Sedges and Forbs in Forest and Wooded Savanna

Much of the vigorous debate about the effect of timing of burning is traceable to the divergent requirements of foresters and pastoralists (Phillips, 1965 and references therein). The effects of early and late fires on grass communities are summarized in Table VII. The responses within the several ecosystems are as follows:

Climax forest communities. Owing to the low light intensity near the ground, grass is absent to sparse: plants are shade demanding or shade bearing such as *Oplismenus, Potamophila, Leptaspis, Streptogyne,* and *Olyra.* Where the canopy is much lighter as near margins and in ecotones, these and some woodland grasses occur, with forest grasses gradually dropping out and wooded savanna species increasing. While more robust grass communities exist only in ecotones, fire during the early, wetter portions of the year has no or very slight effect. Late burns do slightly more harm, particularly in less moist faciations.

Secondary communities are affected to varying degree by early and late fires, or in the moist or less moist seasons, respectively. The effects include the following:

a. Secondary Highly Humid Forest. Early fires have no or very slight effect, but those during the very short ecologically dry spells slightly stimulate growth of grasses and forbs.

b. In other types of secondary forest ecosystems the effects of early burning on grasses and forbs may be slight to moderate to moderately severe. Locally, however, fires early in the less moist spells may retard grasses but stimulate flowering in forb autumnal aspect socies. Burning during ecologically dry spells has variable effects: it sometimes stimulates both grasses and forbs in ecotones. Forb vernal aspect socies occasionally appear in such grass communities.

c. *Macchia.* Secondary vegetation is fairly rich in species of grasses but poor in number and volume. It is, however, rich in species, vigor, and abundance of forbs. Burns during the moister months are slightly to moderately harmful to grasses and forbs, while during the dry spells these stimulate growth overall. Autumnal and vernal forb *socies* may be floristically luxuriant.

d. Subhumid Wooded Savanna—Tropics and Subtropics. Since the grass components are vigorous and extensive, with varying proportions of scattered forbs and forb socies interspersed, the effects of burning are obvious.

Fires during the early part of the ecologically dry season frequently have a slightly to moderately harmful effect on vigor, volume, and basal area of grasses, particularly in areas grazed by wild and domesticated animals. If continued for several years, such early burning and grazing reduce the volume of the more palatable grasses but increase the proportion of grasses and forbs which are not grazed or only occasionally grazed. Early fires therefore are not favored by pastoralists but, because they do less damage to woody plants, are favored by foresters anxious to establish a suitable distribution of age and size classes of valuable species of woodland trees. Late burning during the dry season usually stimulates vigor and the potential volume of grasses, but may reduce the proportion of forbs because of the high temperature generated. Hence, moderately late fires are favored by pastoralists, but not by foresters.

Although the latter fires in the subtropical facies often are hotter than those in the tropics, and notably so where frost occurs, the harm done by early and late burning to grasses, forbs, and woody elements is somewhat similar to that in the tropics.

5. *Overall Ultimate Effect of Unaided Fire on Woody Growth*

Among points of practical interest summarized in Table VII are the following:

(a) Despite general low flammability and combustibility of climax stages in all the types of forest ecosystems, marginal and ecotonal woody elements are subject to damage from fires during ecologically dry spells. Injury ranges from slight to moderately severe and, locally and occasionally, to severe. Such fires are defined as late, contrasted with those far less harmful ones experienced during the last phases of the moist season and earlier parts of the dry season or the earlier EDM. Early and late burning alike take a higher toll as the humidity of the ecosystem decreases: the least susceptible being the highly humid, the most susceptible the subtropical variants of forests lowland, upland, and montane. This results because of their smaller dimensions and relatively isolated occurrence and often also because of their somewhat less humid setting.

(b) In the secondary forest communities, woodland, open woodland, wooded and open grassland, and *macchia,* early fire takes a higher toll than in the climax, even in the highly humid, humid, and humid montane forest successions and definitely more so in *macchia* and other communities in the temperate-subtropical forest succession at Knysna. Late fire generally is more harmful than early fire in secondary communities of all the forest successions reviewed, but the degree of severity increases from the highly humid to the humid and is maintained at this degree in the montane forests. In the temperate-subtropical *macchia* at the Knysna, reduction in vigor of woody growth is even more marked, with forbs and grass tending to be encouraged.

(c) In the subhumid wooded savanna, both tropical and subtropical, early burning is responsible for slight to moderate harm, with that in the subtropics being somewhat more severe. Late burning is clearly more harmful in both these variants, but the effects in the subtropics tend to be somewhat more severe, owing to the less humid setting and greater accumulation of highly flammable and combustible grasses and forbs.

6. *Effect of Fire on Sites Prepared and for a Time Maintained for Shifting Cultivation or Fired after Timber Extraction*

It is academic to think of fire today as being unaided by man except possibly in a few isolated localities. Therefore, the role of fire in two of man's most influential activities in the tropical and subtropical forests of Africa, shifting cultivation and crude exploitation of timber, is discussed briefly.

Notwithstanding the vast extent of closed forest converted to woodland, wooded grassland, and grassland, and the consequent profound alteration in the aerial and edaphic factors induced by long-continued use for crop and animal production, it is inappropriate to discuss the history, practice, advantages, and deteriorating influences of what is widely known as shifting cultivation. Such cultivation in tropical countries is a subject of vigorous controversy, much of it ill-informed, misinformed, and subjective. Attempts to examine objectively the salient features of shifting cultivation in Africa are outlined by de Schlippe (1956), Nye and Greenland (1960), and Phillips (1959, 1964, 1961, 1966).

Areas of seemingly "virgin or near virgin" forest, as in parts of the Gabon, the several Congo states, Cameroon, Nigeria, Ghana, and westward to Guinea and the Gambia are in many places not virgin, but rather high forest regenerated in localities which, through the centuries, have been subjected to selective shifting cultivation. While locally experienced forest ecologists and pedologists can reconstruct the history of change induced by shifting cultivation, the luxuriant and lofty ensemble may readily mislead other observers.

Where forest has been exploited, without attention to the selection system initiated a century ago by von Brandis in India and introduced, at various times, in a modified form in South, East, and West Africa by the British and French, fires have caused local to moderately extensive change. Where exploitation of timber has been followed by shifting cultivation the change has been more profound, because fire has been applied intensively and extensively and, in some localities, consistently for the 2 to 5 or more years of the cultivation phase.

As may be seen in Tables V and VI, changes are intensified when cutting and fire are continued for several years and, where repeated in short cycles, the changes are more profound. Some understanding of the nature of the harm inflicted on the vegetation and soil by felling, burning, and cultivation is obtained from the final column of Table VII.

Despite alterations in the forest ecosystem, ecologically shifting cultivation is a better adaptation to potential use and conservation of forest soil than permanent farming could ever be. Nye and Greenland (1960) and Phillips (1959, 1964, 1961, 1966) discussed the physicochemical–biotic aspects of change wrought in the soil of the forest and savanna through burning of debris in the course of this practice.

Exploitation of timber in association with burning does result in considerable damage to the biotic and other components of the ecosystem, but were not followed by cultivation does no inflict changes lasting as long as those induced by shifting cultivation.

G. EFFECTS ON VERTEBRATES

It is not possible to discuss the influence of fire on vertebrates in each of the major vegetation types because records are unavailable. However, the following generalizations can be made:

Highly humid forest. As fire seldom occurs in the type proper but is confined to clearing and margins, it cannot be argued that animals which restrict themselves to forest and tall thicket communities are directly affected by burning. Because fire does occur in secondary successional stages from grassland to woodland, however, it is logical to assume that unless the area of dense forest is extensive, burning of these facies may indirectly influence browsers and grazers. Elephant and buffalo, formerly typical of this bioclimatic unit, but today occasionally to rarely present in parts of the Congo Basin and West Africa, benefit from the fresh material produced by grasses and edible woody plants on burned sites. As the high humidity, general dampness, and rarely more-than-moderate wind do not encourage fast, fierce fires, damage to young mammals and to birds and their nests is negligible to slight.

Humid forest. Much of the foregoing refers also to humid forest. This unit occurs not only in West Africa where wild mammals are today comparatively rare to local, but also in parts of East Africa where they are most frequent. Hence the beneficial effects of fire on herbivores, elephant, buffalo, antelope, and wild pig, are more significant. In the secondary communities in which grass occurs, fire during the driest months may be somewhat fiercer than in the more humid unit, but seldom destroys young mammals and birds.

Lion and leopard live in or near the now fragmented areas of this unit in East Africa. These carnivores are rarely if ever harmed by fire but rather benefit because herbivores tend to concentrate on burned sites, thus making hunting easier.

Humid montane forest. Elephant, buffalo, large and small antelope, lion, and leopard still occur in this vegetation type in certain parts of East Africa. They have easy access to secondary vegetation, grassland to woodland, and hence are favorably affected. Because of stronger winds and smaller areas, fire often moves faster and is fiercer than in the larger forests at lower elevation. Therefore, the young of the small mammals and birds may be destroyed occasionally.

Humid subhumid coast, coast hinterland, and montane upland forest in the subtropics of South Africa. Owing to man's activities the original large animals, such as elephant and buffalo, have long since disappeared, except for elephant in the Knysna. These were occasionally killed in glades and along the margins but rarely if ever in high forest. Harison (1856–1874) mentioned destruction of these and of bushbuck and smaller antelope in secondary communities, thicket and *macchia,* in the then extensive Knysna forests. The roles of the bush-dove (*Columbia arquatrix* T.&K.), "loerie" (*Turacus corythaix* Wag.), bushpig (*Potamochoerus choeropotamus*), and elephant in the same forests are discussed by Phillips (1925, 1926, 1927, 1928c).

Derived wooded savanna. Because of a wealth of flammable material, this facies may experience fast and fierce fire in dry periods. Mammals, birds, and large and slow reptiles may suffer when caught in a holocaust. Grazers and browsers invariably congregate on the fresh green growth, following the burns, and so do the predators, lion, leopard, cheetah, and many smaller carnivores.

Subhumid wooded savanna. As humidity is lower and the wind often stronger than in the other units, fire may be fast and very hot in dry periods, depending on the vegetation and wind. The burns in the subtropics tend to be faster and hotter because of the drier conditions.

Where frost occurs, the grass and other plant growth is particularly combustible. Young eland, hartebeest, duiker, warthog, and bushpig, among other vertebrates, suffer from time to time, while young birds and eggs are often destroyed. As the fresh growth on the burns is highly palatable and nutritious it is readily sought by herbivores.

It may be concluded that burning locally and occasionally kills or injures young mammals and birds and some reptiles, but this is largely confined to the secondary stages and occurs rarely, if ever, in the forest communities proper. The damage is somewhat greater in the derived and subhumid wooded savanna.

VI. Fire and the Future

It often is impossible to separate the influences of fire from those of other prime activities of man, such as shifting and sedentary cultivation, running of livestock, exploitation of forest products, removal of forest for establishment of exotic tree plantations and for construction of roads, railways, and dams. I have elsewhere predicted (Phillips, 1965) how the natural vegetation, the soil supporting it, and the water resources regulated thereby might be deteriorated by 2000 A.D., assuming that consistent conservation and sound management are not applied immediately. That prediction is a sad one and is repeated here. But I reemphasize that such a state of affairs need not develop: its prevention lies with man (Komarek, 1971).

When the African states most concerned with conservation and forest management (those in West Africa, the Congo, and the Gabon) and those locally concerned (Ethiopia, Uganda, Kenya, and Tanganyika) have gone some distance nearer solving their socioeconomic and population problems, it is feasible that they might attend more ably, wisely, and widely to the welfare of those sectors of forest and subhumid wooded savanna which they should by then have set aside as reserves. Such reserves should embrace forests and other vegetation for economic forestry, nature study, wildlife conservation and management, and the like.

Varying with the country and its special socioeconomic problems, the reserves might total between 15 and 25% of the area of each of the prime vegetation types such as forest and subhumid wooded savanna. Sound management of these types outside the reserves must, of course, be assured according to progressive policies and program. Otherwise the national welfare will inevitably suffer grievously.

Although I am no believer in "ecologloomsterism" or "ecodoomsterism" of the kind far too commonly expounded today, even by some notable ecologists in Britain, America, and elsewhere, I record that while I hope for the best I fear this is not likely to be achieved. Much forest and subhumid wooded savanna will unnecessarily fall before the onset of careless development, which will employ feckless fire as one of its ancillary tools!

References

Adamson, R. S. (1938). "The Vegetation of Soutn Africa." Brit. Emp. Veg. Comm., London.

Adamson, R. S. (1956). Forests of South Africa. In Haden-Guest *et al.* (1956, pp. 385–391).

Aubréville, A. (1936). "La flore forestière de la Côte Invoire." Paris.

Aubréville, A. (1937). Les forêts du Dahomey et du Togo. *Bull. Com. et Hist. et Sc. de l'A.O.F.*

Aubréville, A. (1938). La forêt coloniale: Les forêts de l'Afrique occidentale française. *Ann. Acad. Sc. Colon., Paris* 9, 1–245.

Aubréville, A. (1947a). Les brousses secondaires en Afrique équatoriale (Côte d'Ivoire, Cameroun, A.E.F.). *Bois For. Trop.* 2, 24–35.

Aubréville, A. (1947b). The disappearance of the tropical forests of Africa. *Unasylva, FAO,* Rome, 1, 5–11.

Aubréville, A. (1948). L'Afrique équatoriale française australe. Les régions à longue saison sêche du Cameroun et de l'Oubangui-Chari, Paris.

Aubréville, A. (1949a). "Climates, fôrets et désertification de l'Afrique tropicale. "Soc. l'Édit. Geograph., Marit. Colon, Paris.

Aubréville, A. (1949b). Ancienneté de la destruction de la couverture forestière primitive de l'Afrique tropicale. *Bull. Agr. Congo Belge* 40, 1347–1352.

Aubréville, A. (1950). "La flore forestière soundano-guinéene—A.O.F., Cameroun—A.E.F." Soc. Edit. Geograph., Marit. Colon, Paris.

Aubréville, A. (1956). 16. Forests of tropical Africa *In* Haden-Guest *et al.* (1956) 353–384.

Batchelder, R. B., and Hirt, H. F. (1966). "Fire in Tropical Forests and Grasslands" (mimeo.). Earth Sci. Div. ES-23, U.S. Army Natick Lab., Natick, Massachusetts.

Bews, J. W. (1912 *et seq.*). The vegetation of Natal. *Ann. Natal Mus.* 2, 253–331 (and a number of other papers bearing on forest and fire references in following papers).

Bews, J. W. (1917). The plant ecology of the Drakensberg range. *Ann. Natal Mus.*, 3, 511–565.

Bews, J. W. (1918). "The Grasses and Grasslands of South Africa." Davis, Pietermaritzburg.

Brown, J. C. (1887). "Management of Crown Forests at the Cape." Oliver and Boyd, London and Edinburg.

Busse, W. (1901). "Expedition nach den Deutsch-Ost Afrikanischen Steppen," Bericht, I-VIII. Kolon. Wirtschaftliches Kom., Berlin.

Busse, W. (1902). Forschungereise durch den Sudlichen Teil von Deutsch-Ost Afrika. *Beiheft zum Tropenpflanzer* 5, 93–119.

Busse, W. (1907). Deutsch Ost-Afrika, Zentrales Steppengebit. *In* "Vegetationsbilder" (G. Karsten and H. Schenk, eds.), V. Reihe, Heft 7 Fischer, Jena.

Busse, W. (1908). Die Periodische Grasbrände im Tropischen Afrika, ihr Einfluss auf die Vegetation und ihre Bedeutung fur die Landskultur. *Mitt. Deut. Schutzgeb.* **21**, 113–139.

Chapman, J. D., and White, F. (1970). "The Evergreen Forests of Malawi." Commonw. Forest. Inst., Oxford.

Chipp, T. F. (1927). The Gold Coast Forest. *Oxford Forest. Mem.* **7**.

Clarke, J. D. (1959). "The Prehistory of Southern Africa." Penguin Books, London.

Clayton, W. D. (1961). Derived savanna in Kabba Province, Nigeria. *J. Ecol.* **49**, 596–604.

de Schlippe, P. (1956). "Shifting Cultivation in Africa: The Zande System of Agriculture." Routledge & Kegan, London.

de Wildeman, E. (1920). "Mission forestière et angricole du Comte Jacques de Briey au Mayumbe." Brussels.

de Wildeman, E. (1924). "La fôret congolaise. Sa régressio, sa transformation, sa distribution actuelle." C. R. Ass. Fr. Ao. Sc., Liege.

Engler, A. (1910–1925). "Die Pflanzenwelt Afrika," Vols. I, II, III, V. Leipzig.

Evans, C. C. (1939). Ecological studies on the rain forest of southern Nigeria. II. The atmospheric environmental condition. *J. Ecol.* **27**, 437–482.

Fourcade, H. G. (1889). "Report on the Natal Forests." Govt. Bluebook, Pietermaritzburg.

Gossweiler, J., and Mendonça, F. A. (1939). "Carta fitográfica de Angola." Govt. Printer, Luanda.

Grandvaux Barbosa, L. A. (1970). "Carta fitogeográfica de Angola." Inst. Invest. Cien. Angola. Govt. Printer, Luanda.

Haden-Guest, S., Wright, J. K., and Teclaff, E. M. (1956). "A World Geography of Forest Resources." Ronald Press, New York.

Harison, Capt. C., Conservator of Forests. (1856–1874). Letter book relating to forests, fire and related matters, Tzitizikama forests, South Africa, ms., Dept. Forestry.

Henniker-Gotley, G. R. (1936). A forest fire caused by falling stones. *Indian Forest.* **62**, 422–423.

Hopkins, B. (1961). The role of fire in promoting the sprouting of some savanna species. *J. West Afr. Sci. Ass.* **7**, 154–162.

Hopkins, B. (1965). "Forest and Savanna." Heinemann, London.

Hutchins, D. E. (1909). "Forests of British East Africa." Cmnd. 4723. HM Stationery Office, London.

Johnstone, H. (1906). "Liberia," Vol. 2, p. 894. Hutchinson, London.

Keay, R. W. (1953). An Outline of Nigerian Vegetation." Govt. Printer, Lagos.

Keay, R. W., ed. (1959). "Vegetation Map of Africa South of the Tropic of Cancer." Oxford Univ. Press, London and New York.

Komarek, E. V. (1964). The natural history of lightning. *Proc. 3rd Annu. Tall Timbers Fire Ecol. Conf.*, pp. 139–184.

Komarek, E. V. (1966). The meteorological basis for fire ecology. *Proc. 5th Annu. Tall Timbers Fire Ecol. Conf.*, pp. 85–125.

Komarek, E. V. (1967). Fire—and the ecology of man. *Proc. 6th Annu. Tall Timbers Fire Ecol. Conf.*, pp. 143–170.

Komarek, E. V. (1968). Lightning and lightning fires as ecological forces. *Proc.*

8th Annu. Tall Timbers Fire Ecol. Conf. pp. 169–197.

Komarek, E. V. (1971). Dedication. *Proc. 11th Annu. Tall Timbers Fire Ecol. Conf.*

Laughton, F. S. (1937). "The Sylviculture of the Indigenous Forests of the Union of South Africa, with Special Reference to the Forests of the Knysna Region," Sci. Bull. 157, Forest Ser. 7. Govt. Printer, Pretoria.

Lebrun, J. (1935). "Les essences forestières des régions montagneuses du Congo oriental." INEAC, Brussels.

Lebrun, J. (1936). "Répartition de la fôret équatoriale et des formations végétales limitrophes." Min. des Colonies, Brussels.

Lebrun, J., and Gilbert, G. (1954). "Une classification écologique des fôrets du Congo," Ser. Sci. 63. INEAC, Brussels.

Marloth, R. (1908). "Das Kapland." Fischer, Jena.

Nye, P. H., and Greenland, D. (1960). "The Soil Under Shifting Cultivation," Tech. Commun. 51, Commonwealth Bur. Soils, Harpenden.

Pappe, L. (1862). "Silva capensis." Ward & Co., London.

Phillips, J. (1925). The Knysna elephants: A brief note on their history and habits. *S. Afr. J. Sci.* **22**, 287–293.

Phillips, J. (1926). "Wild pig" (*Potamochoerus choeropotamus*) at the Knysna: Notes by a naturalist. *S. Afr. J. Sci.* **23**, 655–660.

Phillips, J. (1927). The role of the bushdove (*Columbia arquatrix* T.&K.) in fruit dispersal in the Knysna forests. *S. Afr. J. Sci.* **24**, 435–440.

Phillips, J. (1928a). The principal forest types in the Knysna region—An outline *S. Afr. J. Sci.* **25**, 182–201.

Phillips. J. (1928b). Plant indicators in the Knysna region. *S. Afr. J. Sci.* **25**, 202–224.

Phillips, J. (1928c). *Turacus corythaix* Wagl. ("Loerie") in the Knysna forests. *S. Afr. J. Sci.* **25**, 295–299.

Phillips, J. (1930a). Some important vegetation communities in the central province of Tanganyika Territory (formerly German East Africa). *J. Ecol.* **17**, 193–234.

Phillips, J. (1930b). Fire: Its influence on biotic communities and physical factors in South and East Africa. *S. Afr. J. Sci.* **27**, 352–367.

Phillips, J. (1931a). The biotic community. *J. Ecol.* **19**, 1–24.

Phillips, J. (1931b). Forest Succession and Ecology in the Knysna Region. *Bot Surv. Mem.* **14**, 1–326.

Phillips, J. (1934–1935). Succession, development, the climax and the complex organism: An analysis of concepts. Pt. I. *J. Ecol.* **22**, (1), 88–105; Pt. II. **23** (1), 210–223; Pt. III. **23** (2), 554–571.

Phillips, J. (1936). Fire in vegetation: A bad master, a good servant and a national problem. *J. S. Afr. Bot.* **1**, 36–45.

Phillips, J. (1959). "Agriculture and Ecology in Africa (with a bioclimatic map)." Faber, London, and Praeger, New York.

Phillips, J. (1961). "The Development of Agriculture and Forestry in the Tropics: Patterns, Problems and Promise," 1st ed. Faber, London, and Praeger, New York.

Phillips, J. (1963). "The Forests of George, Knysna and the Zitzikama: A Brief History of Their Management: 1778–1939," Bull. 40, Dept. Forestry. Govt. Printer, Pretoria.

Phillips, J. (1964). Shifting cultivation. *International Union for the Conservation of Nature Publ.* [N.S.] **4**, 210–220.

Phillips, J. (1965). Fire as master and servant: Its influence in the bioclimate regions of Trans-Saharan Africa. *Proc. 4th Annu. Tall Timbers Fire Ecol. Conf.,* pp. 1–109.

Phillips, J. (1966). "The Development of Agriculture and Forestry in the Tropics: Patterns, Problems and Promise," 2nd ed. Faber, London, and Praeger, New York.

Phillips, J. (1968a). The ecosystem as a basis for the investigation and development of agriculture, forestry and related industries in the tropics and subtropics. *Proc. Symp. Recent Advan. Trop. Ecol., Int. Soc. Trop. Ecol., 1968* pp. 121–739.

Phillips, J. (1968b). The influence of fire in Trans-Saharan Africa in "Conservation of Vegetation in Africa South of the Sahara." *Acta Phytogeogr. Suecica* **54,** 13–20.

Phillips, J. (1971a). "Physiognomic Classification of the More Common Vegetation Types in Southern Africa, Including Moçambique" (mimeo.). Circulated by R. F. Loxton, Hunting & Associates, Johannesburg.

Phillips, J. (1971b). Re pedo-ecological units. *In* "Report on a Survey of the Natural Resources of the Natal South Coast" (mimeo.). By R. F. Loxton, Hunting & Associates, Johannesburg, for Town and Regional Planning Commission, Natal, Pietermaritzburg.

Phillips, J. (1972). "The Agricultural and Related Development of the Tugela Basin and its Influent Surrounds," Vol. 19. Town and Regional Planning Commission, Natal, Pietermaritzburg.

Rea, R. J. (1935). The forest types of vegetation in Tanganyika Territory. *Emp. Forest. J.* **14,** 202–208.

Richards, P. W. (1939). Ecological studies on the rain forest of southern Nigeria. *J. Ecol.* **27,** 1–61.

Richards, P. W. (1952). "The Tropical Rain Forest." Cambridge Univ. Press, London and New York.

Robyns, W. (1938). Considérations sur les aspects biologiques du problème des feux de brousse au Congo belge et au Ruanda Urundi. *Bull. Sc. Inst. Roy. Col. Belge* **9,** 383–420.

Robyns, W. (1950). "L'Encyclopédie du Congo Belge." Brussels.

Schnell, R. (1970). "La Phytogéographie des Pays Tropicaux—Les Problémes Généraux. 1. Les Flores—Les Structures." Gauthier-Villars, Paris.

Schnell, R. (1971). "La Phytogéographie des Pays Tropicaux—2. Les Milieux—Les Groupements Végétaux." Gauthier-Villars, Paris.

Siebenlist, T. (1914). "Forstwirtschaft im Deutsch-Ostafrika." Parey, Berlin.

Sim, T. R. (1907). "Forests and Forest Flora of the Cape of Good Hope." Taylor & Henderson, Aberdeen.

Sim, T. R. (1909). "Forest Flora and Forest Resources of Portuguese East Africa." Taylor & Henderson, Aberdeen.

Sparrow, J. G., and Ney, E. P. (1971). Lightning observations by satellite. *Nature* (*London*) **232,** 540–541.

Staples, R. R. (1926). Experiments in veld management. First report. *Dep. Agr. S. Afr., Sci. Bull.* **49.**

Story, R. (1952). "A botanical survey of the Keiskammahoek District. *Bot. Surv. S. Afr., mem.* **27.**

Swynnerton, C. F. M. (1918). Some factors in the replacement of the ancient East African forest by wooded pasture land. *S. Afr. J. Sci.* **14,** 493–518.

Swynnerton, C. F. M. (1921). An examination of the Tsetse problem in North Mossurise. *Bull. Entomol. Res.* **11,** 319–385.

Tansley, A. G. (1935). The use and abuse of vegetational concepts and terms. *Ecology* 16, 284–307.

Taylor, C. J. (1952). "Vegetation Zones of the Gold Coast." Govt. Printer, Accra.

Taylor, C. J. (1960). "Synecology and Silviculture in Ghana." Nelson, Edinburgh.

Thomas, W. L., ed. (1956). "Man's Role in Changing the Face of the Earth." Univ. of Chicago Press, Chicago, Illinois.

Trollope, W. S. W. (1971). Fire as a method of eradicating *macchia* vegetation in the Amatole mountains of South Africa—Experimental and field scale results. *Proc. 11th. Annu. Tall Timbers Fire Ecol. Conf.* pp. 99–120.

Troup, R. S. (1922). "Report on Forestry in Kenya Colony." Crown Agents, London.

West, O. (1965). Fire in vegetation and its use in pasture management, with special reference to tropical and subtropical Africa. *Commonw. Agr. Bur. Bur. Pastures Field Crops, Mimeo.* 1.

White, F., ed. (1975). "UNESCO/AETFAT Vegetation Map of Africa." Paris, Oxford Univ. Press, London (in press).

. 14 .

Use of Fire in Land Management

A. J. Kayll

I. Introduction

Preceding chapters have documented in detail the physical, chemical, physiological, ecological, meteorological, and other environmental effects of fire on biomes of the world. The reader has been made aware that fires ignited by natural forces or sources have exerted a profound influence on vegetation (or fuel) types of the world. Wherever or whenever suitable combinations of fuels, burning conditions, and ignition sources have existed concomitantly, vegetation has burned. The corollary is that plants and animals must be adapted in various ways to fire (since they would not exist today if not adapted), and the question arises: "Has man, either ancient or modern, been able to adapt or utilize effects of fire on vegetation to his advantage?" Obviously, "Yes," but asking unqualified questions and obtaining unqualified answers may not produce much new information. The more difficult question becomes: "Has man been able to use fire for his purposes without degrading the particular ecosystem he is trying to manage?" Here the answer equivocates,

"Yes and no," depending on objectives of management and burning techniques employed.

Ancient man of course was not concerned with management objectives. If the fire he lit to facilitate gathering of honey contributed to his survival, then concomitant forest destruction was irrelevant. The need by modern man for limited (versus unconstrained) management objectives gave rise to the concept of using controlled or "prescribed" fires to attain the beneficial effects of fire, while avoiding the harm of uncontrolled conflagrations. Prescribed burning has been defined by the Society of American Foresters (1958) as: "Skillful application of fire to natural fuels under conditions of weather, fuel moisture, soil moisture, etc., that will allow confinement of the fire to a predetermined area and at the same time will produce the intensity of heat and rate of spread required to accomplish certain planned benefits to one or more objectives of silviculture, wildlife management, grazing, hazard reduction, etc." As such, this definition includes "controlled burning" as used in North America, "control burning" in Australia (Hodgson, 1968), "veld burning" in Africa (Phillips, 1965), "muirburn" in Scotland (Gimingham, 1956), or "swaling" in Finland (Viro, 1969). The definition is sufficiently broad to cover all aspects of land management with fire, and the terms are used interchangeably throughout the text.

Since the ecosystems man wishes to manage or manipulate are exceedingly complex, it follows that prescriptions for wise and intelligent use of fire will also be complex, utilizing a host of interacting factors which are not well understood singly, let alone collectively. This understanding will undoubtedly increase with appearance of articles on the utility of properly managed fire in popular magazines (cf. Cooper, 1961; Kilgore and Briggs, 1972; Ternes, 1970) and the commissioning of reports by municipal governments (Boughton, 1970).

In technical journals, a different circumstance exists. As observed by Smith (1970): ". . . There is no shortage of literature on controlled or prescribed burning. . . ." Comprehensive documents or lists have been prepared by many authors for various geographical locations or management objectives. In North America literature on the effects of forest and range fires has been assembled by the Canadian Forestry Service (1950–1972, 1969), Cushwa (1968), Hostetter (1966), Ramsey (1966–1971), and Shipman (1970). Many symposia on fire, including prescribed fire, have been held recently, including "The Role of Fire in the Intermountain West" (Intermountain Fire Research Council, 1970), the "Prescribed Burning Symposium" (U.S.D.A. Forest Service, 1971), "Fire in the Northern Environment" (Slaughter et al., 1971), and "Fire in the Environment" (U.S.D.A. Forest Service, 1972).

Proceedings of the eleven Tall Timbers Fire Ecology Conferences have contributed much to the understanding and use of prescribed fire in American ecosystems, and lately, in Canadian (1970) and African (1971) contexts. In Africa, literature has been collated and reviewed by Phillips (1965) and West (1965, 1971), in Australia by Hodgson (1967), McArthur (1962), and Vines (1968), in tropical ecosystems by Bartlett (1955, 1957, 1961), and in Europe, with particular reference to the management of *Calluna* heathland, by Gimingham (1972). Obviously the foregoing is incomplete: both for the areas or management objectives listed and more particularly, for the areas not mentioned, Asia and South America where no doubt fires occur and are used. Shostakovitch (1925) outlined some forest conflagrations in Siberia in 1915 and mentioned the use of fire for manipulating grass meadows and driving game, and Batchelder (1967) remarks on the ubiquity of fire in South America.

Similarities in the nature and use of fire in various fuel types of the world enable general descriptions and evaluations to be made, but the reader is cautioned against taking the general and applying it to the specific without first thoroughly analyzing local conditions and objectives.

II. Historical Uses of Fire

Anthropologists differ slightly on their estimates of when man first started to use fire. Ardrey (1961) reported some inconclusive evidence from Central Africa indicating that man used fire 800,000 years ago, while Stewart (1963) and Johnston (1970) indicated that man probably used, "kept," and controlled fire for more than 500,000 years. The differences may not be too significant because, more importantly, primitive man did not exert his maximum influence on the vegetation of the world until he learned to produce and use fire 10 to 20 thousand years ago (Johnston, 1970; Phillips, 1965; Stewart, 1963; West, 1971). Stewart (1963) suggested there is a massive amount of evidence that primitive man, with fire as a tool, has been the deciding factor in determining and maintaining the fire subclimax types of vegetation covering one-quarter of the globe. Komarek (1967) disagreed, stating that lightning fires over long periods of time created vegetation "fire mosaics" that man undoubtedly changed by his activities, but such mosaics existed before the advent of man and would continue to exist in his absence. Of course, ancient or aboriginal man was not concerned with, nor interested in, vegetational mosaics. His task immediately at hand was to provide fire for warmth, cooking, attracting game, improving pasture for wild and domestic animals, improving visibility, and aiding travel in grassland, savanna, and forest, hunting and safety, and for promoting growth of nuts, berries, seeds, etc., for food. Fire was also an indispensable adjunct to shifting tropical agriculture, and occasionally fire was used in warfare.

Accounts of early explorers frequently describe the prevalence of vegetation which had been or was being burned by native peoples. In eastern North America, various accounts are available on the openness of eastern forests and the existence of prairie islands surrounded by forest cover (Johnston, 1970). Stoddard (1962) stated that native peoples and early settlers alike used fire liberally and had no doubts as to its necessity and effectiveness for raising livestock and farming activities. Komarek (1965), in support of his thesis that man is a grassland hominid, drew attention to the "popping" characteristic of certain grasses, the popped kernel of *Tripsacum* spp. being almost indistinguishable from open pollinated strawberry popcorn (Komarek, 1965, p. 216). This characteristic could not have gone unnoticed to primitive peoples. On the Gulf of Mexico coast, a Spanish soldier who survived shipwreck in 1528, told of the Indians firing the plains to destroy mosquitoes and to compel deer and other animals to forage within range of hunters (Johnston, 1970). Indians also set large circular fires in flat grassy spots, letting the fires burn slowly toward the center where a hole had been dug to catch the roasted grasshoppers (Johnston, 1970).

In Africa, West (1965) found the earliest reference on occurrence and use of fire to be apparently the "Periplus," an account of a voyage by Hanna the Carthaginian along the west coast in 600 B.C. West (1965) continued with an interesting account of observations by various early explorers, missionaries, and hunters. In the latter part of the nineteenth century, almost every country which was visited and described by the many explorers of the "dark continent" showed evidence of deliberate burning by the natives.

Stokes (1846) in "The Voyage of H.M.S. Beagle" gave this description of aborigines intentionally firing the country near Albany, western Australia:

> . . . we met . . . natives engaged in burning the bush, which they do in sections every year. The dexterity with which they manage so proverbially a dangerous agent as fire is indeed astonishing. Those to whom this duty is especially entrusted, and who guide or stop the running flame, are armed with large green boughs with which, if it moves in a wrong direction, they beat it out . . . I can conceive no finer subject for a picture than a party of the swarthy beings engaged in kindling, *moderating*, and *directing* the destructive element, which under their care seems almost to change its nature, acquiring, as it were, complete docility, instead of the ungovernable fury we are accustomed to ascribe to it . . . [italics by A. J. K.].

Wharton (1966) wrote of Cambodia:

> . . . About 500 years ago a civilization . . . said to dwarf the wonders of Egypt, Greece and Rome laid down its arms and entered a non-martial and non-material period . . . much of the environment that once supported immense cities and armies was abandoned to a few scattered villagers who, with the aid of the agency of fire, have since maintained . . . one of the last great refuges for herbivorous mammals in all of southeast Asia. . . .

Techniques of slash and burn used in Cambodia are similar to those used in the Central Americas as described by Budowski (1966) and covered exhaustively for all the tropics by Bartlett (1955, 1957, 1961).

It is apparent that fire has been used by primitive man whenever and wherever natural fuels would burn and firing techniques must have been simple. The relatively low intensity single purpose grassland and bushland fires would have required little preparation, and if the fires became intense or burned into adjacent forest, the effects may have been considered beneficial, albeit unplanned. But with advent of modern multiple-use and multiple-value concepts, and the increased pressures of man on a limited, finite resource, it has become essential not only to employ fire, but also to do so in a controlled fashion on a predetermined area for one or more specified objectives.

III. Current Uses of Fire

In applying fire as a management tool in complex biological systems, increasing attention has been given to analyzing the various components of the system, evaluating the consequences of invoking steps to achieve management goals, and then devising procedures to attain those goals. The analogy has been drawn with a patient being diagnosed for an illness and having the ingredients of a cure being put together in the form of a prescription. Prescribed burning has been used in an attempt to encompass all of the factors involved in using fire effectively, efficiently, and safely in the management of natural ecosystems.

A. Purposes of Prescribed Burning

Reasons for using prescribed burning are almost as varied as the fuel types in which fire may be a suitable management tool, but in the elaboration of these reasons by various authors (Campbell, 1960; Davis,

1959; Foster *et al.*, 1967; McArthur, 1962; Scott, 1947; West, 1965; Whyte, 1957) there is a marked thread of similarity which is almost independent of fuel type or geographical location. These purposes can be related to the various fuel types being considered, i.e., natural pastures, grasslands, savannas, shrublands, tropical or temperate forests, etc., and include the following:

1. To remove unpalatable growth remaining from previous seasons

2. In some limited instances, to stimulate growth during seasons when there is little green grazing (a practice strongly criticized by Scott, 1947)

3. To control or destroy insects and diseases

4. To control the encroachment or development of undesirable plants and encourage desirable food plants such as legumes for both forage and soil improvement, or shrubs for berry production

5. To aid in the better distribution of animals on a range or management unit, including bird habitat

6. To remove accumulated fuels occurring naturally or as a consequence of logging or cultivation

7. To stimulate seed production or opening of cones and prepare seedbeds for seeding, either naturally or artificially

8. To establish fire breaks in a system of protection from wildfire

9. To provide training for fire fighters and fire researchers

B. TECHNIQUES OF PRESCRIBED BURNING

Techniques of prescribed burning or "fire management" (the title of this chapter is something of a misnomer in that it is application of fire to manage what is on the land, rather than the land itself) vary tremendously depending on the purposes of management and the many factors considered prior to actually igniting a blaze. The necessity for planning is emphasized by all authors (Beaufait, 1966; Davis, 1959; Bonninghausen, 1962; Haddon, 1967; Hodgson, 1967; McArthur, 1962; Vines, 1968; West, 1965). The amount of planning required may be roughly proportional to the intensity of fire expected. Thus a knowledge of fire behavior is intrinsic to developing and applying prescribed burning techniques. Included are data on available fuels, their size and spatial distribution, meteorological controls, use of head-fires (burning with the wind,) or back-fires (burning against the wind), and data on the physical, physiological, and ecological effects of fire. For example, use of fire in Africa to manage a mixed forest–grassland for grazing and browsing by domestic or wild animals could be at the expense of the forest (West, 1965), and must be recognized in the burning prescription.

1. *Grasslands and Savannas*

Although there is not a balanced amount of world literature on the subject, fire has been used in the management of grasslands and savannas in all tropical and temperate regions of the world (Batchelder and Hirt, 1966; Budowski, 1966; Daubenmire, 1968; Humphrey, 1963; Phillips, 1965; Smith, 1960; Vogl, 1965; Wharton, 1966; West, 1965, 1971). Batchelder (1967) pointed out ". . . in South America, so unbiquitous is fire, that very few specific references describe in detail the times or purposes of burning. . . ." Detailed descriptions of burning techniques are almost equally scarce, but the following outline is representative.

Using suitable ignition sources, including matches, fire brands, or diesel oil burners (pressurized or gravity feed), areas are lit at suitable intervals (annual, biannual, half-decade) during the dormant season. Burning late or early in the season will depend on management objectives. For example, in Zambia where rainfall is high (40–50 inches/year), annual, late, high-intensity burning near the end of the dry season has transformed well-developed woodland and coppice woodland into grassland with the woody species persisting below the level of grasses. Annual, early, low intensity burning as early as possible in the dry season has maintained the woodlands in a slightly thickened condition (West, 1965) and similar effects have been noted in Nigeria (West, 1965). This type of burning removes the top layer of litter but heat does not penetrate into the ground to any appreciable depth to kill the roots. However, burning too frequently can engender depletion of root reserves, leading to death of the desirable grasses and invasion of the site by less palatable, but more fire-resistant or heat-tolerant species. Both back-firing and head-firing techniques are used, the latter probably being more common because burning of a specific area can be accomplished more quickly. Where graziers or ranchers are apprehensive about controlling rapidly advancing high intensity head-fires, nighttime ignition may be used.

In tropical American lowlands fire is used in management of natural grassland and in "slash and burn" agriculture, but little information on techniques is available (Budowski, 1966). Similar circumstances exist for such areas as Cambodia (Wharton, 1966) and northern Australia (Smith, 1960). Burning techniques similar to those employed in Africa are likely used for various stated or implied objectives.

Grasslands in the United States have been extensively investigated (cf. Chapter 5). Daubenmire (1968) reviewed characteristics of grassland fires and their effects on the ecosystem, but the simple nature of grassland-burning techniques understandably has not sponsored a similar review. Despite careful definition of grassland-burning objectives, detailed plans and procedures are not as common as with prescribed burning of forest lands. This lack is probably a reflection of the relatively lower intensity of grassland and shrubland fires and the ease with which they may be lit and controlled. However, more planning and research would help preclude "excursions" by prescribed fires. Furthermore, research is needed not only on the effects and use of fire, but also on techniques of burning, i.e., management *of* fire as well as management *with* fire.

2. *Shrublands*

Burning techniques in dwarf shrub communities are similar to those employed in grasslands and have been briefly described by Kayll (1967) and Miller (1964). An estate gamekeeper, using a diesel oil burner to start the fire and a traditional "broom" (usually a birch stem about 10 ft long, weighing up to 20 lb with a loop of chicken wire at one end) to control its edge, attempts to burn on a cycle of 12 to 15 years. Head-fires are commonly used with previous burns, woods, trails, waterways, etc., serving as firebreaks. The management objective is to obtain a patchwork quilt effect of regenerating (recently burned) and maturing heather. Where stands become older than 20 to 25 years, back-burning may be needed to consume the somewhat clumped and discontinuous fuels. Similar burning is practiced by commercial blueberry growers, but the cycle is shorter. Ocassionally straw is scattered to facilitate fire spread.

3. *Forest Lands*

The greater significance attached to forest fires is reflected in the number and scope of publications dealing generally with the subject and dealing specifically with planning and use of prescribed fires. In the United States Bonninghausen (1962), Beaufait (1966), and Davis (1959) gave detailed descriptions. Similar works have been produced in Canada by the British Columbia Forest Service (1969) and Haddon (1967), and in Australia by McArthur (1962) and Vines (1968). All emphasize the necessity of coupling good planning with experience and sound common sense in order to achieve effective and efficient prescribed fires.

a. SOUTHEASTERN UNITED STATES. In the United States, prescribed burning may have had its start in the Southeast (Riebold, 1971). Over the years, the practice has produced many detailed but flexible plans for implementing burning prescriptions. Assuming management objectives can be achieved by using fire, existing barriers such as roads, waterways, and significant fuel discontinuities (e.g., a recently plowed field) are utilized in preparing an area for burning. Some areas may require parallel fire breaks (plowed or existing lines) within the compartment to facilitate ignition (Bonninghausen, 1962; Davis, 1959). Deciding when to burn, including time of day, season of year, and frequency, depends on the stated management objectives, fuel characteristics (e.g., grass under a forest overstory versus slash on a cutover), and meteorological factors. For example, burning for wildfire control (hazard reduction) purposes may be done at 3- to 6-year intervals (sometimes longer), back-fires or head-fires may be used alone or in combination, and winter burning may be preferred to take advantage of the prevailing lower ambient air temperatures (Wade, 1969).

Using simple drip torches for ignition from established fire lines or occasionally spotted at intervals by walking into a steady wind of 5 to 10 mph, small crews of 3 to 6 experienced men can burn 300 to 1000 acres in one day. For control, water backpacks and hand tools are usually immediately available with heavier mechanical equipment on local standby.

b. EASTERN CANADA. With superficially similar forests, but distinctly different climatic and silvicultural conditions, prescribed burning in jack pine (*Pinus banksiana*) management in Canada has the following characteristics (Adams, 1966; Foster *et al.*, 1967). Burns are on cutover compartments subdivided by plowed furrows into blocks of 6 to 20 acres. Preburn planning includes preparing guidelines, preparing sketch maps showing existing roads and water courses ("safe edges"), locating new 10-ft-wide fire lines, and felling of all trees up to 100 ft inside the perimeter. Fuels should be more or less continuous and have a moisture content of less than 20%. Burning in midafternoon with a relative humidity less than 40% and a wind of 8 to 10 mph is preferable. Ignition procedures are similar to those used in the southeastern United States, but rather more men and equipment are utilized because of the generally higher fuel loadings and relative infrequency of burning, perhaps once every 30 to 50 years. (As a general rule, the higher one's latitude, the less frequent is the occurence of natural fire on any given area, the less frequent should prescribed fire be used, and days suitable for prescribed fires are also less frequent.) Where a mineral seedbed is desired, an intense fire is needed, which in turn requires a lengthy drying period. Thus fuels outside the perimeter of a prescribed burn may be hazardous to the point of needing wetting down. Crews of 16 to 20 men, with two-way radio communication and equipped as torchmen, patrol-men, and tanker crews, may burn 200 to 300 acres in a day.

c. Western United States and Canada. Where fuel concentrations become heavy and slopes steep, comprehensive planning for the use of fire becomes critical, especially if benefits are to be commensurate with costs. Burning in the mountainous regions of the western United States and Canada has been largely for reduction of wildfire hazard and site preparation after the merchantable timber has been removed. Removal of residual logging debris (slash) by prescribed burning (slash burning) has been used in western North America since 1910 (LeBarron, 1957), has been recommended in the Vancouver Forest District of British Columbia since 1912 (Smith, 1970), and has been required by law in the District since 1939 (Haddon, 1967). Although first used for hazard reduction purposes, prescribed fire is being increasingly used in silviculture, in habitat manipulation, and in maintenance of wilderness and parklands. Details on burning in western North America have been provided by Beaufait (1966), British Columbia Forest Service (1969), Davis (1959), Emrick (1967), Murphy (1967), and Zwolinski and Ehrenreich (1967).

Slash is usually "broadcast" burned, i.e., burned where felled or lying on the ground after logging. Occasionally, in particularly difficult circumstances, slash may be machine piled or windrowed before burning. Burns are usually conducted in the autumn while the slash still has needles, with compartments laid out prior to logging to take advantage, where practicable, of natural barriers. These barriers are supplemented by bulldozed fire lines one or two blade-widths wide to create blocks of 10 to 400 acres, with an optimum of 100 to 200 acres. Detailed maps and plans are prepared and all crewmen instructed prior to ignition. Ignition times ". . . should be determined by the job, not the clock . . ." (Beaufait, 1966; Beaufait and Fischer, 1969), and thus according to localities and fuel types, ignition time may range from early morning to late afternoon. Desirable winds range from calm to 10 mph, depending on ignition techniques and devices being used, the latter being selected for speed and flexibility. Strip and center ignition patterns are used, the latter providing a strong central convection column which draws air to the center of the fire, enables area ignition (lighting up the area more or less at once), and lifts the smoke to high elevations. Plans are made such that compartments may be burned in one day. Mop-up and patrol after the fire are essential.

d. Australia. Burning techniques similar to those used in western North America have developed in Australia and are thoroughly described by McArthur (1965), although the emphasis in this case is on the use of prescribed burning for reduction of wildfire hazard on a broad area basis. Developing techniques have led to the use of low flying aircraft and small cheap incendiary capsules to ignite up to 10 thousand to 15 thousand acres in one afternoon (Vines, 1968). The object is to reduce the amount of flammable fuel over a broad area and thus lessen the intensity of a wildfire burning during hot, dry, windy weather. On the flat terrain of western Australia, grid flight patterns are flown, but over more mountainous areas, aircraft fly along contours, igniting the ridgetops first in the sequence (a pattern often used in North American forests). Properly executed, the resulting low intensity fires do little damage to the overstory, protection against damaging wildfires is effective, and costs are commensurate with benefits: about 10 cents per acre, including total costs of preparation of fire guards, burning plans, mobile radio beacons, and aircraft time. Each area is burned on a rotation of about 5 years.

IV. Examples of Use of Prescribed Burning

In eucalypts and in other fuel types of the world, it is more difficult to begin controlled burning than to continue the practice. But irrespective of fuel type, as more is learned about managing fire, increasingly sophisticated and subtle burning techniques are being employed to manage natural resources. But as several authors (Austen, 1971; Hodgson and Heislers, 1972; Scotter, 1971) pointed out, fires are not equally useful for all species and the choice of using fire infers disadvantage as well as advantage (as does any other management technique).

The underlying principle in the use of fire in land management is the manipulation of a biological system on a particular land form. Management of grasslands, savannas, and shrublands with fire is usually for a secondary purpose, i.e., to provide browse and cover for domestic or wild animals. Using fire in the management of a forest for fiber or wood production differs basically because of the intrinsic value of the wood. Obviously, the integration of all interactions cannot be achieved with single purpose management concepts. It is only for ease of discussion (or administrative reasons) that we categorize types of prescribed fire use. But just as fire crosses all the artificial and somewhat arbitrary disciplinary boundaries we sometimes establish, so must its uses and implications be examined in what Phillips (1965) has termed a "holistic" approach. Advocates of the wise use of fire as a management tool maximize the advantages and keep the disadvantages at a level tolerable to the ecosystem in question. Unacceptable disadvantages, or alternatives which maximize advantages better, may preclude the use of prescribed fire. On the other hand, fire should not be rejected as a suitable management tool without thorough, comprehensive analysis.

A. Grasslands and Savannas

Whether in the Northern or Southern Hemisphere, most authors agree that late spring is the most beneficial time to burn for stimulation of grass production. In the United States in Kansas, Anderson (1964) makes the following general recommendations:

1. Burn only in late spring (when plants are best able to recover quickly).

2. Burn only when the soil and plant crowns are damp after a rain (to minimize heat penetration to protected growing points).

3. Burn when there is a breeze to move the fire along quickly.

4. Avoid close and early grazing after burning (to allow plant stocks to establish food reserves).

The last point is particularly important because the interaction of fire and grazing can have a profound effect on the health and vigor of rangeland or "veld." Clements (1949) stated most clearly ". . . If not too frequent, it [fire] affects grassland little, but the reaction value of grass may be seriously reduced or almost destroyed by overgrazing . . ."

In Wisconsin, Vogl (1965) recommended spring burning of the tall grass brush–prairie savanna. The resultant green herbage with its higher moisture content is more productive, palatable, and desirable to herbivores. Periodic burning, up to once every other year, prevents the prairie savanna from becoming decadent, helps maintain maximum productivity, and is important in retarding the woody growth which otherwise enables the savanna to succeed to forest.

In Oregon, Hardison (1957) described the use of fire in sanitation of fields used for commercial production of grass seeds. Following combining, burning the straw and stubble destroys old dry leaves together with the spore bodies capable of producing the countless spores which would otherwise be disseminated to reinfect new leaves and plants. Burning has also given good control of several seed disorders by destroying those seeds which have been modified into galls, sclerotia, or other aberrations. In Canada, similar practice (spring burning) in Saskatchewan gives good control of black stem disease in alfalfa fields.

In the less pronounced seasonal changes of Louisiana, Duvall and Whitaker (1964) recommended a system of rotationally burning one-third of the longleaf pine (*Pinus palustris*)–bluestem (*Andropogon tener* and *A. divergens*) ranges in the winter or early spring, both on cutover and timbered lands. By integrating range- and forest-burning programs, economies of operation as well as desired management objectives are achieved. Wildfire hazard in the forest is reduced, range vegetation and grazing distribution are improved, scrub hardwoods are top killed, and unpalatable perennials are curtailed.

In California, Burma (1967) discussed controlled burning in the "public domain" where general rules and detailed procedures have been developed to ensure that not only is burning safe and efficient, but also that due attention is paid to integrating all uses of a particular parcel of land. The provisions made for legal and technical authority to burn and to minimize escape risks are followed by considerations of potential erosion, of aesthetic, habitat, or watershed qualities, and of subsequent land use. For example, constraints on burning may be accompanied by requirements for seeding, spraying, and grazing. Thus burning becomes one step in implementing a management plan. Emrick (1967) provided detailed steps for successful brush-control burning in California. To some readers the amount of planning and organizing considered essential may seem excessive, but such requirements may in part be a function of increasingly high population densities in California and concomitant splintered ownership patterns.

The role of fire in desert grassland is in part determined by the intensity of domestic livestock grazing (Humphrey, 1963). Where grazing has been too heavy, residual, weakened grasses may not have the density or volume to carry a fire. Thus, where in former times hot fires precluded shrub invasion, the mature treelike shrubs which have now grown up are not killed by weak fires. Building up suitable fuel quantities through grazing control seems mandatory if fire is to be a practical and effective tool.

West (1965, 1971) described the effects and uses of fire in Africa. In areas where forest would be the "climax" cover in the absence of fire, controlled burning may be employed to convert the forest to savanna (open woodland and grass understory), and with continued application, perhaps ultimately to grassland. Van Rensburg (1971) outlined characteristics of burning early or late in the seasons in various fuel types. In southern, central, and eastern Africa, the best time for burning to promote vigorous grass growth is the end of the dry season, i.e., as late as possible in the dormant season, just before the grasses begin to grow. Burning in early or mid-dry season encourages brush encroachment at the expense of grasses and causes exposure that may further damage and weaken the sward. In discussing burning as a management practice, West (1965) noted the strong interaction between the intensity of grazing and the efficiency of fires in controlling the encroachment of woody species (cf. Humphrey, 1963). Because late season hot fires are required to suppress and control bush encroachment, it follows that on heavily grazed land efficacious fires can be obtained only after resting. Some postburn resting may also be necessary to enable the grasses to recuperate. The interval between burns depends on management objectives, rates of litter accumulation, and grazing intensities, and may range from annually to every fourth year or more (West, 1965). Scott (1970) commented on the pros and cons of eliminating veld burning and substituting mowing (not practicable over large areas) or "complete utilization" (high grazing pressure). Scott (1970) recounted that the ill effects of veld burning are mainly due to exposure of the soil to insolation and wind, and the "complete utilization" may do the same damage and more, because of heavy trampling. Greater utilization of grass would no doubt be beneficial, but not at the expense of weakening the sward and subsequent site deterioration. Integrated rotational grazing and burning management is suggested by West (1971) in Africa, but because of the different vegetation types, does not include the forest management component suggested by Duvall and Whitaker (1964) for southeastern United States.

B. Shrublands

Integration of management objectives is an essential feature of all effective burning programs. Just as in grasslands and savannas, the use of fire in the more limited "shrubland" fuel type has its interacting effects on wildlife, vegetation, watersheds, aesthetics, and domestic animal husbandry.

In Europe, fire is used extensively in management of dwarf shrub heath communities and has its attendant effects on grass and forest communities (Gimingham, 1970; Kayll and Gimingham, 1965; Grant et al., 1963; Hansen, 1964; Uggla, 1958). Although the effects of large uncontrolled fires may be detrimental (Radley, 1965), the prevailing advocacy is the use of properly controlled fires of suitable intensity on a 10- to 15-year rotation (Allen, 1964; Robertson, 1957; McVean, 1959). Ward (1972) outlined the uses of prescribed fire by farmers and gamekeepers in management of heather, grass, and gorse. Essentially the objectives are the same: provision of food and cover for grouse and sheep. In England and Wales, heather is not as well regarded for sheep grazing as in Scotland, and thus fire management practices tend to favor grass at the expense of heather.

In the United States, brush control and encouragement of grass with prescribed fire are topics of both technical and popular writings, but the operational constraints are distinctly different from the European case. Raymond (1967) showed that the complexity of the problem did not relate merely to the physical and ecological effects of prescribed fire, but also involved questions of legal constraints (complex land tenure) and liabilities associated with fire escapes. Nevertheless, prescribed fire is used as an effective tool in range improvement for both wild and domestic animals, and in reduction of wildfire hazard. Baldwin (1968), Doman (1967), Murphy (1967), and Pase and Glendening (1965) outlined brush conversion programs which include crushing with a tractor, ignition in late fall through to early spring, seeding with grass and/or legumes, control of brush sprouts and seedlings with chemical herbicides, and continued active management (grazing control) of the treated area. Costs are high ($40 to $70 per acre) but may be commensurate with resultant benefits to wildfire control, wildlife habitat, domestic livestock grazing, and watershed management.

C. FOREST LANDS

Well-developed prescribed burning techniques for major forest types usually exist if the relationship of fire to the commercially valuable species is pronounced, if safe burning is achieved relatively easily, or if the wildfire hazard associated with cutting practices is so high as to make fuel-reduction burning almost mandatory. Thus, in many parts of North America, prescribed burning has been extensively developed and applied.

In the southeastern United States, as in other areas, details of prescribed burning in forest management vary according to species, site, and management objectives. Riebold (1964, 1971) outlined its history of use and other authors (Bonninghausen, 1962; R. W. Cooper, 1963; Lotti, 1959, 1962; Neel, 1965; Stoddard, 1962) presented details. It is dangerous to generalize, but low intensity head- or back-fires are prescribed, principally in the dormant or winter season. Dieterich (1971) estimated that 2.5 million acres are burned annually, with costs ranging from a few cents to $1.00 per acre (Neel, 1965), depending on area burned, time since last burn (fuel accumulation), proximity of structures or areas requiring special fire guards or precautions, and other factors related to controlling and executing the burn.

Little and Somes (1961) summarized the extensive use and effects of prescribed burning in the pine region of southern New Jersey, but in the north central part of the continent, prescribed burning has not been as common. Recommendations for the Lake States by Dieterich in 1963 included using fire for site preparation on clear-cut jack pine and black spruce (*Picea mariana*) areas, on areas of jack pine–hardwood mixtures, and under mature red pine (*Pinus resinosa*). Costs at that time ranged as high as $43 per acre. Sando (1969) summarized the status in 1968 with about 15,000 acres being treated at costs ranging from $0.15 to $19.00 per acre. For northeastern Minnesota, Ahlgren (1970) did not include costs of burning in his experimental studies, but he did elaborate on the efficiency of prescribed burning in removing slash, reducing humus, retarding competing shrubs, and establishing a new jack pine stand. Ahlgren (1970) also highlighted the necessity of allowing for natural cycles (time) when evaluating the success or failure of prescribed burning. After fire, at least 2 or 3 years are needed because survival of germinants depends not only on the nature and extent of the fire, but also on a good seedbed, adequate rainfall, seed supply, rodent activity, shrub competition, etc. (Some indications that poplar was not reduced by fire, but occupied about the same position as in a cut, unburned stand, were based on 9-year results.)

Although several researchers have explored the utility of fire in the silviculture of jack pine in Canada (Cayford, 1963, 1970; Chrosciewicz, 1967, 1968; VanWagner, 1966; Williams, 1960) and the results have been promising, prescribed burning nevertheless plays a relatively small role on an operational basis (Cayford, 1970). The few days in each year when prescribed burning can be safely undertaken and the mechanical site treatment techniques available at comparable costs (ca. $20 per acre) militate against initiating burning programs. A psychological reluctance to burn on the part of foresters is aggravated by the relatively high fuel loadings extant initially, and as has been pointed out, it is more difficult to begin controlled burning than to continue the practice.

To obtain the mineral soil essential for germination and survival of conifer seedlings, high intensity, high hazard summer fires are needed. Spring hazard-reduction burns on cutover areas do not consume the wet or perhaps still frozen organic layer (Foster *et al.*, 1967). A sequence of low intensity fires coupled with mechanical scarification may be needed to attain the desired seedbeds (Jarvis and Tucker, 1968). Use of fire in northern hardwood silviculture is in the experimental stages only, but some applications seem possible (Niering *et al.*, 1970; Sykes, 1964).

In Scandinavia, prescribed burning is used to release immobile nitrogen in the cold moist soils, care being taken to remove only the slash and upper portions of the humus layer. Fire is not used on dry sites with thin humus layers (Weetman and Nykvist, 1963). Cutovers of 200 acres or more are burned all through the summer period whenever conditions of settled weather, light winds, and dry surface humus exist. As well as activating the humus, planting and seeding are facilitated by burning (Uggla, 1958). Viro (1969) suggested that prescribed burning in Finland was best utilized on sites with thick humus accumulations, cold soils, and bound nutrients. After burning, pines should be established on such sites.

Certain parallels have been drawn with the Scandinavian experience by Hardy and Franks (1963) in Alaska by their suggesting that prescribed fire had yet to be effectively used as a forest management tool. Experience subsequently gained through research (Slaughter *et al.*, 1971) and long-term studies such as Viro's (1969) in Finland will aid in clarifying the role and potential uses of pescribed fire in the far north. For instance, slash burning may have aided sitka spruce (*Picea sitchensis*) regeneration in southeast Alaska (Harris, 1966), but advantages over burned areas seemed marginal.

In western North America, intentional burning probably was conducted in one form or another since the beginning of the twentieth century. LeBarron (1957) reported that slash burning has been conducted in the intermountain west since 1910. From 1920 onwards, burning of hand-piled logging slash and debris gradually extended to include preparation of sites for regeneration, range improvement, wildlife habitat improvement, and reduction of natural hazards (DeSilvia, 1965). In the rough mountainous country of Montana and Idaho, much of the burning is on cutovers ranging from 10 to 500 acres with practically all burning done in the autumn after rains have broken the fire season. Piling and burning may adversely affect conifer reproduction (Vogl and Ryder, 1969) and it is suggested that broadcast burning, in more closely simulating wildfire, may lead to more uniform restocking of cutovers.

Use of fire in management of ponderosa pine (*Pinus ponderosa*) has been extensively covered (cf. Lindenmuth, 1960; see also Chapter 9 of this volume). Briefly, management should be directed toward achieving a mosaic of even-aged groups (maximum area 1.5 acres) of trees with the debris being burned following cutting. Following regeneration, areas should be periodically burned to thin the stands and maintain vigorous growth. Burning should be done when trees are dormant, soils are moist, and winds are low. As elsewhere, first fires in heavy fuels are most difficult, but easier conditions follow (Biswell, 1970).

In Alberta, effects of prescribed fire in subalpine spruce-fir slash have been described by Kiil (1970, 1971). In partially cut stands, low intensity fires are not effective in creating site conditions suitable for survival of conifer seedlings, but do show promise for hazard reduction and improvement of wildlife habitat. Single fires on clear-cut areas are not effective in exposing mineral soil because of deep organic layer accumulations. With late September or early October being the periods of safe burning, smoldering fires, a sequence of burns, or mechanical scarification in conjunction with burning, may be required to achieve satisfactory conifer regeneration.

The use of fire in the Douglas-fir (*Pseudotsuga menziesii*) region of western North America has been extensive for half a century. However, Isaac (1963) pointed out that fire is a tool and not a blanket rule in Douglas-fir ecology and he listed the times to burn as (a) when slash and weather conditions are safe, (b) when slash areas become so large or continuous that fire control is impracticable, (c) when slash accumulations are extremely heavy, (d) when competing, undesirable species invade, (e) when it is necessary to prepare sites for seeding or planting, (f) when insect or disease infestations threaten, or (g) when there are neither seeds nor seedlings on a cutover area and there is a seed crop in prospect on nearby seed sources. Times when not to burn include: (a) when fire cannot be safely controlled, (b) when cutting has left a good residual stand, or one has become established, (c) when slash is light and unlikely to be dangerous even during periods of high hazard or the slash provides cover on exposed slopes, or (d) when burning conditions will produce an extremely intense fire that may seriously affect the soil or the habitat generally. Long-term studies are altering some of the details of burning (Morris, 1970), but its role in wildfire hazard abatement has been clearly shown (Smith, 1970). In other forest types of western North America, Muraro (1968, 1971) has shown hazard reduction in cedar–hemlock logging slash can be successful within narrow constraints, but continuing use of fire is expected.

Perhaps one of the narrowest sets of burning constraints is the special circumstance where a "let-burn" policy is being developed and applied to high elevation, low intensity, lightning started fires (Kilgore, 1971b, 1972). In giant sequoia (*Sequoiadendron giganteum*) and red fir (*Abies magnifica*) forests where other economic values are not threatened, fires are being allowed to follow their natural course in maintaining the appearance and substance of the forest complex prior to the advent of fire exclusion policies. The management implications are far reaching; similar plans are being considered for other areas of western North America (e.g., Prasil, 1971) and implementation of such policies will require effective and continued dialogue between resource scientists, managers, administrators, and the general public.

In Australia, a good public understanding has been one of the contributing factors to the widespread success of the hazard reduction control burning program, which utilizes low intensity fires (<100 BTU/sec-ft) to remove fuel accumulations and facilitate wildfire control (Hodgson, 1968, 1970). As mentioned, large areas are treated annually using aerial incendiaries, but ground ignition techniques are also used, for example, in Tasmanian eucalypt forests (Mount, 1965). The use of fire in radiata pine (*Pinus radiata*) plantations has been explored experimentally by Gilmour and Cheney (1968) and used operationally as a hazard reduction tool. With an upper limit of 100 BTU/sec-ft, 25 to 50 BTU/sec-ft is considered an optimum intensity.

D. WILDLIFE

In North America, the first principal work on use of fire in management of wildlife populations was that of Stoddard (1931) for bobwhite quail. Since that time, many researchers and managers (Komarek, 1963) have developed and used techniques for manipulating forests, grasslands, and savannas for the benefit of wildlife. Miller (1963) has described fire as a tool for maintaining the subclimax vegetation (the usual preferred habitat of most species) in vigorous and proper condition, density, and composition. Komarek (1971), in summarizing the effects of fire on wildlife habitats in southeastern United States, pointed out that grasslands and early stages of brushland, maintained by controlled burning, provide the diversity of flora necessary for healthy wildlife populations. Czuhai and Cushwa (1968) found similar evidence for the more upland areas of the Southeast. Miller (1963) was referring largely to "upland" game, but his statement is applicable to many other forms of wildlife. Thus habitat maintenance with knowledgeably applied fire has benefitted various species of grouse, prairie chicken, pheasant, turkey, quail, woodcock, snipe, ducks, geese, songbirds, birds of prey, deer, elk, moose, muskrats, and others (Vogl, 1967). The use of fire for habitat manipulation for both wild and domestic animals is common, e.g., deer and cattle in California (Hendricks, 1968) and sheep and grouse in Britain (Ward, 1972).

Marsh burning is an accepted management practice in most waterfowl refuges on the East Coast of the United States (Givens, 1962; Zontec, 1966) and burning on inland, freshwater wildfowl habitats has been undertaken both in Canada (Ward, 1968) and the United States (Schlichtemeier, 1967). The dual benefits on wildfowl and muskrat populations have also been described (Perkins, 1968). In the giant sequoia forests of California, Kilgore (1971a) recorded the effects of habitat manipulation, including cutting and burning of brush and saplings on breeding bird populations. As he expected, habitat requirements of the various species engendered varying responses to the treatments, but changes were not substantial because of the limited areas and degree of change. Small mammals likely have similar responses, e.g., burning of jack pine areas in Minnesota gave rise to increases in seed-eating mouse species while other species with a more varied diet remained relatively constant in number (Ahlgren, 1966). Studies in Australia are exploring the effects of the widespread control-burning policies on vegetation (C. F. Cooper, 1963), wildlife populations (Butcher and Dempster, 1970), and on small mammals (Leonard, 1970), but definitive statements are not yet available.

Management of big game animal habitats with fire is common in Africa and becoming so in North America. In the former, Brynard (1964, 1971), Austen (1971), Owen (1971), and others have outlined the use of "veld" and savanna burning for management of various animals in African national parks. Under the premise that grass should be eaten, mowed, or burned (Austen, 1971), fire management plans have been developed which incorporate not only burning regimes favorable to wildlife (i.e., promotion of grasslands), but also incorporate long-term principles of management related to forest production (early versus late burning), watershed protection, and aesthetics. Integration of management objectives is clearly demonstrated.

In the United States, Leege (1968) outlined burning procedures favoring elk habitats in northern Idaho. Spring fires after the snow recedes, but before sprouting of new growth occurs, seem most efficient in reducing height of existing browse and stimulating seed germination, both of which are beneficial to the elk habitat. In northern Canada, limited controlled burning may favor moose populations (Spencer and Hakala, 1964) but be detrimental to barren-ground caribou (Scotter, 1970, 1971). Since the effects of fire in the slowly cycling northern environment can be profound and long lasting, an intensification of research and development effort seems imperative, and suggestions for a northern research center have been made (Slaughter et al., 1971).

E. WATERSHEDS, AIR QUALITY, AND RECREATION

Public concern and awareness of ecological subjects have prompted the writer to clump the uses of fire on watershed, air quality, and recreation management because of their "political" visibility, rather than intrinsic relationships. The use of fire in manipulation of watershed areas in Arizona has had the multiple-use objectives of increasing yield of water, timber products, forage for livestock and game, improving conditions for recreation, reducing the adverse effects of wildfires, and reducing soil erosion (Arnold, 1963; Kallander, 1969). Burning practices have been applied in watersheds with cover types of spruce-fir, ponderosa pine, pinyon-juniper, chaparral, and desert grassland (Zwolinski and Ehrenreich, 1967). Conversion from brush to grass has increased water yields in California (Burma, 1967), but in smog-sensitive California, the use of broadcast burning in watershed management may be restricted for aesthetic, atmospheric, and economic reasons (Zivnuska, 1968). More intensive and carefully controlled environmental management may preclude use of prescribed fire in certain circumstances, but alternatives will need to be developed.

In the Douglas-fir region of western North America, multidisciplinary studies have been initiated to determine prescribed fire smoke constituents (Fritschen *et al.*, 1970) as well as effective smoke dispersal in burning for forest, grassland, and watershed management (Beaufait, 1968; Dell and Green, 1968). Vines *et al.* (1971) have undertaken similar studies in Australia. It is interesting to note that seven individuals cooperated with Fritschen and six with Vines, indicating perhaps not only a diversity of interests but also of the effects engendered by prescribed burning.

Intentional burning on campgrounds and intensively used recreation areas has not been widely applied, but the finesse required for this type of operation is being developed (Cumming, 1969) with low intensity, carefully controlled fires being prescribed to lower wildfire hazards. Secondary benefits have accrued in the form of additional wildlife and improved aesthetics. The latter has been partly brought about by the creation of "edges," considered a prime requisite by Meskimen (1971) in the production of forest landscapes. The creation of edge leads to diversity and variety, and Meskimen (1971) presents the thesis that, irrespective of geographical or vegetational location, only three building blocks provide all the characteristic landscapes, viz., meadow, shrub thicket, and forest stand. Perkins (1971) has indicated several forms of outdoor recreation that are compatible with use of prescribed fire, including hunting, camping, picnicking, hiking, bird watching, and outdoor photography.

According to Hoffman (1971), one purpose of natural parks is to enable people to enjoy the features of a natural environment and this environment must include fire. Robinson (1970) summarized the future direction of fire management by identifying the following trends: an increasing use of fire in managing natural resources, an increasing requirement for demonstration of favorable benefit/cost ratios, a recognition of the role of fire as a natural component of wilderness, and an allowance for fire to follow its natural course under carefully specified conditions.

Prescribed fire will never become a management tool to the total exclusion of other techniques, but it will continue to be effective and its uses will diversify in management of the world's natural resources for the benefit of man.

References

Adams, J. L. (1966). Prescribed burning techniques for site preparation in cut-over jack pine in southeastern Manitoba. *Pulp Pap. Mag. Can., Woodl. Rev. Sect. Index* 2393 (F-3), 1–7.

Ahlgren, C. E. (1966). Small mammals and reforestation following prescribed burning. *J. Forest.* 64, 614–618.

Ahlgren, C. E. (1970). Some effects of prescribed burning on jack pine reproduction in northeastern Minnesota. *Agr. Exp. Sta., Minn., Misc. Rept.* 94, 1–14.

Allen, S. E. (1964). Chemical aspects of heather burning. *J. Appl. Ecol.* 1, 347–367.

Anderson, K. L. (1964). Burning Flint Hills bluestem ranges. *Proc. 3rd Annu. Tall Timbers Fire Ecol. Conf.* pp. 89–104.

Ardrey, R. (1961). "African Genesis" (reprinted by Collins Fontana Books, London, 1970) (in West, 1971).

Arnold, J. F. (1963). Use of fire in the management of Arizona watersheds. *Proc. 2nd Annu. Tall Timbers Fire Ecol. Conf.* pp. 99–112.

Austen, B. (1971). The history of veld burning in the Wankie National Park, Rhodesia. *Proc. 11th Annu. Tall Timbers Fire Ecol. Conf.* pp. 277–296.

Baldwin, J. J. (1968). Chaparral conversion on the Tonto National Forest. *Proc. 8th Annu. Tall Timbers Fire Ecol. Conf.* pp. 203–208.

Bartlett, H. H. (1955, 1957, 1961). "Fire in Relation to Primitive Agriculture and Grazing in the Tropics," Annotated Bibliography, Vols. I–III. University of Michigan, Botanical Gardens, Ann Arbor, Michigan.

Batchelder, R. B. (1967). Spatial and temporal patterns of fire in the tropical world. *Proc. 6th Annu. Tall Timbers Fire Ecol. Conf.* pp. 171–207.

Batchelder, R. B., and Hirt, H. F. (1966). "Fire in Tropical Forests and Grasslands," Tech. Rep. 67-41-ES. U.S. Army Natick Lab., Natick, Massachusetts.

Beaufait, W. R. (1966). Prescribed fire planning in the intermountain west. *U.S., Intermt. Forest Range Exp. Sta., Res. Pap.* INT-26, 1–27.

Beaufait, W. R. (1968). "Scheduling Prescribed Fires to Alter Smoke Production and Dispersion," Presented paper: Seminar on Prescribed Burning and Management of Air Quality, pp. 33–42. Southwest Interagency Fire Council, Tucson, Arizona.

Beaufait, W. R., and Fischer, W. C. (1969). Identifying weather suitable for prescribed burning. *U.S., Forest Serv., Res. Note* INT-94, 1–7.

Biswell, H. H. (1970). Ponderosa pine management with fire as a tool. *Symposium. The Role of Fire in the Intermountain West, Missoula, Mont. 1970,* pp. 130–136.

Bonninghausen, R. A. (1962). Forest land management and the use of fire. *Proc. 1st Annu. Tall Timbers Fire Ecol. Conf.* pp. 127–132.

Boughton, V. H. (1970). "A Survey of the Literature Concerning the Effects of Fire on the Forests of Australia." Municipal Council of Ku-ring-gai, Gordon, New South Wales.

British Columbia Forest Service. (1969). "A Guide to Broadcast Burning of Logging Slash in British Columbia," Forest Prot. Handb. Ser. No. 2. B.C. Dept. Lands, Forests, Water Resources.

Brynard, A. M. (1964). The influence of veld burning on the vegetation and game of the Kruger National Park. *Monogr. Biol.* 14, 371–393.

Brynard, A. M. (1971). Controlled burning in the Kruger National Park—history and development. *Proc. 11th Annu. Tall Timbers Fire Ecol. Conf.* pp. 219–231.

Budowski, G. (1966). Fire in tropical American lowland areas. *Proc. 5th Annu. Tall Timbers Fire Ecol. Conf.* pp. 5–22.

Burma, G. D. (1967). Controlled burning in the public domain in California. *Proc. 7th Annu. Tall Timbers Fire Ecol. Conf.* pp. 235–244.

Butcher, A. D., and Dempster, J. K. (1970). Fire and the management of wildlife. *Proc. Fire Ecol. Symp., 2nd, 1970* pp. 1–10.

Campbell, R. S. (1960). Use of fire in grassland management. Paper delivered at FAO Work. Party on Pasture and Fodder Development in Tropical America, Maracay, Venezuela (in West, 1965, p. 32).

Canadian Forestry Service. (1950–1972). "Forest Fire Protection Abstracts," Vols. 1–23. Prepared by Forest Fire Res. Inst., Canada Dept. Environment, Ottawa.

Canadian Forestry Service. (1969). "Document List of Forest Fire Technical Information." Can. Forest Serv., Forest Fire Res. Inst. Inform. Cent., Canada Dept. Environment, Ottawa.

Cayford, J. H. (1963). Some factors influencing jack pine regeneration after fire in southeastern Manitoba. *Can. Dep. Forest., Forest Res. Branch Publ.* **1016**, 1–16.

Cayford, J. H. (1970). The role of fire in the ecology and silviculture of jack pine. *Proc. 10th Annu. Tall Timbers Fire Ecol. Conf.* pp. 221–244.

Chrosciewicz, Z. (1967). Experimental burning for humus disposal on clear-cut jack pine sites in central Ontario. *Can. Dep. Forest Rural Develop., Forest Branch, Publ.* **1181**, 1–23.

Chrosciewicz, Z. (1968). Drought conditions for burning raw humus on clear-cut jack pine sites in central Ontario. *Forest. Chron.* 44, 30–31.

Clements, F. E. (1949). "Dynamics of Vegetation" (selections from the writings of Clements by B. W. Alred and E. S. Clements). Wilson Co., New York.

Cooper, C. F. (1961). The ecology of fire. *Sci. Amer.* 204, 150–160.

Cooper, C. F. (1963). "An Annotated Bibliography of the Effects of Fire on Australian Vegetation" (Mimeo.). Soil Conserv. Auth. of Victoria, Australia.

Cooper, R. W. (1963). Knowing when to burn. *Proc. 2nd Annu. Tall Timbers Fire Ecol. Conf.* pp. 31–35.

Cumming, J. A. (1969). Prescribed burning on recreation areas in New Jersey: History, objectives, influence, and technique. *Proc. 9th Annu. Tall Timbers Fire Ecol. Conf.* pp. 251–269.

Cushwa, C. T. (1968). Fire: A summary of literature in the United States from the mid-1920's to 1966. *U.S., Forest Serv., Southeast. Forest Exp. Sta., Mimeo.* pp. 1–117.

Czuhai, E., and Cushwa, C. T. (1968). A resume of prescribed burnings on the Piedmont National Wildlife Refuge, *U.S., Forest Serv., Res. Note* **SE-86**, 1–4.

Daubenmire, R. (1968). Ecology of fire in grasslands. *Advan. Ecol. Res.* 5, 209–266.

Davis, K. P. (1959). "Forest fire: Control and Use." McGraw-Hill, New York.

Dell, J. D., and Green, L. R. (1968). Slash treatment in the Douglas-fir region— Trends in the Pacific Northwest. *J. Forest.* 66, 610–614.

DeSilvia, E. R. (1965). Prescribed burning in the northern Rocky Mountain area. *Proc. 4th Annu. Tall Timbers Fire Ecol. Conf.* pp. 221–230.

Dieterich, J. H. (1971). Air-quality aspects of prescribed burning. *Proc. Prescribed Burning Symp.* pp. 139–151.

Dieterich, J. H. (1963). Use of fire in planting site preparation. *Proc. Lake States Forest Tree Improv. Conf. 6th, 1963* pp. 22–32.

Doman, E. R. (1967). Prescribed burning and brush type conversion in California National Forests. *Proc. 7th Annu. Tall Timbers Fire Ecol. Conf.* pp. 225–234.

Duvall, V. L., and Whitaker, L. B. (1964). Rotation burning: A forage management system for longleaf pine–bluestem ranges. *J. Range Manage.* **17**, 322–326.

Emrick, W. E. (1967). Organize, plan and prepare for control bush burning. *Proc. 7th Annu. Tall Timbers Fire Ecol. Conf.* pp. 163–178.

Foster, W. T., Hubert, G. A., Cayford, J., Dickson, W. A., and MacBean, A. P. (1967). Symposium on prescribed burning. *Pulp. Pap. Mag. Can., Woodl. Rev. Sect. Index* **2407**, (F-2), 1–11.

Fritschen, L., Bovee, H., Buettner, K., Charlson, R., Monteith, L., Pickford, S., Murphy, J., and Darley, E. (1970). Slash fire atmospheric pollution. *U.S. Forest Serv., Res. Pap.* **PNW-97**, 1–42.

Gilmour, D. A., and Cheney, N. P. (1968). Experimental prescribed burn in radiata pine. *Aust. Forest.* **32**, 171–178.

Gimingham, C. H. (1956). Fire on the hills. *J. Brit. Forest. Comm.* **25**, 148–150.

Gimingham, C. H. (1970). British heathland ecosystems: The outcome of many years of management by fire. *Proc. 10th Annu. Tall Timbers Fire Ecol. Conf.* pp. 293–322.

Gimingham, C. H. (1972). "Ecology of Heathlands." Chapman & Hall. London.

Givens, L. S. (1962). Use of fire on southeastern wildlife refuges. *Proc. 1st Annu. Tall Timbers Fire Ecol. Conf.* pp. 121–126.

Grant, S. M., Hunter, R. F., and Cross, C. (1963). The effects of muir burning Molinia—dominant communities. *J. Brit. Grassl. Soc.* **18**, 249–257.

Haddon, C. D. (1967). "Slash Disposal. Vancouver Forest District." B.C. Forest Serv., Prot. Div., Victoria, British Columbia, Canada.

Hansen, K. (1964). Studies on the regeneration of heath vegetation after burning-off. *Saertryk Bot. Tiddskr.* **60**, 1–41.

Hardison, J. R. (1957). Disease problems in forage seed production and distribution. *In* "Grasslands Seeds" (W. A. Wheeler and D. D. Hill, eds.), Chapter VII, Van Nostrand-Reinhold, Princeton, New Jersey (In Komarek, 1965).

Hardy, C. E., and Franks, J. W. (1963). Forest Fires in Alaska. *U.S., Forest Serv., Intermt. Forest Range Exp. Sta., Res. Pap.* **INT-5**, 1–163.

Harris, A. S. (1966). Effects of slash burning on conifer regeneration in southeast Alaska, *U.S., Forest Serv., Northern Forest Exp. Sta., Res. Note* **NOR-18**, 1–5.

Hendricks, J. H. (1968). Control burning for deer management in chaparral in California. *Proc. 8th Annu. Tall Timbers Fire Ecol. Conf.* pp. 219–233.

Hodgson, A. (1967). Fire management in eucalypt forests. *Proc. 6th Annu. Tall Timbers Fire Ecol. Conf.* pp. 97–112.

Hodgson, A. (1968). Control burning in eucalypt forests in Victoria, Australia. *J. Forest.* **66**, 601–605.

Hodgson, A. (1970). Fire as a forest management tool. *Proc. Fire Ecol. Symp., 2nd, 1970* pp. 1–5.

Hodgson, A., and Heislers, A. (1972). Some aspects of the role of forest fire in southeastern Australia. *Proc. World Forest. Congr., 7th, 1972, Mimeo.* pp. 1–26.

Hoffman, J. E. (1971). Fire in park management. *Proc. Symp. Fire Northern Environ. U.S., Forest Serv., Pac. Northwest Forest Range Exp. Sta.* pp. 73–78.

Hostetter, A. (1966). Annotated bibliography of publications by U.S. Forest Service forest fire research staff, their colleagues, and co-operators in California. *U.S., Forest Serv., Pac. Southwest Forest Range Exp. Sta., Mimeo.* pp. 1–24.

Humphrey, R. R. (1963). The role of fire in the desert and desert grassland areas of Arizona. *Proc. 2nd Annu. Tall Timbers Fire Ecol. Conf.* pp. 45–62.

Intermountain Fire Research Council. (1970). "Symposium on the Role of fire in the Intermountain West." IFRC, Missoula, Montana.

Isaac, L. A. (1963). Fire—a tool not a blanket rule in Douglas-fir ecology. *Proc. 2nd Annu. Tall Timbers Fire Ecol. Conf.* pp. 1–18.

Jarvis, J. M., and Tucker, R. E. (1968). Prescribed burning after barrel-scarifying on a white spruce-trembling aspen cut-over. *Pulp Pap. Mag. Can.* pp. 70–72.

Johnston, V. R. (1970). The ecology of fire. *Audubon* **72**, 76–119.

Kallander, H. (1969). Controlled burning on the Fort Apache Indian Reservation, Arizona. *Proc. 9th Annu. Tall Timbers Fire Ecol. Conf.* pp. 241–249.

Kayll, A. J. (1967). Moor burning in Scotland. *Proc. 6th Annu. Tall Timbers Fire Ecol. Conf.* pp. 29–40.

Kayll, A. J., and Gimingham, C. H. (1965). Vegetative regeneration of *Calluna vulgaris* after fire. *J. Ecol.* **53**, 729–734.

Kiil, A. D. (1970). Effects of spring burning on vegetation in old partially cut spruce–aspen stands in east-central Alberta. *Can. Dep. Fish. Forest., Can. Forest Serv. Inform. Rep.* **A-X-33**, 1–12.

Kiil, A. D. (1971). Prescribed fire effects in subalpine spruce-fir slash. *Can. Dep. Environ., Can. Forest Serv. Inform. Rep.* **NOR-X-3**, 1–30.

Kilgore, B. M. (1971a). Response of breeding bird populations to habitat changes in a giant sequoia forest. *Amer. Midl. Natur.* **85**, 135–152.

Kilgore, B. M. (1971b). The role of fire in managing red fir forests. *Trans. N. Amer. Wildl. Natur. Resour. Conf.* **36**, 405–416.

Kilgore, B. M. (1972). The role of fire in a giant sequoia–mixed conifer forest. *Proc. AAAS Symp., 1971* pp. 1–38.

Kilgore, B. M., and Briggs, G. S. (1972). Restoring fire to high elevation forests in California. *J. Forest.* **70**, 266–271.

Komarek, E. V. (1965). Fire ecology—grasslands and man. *Proc. 4th Annu. Tall Timbers Fire Ecol. Conf.* pp. 169–220.

Komarek, E. V. (1967). Fire and the ecology of man. *Proc. 6th Annu. Tall Timbers Fire Ecol. Conf.* pp. 143–170.

Komarek, E. V. (1971). Effects of fire on wildlife and range habitats. *Proc. Prescribed Burn. Symp., 1971* pp. 46–52.

Komarek, R. (1963). Fire and the changing wildlife habitat. *Proc. 2nd Annu. Tall Timbers Fire Ecol. Conf.* pp. 35–44.

LeBarron, R. K. (1957). Silviculture possibilities of fire in northeastern Washington. *J. Forest.* **55**, 627–630.

Leege, T. A. (1968). Prescribed burning for elk in northern Idaho. *Proc. 8th Annu. Tall Timbers Fire Ecol. Conf.* pp. 235–254.

Leonard, B. (1970). Effects of control burning on the ecology of small mammal populations. *Proc. Fire Ecol. Symp., 2nd, 1970* pp. 1–7.

Lindenmuth, A. W. (1960). A survey of effects of international burning on fuels and timber stands of ponderosa pine in Arizona. *U.S., Forest Serv., Rocky Mt. Forest Range Exp. Sta., Pap.* **54**, 1–22.

Little, S., and Somes, H. A. (1961). Prescribed burning in the pine region of southern New Jersey and eastern shore Maryland. A summary of present knowledge. *U.S., Forest Serv., Northeast Forest Exp. Sta., Pap.* **151**, 1–21.

Lotti, T. (1959). The use of fire in the management of coastal plain loblolly pine. *Proc. Soc. Amer. Foresters, 1959* pp. 1–3.

Lotti, T. (1962). The use of prescribed fire in the silviculture of loblolly pine. *Proc. 1st Annu. Tall Timbers Fire Ecol. Conf.* pp. 109–120.

McArthur, A. G. (1962). Control burning in eucalypt forests. *Forest. Timber Bur. Aust., Leafl.* **80**, 1–31.

McArthur, A. G. (1965). Prescribed burning in Australian fire control. *Aust. Forest.* **30**, 4–11.

McVean, D. N. (1959). Muir burning and conservation. *Scot. Agr.* **39**, 79–82.

Meskimen, G. (1971). Managing forest land*scapes:* Is prescribed burning in the picture? *Proc. Prescribed Burning Symp., 1971* pp. 54–58.

Miller, G. R. (1964). The management of heather moors. *Advan. Sci.* **21**, 163–169.

Miller, H. A. (1963). Use of fire in wildlife management. *Proc. 2nd Annu. Tall Timbers Fire Ecol. Conf.* pp. 19–30.

Morris, W. G. (1970). Effects of slash burning in overmature stands of the Douglas-fir region. *Forest Sci.* **16**, 258–270.

Mount, A. B. (1965). The vegetation as a guide to prescribed burning in Tasmania. *Pap., Inst. Forest. Aust. Hobart Conf.* pp. 1–13 (revised, 1968).

Muraro, S. J. (1968). Prescribed fire–evaluation of hazard abatement. *Can. Dep. Forest. Rural Develop. Forest Branch, Publ.* **1231**, 1–28.

Muraro, S. J. (1971). Prescribed-fire impact in cedar–hemlock logging slash. *Can. Dep. Environ., Can. Forest. Serv., Publ.* **1295**, 1–20.

Murphy, A. H. (1967). Controlled burning in chamise chaparral. *Proc. 7th Annu. Tall Timbers Fire Ecol. Conf.* pp. 245–255.

Neel, L. (1965). Head fires in southeastern pines. *Proc. 4th Annu. Tall Timbers Fire Ecol. Conf.* pp. 231–240.

Niering, W. A., Goodwin, R. H., and Taylor, S. (1970). Prescribed burning in southern New England: Introduction to long-range studies. *Proc. 10th Annu. Tall Timbers Fire Ecol. Conf.* pp. 267–286.

Owen, J. S. (1971). Fire management in the Tanzania national parks. *Proc. 11th Annu. Tall Timbers Fire Ecol. Conf.* pp. 233–241.

Pase, C. P., and Glendening, G. E. (1965). Reduction of litter and shrub crowns by planned fall burning of oak–mountain mahogany chaparral. *U.S., Forest Serv., Rocky Mt. Forest Range Exp. Sta., Res. Notes* **RM-49**, 1–2.

Perkins, C. J. (1968). Controlled burning in the management of muskrats and waterfowl in Louisiana coastal marshes. *Proc. 8th Annu. Tall Timbers Fire Ecol. Conf.* pp. 269–280.

Perkins, C. J. (1971). The effects of prescribed fire on outdoor recreation. *Proc. Prescribed Burning Symp., 1971* pp. 59–63.

Phillips, J. (1965). Fire—as master and servant: Its influence in the bioclimatic regions of Trans-Saharan Africa. *Proc. 4th Annu. Tall Timbers Fire Ecol. Conf.* pp. 6–109.

Prasil, R. G. (1971). National Park Service fire policy in National Parks and Monuments. *Proc. Symp. Fire Northern Environ., 1971* pp. 79–81.

Radley, J. (1965). Significance of major moorland fires. *Nature (London)* **205**, 1254–1259.

Ramsey, G. S. (1966–1971). Bibliography of departmental forest fire research literature. *Can. Dep. Forest., Rural Develop. Inform. Rep.* **FF-x-2,** Suppl. 1 (1967);
Suppl. 2 (1968); Suppl. 3 (1969); Suppl. 4 (1970); Suppl. 5 (1971).

Raymond, F. H (1967). Controlled burning on California wildlands. *Proc. 7th
Annu. Tall Timbers Fire Ecol. Conf.* pp. 151–162.

Riebold, R. J. (1964). Large scale prescribed burning. *Proc. 3rd Annu. Tall Timbers
Fire Ecol. Conf.* pp. 131–138.

Riebold, R. J. (1971). The early history of wildfires and prescribed burning. *Proc.
Prescribed Burning Symp., 1971* pp. 11–20.

Robertson, R. A. (1957). Heather management. *Scot. Agr.* **37,** 126–129.

Robinson, R. R. (1970). Fire management direction. *Symp., Role Fire Intermt.
West, Missoula, Mont., 1970* pp. 143–152.

Sando, R. W. (1969). The current status of prescribed burning in the Lake States.
U.S., Forest Serv., Res. Notes **NC-81,** 1–2.

Schlichtemeier, G. (1967). Marsh burning for waterfowl. *Proc. 6th Annu. Tall
Timbers Fire Ecol. Conf.* pp. 41–46.

Scott, J. D. (1947). Veld management in South Africa. *Repub. S. Afr., Dep. Agr.
Tech. Serv., Sci. Bull.* **278.**

Scott, J. D. (1970). Pros and cons of eliminating veld burning. *Proc. Grassl. Soc.
S. Afr.* **5,** 23–26.

Scotter, G. W. (1970). Wildfires in relation to the habitat of barren-ground caribou
in the taiga of northern Canada. *Proc. 10th Annu. Tall Timbers Fire Ecol.
Conf.* pp. 85–105.

Scotter, G. W. (1971). Fire, vegetation, soil and barren-ground caribou relations
in northern Canada. *Proc. Symp. Fire Northern Environ., 1971* 209–230.

Shipman, R. D. (1970). Abstracts of publications relating to forest fires research
and use in resource management. *Pa. State Agr. Sch. Forest Resour., Mimeo.* pp.
1–110.

Shostakovitch, V. B. (1925). Forest conflagrations in Siberia with special reference
to the fires of 1915. *J. Forest.* **23,** 365–371.

Slaughter, C. W., Barney, R. J., and Hansen, G. M., eds. (1971). Fire in the
Northern environment—A symposium. U.S. Forest Serv., Pacific Northwest
Forest Range Exp. Sta., Portland, Oregon.

Smith, E. L. (1960). Effects of burning and clipping at various times during the
wet season on tropical tall grass range in northern Australia. *J. Range Manage.*
13, 197–203.

Smith, J. H. G. (1970). A British Columbian view of the future of prescribed
burning in western North America. *Commonw. Forest Rev.* **49,** 365–367.

Society of American Foresters. (1958). "Forestry Terminology." SAF, Washington,
D.C.

Spencer, D. L., and Hakala, J. B. (1964). Moose and fir on the Kenai. *Proc. 3rd
Annu. Tall Timbers Fire Ecol. Conf.* pp. 10–33.

Stewart, O. C. (1963). Barriers to understanding the influence of use of fire by
aborigines on vegetation. *Proc. 2nd Annu. Tall Timbers Fire Ecol. Conf.* pp.
117–126.

Stoddard, H. L. (1931). "The Bobwhite Quail, Its Habits, Preservation, and Increase." Scribner's, New York.

Stoddard, H. L. (1962). Some techniques of controlled burning in the deep southeast.
Proc. 1st Annu. Tall Timbers Fire Ecol. Conf. pp. 133–144.

Stokes, L. (1846). "Discoveries in Australia—Voyage of the HMS Beagle During 1837–43" (in Vines, 1968).

Sykes, J. M. (1964). Report on the interim results of prescribed spring burning in a poor quality hardwood stand. *Ont., Dep. Lands Forests, Sect. Rep.* **49**, 1–13.

Ternes, A. (1970). Do we need forest fires? *True Mag.* Sept., pp. 33–38.

Uggla, E. (1958). "Ecological Effects of fire on North Swedish Forests." Almqvist & Wiksell, Stockholm.

U.S.D.A. Forest Service. (1971). "Prescribed Burning Symposium Proceedings." Southeast. Forest Exp. Sta., Ashville, North Carolina.

U.S.D.A. Forest Service. (1972). "Fire in the Environment." Symp. Procs. N. Amer. Forest. Comm. F.A.O., Denver, Colorado.

Van Rensburg, H. J. (1971). Fire: Its effect on grasslands, including swamps—Southern, Central and Eastern Africa. *Proc. 11th Annu. Tall Timbers Fire Ecol. Conf.* pp. 175–199.

VanWagner, C. E. (1966). Three experimental fires in jack pine slash. *Can., Dep. Forest., Publ.* **1146**, 1–22.

Vines, R. G. (1968). The forest fire problem in Australia—A survey of past attitudes and modern practice. *Aust. Sci. Teacher's J.* pp. 1–11.

Vines, R. G., Gibson, L., Hutch, A. B., King, N. K., MacArthur, D. A., Packham, D. R., and Taylor, R. J. (1971). On the nature, properties and behaviour of bush-fire smoke. *Aust. CSIRO, Div. Appl. Chem., Tech. Pap.* **1**, 1–32.

Viro, P. J. (1969). Prescribed burning in forestry. *Commun. Inst. Forest. Fenn.* **67.7**, 1–49.

Vogl, R. J. (1965). Effects of spring burning on yields of brush prairie savanna. *J. Range Manage.* **18**, 202–205.

Vogl, R. J. (1967). Controlled burning for wildlife in Wisconsin. *Proc. 6th Annu. Tall Timbers Fire Ecol. Conf.* pp. 47–96.

Vogl, R. J., and Ryder, C. (1969). Effects of slash burning on conifer reproduction in Montana's Mission Range. *Northwest Sci.* **43**, 135–147.

Wade, D. D. (1969). Research on logging slash disposal by fire. *Proc. 9th Annu. Tall Timbers Fire Ecol. Conf.* pp. 229–234.

Ward, P. (1968). Fire in relation to waterfowl habitat of the Delta Marshes (Manitoba). *Proc. 8th Annu. Tall Timbers Fire Ecol. Conf.* pp. 255–268.

Ward, S. D. (1972). The controlled burning of heather, grass and gorse. *Nature Wales* **13**, 24–32.

Weetman, G. F., and Nykvist, N. B. (1963). Some mor humus, regeneration and nutrition problems and practices in North Sweden. *Pulp Pap. Res. Inst. Can., Tech. Rep. Woodl. Res. Index* **139**, 1–15.

West, O. (1965). Fire in vegetation and its use in pasture management with special reference to tropical and subtropical Africa. *Commonw. Bur. Pastures Field Crops, Mimeo Publ.* **1**, 1–53.

West, O. (1971). Fire, man and wildlife as interacting factors limiting the development of vegetation in Rhodesia. *Proc. 11th Annu. Tall Timbers Fire Ecol. Conf.* pp. 121–146.

Wharton, C. H. (1966). Man, fire and wild cattle in northern Cambodia. *Proc. 5th Annu. Tall Timbers Fire Ecol. Conf.* pp. 23–66.

Whyte, R. O. (1957). The grasslands and fodder resources of India. *Comm. Agr. Res. India, New Delhi, Sci. Monogr.* **22** (in West, 1965).

Williams, D. E. (1960). Prescribed burning for seedbed preparation in jack pine types. *Can. Pulp Pap. Ass., Woodl. Sect. Index* **1959** (F-2), Annu. Meet. Release No. 9, 1–5.

Zivnuska, J. A. (1968). An economic view of the role of fire in watershed management. *J. Forest.* **66**, 596–600.

Zontek, F. (1966). Prescribed burning on the St. Marks National Wildlife Refuge. *Proc. 5th Annu. Tall Timbers Fire Ecol. Conf.* pp. 195–202.

Zwolinski, M. J., and Ehrenreich, J. H. (1967). Prescribed burning on Arizona watersheds. *Proc. 7th Annu. Tall Timbers Fire Ecol. Conf.* pp. 195–206.

Author Index

Numbers in italics refer to the page on which the complete references are listed.

A

Aaltonen, V. T., 10, 22, *44*, 210, *219*
Adams, J. L., 491, *504*
Adams, M. S., 102, *125*
Adams, S., 102, *125*
Adamson, R. S., 439, *477*
Agnostopoulos, G. D., 405, *431*
Ahlen, I., 120, *125*
Ahlgren, C. E., 32, 33, *44*, 49, 50, 51, 53, 56, 57, 59, *69*, 75, 76, 78, 79, 97, 99, 102, 105, 106, 108, 119, 121, *125*, 168, *182*, 197, 200, 201, 202, 204, 205, 207, 208, 209, *219*, 254, *272*, 498, 502, *504*
Ahlgren, I. F., 33, *44*, 49, 50, 51, 53, 56, 57, *69*, 75, 76, 78, 97, 99, 102, 119, 121, *125*, 168, *182*, 254, *272*
Aikman, J. M., 139, 151, 156, 158, 159, 161, 163, 174, *182*, *186*
Albertson, F. W., 139, 148, 149, 150, 153, 158, 165, *182*, *187*, *193*
Aldous, A. E., 139, 160, *182*
Allen, S. E., 100, *125*, 496, *504*
Amman, G. A., 104, *125*
Anderlind, A., 409, *432*
Anderson, H. W., 232, *248*, 352, *360*
Anderson, K. L., 161, 170, *182*, *191*, 494, *504*
Anderson, R. C., 78, *125*, 180, 181, *182*
Andresen, J. W., 241, *247*
Applegate, J., 308, *317*
Ardrey, R., 485, *504*
Armstrong, C. G., 175, *189*
Arnold, J. F., 102, *125*, 170, 176, *182*, 502, *504*
Arnold, F., 252, *272*
Ashby, W. R., 431, *432*
Aubréville, A., 439, 441, 448, 449, *477*
Aumann, G. D., 115, *125*
Austen, B., 84, *125*, 493, 502, *504*
Austin, R. C., 100, *125*
Axelrod, D. I., 407, *432*

B

Backus, M. P., 52, 56, *72*
Baerreis, D. A., 252, *273*
Bailey, A. W., 120, *125*
Baisinger, D. H., 100, *125*
Baker, G. O., 52, 56, *72*
Baker, W. W., 266, *272*
Bakuzis, E. V., 204, 206, *219*
Balci, A. N., 100, *128*
Baldanzi, G., 159, 160, *182*
Baldwin, J. J., 497, *504*
Banfield, A. W. F., 84, *125*
Barden, L., 270, *272*
Barick, F., 91, *125*
Barnette, R. M., 160, *187*
Barney, R. J., 484, 499, 502, *509*
Bartholomew, B., 345, *360*
Bartlett, H. H., 145, 146, *183*, 485, 487, *504*
Bar Yosef, O., 430, *434*
Basham, J. T., 61, *69*
Batchelder, R. B., 140, 148, 149, *183*, 443, *477*, 485, 489, *504*
Bate, D. M. A., 407, *432*
Batzli, G. O., 114, *125*
Beadel, H. L., 261, *276*
Beale, E. F., 285, *317*
Beaton, J. P., 197, *219*
Beaufait, W. R., 197, 201, 202, *219*, 488, 490, 492, 503, *504*
Beck, A. M., 79, 105, 108, *125*, 145, 161, 179, *183*, *193*
Bedell, G. H. D., 205, *219*, 232, *248*
Beeson, K. C., 100, *125*, *276*
Beetle, A. A., 142, *183*
Bendell, J. F., 81, 84, 85, 89, 90, 91, 92, 94, 102, 104, 116, 117, 118, 122, 123, *125*, *126*, *135*, *138*
Benson, L., 390, *399*
Bentley, J. R., 325, *360*, 411, *432*
Bergerud, A. T., 98, 103, 120, *126*
Berry, F. H., 236, *247*

Subject Index

A

Abert Squirrel, 119, 120
Aborigines, 145, 284, 313, 485
Abscission, 253, 270
Absorption, 15, 28, 29, 160, 199
Acetic acid, 24, 25
Acidity, 10, 20–22, 28, 30, 31, 33, 34, 37, 56, 58, 59, 141, 160, *see also* Alkalinity, pH
Actinomycetes, 31, 53–55
Adaptation, 84, 95, 106, 122, 142, 282, 311, *see also* Adaptation to fire, Fire resistance, Fire Tolerance
Adaptation to fire, 4, 64, 119, 122–124, 143, 150, 152, 162, 169, 172, 175, 199–204, 237, 255, 259, 262, 265, 270, 334–339, 356, 426, 429, 483, *see also* Fire resistance, Fire tolerance
Adiabatic heating, 428, *see also* Heat
Adsorption, 27
Adventitious buds, 313, 414, *see also* Buds
Adventitious roots, 414, 415
Aeration, 11, 18, 200, 454, 455
Aesthetics, 294, 495, 496
African ecosystems, 435–481
African elephant, *see* Elephant
Aging, *see* Longevity
Air currents, 59, *see also* Wind
Alaska, 316, 317
Albite, 10
Alkali, 21, 25, 33
Alkaline earths, 21, 159
Alkalinity, 30, 34, 97, 141, 160, 368, *see also* pH
Alleles, 118
Allelopathy, 144, 158, 164, 167, 204, 345, 414
Alluvium, 377, 387, 447
Aluminum, 38, 241
Ammonia, 18, 24, 28, 33–37, 39, 52
Ammonification, 34, 39, 53
Ammonium acetate, 26
Amphiboles, 10

Animal migration, *see* Migration
Annelids, 47, 63, 64
Antelope, 75, 374, 398, 449, 475
Anthracobionte, 59
Anthracophile, 59
Anthracophyte, 426
Anthrakophobe, 59
Anthrakozene, 59
Anthropogenic grassland, 143, 144, *see also* Grassland
Antibiotics, 167, 414
Ants, 64, 65
Apatite, 10
Apomixis, 143
Arachnids, 66, 67
Arboricides, 411, *see also* Herbicides
Archeology, 407, 430
Aridity, 170, 281, 368, 375, 376, 407, 421, 428–430, *see also* Drought, Soil moisture, Water supply
Arthropods, 62–64
Ascomycotes, 43, 55
Ascospores, 61, *see also* Spores
Ash, 19, 21, 22, 24, 29, 39, 49, 52, 53, 55, 58, 59, 75, 84, 99, 100, 102, 146, 153, 154, 156–160, 168, 206, 208, 263, 289, 291, 293, 297, 298, 337, 349, 372, 407, 408, 421, 443, 451, 453, *see also* Fly ash
Asphodelus deserts, 410
Asphyxiation, 77
Awns, 428, 429
Axillary buds, 420, *see also* Buds

B

Backfire, 153, 254, 488, 489, 497
Bacteria, 48–55, 62, 151, 160, 168, *see also* Microbes
Bacterial rot, 384
Bark, 4, 175, 200, 231, 233, 237, 245, 257, 258, 307, 314, 383, 399, 420
Bark beetles, 65, 246, *see also* Beetles
Basal crooks, 237, 238, 259
Basalt, 428